Mobile Antenna Systems Handbook

Second Edition

For a listing of recent titles in the *Artech House Antennas and Propagation Library*, turn to the back of this book.

Mobile Antenna Systems Handbook

Second Edition

K. Fujimoto
J. R. James

Editors

Artech House
Boston • London
www.artechhouse.com

Library of Congress Cataloging-in-Publication Data
Mobile antenna systems handbook / K. Fujimoto, J. R. James, editors.—2nd ed.
   p. cm. — (Artech House antennas and propagation library)
   Includes bibliographical references and index.
   ISBN 1-58053-007-9 (alk. paper)
   1. Mobile communication systems. 2. Antennas (Electronics). I. Fujimoto, K. (Kyohei),
   1929– II. James, James R. III. Series.

TK6570.M6 M57   2000   00-050818
621.3845—dc21   CIP

British Library Cataloguing in Publication Data
Mobile antenna systems handbook.—2nd ed. — (Artech House
   antennas and propagation library)
   1. Mobile communication systems 2. Antennas (Electronics)
   I. Fujimoto, K. (Kyohei), 1929– II. James, J. R. (James Rodrick), 1933–
   621.3'845

   ISBN 1-58053-007-9

Cover design by Igor Valdman

© 2001 ARTECH HOUSE, INC.
685 Canton Street
Norwood, MA 02062

International Standard Book Number: 1-58053-007-9
Library of Congress Catalog Card Number: 00-050818

10 9 8 7 6 5 4 3 2 1

Contents

Authorship by Chapter

Chapter 1   J. R. James and K. Fujimoto

Chapter 2   K. Kagoshima: 2.1.1, 2.1.2
W. C. Y. Lee: 2.1.3, 2.2, 2.3.2
K. Fujimoto: 2.3.1
T. Taga: 2.4

Chapter 3   S. Saunders

Chapter 4   K. Kagoshima: 4.1, 4.2, 4.3
T. Taga: 4.4

Chapter 5   Y. Yamada: 5.1
Y. Karasawa: 5.2

Chapter 6   R. Mumford: 6.1.1, 6.1.2
Q. Balzano: 6.1.3, 6.1.4, 6.1.5, 6.3
Q. Balzano and C. K. Chou: 6.5
T. Taga: 6.2, 6.4

Chapter 7   Y. Rahmat-Samii, K. W. Kim, and M. A. Jensen: 7.1
K. Fujimoto: 7.2.1, 7.2.3, 7.2.4, 7.2.5
O. Edvardsson: 7.2.2

Chapter 8   H. K. Lindenmeier, L. Reiter, and J. Hopf: 8.1
K. Fujimoto: 8.2, 8.4
K. Hirasawa: 8.3

Chapter 9   T. Shiokawa and S. Ohmori: 9.1, 9.2, 9.3, 9.4
                           T. Teshirogi: 9.5
                           M. Williamson: 9.6.1
                           O. Edvardsson: 9.6.2

Chapter 10   Y. Suzuki

Appendix   K. Fujimoto and J. R. James

Preface to the Second Edition

In the past decade the communications field, and particularly mobile communications, has emerged as a dominant influence on antenna design. *The Mobile Antenna Systems Handbook* was conceived to present these important developments. At the same time we also set out to produce a rather different book about antenna systems where applications are brought to the forefront in a typical top-down system approach. In parallel with this, we ensured that the text contained adequate details and references concerning the impact of numerical analysis and computational techniques on practical antenna design. The worldwide response to the book has indeed been very gratifying and we understand that this systems approach to antennas has been much appreciated by a wide spectrum of people who have felt the need for enlightenment and a broader perspective on some or all aspects of mobile communication antennas.

We stated in the preface to the first edition that "... the potential of mobile communication is so great that it may change the infrastructure and scale of communications worldwide. ..." We did not imagine, however, that the speed of change in some areas of mobile communications would be so fast and the influence on antenna design so profound. Even now, new communication concepts are appearing on the horizon such as the seamless connectivity of Bluetooth technology, the realms of optical switching with prospects for ultra-wide-band all-optical communication technology and so on. Already one can conjecture on the new demands that will be placed on mobile antenna designers. We are therefore very pleased to be invited by the publishers to produce this second edition, which is an opportunity both to update the handbook and to further consolidate it as a recognized contribution to the international antenna engineering literature.

The success and rapid growth of cellular systems has been foremost in establishing a critical need for design techniques that will greatly increase mobile communication capacity and flexibility, to deliver the now much sought-after services. To meet these critical needs, system designers have made advances on many research fronts such as improved techniques for efficient signal processing, more precise propagation prediction methods, and physically compact higher performance antennas. These new techniques are now inextricably interlinked in a system sense to create antenna systems for base stations, personal handsets, and mobile satellite terminals. We have embodied these advances by inserting new sections and chapters in this second edition while retaining as much of the previous foundation material as the book size allowed.

The subdivision of service areas into smaller cellular zones has led to new prediction models for macro-, micro-, and picocells and Chapter 3 describes the progress being made and the greater precision achieved in system trade-offs to achieve a given level of service. The concept of picocells for close-range urban and in-building propagation has stimulated much thought on the design of inconspicuous, environmentally friendly, easy-to-mount base station antennas, while many design advances associated with large cell base station antennas are evident and are addressed in Section 5.1 in conjunction with advances in adaptive antenna arrays (Section 5.2). These latter arrays are now commonly referred to as "smart," inferring some built-in "intelligence" property and their practical realization and merit has now been demonstrated, as reported in the many extensive practical trials. In essence, these adaptive arrays provide a valuable degree of spatial discrimination to supplement software time and frequency-domain signal processing techniques and emphasizes this new role for mobile antennas; a complex role that was previously the preserve of defense electronic systems.

The public awareness worldwide of electromagnetic radiation is now so significant that communication system designers are carrying out extensive computations and measurements to ensure safety and allay unnecessary public concern. Every opportunity is being taken by designers to reduce transmitted powers, and personal handset antennas have themselves been intensively investigated in this respect. Section 6.5 presents up-to-date information on these safety aspects, while Chapter 7 describes the application of computer modeling and measurement together with examples of recent handset antennas.

Some interesting developments in Intelligent Transportation Systems (ITS) are presented in Section 8.4, while Section 9.6.1 summarizes the revolutionary constellations of mobile satellite systems (MSS) that are now just coming into exploratory service. Section 9.6.2 gives an appreciation of the design opportunities presented to the MSS handset antenna designer. Network planners predict that the megacell propagation zones afforded by the MSS will complement the macro-, micro-, and picocell topologies to achieve Global Village Communication but the timescale and final services will clearly depend on both financial resourcing and the ultimate subscriber demand.

Much of the glossary is in the nature of foundation material but some new antenna examples have been added where the design is considered to be of significant interest, such as showing evidence of new fundamental properties or manufacturing advantages. We believe that the projected book readership remains essentially the same, being largely made up of specialists and some generalists in antenna engineering, propagation prediction, mobile communication system planning, electronic packaging for communication together with supporting mathematicians, physicists, and postgraduates doing research on antennas.

The latter readership is likely to increase, however, due to the continuing international research emphasis on mobile communications.

Once again our sincere thanks are due to the previous authors for establishing the handbook and to the new authors who have contributed to this second edition. A multiauthor text naturally increases the editorial work and we are grateful to the reviewers for their skilful help, to the new authors for their patient responses, and to Dr. Julie Lancashire and Ruth Young of the publishing staff, for their enthusiasm and assistance. Last but not least we acknowledge the many colleagues who have provided technical material throughout, most of which is disclosed for the first time, and who have contributed to this work in various ways.

K. Fujimoto
J. R. James

Chapter 1

Importance of Antennas in Mobile Communication Systems and Recent Trends

J. R. James and K. Fujimoto

1.1 INTRODUCTION

The evolution and development of antennas have a long and fascinating history [1] stretching back to Hertz's radiation experiments in 1887, confirming Maxwell's theories of 1864 although Henry is considered to have done initial radiation experiments with discharges in 1842. It seems remarkable that a parabolic dish at 470 MHz (Hertz, 1893), millimeter radiation at 60 GHz (Bose, 1895), wireless signaling (Marconi, 1896), and phased-array principles (1906) were demonstrated soon after, but the ideas and concepts needed supporting technology to accelerate and endorse design progress, as is still true today.

The first mobile communication was initiated in 1885 with wireless telegraph between trains and stations, which was developed by Thomas Edison [2]. Telegraph signals were conveyed through the trolley wires, which were electrostatically coupled with a metal plate installed on the ceiling of the train. Edison also experimented with communication on a vehicle in 1901 [3] using a thick cylindrical antenna placed on the roof of the vehicle. The real mobile communication services started with wireless telegraph on ships in 1898 developed by Guglielmo Marconi using long vertical wire antennas in various forms such as T, inverted L, and umbrella shapes. Portable equipment appeared in 1910 [4].

Both world wars provided the need for advanced antenna design and the technology surge [5]; wire antennas were firmly established in the 1920s, while present-day microwave antenna design and technology was commonplace in the 1950s. In the 1960s a new antenna era emerged, triggered by the revolutionary progress in semiconductor integrated circuits, attributed initially to the Cold War defense industry but subsequently carried forward into the commercial equipment sector. Quite simply, the demand opened up designers to the possibilities of redesign, recreation, and transformation of known antenna types into less bulky, lightweight, low-cost, easy-to-manufacture radiating structures, compatible with the newly conceived integrated electronic packages. Most notable has been the creation of printed antenna technology. which lends itself to multifunction antenna devices [6]. Some of the salient factors that have increasingly influenced antenna design in this era [7] and continue to do so today are noted in Table 1.1, which clearly emphasizes that communications, and particularly mobile communication systems, are the most significant drivers of antenna technology at present.

Many other information-relaying systems are now emerging that have much in common with mobile communication systems, but do not necessarily involve interactive speech or vision; these communication related systems are noted in Table 1.2 and although they demand much ingenuity from the antenna designer, these systems generally do not as yet have the massive customer base and resourcing associated with mobile communication systems. The exception is GPS, which is being deployed increasingly in conjunction with mobile communication systems.

Table 1.1
Factors Influencing Recent Antenna Technology and Design

Factor	Trends
Spectral congestion and utilization	Wider bandwith operation, improved performance, interference rejection, use of millimeter and submillimeter antennas
Explosive growth in mobile/ personal communication systems	New compact user-friendly higher performing antennas for cellular terrestrial operation of handsets and vehicles
Escalating information and processing speeds	Wider bandwidth "smart" antennas with pattern agility and fast scan acquisition, antennas for microwaves
Growth in SATCOMS	Higher performance space-borne antennas offering multifunction operation and reduced payloads, small high-performance handset antennas
Link with IN and ATM networks	Small high-performance antennas for mobile terminals
Traffic information and control	Specifically designed antennas for systems
New materials	Redesign of existing, and creation of new, robust antenna structures to simplify manufacture and operation
Impact of computer modeling and computer-controlled measurement	Strengthens design methods to create higher performing equipment, compatible antennas at lower cost
Public awareness of electromagnetic radiation	Preference for lower transmitted powers, environmentally friendly antennas, and antenna platforms

Table 1.2
Some Applications Related to Mobile Communications Demanding Innovative Antenna Design

Application	Requirements
Animal tracking	Inconspicuous robust body-mounted antennas for satellite or terrestrial monitoring
Product information	Remotely interrogated "barcode" system mainly utilizing electrically small antenna concpets
Automatic meter telemetry for utility industries	Enables meter reading to be interrogated remotely; compact efficient antenna needed, possibly directional
Traffic information, control, and management systems	Vehicle-mounted inconspicuous antennas for short-range communications with pattern shaped base station antennas
Security systems	Inconspicuous antennas for smart cards, car door operation, and personal identification
Environmental monitoring	Remote recording of terrestrial environment and weather data
Navigation	Inconspicuous user-friendly antennas for both Global Positioning Satellite (GPS) handsets and vehicle-mounted terrestrial radio location systems
Information and data transmission system at home/office	Short-range communication (data and video) and control systems at frequencies >2 GHz

This brief historical perspective shows how the creation and development of antennas have accelerated rapidly in response to worldwide demands that were previously dominated primarily by defense requirements. The worldwide impact of mobile communication and related systems on antennas during the past decade is significant in many ways:

- It arises mainly from the commercial sector.
- The period of accelerated antenna design activity is already at a record length, shows no sign at present of diminishing, and is associated with massive and increasing resourcing by the public.
- The infusion of mobile communications into all remote corners of the world community is bringing about both sociological changes [8] and an increased public awareness of antennas in their everyday environment. The latter awareness of electromagnetic radiation is already influencing base station and handset antenna design specifications.

It is appreciated that future forecasts are seldom very accurate, and the influence of mobile communications on antennas was certainly not foreseen. As such, the present authors will not attempt predictions beyond the millennium but will instead examine the many facets of contemporary mobile communications and how they in turn influence current antenna design and manufacture. There is much that can still be learned from the immediate views and aspirations of this now vibrant communications industry as a whole.

During recent years the sales of mobile communication equipment have outstripped predictions and this is likely to continue into the foreseeable future, given the overwhelming public acceptance of cellular handsets as a necessary personal item. Pager subscribers have been growing an order magnitude per decade but have already been caught up to by cellular and PCS subscribers, whose numbers had grown at two orders magnitude per decade [9], leading to a prediction that about 5% of the world population will have subscribed to these two components of the mobile communication market alone at the millennium. The Asia-Pacific region is projected [10] to account for just under half of worldwide cellular/PCS subscribers by the year 2004, driven by wireless growth in China [11] and Japan [12]. The increase in the number of cellular subscribers in Europe has been dramatic and Figure 1.1 shows these statistics.

There has been much discussion of the Global Communications Village concept comprised of large-scale satellite-served terrestrial megacells, which are resolved into nested geometrical regions of macro, micro, and pico propagation cells, the latter including in-building communication. Such aspirations have led to the conception of complex network configurations of integrated mobile communication systems; however, it is widely appreciated that the global standardization of signal processing methods, network topologies, and equipment is the key to successful global realization.

An outline of some of the milestones, both achieved and planned in relation to the evolution of Global Mobile Communication Systems, is given in Figure 1.2. The early analog systems of the 1980s now described as "first generation," laid the foundations for the subscriber market, demonstrating the benefits of communication on the move. In comparison with present-day equipment, the systems were relatively simple and were developed to suit local requirements. The "second-generation" digital systems have shown not only the advantages of digital over analog processing with respect to increasing the available capacity and services provided, but also the way forward to achieve the vital global standardization.

The very successful pan-European Global System for Mobile Communications (GSM) has been instrumental in this and has paved the way for the proposed "third-

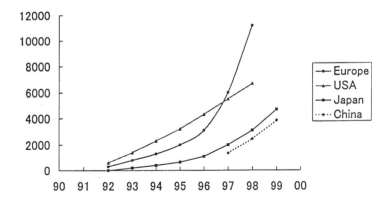

Figure 1.1 Number of cellular system subscribers in Europe, the United States, and Japan.

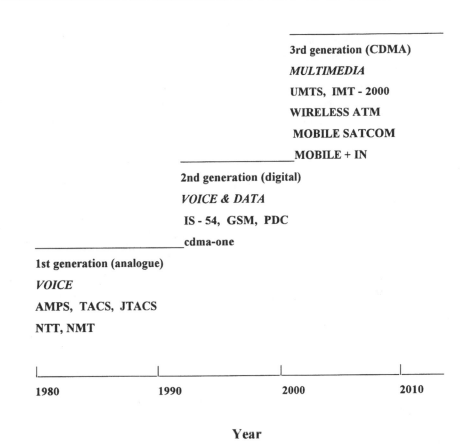

3rd generation (CDMA)

MULTIMEDIA

UMTS, IMT - 2000

WIRELESS ATM

MOBILE SATCOM

MOBILE + IN

2nd generation (digital)

VOICE & DATA

IS - 54, GSM, PDC

cdma-one

1st generation (analogue)

VOICE

AMPS, TACS, JTACS

NTT, NMT

| 1980 | 1990 | 2000 | 2010 |

Year

Figure 1.2 Mobile communication systems evolution showing some widely used systems and future concepts.

generation'' European Universal Mobile Telecommunications Systems (UMTS), which will occupy bands up to 2.2 GHz. It is proposed [13] that global integration will take place under the International Mobile Telecommunications (IMT-2000) standards; the anticipated lower cost per bit capacity demands for multimedia, including the Internet, can only be realized by employing code division multiple access (CDMA) technology [14,15], further signal compression techniques particularly concentrating on speech coding [16] and video coding, ''smart'' adaptive antenna array methodologies [17], and a more quantifiable understanding of the propagation phenomena taking place in the various cell regions.

The complexity facing mobile communication systems designers is further compounded by the rapid advance of both wireless technology for the provision of in-building and close-range urban links [14,18] and mobile satellite systems (MSS) [19], the latter opening up communication to any remote location in the world. In addition high-altitude platform stations (HAPSs) in the form of balloons or circulating aircraft are seen as attractive and flexible alternatives [20] where rapid deployment of mobile services are required over areas from tens of kilometers to hundreds of kilometers in diameter. The enormous task embodied in IMT-2000 is thus the harmonious integration of all of these types of mobile communication systems, and antennas will continue to play an important, and often pivotal, role.

Attention should also be paid to significant developments in intelligent transportation systems (ITS). Typical mobile communications in ITS are categorized into two systems: first, communication between vehicles and roadside facilities (CVR), and second, between vehicles (CVV). These systems also provide control facilities. The dedicated short-range communication system (DSRC) is one type of communication system used in ITS. In ITS circular or elliptical or elongated spot beams are projected onto one or perhaps several highways, depending on the environmental conditions and requirements. Antennas for ITS are of various designs to suit the application requirements for the link budgets, radiation patterns, and propagation environment.

Like many emerging areas in electronics, these new equipment concepts have been made possible by the revolutionary semiconductor chip products that are now freely available to exploit; the electronics packed into a portable telephone headset or pager are indeed an impressive sight. This of course is a familiar theme [7,21], whereby electronics equipment is now so reduced in size that the use of a conventional antenna would not be acceptable to the user and would in any case make equipment miniaturization rather pointless.

However, the challenge for the mobile antenna designer goes further because there is now an awareness that with clever design the antenna can give added value by embodying additional system functions such as diversity reception capability, reduction of multipath fading, selectivity of polarization characteristics, or adaptiveness to environmental conditions. Mobile antenna design is no longer confined to small, lightweight, low-profile or flush-mounted, omnidirectional radiators on a well-defined flat ground plane, but is rather the creation of a sophisticated electromagnetic configuration that plays a significant role in signal processing while operating in a generally ill-defined time-varying environment. The antenna is now an integral part of the overall system design as described pictorially in Figure 1.3.

The nature of the mobile system itself greatly influences the ultimate antenna design, and several distinctions can be made between land, maritime, aeronautical, and satellite mobile systems, for instance, and the type of mobile platforms such as vehicles, ships, aircraft, and portable equipment. Frequency reuse capabilities, the type of information, modulation, and personalization of mobile terminals are some of the many factors that are of concern to antenna designers. In zoned systems, radiation patterns have to match the zoned patterns to avoid interference, and performance is also subject to variations in the field strength according to the movement of the mobile terminals and environmental conditions in the propagation path. Also the equipment onto which an antenna element

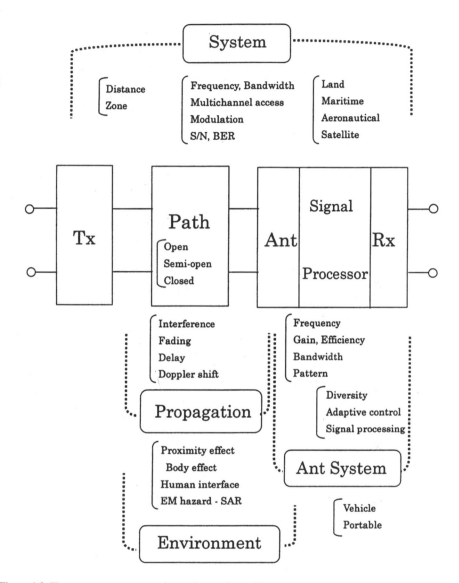

Figure 1.3 The antenna system as an integral part of a mobile system.

is mounted may itself act as a radiator, so that the antenna element and the body of the equipment must be treated together as an antenna system. Proximity effects caused by obstacles near an antenna element affect the antenna performance and must be allowed for in the design. The operator who holds a portable mobile terminal can significantly perturb antenna performance and the human hazard problem must also be kept in mind. Extreme examples are pendant-size pagers and miniature portable telephones, which are now feasible using high-frequency materials with low-loss characteristics. However, the human interface remains one of the most important issues in operating mobile terminals where ease and safety in handling are essential. Recent investigations have shown [22] that the use of a balanced antenna feed arrangement can significantly reduce the influence of the operator's hand and head. Table 1.3 summarizes the influence of system requirements on antenna design.

1.2 TRENDS

Mobile systems are being advanced toward the third-generation systems IMT-2000/UMTS, which offer various new services with higher data rate, video transmission capability, and other advanced facilities. Main regional standard bodies have decided the preferred technology for IMT-2000, for example:

- UWC-136 is a TDMA-based proposal from the United States.
- W-CDMA is a joint proposal from Japan and Europe.
- cdma2000 has advanced from cdma-One in the United States.

Table 1.3
Mobile Antenna System Design Requirements

Requirement	Implication
Antenna as a system	Not as an isolated receive/transmit terminal
Designed to accommodate propagation effects	Some degree of polarization or pattern diversity control embodied, leading to "smart" adaptive antenna arrays
Compatible with environmental conditions	Pattern characteristics to match zone requirements and allow for nearby obstacles
Integration of antenna with vehicle or platform	To include hand and body effects and possible hazard considerations
Latest manufacturing technology	Exploitation of new composite materials and integrated electronic technology
User-friendly and reliable performance	Minimum of moving parts and switches; high reliability of mechanical design
EMC constraints	Reduction of spurious radiation and coupling
Multimedia applications	Increased bandwidth requirements

The W-CDMA system features high-quality and variable-multirate services, multimedia transmission, and international roaming. Antenna systems with intelligent functions realized by adaptive signal processing and software implementation will appear in future advanced systems. The typical trends in modern mobile communications are listed in Figure 1.4 and are discussed in the following subsections.

Personalization

This has been accelerated by downsizing hardware in both mobile stations (MS) and base stations (BS). In particular the downsizing of mobile terminals' handsets has given an

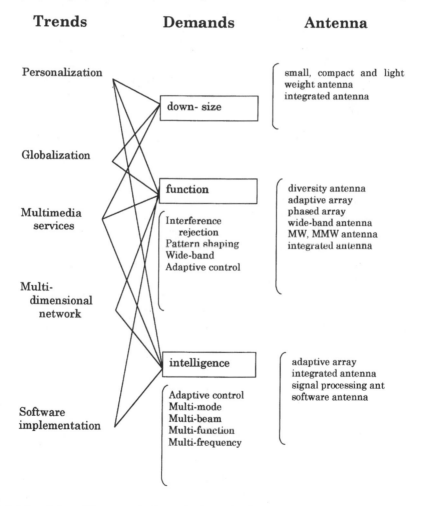

Figure 1.4 Trends in mobile communication and antenna structure.

impetus to the personalization of mobile systems because the smaller handsets are inexpensive, more convenient to carry, and easier to operate. One of the greatest problems encountered is the realization of smaller antennas for downsized handsets without degrading system performance. Downsizing of BSs has also required antennas to be small in size and lightweight as a consequence of the increase in the number of subscribers and need for more efficient use of channels. The size of zones has been made smaller and hence BS antennas are given less space for installation. Lightweight antennas are required for mounting on the wall of a building and MSA arrays are a typical example of downsized and lightweight antennas being deployed. Other advanced antennas, developed in order to save installation space, are dual-beam antennas and dual-frequency antennas.

Adaptive arrays are usually considered as the antenna system which adaptively steers a null, so that the receiving sensitivity toward the interference can be reduced. However, as another important application, an adaptive array can be used for enhancing the channel capacity. The array produces multibeams, which are adaptively controlled toward MSs, so that each mobile is tracked by a narrow beam for both MS-BS and BS-MS transmissions. Wireless multiple access performed by an adaptive array like this is named space division multiple access (SDMA) and is often referred to as a "smart antenna," having an apparent intelligence function.

Globalization

This has been realized for mobile communication by using satellites on the orbits of LEO or MEO as well as GEO. Iridium was the first commercial system, which started providing services in April 1998. Some of the mobile terminals have dual mode, one of which is GSM. The handsets typically employ a quadrifilar helical antenna or a microstrip antenna. Globalization is realized not only by satellite systems but also by wired networks such as the ATM-cored network and the IP-cored network. Cellular network services will be extended from local area to worldwide via connections to these types of networks. Global multimedia services have now become available through the mobile phone. The third-generation mobile systems such as IMT-2000/UMTS aim at the establishment of worldwide multimedia services.

Multimedia

There is a new trend in the application of mobile phone systems that is evolving as their new status evolves; that is, their use as nonvoice systems such as sound delivery, data and video transmission, radio and TV broadcast reception, net-banking, and net-commerce. One example is a type of PHS that is capable of slow video display, playing games, and reception of music delivery as well as speech transmission. Nonvoice services include control systems, which are applied to not only computer systems in offices, but to appliances in the home. Wireless control and short-range communication systems will become an essential aspect of the office/home network and small portable equipment will play an important role. Compact antennas with a wide spectral bandwidth will be sought, thus making operation at microwave and millimeter wavelengths attractive. The trend toward multimedia in ITS is also inevitable to provide drivers with higher information user-friendly 3D displays of motoring data, reinforced with sound prompting.

Multidimensional Networks

Mobile communication systems are now becoming integrated into multidimensional networks that embrace multi-informational media, multitransmission media, and multilayered networks.

Information media is composed of both voice and nonvoice systems including digital voice, sound, still and moving images, and computer data. Transmission media include both wire and wireless lines, radio, and optical links. Land, maritime, aeronautical, and satellite systems will be integrated into complex multilayered networks, thus allowing seamless communication worldwide regardless of time and space. The demand for more intelligent antennas that are integrated with signal processing, adaptive control, and software will continue unabated.

Software Antennas

This is an example of an intelligent antenna that enhances adaptiveness to environmental conditions more dynamically than a conventional adaptive antenna. It consists of both hardware and software functions that sense and evaluate the ambient environmental conditions through the deployment of an appropriate selected algorithm and results in an optimum radiation pattern. In principle, the software antenna concept makes it feasible to adjust to a very wide range of ambient conditions. The antennas for this new concept will be characterized by multimode, multibeam and multifrequency band features.

1.2.1 Communications

One of the biggest changes in electronics has been the onset of digital techniques, which were made possible by the progress in the large-scale integration of semiconductors. Analog components are now in the minority, and this change, in turn, has enabled an order increase in processing power to be packaged in a much reduced volume of equipment space. Electronics designers have been successful in compacting the electronics into wristlet watches, pendants, and so forth, and this has presented a great challenge to antenna designers.

A distinct bonus that we now take for granted is the increased reliability of integrated semiconductor chip technology. A striking feature of communications today is the operational extremes, with satellites providing global coverage for speech, video data transmission, and navigation systems while cellular and microcell arrangements enable mobile telephone services to be highly optimized in dense urban communities. One possible constraint that might limit the growth of mobile systems concerns the ambient level of man-made noise, but electromagnetic compatibility specifications [23] are likely to hold this constraint at bay at least into the near future. More serious issues for mobile systems are likely to be the finite limitation of the frequency spectrum, unification of systems into those with common standards, and security problems.

Table 1.4 itemizes the trends in mobile communications during this century and is subdivided into five generations. In the early days the services were provided by single-channel systems covering unspecified areas on the earth and oceans. Subsequently antennas have evolved into numerous composite and integrated configurations and arrays with sophisticated adaptive control and signal processing facilities. In urban areas, multipath fading has become a serious problem, while zone arrangements and multichannel access networks improve the efficiency of frequency spectrum management. The rapid growth in personal mobile terminals has stimulated the creation of new types of small antennas but further innovative design is needed.

1.2.2 Increasing Information Flow

Mobile phone services are now being expanded to include nonvoice messages. For example, mobile phone handsets are being designed for the transmission and reception of a variety of information media, including sound, graphics, characters, computer data, and video images. Graphics, characters, and both still and moving video can be displayed in color on the LCD (liquid crystal display) panel of a handset. Services such as delivery of music, video games, reception of radio and/or TV, e-banking, e-commerce, and so forth, are becoming available.

Table 1.4
Trends in Mobile Communication Concepts

Year	*1900–*	*1950–*	*1970–*	*1990–*	*2000–*
	Over earth/sea	Urban, highway, rail		Indoors/tunnels	satellites
Propagation	Reflection/diffraction	Multipath		Delay spread	
Radiation	Omnidirectional		Shaped	Circular polarization	
Antenna	Single element	Composite	Phased array	Sig. proc.	Microwave
Size	Electrically small	Low profile	Built-in	Downsizing	MM-wave
Function			Diversity	Adaptive control, intelligence	
Service	Wide coverage	Zone		Microzone	Short range
Channel	Single	Multizoned, multichannel access, digital modulation			
Problems		Intermodulation, multipath fading, EM noise, delay, proximity effects			

In addition, mobile terminals can be used for the purpose of control and telemetering. There is a growing tendency to replace wireline with wireless systems for the control and management of computers and processors in small offices and for the control of electronic appliances at home. For that purpose, short-distance wireless systems are required and small portable terminals are needed. Small antennas are essential for these systems.

Mobile services are expected to be integrated within integrated services digital networks (ISDN) in the future, and in Europe the integration will embody GSM standards. The integration of mobile systems with Internet-cored or ATM-cored networks, and optical fiber networks, is also proceeding. The global communication services, including worldwide multimedia mobile services, will become available in the very near future through these types of networks.

All this points to the use of even higher frequencies to obtain the bandwidth and greater use of mobile satellite links to serve wider areas without the attendant problems of propagation in urban environments. The design of antennas for higher frequency bands introduces both additional problems and increased design freedom, so the techniques for incorporating system functions such as diversity will be somewhat different to those at lower frequencies.

1.2.3 Propagation Challenges

The upsurge in mobile communications has revitalized many topics in propagation and much effort is now devoted to refining previously established propagation mathematical models, as in Table 1.5, for urban and other scenarios, together with substantiation by field trials. In most cases the field structure becomes very complicated because it is often composed of reflected and diffracted waves produced by multipath propagation environments. Propagation in open areas free from obstacles is the simplest to treat but, in general, propagation over the earth and the sea invokes at least one reflected wave.

Before full-wave theory was developed, ray theory was used and Sommerfeld [24] and van der Pol and others [25] contributed greatly to the understanding of propagation. For land mobile propagation, charts developed by Bullington [26] based on the theory of Burrows [27] were used for estimating the VHF/UHF field strength. Various models for land mobile, maritime, aeronautical, and mobile satellite systems have subsequently been developed [28–31] for which theory is well supported by experimental data.

In urban areas, the field is commonly composed of a multiplicity of waves and multipath propagation problems are significant. Analytical solution of the urban propagation scenario is generally impossible and reliance is placed on statistical methods. Okumura provided useful charts, known as *Okumura curves* and *Hata's empirical formulas,* for estimating propagation loss in both urban and suburban regions [32–39].

Propagation in maritime and aeronautical systems is associated with similar problematical phenomena. In closed indoor areas such as tunnels and underground passages, no established models exist due to the complicated field behavior [40,41]. However, when

Table 1.5

Established Propagation Models

Environment	Incident Waves	Models	Mobiles
Open area			
Free space	D	FS; van der Pol	Vehicles, portables
Earth	D+R+DT*	Sommerfeld, Burrows	Vehicles, portables
Rural	D+R	RT; Okumura; Nakagami-Rice	Vehicles, portables
Suburban	D+R	RT; Okumura; Nakagami-Rice	Vehicles, portables
Urban	D+R+DT	Okumura; Rayleigh log-normal	Vehicles, portables
Ocean	D+R+DT	FS-Reflection; Nakagami-Rice	Ships
Air	D+R+DT	FS-Reflection; Nakagami-Rice	Aircraft
Semi-Open Area			
Mountain	D+R	Diffraction	Vehicles
Highway	D+R	FS-LOS; Nakagami-Rice	Cars
Rail	D+R	FS-LOS; Nakagami-Rice	Train
Underground street	D+R+DT	FS-LOS-RT; Rayleigh	Portables
Indoor to outdoor	D+R+DT	FS-LOS-RT; Rayleigh	Portables
Closed Area (short range)			
Indoor	D+R+DT	TLW; Nakagami-Rice	Portables
Underground passage	D+R+DT	TLW; Nakagami-Rice	Portables
Tunnel	S,G	TLW; Nakagami-Rice	Vehicles

Note: LOS = line of sight; D = direct wave; R = reflected wave; DT = diffracted wave; S = standing wave; G = guided wave; FS = free space; RT = ray theory; TLW = transmission line waveguide.
*Single wave

the field structure is random the Rayleigh model, as used in urban regions, may be applied. When the propagation path is on the line of sight, as in tunnels and underground passages, the field may be treated by lossy transmission line or waveguide theory. Direct-wave models may be used for propagation in a corridor.

Multipath propagation can occur in satellite communication systems if the receiver is near a reflecting surface(e.g., an aircraft flying over the ocean) but in urban mobile communications it is the dominant effect. Both tall and small buildings can totally block the view of a satellite while the use of lower frequencies below 2 GHz, and typically below 1 GHz, leads to a high multipath content in urban areas and the fading characteristics are highly dependent on the nature of the local environment, time of day, and other factors. Not surprisingly there is a demand for adaptive systems to cope with ambient conditions as they occur and this has been an opportunity for the antenna designer to create new diversity functions based on ingenious algorithms, discrimination in polarization, and control of patterns. These advancements in antenna design clearly rely on advances in propagation knowledge.

With the enormous demand for mobile communications, radio environments have been changed to include smaller zones such as micro- and picocells and lower antenna height as the trend continues to place antennas at less visible lower levels. Much research is now directed toward methods of precisely locating the site coordinates of a subscriber call within a small cell region. The use of indoor mobile systems is increasing and the problems of small cell propagation are well appreciated [40–42].

1.2.4 Maritime Systems

Traditional high-frequency telegraphy and voice has been supplemented by global satellite INMARSAT-type systems [43] offering superior voice and data transmission. An international paging service is also planned. Global positioning systems (GPS) provide precise navigation data, and much work is being carried out on reliable distress and rescue systems. The Global Maritime Distress and Safety System (GMDSS) implemented in 1992 is one such example. GPS is also being applied to land vehicle navigation and fleet vehicle management in Japan, Europe, and the United States. Rugged antenna installations with added-value system functions compatible with shipborne operation and compact antennas for small-boat operation are some of the fascinating requirements for the antenna designer. Mobile phone services have been extended by the Iridium system, which has enabled use of handsets at any time in almost all areas in the world.

1.2.5 Aeronautical Systems

The aerodynamic constraints are significant, and antennas for both satellite and radio systems must conform to minimum drag and reliability requirements. International services are currently available that carry data and voice communication, together with navigation information on a global basis, but higher data rates to accommodate image and computer data are necessary and have been developed. ETS/V (Engineering Test Satellite-V) experiments [44] for the transmission of telephone, image, and low- and high-speed digital data between satellite and land vehicles, aircraft, and ships have been successful.

1.3 MODERN MOBILE ANTENNA DESIGN

The progress in mobile antenna design is listed in Table 1.6, which shows the related technical issues over the five generations in the time periods cited in Table 1.4. As the system capacity reached its limit with the rapid increase in mobile system users, the frequencies allocated for mobile communications were gradually raised from 30 MHz to 50, 150, 250, and then 450 MHz. Currently, frequencies from 800 MHz to 1.9 GHz have been assigned for mobile telephones, and an allocation of even higher frequency bands such as 2.4 and 5.8 GHz is being considered.

Various new kinds of antennas, such as inverted-F (IFA) and very small rectangular loops, have been developed and applied to pagers and portable telephones. Since the early 1980s, further advances in antenna design were observed as a consequence of the

Table 1.6

Progress in Antennas for Mobile Systems

	1900–	1950–	1970–	1990–	2000–
Frequency		150 MHz<	800 MHz<	1.9 GHz<	2 GHz<
System	Telegraph/telephone for train, ship, aircraft, police cars, portable receiving and transmitting	Voice system for business, navigation, taxis, tone pagers		Satcom, voice, and data channel for aircraft, personal phones, microzones, facsimile, TV-type images, wireless local-area network	
Antenna	Monopole/dipole, whip, top-loaded monopole, inverted-L, loop	Blades, coil loaded, ferrite, helical	Corner reflector, leaky coaxial cable, diversity configurations, body integrated planar inverted-F, bililar helix, microstrip arrays, parallel plate pager, base station, printed wire on glass	Adaptive signal processing	Intelligent software algorithms

introduction of cordless phones, multichannel access systems, and navigation systems. In urban mobile communications, diversity antennas with space or polarization schemes have evolved to reduce multipath fading, while the personalization of mobile terminals has demanded electrically small antennas. The concept of integrated antenna systems [45] has been applied to antenna systems for portable equipment.

The development of even smaller antennas has been required for more advanced personal communication systems, particularly like the personal handy-phone systems (PHS) and PDC in Japan. Other requirements are antennas for the handsets of hybrid systems; for example, the combined handsets of GSM and DECT and for PHS and GPS need dual- or wideband small antennas. In designing a built-in antenna for handsets, the basic concept is to choose an element that is small, low-profile, compact, and lightweight. A PIFA is commonly used and recently a normal mode helical antenna (NMHA) or a meander line element has been combined with a monopole element to act as a composite main antenna for a handset.

A tiny chip antenna which is essentially a NMHA encapsulated in the high-permeability ceramic material, has also been employed in the PHS. Note that when a small antenna element is built into the handset, antenna performance is generally enhanced because currents induced on the conducting surfaces of the handset reradiate. Increased gain and bandwidth was observed when a small PIFA element was deployed in a handset. However, the handset radiation performance will, in most cases, be noticeably degraded when the operator holds the handset due to induced hand and body losses. Ways of reducing this unwanted effect have been reported [22] and may change the design concept of handset antennas. An illustration of the size and weight reduction achieved for pagers and portable phones is shown in Figure 1.5. The smallest portable phone now has a volume of about 57 cm^3 while pendant pagers have dimensions of the order $7 \times 7 \times 9$ mm^3.

The above antenna design is increasingly dependent on computer-aided design (CAD) based on well-known mathematical methods. In particular the finite difference time domain (FDTD) method has proven to be valuable for computing radiating structures embedded in layers of heterogeneous material such as a handset close to the human head. Some examples of computational methods applied to antennas are as follows.

1. *Moment methods:* monopole or IFA on portable equipment [46–48] and a monopole on a vehicle at VHF [49];
2. *Geometric theory of diffraction (GTD):* monopole on a vehicle at UHF [50];
3. *Hybrid methods:* moment methods and GTD [51];
4. *FDTD:* an IFA on portable equipment [52] and simple antennas [53];
5. *Spatial network method:* impedance characteristics of an IFA on an infinite ground plane [54];
6. *IE3D:* microstrip antenna [55] and current distributions on the body of handsets [56]; and
7. *Genetic algorithm* [57]: loaded wire antennas [58] and array thinning [59].

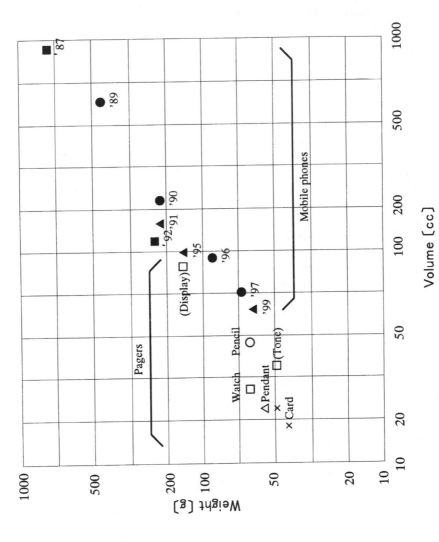

Figure 1.5 Volume and weight of pagers and portable telephones.

The IE3D method is based on a full-wave integral equation and the method of moments in the frequency domain and is applied to three-dimensional multilayered structures of arbitrary shape.

1.4 OBJECTIVES OF THIS BOOK

The myriad of antennas and antenna arrays that are already associated with the mobile communications industry presents those engaged in the industry with a bewildering collection of seemingly different antenna types, electromagnetic theory concepts, and subsequent design techniques. There is a clear requirement to classify and catalog the antennas into generic types so that the fundamental principles and reasons for creating design variants are evident. Such a classification is seldom attempted in antenna texts and as such we have devoted an entire glossary to this important task. Placing this glossary at the end of the book in no way subtracts from its importance but rather leaves the editors free to address the main book text to the wider task of examining antenna systems for numerous mobile applications; our scope necessarily encompasses all relevant aspects as listed in Tables 1.1 to 1.6.

Since we are dealing with a system, it was felt inappropriate to break the text down into headings as in these tables, but rather address the main application areas of land, maritime, satellite, and aeronautical mobile systems. These main areas are in themselves further subdivided into numerous ways to reflect current developments. Most antenna systems described arise from communication requirements, but there are exceptions, where the acquisition of data relates directly to navigation, identification, and so on.

Another decision that confronted the editors was how to address the rapid advances of techniques now being deployed in certain aspects of mobile communications and it was decided that these very recent developments justified separate sections or chapters, naturally following from foundation topics and established techniques.

For completeness, Chapter 2 collates background material on technology, propagation, and antennas to support the subsequent chapters. Because propagation is one aspect of mobile communications that continues to be intensively researched worldwide it was considered essential to report these new developments in Chapter 3. Here, recent advances in methods of predicting the coverage and capacity of mobile systems are described. The propagation behavior and prediction methods relate to the type of propagation cell, that is, pico, micro, and so on, and enable important trade-offs to be made between equipment complexity and the communication services sought.

Land communication systems outnumber those in other global sectors and occupy several chapters commencing with Chapter 4 where basic techniques for base station and vehicle antennas are described; recent advances in base station design and research into adaptive antenna arrays then follow in Chapter 5 and consolidate much of the contemporary research thrust into land systems. Examples of pagers and portable phones are then presented in Chapter 6 together with some early work on the exposure of operators to electromagnetic radiation.

A topic of continued enormous interest concerns the exposure of operators to the electromagnetic radiation emanating from their handsets, and exhaustive research activity has been committed to obtaining more precise mathematical modeling methods and reliable field measurement to assess the operator safety. Not surprisingly, this has influenced handset antenna design and the latest aspects of safety are presented in Chapter 6 while an appreciation of the progress in modeling, measurement and antenna development is given in Chapter 7.

The demand for quality in-car entertainment systems and reliable communication for trains and road traffic in general by way of ITS continues to grow and several disparate applications are described in Chapter 8 to complete the coverage of land systems.

Chapter 9 is devoted to satellite communication systems, which are mainly the well-established geostationary systems embracing vehicle, shipborne, and broadcast applications. The chapter concludes with an outline of recent developments in mobile satellite systems and the impact on antenna design.

Aircraft communication has had a long history and, although well established, it can benefit from advances in antenna technology. Chapter 10 concentrates on the variety of airborne systems for communication, navigation, and other related aspects of avionics.

REFERENCES

[1] O'Hara, J. G., and W. Pricha, *Hertz and the Maxwellians*, Peter Perigrinus Ltd., 1987.

[2] Hawks, E., *Pioneers of Wireless*, Methuen, 1927.

[3] Isobe, S., "From Chappe's Semaphore Telegraph to Satellite Communications," p. 133, KDD Data Centre, 1968 (in Japanese).

[4] Ibid, p. 134.

[5] Historical papers on developments in antennas and propagation, 1912 to 1962, published in *The Fiftieth Anniversary Issue of the Proceedings of the Institute of Radio Engineers*, May 1962, pp. 679–717.

[6] James, J. R., and G. Andrasic, "Multifunction Printed Antennas," Ch. 6 in *Advances in Microstrip and Printed Antennas*, H. F. Lee and W. Chen. eds., John Wiley and Sons Ltd., 1997, pp. 273–324.

[7] James, J. R., "What's New in Antennas," *IEEE Antennas and Propagation Society Magazine*, Feb. 1990, pp. 6–18.

[8] Lucky, R. W., "New Communication Services—What Does Society Want," *Proc. IEEE*, Vol. 85, No. 10, Oct. 1997, pp. 1536–1543.

[9] Siwiak, K., *Radiowave Propagation and Antennas for Personal Communications*, Artech House, 1998, pp. 2–3.

[10] "World Cellular and PCS Markets, 1999," The Strategis Group.

[11] Kneltsch, W., and P. Berrimon, "Inside Chinese Telecom," *Telecommunications*, Aug. 1999, p. 19.

[12] Ohta, K., "75 Million Mobile Phone Subscribers Would Be Near; Mobile Communication Market Review," Nikkei Electronics, 3 Mar. 1999, p. 155 (in Japanese).

[13] Clarke, P., "CDMA and the Race Towards 3G," *Microwave Engineering Europe*, July 1999, pp. 17, 18, and 20.

[14] Prasad, R., "*Universal Wireless Personal Communications*," Artech House, 1998, Ch. 8 and Ch. 9.

[15] Woodward, G., and B. S. Vucetic, "Adaptive Detection for DS-CDMA," *Proc. IEEE*, Vol. 86, No. 7, July 1998, p. 1413.

[16] Budagavi, M., and J. D. Gibson, "Speech Coding in Mobile Radio Communications," *Proc. IEEE*, Vol. 86, No. 7, July 1998, pp. 1402–1412.

[17] Godara, L. C., "Applications of Antenna Arrays to Mobile Communications, Part 1: Performance Improvement, Feasibility and System Considerations," *Proc. IEEE*, Vol. 85, No. 7, July 1997, pp. 1031–1060.

[18] Shafi, M., A. Hashimoto, M. Umehira, S. Ogose, and T. Murase, "Wireless Communication in the Twenty-First Century: A Perspective," *Proc. IEEE*, Vol. 85, No. 10, Oct. 1997, pp. 1622–1638.

[19] Evans, J. V., "Satellite Systems for Personal Communications," *Proc. IEEE*, Vol. 86, No. 7, July 1998, pp. 1325–1341.

[20] Wakling, J., "Antenna System Requirements—Pointing in a New Direction," presented at IEE Conference on Antennas and Propagation, University of York, March–April 1999.

[21] Fujimoto, K., A. Henderson, K. Hirasawa, and J. R. James, *Small Antennas*, Research Studies Press, 1987, distributed by Wiley & Sons.

[22] Furuuchi, H., H. Morishita, Z. Ide, Z. Tanaka, and K. Fujimoto, "A Balance-Fed Loop Antenna System for Handsets," *IEEE Int. Symp. on Antenna & Propagation*, Aug. 1999, pp. 6–9.

[23] European Community's Directive on Electromagnetic Compatibility starting 1 Jan. 1992.

[24] Sommerfield, A., "Uber die Ausbreitung der Wellen in der Drahtlosen Telegraphic," *Ann. Phys.*, Vol. 28, 1909, p. 665.

[25] van der Pol, B., and H. Bremmer, "The Diffraction of Electromagnetic Waves From an Electrical Point Source Round a Finitely Conducting Sphere With Application to Radiotelegraphy and the Theory of Rainbow," *Phil. Mag.*, Vol. 24, 1937, pp. 141–825.

[26] Bullington, K., "Radio Propagation at Frequencies Above 30 MC," *IRE*, Vol. 35, 1947, pp. 1122–1136.

[27] Burrows, C. R., and S. S. Attwood, *Radio Wave Propagation*, New York, 1949.

[28] Akeyama, T., S. Sakagami, and K. Yoshizawa, "Propagation Characteristics of Air-Ground Paths, 900 MHz Band," *Trans. IECIE*, Vol. J73-B-11, No. 8, Aug. 1990, pp. 383–389.

[29] CCIR Rept. 567-3, "Methods and Statistics for Estimating Field-Strength Values in the Land Mobile Services Using the Frequency Range 30 MHz to 1 GHz" *Rec. and Reports of the CCIR*, Vol. 5, ITU, Geneva, 1986.

[30] Karasawa, Y., "Complex Frequency Correlation Characteristics of L-Band Multipath Fading Due to Sea Surface Scattering," *Trans. IECIE*, Vol. J72-B-11, No. 12, Dec. 1989, pp. 633–639.

[31] CCIR Rept AL/5, "Propagation Data for Aeronautical Mobile-Satellite Systems for Frequency Above 100 MHz," *Rec. and Reports of the CCIR*, Vol. 5, ITU, Geneva, 1991.

[32] Suzuki, H., "A Statistical Model for Urban Radio Propagation," *IEEE Trans. Comm.*, Vol. 25, 1977, pp. 673–679.

[33] Turin, G. L., et al., "A Statistical Model for Urban Multipath Propagation" *IEEE Trans.*, Vol. VT-21, 1997, pp. 1–8.

[34] Hata, M., "Empirical Formula for Propagation Loss in Land Mobile Radio Services" *IEEE Trans.*, Vol. VT-29, No. 3, 1980, pp. 317–325.

[35] Okumura, Y., et al., "Field Strength and its Variability in VHF and UHF Land-Mobile Radio Services," *Rev. Elec. Comm. Lab.*, Vol. 16, Sep./Oct. 1968, pp. 825–873.

[36] Tsuruhara, T., et al., "Mobile Radio Propagation Characteristics in Urban Areas in UHF Bands" *Trans. IECE*, Vol. E66, 1983, pp. 724–725.

[37] Okumura, Y., and T. Akeyama, "Propagation in Mobile Communications," Ch. 2 in *Fundamentals of Mobile Communications*. Y. Okumura and M. Shinji, eds., IEICE, 1986.

[38] Akeyama, T., "Propagation in Mobile Communication Networks," Ch. 12 in *Radiowave Propagation*, M. Shinji, ed., 1992, pp. 203–248.

[39] Lee, W. C. Y., *Mobile Cellular Communications Systems*, McGraw-Hill, 1990.

[40] Hashemi, H., "The Indoor Radio Propagation Channel," *Proc. IEEE*, Vol. 81, No. 7, 1993, pp. 943–967.

[41] Dresch, U., and E. Zollinger, "Propagation Mechanism in Microcell and Indoor Environments" *IEEE Trans. Vehicular Technology*, Vol. 43, No. 4, 1994, pp. 1058–1066.

[42] Feuerstein, M. J., et al., "Path Loss, Delay Spread and Outage Models as Function of Antenna Height for Microcellular System Design," *IEEE Trans Vehicular Technology*, Vol. 43, No. 3, Pt. 2, 1994, pp. 487–498.

[43] Miya, K., ed., *Satellite Communication Technology*, KDD Engineering and Consulting, 1982, pp. 5, 37.

[44] Hamamoto, N., et al., "Results on CRL's Mobile Communication Experiments Using ETS-V Satellite," *Space Communications*, No. 7, 1990, pp. 483–493.

[45] Fujimoto K., "A Treatment of Integrated Antenna Systems," *IEEE AP-S Int. Symp.*, 1970, pp. 120–123.

[46] Nakano, H., and S. R. Kemer, "The Moment Method Solution for Printed Wire Antennas of Arbitrary Configuration," *IEEE Trans. on Antennas and Propagation*, Vol. AP-36, 1988, p. 1667.

[47] Taga, T., and K. Tsunekawa, "Performance Analysis of a Built-in Planar Inverted-F Antenna for 800MHz Band Portable Radio Units," *IEEE Trans. Selected Areas in Communication*, Vol. SAC-5, No. 5, June 1987, pp. 921–929.

[48] Sato, K., et al., "Characteristics of a Planar Inverted-F Antenna on a Rectangular Conducting Body," *Electronics and Communications in Japan*, Vol. 71-B, Scripta Publishing, Aug. 1989, pp. 43–51.

[49] Nishikawa, K., "Effects of Automobile Body and Earth on Radiation Patterns of Antennas for FM Radio," *Trans. IECE Japan*, Vol. E67, Oct. 1984, pp. 555–562.

[50] Nishikawa, K., and Y. Asano, "Vertical Radiation Patterns of Mobile Antenna in UHF Band," *IEEE Trans. Vehicular Technology*, Vol. VT-35, May 1986, pp. 57–62.

[51] Thiele, G. A., and T. H. Newhouse, "A Hybrid Technique for Combining Moment Methods with the Geometrical Theory of Diffraction," *IEEE Trans.* Vol. AP-17, 1969, pp. 62–69.

[52] Kagoshima, K., A. Ando, and K. Tsunekawa, "FD-TD Analysis of a Planar Inverted-F Antenna Mounted on a Conducting Box," *Proc. Int. Symp. Ant. Propag.*, Sapporo, Japan, Sep. 1992, pp. 713–716.

[53] Maloney, J. G., et al., "Accurate Computation of the Radiation From Simple Antennas Using the Finite-Difference Time-Domain Method," *IEEE Trans. Antennas and Propagation*, Vol. AP-38, 1990, p. 1059.

[54] Taga, T., "Analysis of Planar Inverted-F Antennas and Antenna Design for Portable Radio Equipment," Ch. 5 in *Analysis, Design, and Measurement of Small and Low-Profile Antennas*, K. Hirasawa and M. Haneishi, eds., Artech House, 1991.

[55] Harscovici, H., "A Wide-Band Single-Layer Patch Antenna," *IEEE Trans. Antennas and Propagation*, Vol. 46, No. 4, Apr. 1998, pp. 471–474.

[56] Morishita, H., H. Furuuchi, and K. Fujimoto, "Balance-fed L-type Loop Antenna Systems Handset" *IEEE VTC-Fall Proc.*, Vol. 3, 1999, pp. 1346–1350.

[57] Goldberg D E., *Genetic Algorithms*, New York, Addison-Wesley, 1989.

[58] Boag, A., et al., "Design of Electrically Loaded Wire Antennas Using Genetic Algorithms," *IEEE Trans. Antennas and Propagation*, Vol. 44, May 1996, pp. 687–695.

[59] Haupt, R. L., "Thinned Array using Genetic Algorithms," *IEEE Trans. Antennas and Propagation*, Vol. 42, July 1994, pp. 993–999.

Chapter 2

Essential Techniques in Mobile Antenna Systems Design

K. Kagoshima, W. C. Y. Lee, K. Fujimoto, and T. Taga

2.1 MOBILE COMMUNICATION SYSTEMS

2.1.1 Technologies in Mobile Communications

Throughout the history of mobile communication systems, various kinds of systems have been developed and commercialized. The depth and breadth of the technologies supporting the systems reflect system size and complexity. For example, public mobile phone systems, such as the automobile telephone system, are some of the largest and most complex systems; the outline of the system configuration is shown in Figure 2.1. A complete understanding of wave propagation, radio transmission, channel control, and hardware devices is essential to construct a mobile phone system [1].

To create antenna hardware, we must first consider propagation and radio transmission characteristics. For example, the antenna pattern used at the base station affects the propagation characteristics, and the arrangement of the antenna pair in a reception diversity system virtually determines the transmission characteristics of the radio channels. Antenna engineers must consider all related factors in order to realize the full potential of any antenna hardware.

The main features and recent trends of the related factors are summarized in the following sections. Readers interested in more detail may refer to individual textbooks [2,3] or the papers appearing in the special issue of [4].

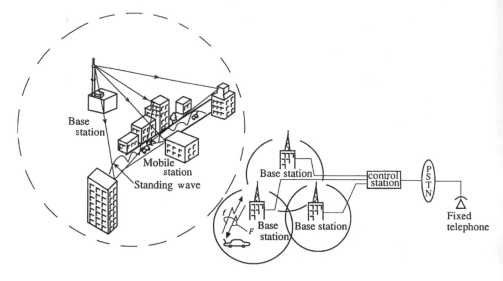

Figure 2.1 System configuration of mobile communications (cellular radio telephone systems).

Propagation

In all radio communication systems, the study of radio wave propagation is inevitable, especially in mobile communication systems, where the propagation path is seldom line-of-sight within the propagation region. This means that transmitted waves are much affected by buildings, towers, and objects within the spatial environment before they reach the receiver. Their propagation characteristics differ greatly from those observed in free space. Although much research concerning mobile propagation has been published, Okumura was first to present the entire range of mobile propagation characteristics needed in mobile communication system design [5]. In 1970, he clarified propagation-path loss characteristics in the propagation environments of urban, suburban, and open areas across wide-frequency ranges from the 200-MHz band to the 2.0-GHz band. In his monumental paper, he also demonstrated the influence of antenna height on the path loss characteristics. He summarized his results in graphic form as *Okumura's curve,* which was adopted by CCIR Recommendation 370 of SG5 [6]. This has become the basis of mobile communication system design throughout the world. Following his seminal research, propagation studies were continuously conducted to increase prediction accuracy, and some improvements in propagation characteristics have been achieved. The predicting formulas for path loss, which took into account building density in the horizontal plane, were obtained [7]. Hata derived the predicting formulas for propagation-path loss, which are now used by system designers [8]. Recently, theoretical predictions for determining field strength were presented based on the data of buildings and geographical features using geometrical optics and/or geometrical theory of diffraction [9,10]. In the future, it will be possible to estimate propagation characteristics by using computer simulations.

One of the significant achievements in the field of radio wave propagation concerns reception diversity. The correlation coefficient between two antennas was measured in various propagation environments, and the received levels achievable through diversity reception were clarified. Digital mobile systems are being developed throughout the world to provide enhanced services, but the problems facing wide-frequency-band transmission within a severe fading environment are formidable. For example, measurements of wideband mobile propagation were carried out for the development of the GSM, and the important results obtained through these measurements and evaluations are summarized by Lorenz [11]. Diversity reception or equalization to reduce multipath delay is being actively persued throughout the world. In the study of propagation delay, the characteristics of propagation and transmission cannot be considered in isolation. Antenna characteristics such as the radiation pattern and correlation coefficient are related to propagation delay. Therefore, it should be pointed out that a total study, taking into account propagation, transmission (system), and the antenna, is necessary [12].

Radio Transmission

Frequency modulation/demodulation (FM) is the most used modulation scheme in mobile communication systems, because it is robust against thermal noise and/or interference and is easy to realize as hardware. From the viewpoint of frequency-effective use, an important point in developing an appropriate FM transmission technique is determining how to narrow the bandwidth. In the Japanese 400-MHz band mobile communication system, channel separation was successfully narrowed from 50 to 25 kHz and then from 25 Hz to 12.5 kHz. Moreover, an 800-MHz automobile telephone system has been created around the interleave channel allocation method and offers the narrow channel separation of 6.25 kHz. The achievement of this narrow channel separation is due to the development of a highly stabilized oscillator and high-performance filter, as well as a syllabic compandor [13].

For analog transmission, single-sideband (SSB) is an attractive modulation scheme, which can realize frequency-effective use. SSB equipment has been substantially improved, but this has not prevented the emergence of digital systems.

Digital mobile systems have been actively pursued in order to realize enhanced mobile communication services. Commercial services started in Europe and Japan in 1991 and 1993, respectively [14]. For these systems, the technologies of voice codecs and narrow-band digital modulation were greatly advanced. Diversity reception, as well as equalization, continues to be studied to improve the bit error rate; error correction also plays an important role in digital transmission systems. Moreover, the effect of all the above techniques on transmission system performance was investigated.

Control

In mobile communication systems, numerous control technologies such as channel connection, register of position, and zone switching are required to establish a link between the transmitter and receiver. Special switching technology is also necessary to link the mobile radio network to public telephone networks. These control technologies differ widely in many points from those of the ordinary fixed radio communication system. Although the quality of the control technology significantly affects the performance of mobile communication, it is peripheral to the subject of this book and readers interested in control technologies should refer to [15].

Hardware

The hardware technologies needed for establishing the radio transmission network cover a wide area, including antennas, active circuits, and batteries. Nippon Telegraph and Telephone Corporation's (NTT) automobile telephone service was put into commercial use in 1979, and its portable telephone service began in 1987. Over two decades, the volume and weight of the portable equipment unit have been decreased as shown in Figure 2.2. Cutting size and weight by 75% has been achieved in under five years. These results are due mainly to the use of large-scale integration (LSI) parts, especially in the intermediate frequency (IF) circuits, baseband circuits, and control circuits. Reducing the power consumption of RF circuits is also an important breakthrough, which has lengthened the operation time of radio units, as have highly effective batteries.

    As for the passive circuits, the high ϵ dielectric filter with relative permittivity of more than 90 and surface acoustic wave (SAW) filters contribute to the downsizing of

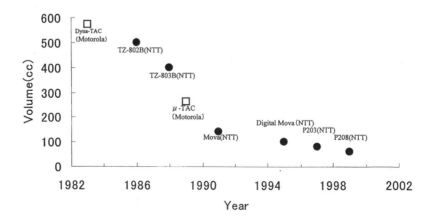

Figure 2.2 Change in volume of mobile phone.

radio units [16]. As the radio units become smaller, the antennas of the mobile units should be equally small or even integrated within the units' bodies. Research and development are being carried out to obtain a suitably small antenna with high performance [17].

Major research items concerning the four technologies mentioned above are summarized in Table 2.1. The development of these technologies are aimed to achieve three goals: (1) effective use of radio frequency resources, (2) a wider variety of enhanced services, and (3) cost effectiveness. Even when we pursue individual technical developments, we should keep these goals clearly in mind to ensure that the developments will contribute to an effective and successful system.

By downsizing and increasing the gain of the portable radio antenna, very effective radio units can be created that will have long operation times and will be easy to use. Increasing the gain and lowering the sidelobe levels of the base station antenna [17] will enhance system economy, because the transmitting power of the base station can be reduced and frequency reuse can be pursued more aggressively. The creation of a more sophisticated system that can handle dramatically increased levels of service requires the development of a base station antenna that offers higher performance and expanded functions. The relationship between antenna technology and its impact on mobile communication systems is shown in Figure 2.3.

The above discussion focuses mainly on land mobile telephone systems. The aeronautic and maritime mobile telephone systems need almost the same technologies, since only the service areas are different from that of the land mobile systems. However, the antenna patterns required for each system depend on the expanse and shape of the service area. Thus, different types of antennas are needed, as discussed elsewhere in this book. For example, in aeronautic mobile telephone systems, the service area exists from ground level up, and the base station antenna, which is located on the ground, should radiate upwards. To compensate for differences in field intensity, cosecant squared shaped beams are used.

Private mobile radio systems like the multichannel access (MCA) system in Japan [18] (which were common in the early period of mobile communication systems) have

Table 2.1
Major Research Items for Mobile Communications

	Research Items
Propagation	Propagation-path loss prediction, delay profile measurement, diversity, microwave and millimeter wavelength bands
Transmission	Modulation/demodulation, signal processing, code/decode, error correction, diversity reception, multichannel access, rake reception
Control	Power control, channel allocation, channel switching, location registration, file access, adaptive processing, and beamforming
Hardware	Antennas, filter, oscillator, synthesizer, amplifier, codec, LSI, battery, MMIC, DSP, and A/D and D/A in RF regions

28

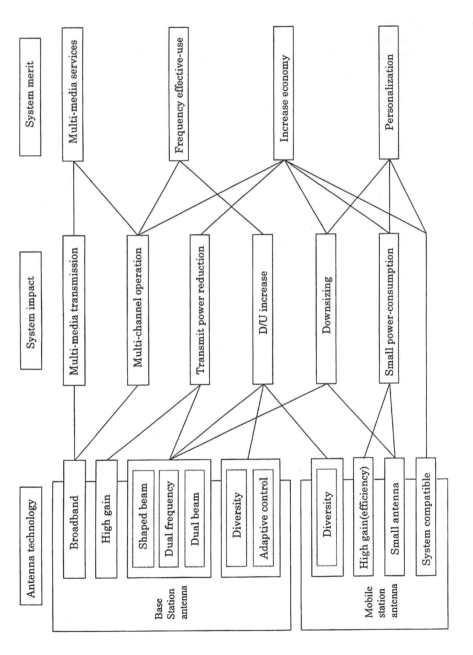

Figure 2.3 Relation between antenna technology and its system impact.

much simpler control station functions than cellular systems because they (the private systems) are not connected to the public service telephone network (PSTN). However, they require virtually identical technologies for propagation, radio transmission, and hardware.

As will be seen in Chapter 9, mobile satellite communication systems have been commercialized through INMARSAT. In the near future, advanced mobile satellite systems will be realized in some countries using domestic satellites. In these systems, the antennas for mobile stations usually need higher gain and satellite tracking. Moreover, they should be economical. Therefore, the design of those antennas becomes more complicated, and notable progress has been made in many countries.

2.1.2 Frequencies Used in Mobile Systems

Technical Aspects

Frequencies used in mobile communication systems have undergone a slow shift from the lower frequency bands to the higher frequency bands. The optimum frequency band is basically determined by the following technological demands: (1) Frequencies at which small and light mobile terminals can operate, (2) frequencies yielding the largest propagation distances, and (3) frequencies offering adequate bandwidths to meet for system requirements.

The frequencies are not permanently fixed, but change with system enhancement or the appearance of new systems. Figure 2.4 shows the changes in frequencies used in major Japanese mobile communications systems [19] up to 1994, and Figure 2.5 illustrates the change process. If a system is introduced for commercial service and grows rapidly, it soon exhausts its frequency resources. The typical response is to develop technologies that maximize the utilization of frequency resources. If the frequency shortage cannot be overcome, new frequency bands are employed. This generally requires the establishment of a new system based on the largest technologies. Frequencies used by the mobile

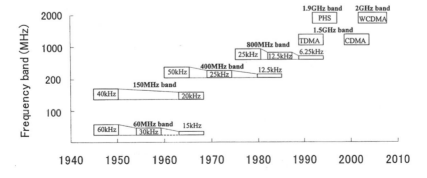

Figure 2.4 Frequency band and channel separation for major mobile systems in Japan.

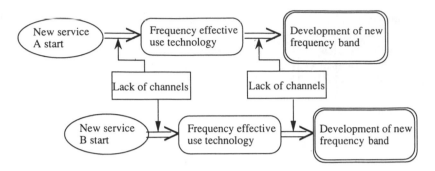

Figure 2.5 Channel increase process due to frequency-effective use technology and higher frequency band development.

communication systems have expanded to almost completely occupy the VHF and UHF bands, because these frequencies best satisfy the above three demands, as shown in Figure 2.6 [19]. This continued development mirrors the development experienced by fixed radio relay systems, which use frequencies below 10 GHz, the so-called *microwave frequency band.*

The first mobile communication systems were private networks, and service demands were not so high. The recent introduction of public systems and convenient equipment such as automobile telephones, portable telephones, cordless telephone, and pager systems has led to a dramatic increase in demand such that frequency resources are rapidly exhausted. The response is to utilize new bands; for example, because the 800-MHz band is becoming congested, the 1.5-GHz band or 2-GHz band and higher are now being developed for mobile communication system used. As the frequency increass, the radio waves attenuate more with distance. It is necessary therefore to increase antenna gain or to narrow the cell size to maximize system economy. If the higher frequency bands are to be used, the power consumption and size of the mobile station, especially the RF circuit, must be decreased to achieve a convenient and economical mobile station.

Regulation Aspects

The frequency band adopted depends on the frequencies permitted by Radio Regulation (RR), as well as technical considerations. The VHF and UHF bands, which are commonly used in mobile communication systems, are also allocated for FM broadcasting, TV broadcasting and various private fixed radio communication systems. The principle of frequency sharing is determined by the RR of each country. Under RR, an administrative body in each country determines allocation of the radio frequency within the country.

The 1992 and 1995 World Administrative Radio Committee (WARC) and the International Telephony Union (ITU) allocated new frequency bands for mobile communi-

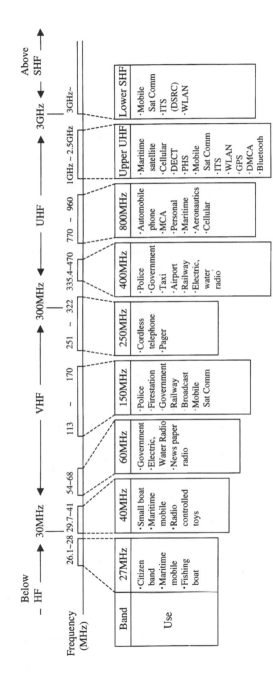

Figure 2.6 Use of frequency band for mobile communications in Japan. (After [19].)

cation systems, as shown in Table 2.2 [20]. This resulted in a rapid increase of mobile services all over the world.

2.1.3 System Design and Antennas

The word *system* is commonly used in engineering today and through this book. The word implies that many factors are considered in the design, and it is instructive here to briefly discuss system design in relation to antennas. Several decades ago, equipment was often designed by assembling a collection of component parts that had been optimized in isolation from one another. For instance, communication apparatus would be fitted with an antenna of standard optimized design and subsequent adjustments made in situ. As equipment complexity increased, this bottom-up design process gave way to today's top-down system approach, where component parts are optimized in relation to all factors that influence the equipment function. An obvious yet seldom appreciated fact is that a component that is highly optimized in performance may not be the best choice indicated by the system approach. A good example is the printed patch antenna, which can have a lower efficiency than a conventional wire monopole antenna, yet it is the printed patch with its low profile and compatibility with printed technology that has made many new types of systems possible, particularly in mobile communications, radar, and navigation equipment. Antennas today cannot be designed in isolation from their host equipment, and systems design is an essential technological approach in the realization of high-performance radio equipment operating to a critical specification. System design is amply demonstrated by the various antenna systems described in this book, and it serves no useful purpose to generalize further. However, a few comments on how system design is carried out are in order and illustrated by the following example of an urban mobile communication system for vehicles and portable units.

The factors that a system designer would list include the following.

- Zone configurations—defining signal coverage and antenna patterns;
- Base station antennas—height, physical constraints, and requirements for beam down tilting;
- Noise levels—thermal and environmental;
- Interference—its level and nature and cochannel and adjacent channel effects;
- Signal requirements—optimal frequency of operation, bandwidth, intermodulation effects, and effective utilization of frequency spectrum;
- Cost of development and subsequent manufacture;
- Reliability—servicing required and ease of access, and costs;
- Vulnerability to damage—exposure to weather, corrosion, and wear and tear;
- Network operation requirements;
- Customer appeal.

Table 2.2

Frequency Allocations and Bands Identified for Mobile Services by WARC-92 and WARC-95

	Allocation	*Frequency (MHz)*
Mobile satellite:		
Low earth orbit	New	137–138 (down)*
		148–149.9 (up)
		149.9–150.05 (up); land mobile satellite
		312–315 (up) +
		312–315 (down) +
		400.15–401 (down)
1.6/2.4-GHz band	New	1610–1625.5 (up)
		2483.5–2560 (down)
L band	Extended	1525–1530 (down)
		Region 1: maritime and land mobile satellite
		Regions 2, 3: mobile satellite
S band	Extended	2500–2520 (down)
	(region)	2670–2690 (up)
IMT-2000		
Terrestrial	ITU	1885–1980 (up)
		2110–2170 (down)
MSS		1980–2010 (up)
		2170–2200 (down)
Terrestrial	Japan	1920–1980 (up)
		2110–2170 (down) (WCDMA)
MSS		1980–2010 (up)
		2170–2200 (down)
Terrestrial		2010–2025 (TD-CDMA)
UMTS		
Terrestrial	Europe	1910–1980 (up)
		2110–2170 (down) (WCDMA)
MSS		1980–2010 (up)
		2170–2200 (down)
Terrestrial		2010–2025 (TD-CDMA)
PCS**		
Terrestrial	Americas and	1850–1990
MSS	Caribbean	1990–2025 (down) ++
		2160–2200 (up) ++

Note: down = downlink from satellite to earth; up = uplink from earth to satellite.
*Partly including secondary service.
+Including secondary service.
**Including other systems not related to IMT-2000.
++Revised WARC-95.

There arc, no doubt, many other considerations, but the list is sufficient to illustrate the various factors that will influence the antenna design to some greater or lesser degree. The point we are leading to here is the actual process of design that enables the designer to embody these factors and translate them into physical constraints on the hardware. It is tempting to think that the constraints can be translated into equations and the requirements computed, but this is most unlikely. It is true that in academic circles there are topic areas called *system analysis,* where every facet is described by equations and numerical optimization methods, but this assumes that all unknowns have given functional forms and, likewise, noise and interference effects. In reality, the mobile antenna system designer will have incomplete data, much in subjective form. Furthermore, the system requirements will be time-varying and likely to change significantly with the fashion of customer demand. The process of practical system design amounts to an iterative procedure geared to a commercial business plan with limited time scales. The procedure will embrace computation, measurement, and field trials and will involve a team of individuals responsible for different aspects of the system. One might enquire how such a procedure could guarantee that an optimum design is achieved, and the answer lies in the existence of competition from other manufacturers and the influence of the customer, who has the ultimate experience with the system and the choice. Most of the antenna systems in this book have been through this pragmatic design process, and the work of a variety of manufacturing companies worldwide is presented.

2.2 FUNDAMENTALS AND PREDICTIVE MODELS IN LAND MOBILE PROPAGATION

As already mentioned, the propagation mechanism of radio waves in a mobile environment is a vital aspect of design. A book such as this that concentrates on antennas does not allow the space to give a comprehensive coverage of the subject of propagation, but some outline is necessary to introduce the terminology, physical behavior, mathematical models, and predictive processes encountered by system designers. Both introductory and advanced texts on propagation are readily available in the literature, so we have chosen here to present a compact outline of established techniques for land mobile systems, due mainly to W. C. Y. Lee, to illustrate the role of propagation in antenna system design. Propagation methodologies pertaining specifically to maritime, aerospace, and satellite systems are available in the literature, but are generally not as complex and convoluted as those in land mobile systems.

2.2.1 Propagation Problems in Land Mobile Communications

In the mobile radio environment, there are five unique factors [21]:

- Natural terrain configurations such as flat ground, hills, water, mountain, valley, and desert;

- Manmade structures such as open areas, suburban and urban areas, and metropolitan areas;
- Manmade noise such as automotive ignition noise and machine noise;
- Moving medium brought about by the mobility of the mobile and portable units;
- Dispersive medium causing frequency-selective fading and time-delay spread.

The above five factors are made signifiant because the mobile antenna is very close to the ground. When the mobile antenna is around 1.5 to 3m above the ground, the signal received by the mobile unit comprises a direct path signal and a strong reflected wave due to the closeness with the ground. These two waves, when combined, result in an excessive path loss at the mobile reception. Also, because the mobile antenna is close to the manmade structures and manmade noise sources, the path losses, multipath fading, and interference will have profound effects.

2.2.2 Propagation Models and Field Strength

There are essentially two kinds of progagation models. One is the propagation prediction model and the other is the multipath fading model. The propagation prediction model is used to predict the average field strengths (also called the *local means*) and as a tool to design a mobile radio communication system in different geographical areas. The antenna height, gain, and directivity at the base station play a big role in the model. The multipath fading model is derived from the natural randomness of multipath wave arrival. The model can predict the behavior of instantaneous field strengths in the field.

Value of the Prediction Model

Why is the prediction model so important? The value of the prediction model is to save manpower, cost, and time. Before planning a cellular (or mobile radio) system in an area, selecting the cell site (or base station) locations for signal coverage that are mutually interference-free is a big task. Without prediction tools, the only way is to use cut-and-try methods by actual testing. This involves the measurement of the coverage of the cell sites following each proposed plan and selection of the best one, which is very costly. With a fairly accurate prediction tool and computer manipulation, we can easily pick up the optimal cell site locations of a plan after comparing and evaluating the performances of all plans from the computer outputs. As we all know, anyone can write a prediction model. However, a reasonably accurate model that has been verified by measured data in a mobile radio environment over the years is the one that should be chosen.

Requirements for Selecting the Right Model for Mobile Radio

A mobile radio prediction model has to be able to distinguish various natural terrain contours, such as flat areas, hilly areas, and valleys, from the signal prediction results.

Also, the model has to be able to distinguish various manmade environments, such as open, suburban, urban, and metropolitan areas, from the signal prediction results. There are many variables involved, and they all have their appropriate roles to play. Therefore, a good mobile radio prediction model is very hard to form. The manmade structures are different in different cities; hence, to predict the signal strength received in a city of interest one cannot use the measured data collected and averaged from one or many different cities elsewhere. The radio signal propagating in a mobile radio environment, in general, follows wave propagation theory; but because many environment-generated variables are involved, a statistical methodology is necessary to complete the prediction tool.

A good prediction model should be simple to use. The model should be specified very clearly and should not provide the user with any subjective judgment or interpretation, which could draw different predicted values in the same area. It should also allow a clear explanation of why the theory of wave propagation and the properties of communication statistics are representative of the physical action.

Free-Space Path Loss Formula

$$\text{FSPL} = -20 \, \log_{10}\left(\frac{4\pi R}{\lambda}\right)\text{dB},$$

where λ = wavelength.

The free-space path loss (FSPL) is the minimum loss that can be found at the same distance R. To achieve free-space loss is to place both the transmitting and receiving antennas high above the ground or away from nearby scatterers. Thus, the reflected waves from the ground or scatterers become weak and can be neglected.

Reciprocity Theory in Mobile Radio Environment

If the receiving antenna is made to transmit and the transmitting antenna receive, then the received signal strength will be the same as before, given the same transmitted power and matching conditions. This is known as the *reciprocity theory*. However, the received carrier-to-interference ratio (CIR) is not reciprocal because the interference level I received at the vehicle and the base station is different.

2.2.3 Formula for the Two-Wave Theory on a Flat Ground

The received power at the mobile antenna is obtained by summing up two waves: a direct wave and a reflected wave from a direct path and a reflected path, respectively, as shown in Figure 2.7.

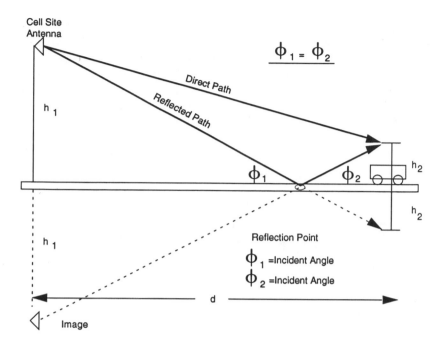

Figure 2.7 A two-wave model.

$$P_r = P_0 \left(\frac{1}{4\pi d/\lambda} \right)^2 |1 + a_v \exp(j\Delta\Phi)|^2 \qquad (2.1)$$

where $P_0 = P_t G_t G_m \alpha_e$; P_t = transmitted power; G_t = gain of the cell site antenna; G_m = gain of the mobile antenna; α_e = additional loss due to propagation in different manmade environments; a_v = reflection coefficient = −1 when the incident angle of the reflected wave is very small;

$$\Delta\Phi = \beta\Delta d \qquad (2.2)$$

equals the phase difference between a direct path and a reflected path; β = wave number = $2\pi/\lambda$; Δd = the difference in path lengths between the direct path and the reflected wave path;

$$\Delta d = \sqrt{(h_1 + h_2)^2 + d^2} - \sqrt{(h_1 - h_2)^2 + d^2} \qquad (2.3)$$

$h_1 h_2$ = antenna height of the cell site and the mobile unit, respectively; d = the distance between the cell site and the mobile unit; and $[1/(4\pi d/\lambda)]$ = the free-space path loss formula.

Equation (2.1) is an equation of the exact solution of the two-wave model, which can be calculated by a computer. This simple model does not represent the close-in mobile radio environment very well, but when the distance from the cell site becomes 1 km or greater, the two-wave model curve fairly matches the measured data in a statistical sense. For this reason, we are interested in a distance greater than 1 km. Under this condition, $h_1 + h_2 \ll d$, and (2.3) can be approximately expressed by

$$\Delta d \approx (2/d)h_1 h_2 \tag{2.4}$$

Also, $\Delta\Phi$ is very small, so that $\sin\Delta\Phi \approx \Delta\Phi$ and $\cos\Delta\Phi \approx 1$. Hence,

$$\exp(j\Delta\Phi) = 1 + j\Delta\Phi = 1 + j\beta(\Delta d) \tag{2.5}$$

Substituting (2.4) into (2.5) and putting the result into (2.1) yields

$$P_r = P_0(h_1^2 h_2^2/d^4) \tag{2.6}$$

which is an approximation with the wave length canceled out by the approximation process, and α_e is an unknown shown in P_0 of (2.1). Without the wavelength information, (2.6) cannot be used for calculating the absolute path loss as the free-space loss formula can. However, (2.6) has a simpler expression than (2.1) and agrees with the measurement of approximately 40 dB/dec (dec = decade) for the path loss and 6 dB/oct (oct = octave) for the cell site antenna height gain for the values of d and h_1 and h_2 of interest. Unfortunately, a value of 3 dB/oct, not 6 dB/oct, for the mobile unit antenna height gain was found from the measurements taken when the mobile antenna height is close to 3m. This finding does not agree with (2.6). Therefore, (2.6) has to be modified as follows.

$$P_r = P_0(h_1^2 h_2^c/d^4) \tag{2.7}$$

Where c is in the range of $1 \leq c \leq 2$. When $h_2 \geq 30$m, $c = 2$; when $h_2 \leq 10$m, $c = 1$. When 10m $\leq h_2 \leq 30$m, the value of c varies because of the different effects caused by the manmade environments. Based on measured data, a linear formula may be used to estimate the value of c, thus: $c = h_2/20 + 1/2$.

Since (2.7) is fairly good at predicting the path loss and the base station antenna height gain, the two-wave model (Figure 2.7) used to derive (2.7) is thus regarded as the correct model. This is because the mobile antenna height is close to the ground and the incident angle of the reflected wave arriving at the reflection point is very small, so the reflection coefficient approaches -1. This means that the direct and reflected waves have approximately the same signal strength, but due to the properties of electromagnetic waves, the reflected wave always has a 180-deg phase shift after reflecting from the ground. Therefore, the two waves are actually subtracting rather than adding. With a slight phase difference due to the different path lengths of the two waves, a loss of 40 dB/dec is obtained, as compared to a free-space loss of 20 dB/dec due to a single-wave path.

2.2.4 Some Established Models

Path Loss Prediction Model Versus Local Mean Prediction Model

Most models predict the path loss over the radio path, which is the distance measured from the mobile site to the base station [21–23]. The standard deviations of the prediction errors against the measured data for these models are about 6 to 18 dB. This indicates that the path loss curves used for the prediction of signal strength are too loose to use to design a cellular system. Another prediction model [24,25] is to predict the local means along the mobile path (i.e., along individual streets and roads).

The Path Loss Model

We present the Okumura Model [5], which was formed from averaging measured data in Japan and expressed as a statistical average as follows:

$$L = 69.55 + 26.16\log_{10}F - 13.82\log_{10}h_b + (44.9 - 65.5\log_{10}h_b)\log_{10}R - A_{b_m}$$

(2.8)

where

L = path loss from the base station to the mobile unit;

F = carrier frequency in megahertz;

h_b = base station antenna height in meters;

R = distance between the base station and mobile unit in kilometers;

A_{b_m} = $(1.1\log_{10}F - 0.7) \cdot h_m - (1.56\log_{10}F - 0.8)$;

h_m = mobile antenna height in meters.

Local-Mean Prediction Model—Lee's Model

Lee's Model is a local-mean prediction model. The predicted local mean is used as a comparison with the measured local mean at any given location. It should be noted that the local-mean prediction is different from the path loss prediction. The philosophy of Lee's model is to try to separate the effects of the received signal on the natural terrain configuration and on the manmade structures.

2.2.5 Effect of Manmade Structures

Since the terrain configuration of each city is different, and the manmade structure of each city is also unique, we have to find a way to separate these two. The way to factor out the effect due to the terrain configuration from the manmade structures is to work out a way to obtain the path loss curve for the area as if the area were flat, even if it is not. The path loss curve obtained on virtually flat ground indicates the effects of the signal loss due to solely manmade structures. This means that the different path loss curves obtained in each city show the different manmade structure in that city. To do this, we may have to measure signal strengths at those high spots and also at the low spots surrounding the cell sites, as shown in Figure 2.8(a). Then the average path loss slope (Figure 2.8(b)), which is a combination of measurements from high spots and low spots along different radio paths in a general area, represents the signal received as if it is from a flat area affected only by a different local manmade structured environment. We are using 1-mi intercepts (or, alternatively, 1-km intercepts) as a starting point for obtaining the path loss curves. The reasons are:

- There are fewer streets within a mile of the cell site. Statistically, the signal-strength data collected within 1 mi of the cell sites are not enough.
- The streets oriented in line with the radio path always have stronger signals than the ones oriented perpendicular to the radio path in the area closer to the cell site, as shown in Figure 2.9. Sometimes the received signals can be different by 20 dB within a 1-mi radius. As soon as the distance is larger than 1 mi, this phenomenon will begin to disappear. Therefore, the near-in data can be biased up or down and should not be used.
- Certain noticeable objects close to the cell site could affect the signal strengths at the near-in areas, and should not be considered from a statistical sense.

Now, a path loss should be obtained from measurement in each city following the rule previously described. The general representation of a flat-area path loss slope is as follows.

$$P_r = P_{r0}(d/d_0)^{-\gamma}\alpha_0 \quad \text{mW} \tag{2.9a}$$

$$= P_{r0} - 10\gamma\log(d/d_0) + \alpha_0 \quad \text{dBm} \tag{2.9b}$$

where P_{r0} is the received power in milliwatts at the distance of d_0, which can be 1 mi or 1 km, and γ is the slope of the path loss in mobile radio environment. γ shown in (2.9a) is closer to 4 rather than 2, which is proper for free-space loss. γ, shown in (2.9b), represents the decibel value per decade. α_0 is the correction factor when the actual condition is different from the reference conditions, which may be specified as:

- Antenna height at the cell site = 100 ft (or 30m):
- Antenna gain at the cell site = 6 dBd;

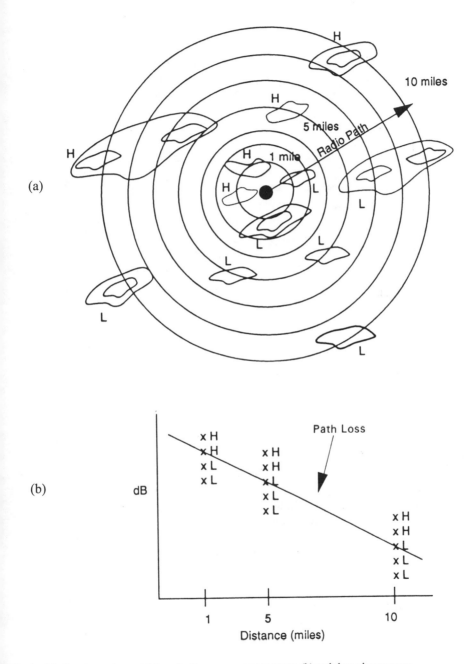

Figure 2.8 Terrain contours: (a) for selecting measurement areas; (b) path loss phenomenon.

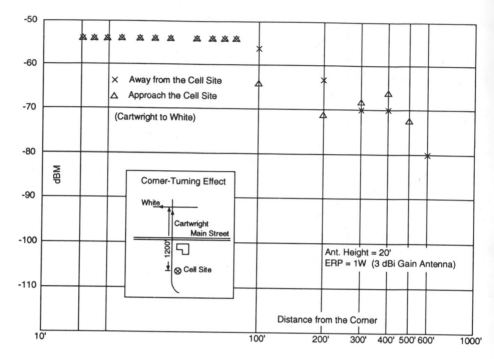

Figure 2.9 Corner-turning effect tested in Irvine, CA [28].

- Transmitter power = 10W;
- Antenna height at the mobile unit = 10 ft (or 3m);
- Antenna gain at the mobile unit = 0 dBd.

All of the curves shown in Figure 2.10 [26] are normalized to this set of reference conditions and illustrated in each geographical area as if they were obtained from flat ground. Their differences are solely due to the manmade structures in their corresponding cities. We call these path loss slopes the *flat-area path loss slopes.* These slopes have to be measured according to the special rule mentioned previously, so the path loss slopes obtained from the other published sources may not be appropriate. Each of the slopes has to be measured in each corresponding city, since no manmade structures are alike.

2.2.6 Multipath Fading Model

2.2.6.1 Long-Term Fading

The raw data shown in Figure 2.11 has been artificially broken down into two parts, the long-term fading and the short-term fading [27]. The long-term fading is affected by

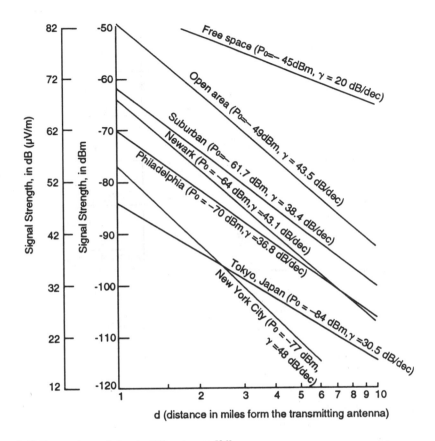

Figure 2.10 Propagation path loss in different areas [26].

the terrain configuration. It is the local-mean (L) variation and follows the log-normal distribution [28]:

$$p(L) = \frac{1}{2\pi\sigma} \exp\left(-\frac{b^2}{2\sigma^2}\right) \qquad (2.10)$$

where L is in dBm and σ is the standard deviation in decibels. σ is different depending on the terrain variation: in a hilly area, σ is large, and in a flat area, σ is small. The value of σ from a log-normal plot shown in Figure 2.12 can be found by the following method. Find the level $\overline{L}$ at a percentile of 90%:

$$P(L \leq \overline{L}) = 90\% \quad \overline{L}/\sigma = 1.29 \qquad (2.11)$$

Then

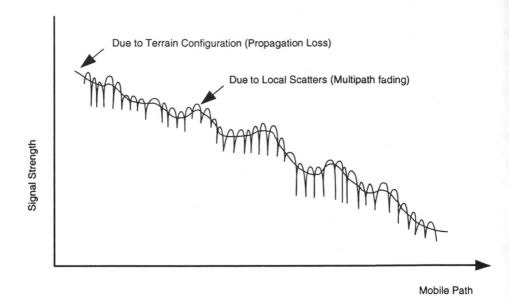

Figure 2.11 The nature of fading.

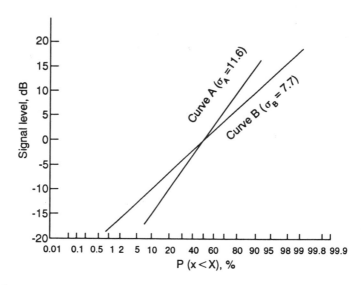

Figure 2.12 Cumulative probability distribution curves for lognormal fading in two different areas.

$$\sigma = \frac{\overline{L}}{1.29} \quad \text{dB} \qquad (2.12)$$

2.2.6.2 Short-Term Fading

Short-term fading is fast fading. It is caused by the structures surrounding the mobile receiver. The signal is transmitted from the base station and arrives at the mobile unit through the multipath reflections. If a strong line-of-sight path also exists, then the signal received by the mobile unit will perform a Rician fading [29]. If there is no line-of-sight path, then the signal received will perform a Rayleigh fading [27]. The Rayleigh fading is formed from many reflected waves. The components of the Rayleigh fading can be expressed as:

$$X_k = \sum_{j=1}^{9} a_j \cos\left[2\pi \frac{k}{100}\cos\theta_j\right] + \sum_{j=1}^{9} b_j \sin\left[2\pi \frac{k}{100}\cos\theta_j\right] \qquad (2.13)$$

$$Y_k = -\sum_{j=1}^{9} a_j \sin\left[2\pi \frac{k}{100}\cos\theta_j\right] + \sum_{j=1}^{9} b_j \cos\left[2\pi \frac{k}{100}\cos\theta_j\right] \qquad (2.14)$$

and

$$r_k = [X_k^2 + Y_k^2]^{1/2} \qquad (2.15)$$

where a_j and b_j are Gaussian variables with mean 0 and variance 1. θ_j has nine values ranging from 0, 40, 80, . . . , 320 deg. The Rayleigh fading can be generated by (2.15) and is shown in Figure 2.13. The distribution of Rayleigh fading is

$$p(r) = (1/\overline{r})\exp[-(r^2/\overline{r}^2)] \qquad (2.16)$$

and

$$P(r \le R) = 1 - \exp[-(R^2/\overline{r}^2)] \qquad (2.17)$$

Fading Depending on Frequency

In (2.16) we are assuming that all the waves are arriving at the mobile unit at the same time. This is called *flat fading*. In reality, the wave arrivals are at different times and cause time-delay spread. However, the flat-fading model is usually applied to analog-modulated signals on voice systems or to a slow transmit bit rate on digital systems.

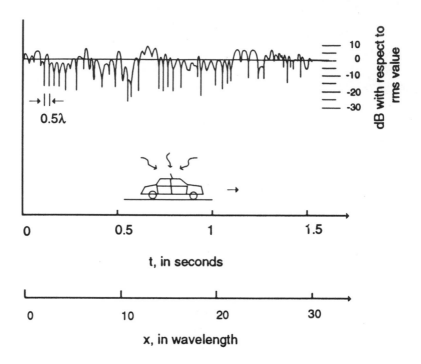

Figure 2.13 Typical signal fading while vehicle is in motion.

Under these circumstances, waves arriving at different times do not cause noticeable degradation in voice quality or bit error rate in data transmission. When the transmit bit rate becomes high, the waves arrive at different times, causing selective frequency fading. Selective frequency fading will affect the digital modulation or signaling transmission over an analog system. We have to indicate the cause of the selective frequency fading as follows.

2.2.6.3 Selective-Fading Model

We can modify the flat model shown in (2.13) and (2.16):

$$X_k = \sum_{j-1}^{9} a_j \cos\left[2\pi \frac{k\Delta_j}{100} \cos\theta_j\right] + \sum_{j-1}^{9} b_j \sin\left[2\pi \frac{k\Delta_j}{100} \cos\theta_j\right] \tag{2.18}$$

$$Y_k = -\sum_{j-1}^{9} a_j \sin\left[2\pi \frac{k\Delta_j}{100} \cos\theta_j\right] + \sum_{j-1}^{9} b_j \cos\left[2\pi \frac{k\Delta_j}{100} \cos\theta_j\right] \tag{2.19}$$

where Δ_j is the time delay of the jth wave. Δ_j will be different due to different environments. The relationship between the interval of Δ_j and the difference in frequency Δf can be expressed as

$$\Delta_j = \frac{1}{2\pi(\Delta f)} \tag{2.20}$$

which indicates that when the frequency changes, the time delay of wave arrival also changes. The time delay of wave arrival forms a frequency-selective fading as shown in Figure 2.14.

Then, in a fading environment, we can use two different frequencies to achieve a frequency diversity if

$$\Delta f \geq \frac{1}{2\pi\Delta} \tag{2.21}$$

where Δf is the separation between two carrier frequencies. In the mobile radio environment:

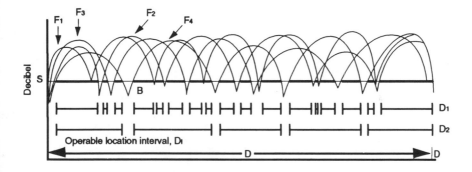

F₁ — Forward link setup channel
F₂ — Reverse link setup channel
F₃ — Forward link voice channel
F₄ — Reverse link voice channel
D₁ Distance interval where the signal is above level S; operable location interval
D Total distance of interest
D₁ Total distance that all four frequencies are above level S
D₂ Total distance that only F₄ is above level S

Figure 2.14 Portable cellular four-frequency selective fading system. (Lee, W. C. Y., "In Cellular Telephone, Complexity Works," *IEEE Trans. on Circuits and Devices,* Vol. 7, No. 1, January 1991, pp. 26–32.)

- Suburban: $\Delta = 0.5$ μs, and $\Delta f = 300$ kHz;
- Urban: $\Delta = 3$ μs, and $\Delta f = 50$ kHz.

Applying frequency diversity in suburban areas requires two frequencies separated by 300 kHz, and in urban areas separated by 50 kHz. If the frequency diversity has to apply in both areas, then take the 300-kHz separation. Selective fading is harmful to system reception.

Selective-Fading Problems

For example, in cellular systems, we need four frequencies to complete a call. Two frequencies (because of a duplex system) are used for setup calls and two for connecting to voice. When the mobile unit is moving, the average signal strengths of all four frequencies are the same. If one frequency signal is strong, all four frequency signals will be strong. The calls do not depend on the condition of any particular frequency. But if the mobile or a portable unit is standing still, then the antenna is located at one spot. Due to frequency-selective fading, all four frequency signals are not always above a certain threshold; then the system only needs the signal strength of one frequency below the threshold to cause dropped calls. In order to calculate the percentage that the signals of all four frequencies are above a certain threshold, the following formula is used.

$$P_4(r > R) = [P_1(r > R)]^4 \qquad (2.22)$$

where $P_1(r > R)$ is the probability that one frequency is above the threshold R. In a Rayleigh fading environment,

$$P_1(r > R) = \exp(-R/R_0) \qquad (2.23)$$

where R_0 is a reference level at which the threshold R can be referred to. Then (2.22) becomes

$$P_4(r > R) = [\exp(-R/R_0)]^4 \qquad (2.24)$$

The four-frequency phenomenon can be illustrated as shown in Figure 2.14.

2.2.7 Maritime and Aeronautical Mobile Systems

Maritime and aeronautical mobile systems basically operate in different environments. As mentioned in Section 2.2.1, ground mobile systems are very much affected by natural terrain configuration and manmade structures.

However, the transmitted media of maritime and aeronautical mobile systems are affected little by the natural terrain configuration and manmade structures. Usually, frequencies below 30 MHz (referred to as HF) are used for maritime and aeronautical mobile systems. For aeronautical mobile satellite systems, the allocated frequency is about 1.5 to 7.3 GHz.

HF is used for communication distances greater than 150 km. The transmission depends on the sky wave reflected from the ionosphere. Depending on the distance, the E-layer, which is at a height of about 110 km, can communicate over a distance of less than 150 km during the day, and less at night. The F_1-layer is at a height of 175 to 250 km, only exists during the day, and does not actively reflect the energy from the incident wave. The F_2-layer is at a height of about 200 to 400 km and can have a greater propagation distance of up to 3,000 km.

Because of constant changes in ionospheric conditions, maritime mobile and aeronautical mobile communication systems should use space and frequency diversities to compensate for fading caused by these ionospheric changes.

Now, maritime satellite and aeronautical mobile satellite systems use satellites to replace the ionospheric layers. There are two kinds of satellites, which are differentiated by their altitudes. One is a stationary satellite (2,200 mi above the earth) and the other is the low-elevation orbit (LEO) satellite (200 to 500 mi above the earth). These satellite systems are also developed for coverage of rural areas on land. The propagation mechanisms for satellite communication are generally less problematical than those for land mobile systems, particularly when the latter is in an urban environment.

2.3 ANTENNA DESIGN

2.3.1 Requirements for Mobile Antennas

In general, antennas for mobile terminals have been required to be small, lightweight, and low-profile, and to have an omnidirectional radiation pattern in the horizontal plane. In addition, antennas must exploit ambient propagation characteristics and be robust against mechanical and environmental hazards encountered while moving. In the early days of mobile communications, wire antennas such as whip antennas, monopole antennas, and inverted-L antennas were commonly used for vehicles and mobile equipment. These antennas can be simply mounted onto the body of the vehicle or mobile unit and still meet the requirements for mobile terminal use.

With the evolution of mobile communication systems, antenna technology has also progressed, and the design concept has changed as well, although the fundamentals essentially remained the same. A typical example is the development of small antennas. The rapid growth in civil applications of mobile communications, particularly the increased use of personal mobile terminals, has generated a need for the development of small mobile terminals and small-size radiating systems. It is well known that the smaller the antenna size, the lower the antenna efficiency and the narrower the bandwidth; but, in addition, the design concept for portable equipment must include the body of the equipment as a part of the radiator [30], since currents flowing on the body contribute to radiation [31,32]. The radiation characteristics of such an antenna differ much from those of the antenna element solely in free space. The radiation pattern varies depending on the size and shape of the portable unit and the location of the antenna element [16,32,33].

Another example is antennas used in zone systems, which were adopted in order to increase the utilization of the frequency spectrum. The radiation pattern of base station antennas in cellular land mobile systems is not always omnidirectional, but designed to conform to the pattern of specified zones in the horizontal plane and to be tilted downward in the vertical plane to minimize the cochannel interference. Base station antennas in aeronautical mobile systems must have either a conical beam to cover a specified zone or a satellite tracking facility.

Today, antennas used in mobile communication systems have been recognized as critical elements that can either enhance or constrain system performance. Accordingly, antenna performance and characteristics have come to be studied extensively, specifically taking mobile environmental and propagation conditions into account.

Requirements for antennas depend on the types of mobile systems, as summarized in Table 2.3 [34–36], where the types are categorized into land, maritime, aeronautical, and satellite mobile systems. There are various types of antennas, from simple ones, such as whip antennas for a single-channel operation, to complicated ones, such as beam-shaped array antennas used in multichannel systems. Table 2.4 [35] shows some typical antennas used in both base and mobile stations [36] in various mobile communication systems. Antennas used in base stations have a different role from those in mobile stations and are designed under different requirements. The electrical performance and mechanical configuration of a base station antenna mainly depends on the size and the shape of the service area.

For instance, in maritime and aeronautical mobile systems, the base station antenna should cover a wide area, and hence high directivity (7 to 11 dBd) is required. In train communication systems, service areas are restricted within long, narrow areas along the tracks. In order to cover such areas, a parabolic reflector antenna has been used. To solve problems in mountainous areas and in tunnels, where the radio wave is obstructed, the LCX communication system has been developed and used. This system has been applied to the train telephone systems in Japan. Since antennas installed on the body of vehicles such as trains, automobiles, and particularly aircraft must be lightweight and tolerate severe aerodynamic conditions, thin, short monopoles or low-profile antennas (such as inverted-L) are usually used. Types of flush-mounted or conformal antennas such as blade and microstrip antennas are preferred for aircraft. In automobile telephone systems, either a sleeve antenna or a monopole antenna is commonly used and is mounted either on the trunk lid or on the roof of a car. They are designed so that the antenna element will not disturb the driver's view and the car body will not have very much of an effect on the radiation pattern. Various other types of antennas, such as wire antennas printed on the

Table 2.3
Requirements for Mobile Antenna Systems

System (Band)	Base Station Antenna		Mobile Station Antenna	
	Requirement	Example	Requirement	Example
Maritime telephone system (250-MHz)	Cover wide area (50 to 100 km)	Corner reflector Plane reflector	Omnidirectional in horizontal plane High gain	Dipole (sleeve)
Aeronautical telephone system (800-MHz)	Cover wider area (400 km) Suppress ground reflection wave	Collinear array Broadside array (shaped beam)	Omnidirectional Extremely low aerodynamic resistance	Blade antenna (flush mount)
Train telephone system (400-MHz)	Cover belt-shape zone Cover inside tunnel	Grid paraboloid reflector Coaxial leaky cable	Low aerodynamic resistance	Monopole
Automobile/portable (800-MHz)	Efficient illumination for cell configuration Multichannel operation Reduction of fading	Collinear array Broadside array with beam tilt and/or shaped beam Yagi	Vehicle: Omnidirectional, high gain, diversity Portable: high efficiency, small volume, diversity	Sleeve Monopole (on glass) Monopole Planar IFA
Pager (250-MHz)	Cover wide area Efficient illumination inside the area	Corner reflector Brown antenna	High efficiency Inhouse antenna	Small loop

Table 2.4
Typical Antennas Used in Practical Mobile Communication Systems

System	Mobile Station		Base Station	
	Antenna Type	Requirements	Antenna Type	Requirements
Pager: 150 MHz, 280 MHz, 450 MHz, 900 MHz	Small square loop Multiturn loop Ferrite coil antenna Micro strip antenna Parallel plate (magnetic current loop)	Limited space to mount antenna element Built-in, lightweight Nondirectional sensitivity by combining H- and E-components produced in antenna system Sensitivity enhanced by using image loop Low cost	Brown antenna Corner reflector Plane reflector Array (broadside, low sidelobe)	Wide-area coverage or specified area coverage with shaped beam Homogeneous field distribution in service area Suppression of field near base station
Mobile telephone: Vehicles, 800 MHz	$\lambda/4$ monopole $\lambda/2$ sleeve dipole Printed dipole 2 monopoles: horizontal, vertical	Omnidirectional pattern in horizontal plane Low-elevation angle in vertical plane Space diversity	Collinear array MSA Cylindrical parabola Broadside array Corner refector 2 dipoles	Same as above Pattern tilt downward Low sidelobe in vertical plane Space or polarization diversity
Portable, 800 MHz	$\lambda/4$ monopole $\lambda/4$ whip normal mode helix Planar IFA	Limited space to mount antenna elements Body or portable unit included in antenna system Space diversity	Same as above Cross dipole	Same as above
Cordless telephone: 280/400 MHz	Short dipole Small loop $\lambda/4$ monopole	Built-in or mounted on body of telephone	Short dipole Short monopole	Inhouse use (20m–100m)
Maritime telephone: 250 MHz	Dipole Brown antenna	Omnidirectional horizontal plane pattern Antenna mounting space not severely limited	Corner reflector Plane reflector	Long-distance transmission (50–100 km)
Train telephone: 400 MHz	$\lambda/4$ monopole ILA	Antennas usually mounted on roof	Parabola with grid reflector	Transmission along railway

Application	Antenna	Characteristics	Remarks
	Monopole buried on train body	Low profile; Rigid structure; Resistant to weather and aerodynamic conditions	
	LCX	Collinear array; Broadside array (shaped beam)	Along railway; Long distance transmission (~400 km); Conical zone coverage
Airplane telephone: 800 MHz	Blade antenna; MSA array	Flush-mounted onto lower side of train; Low-profile, flush-mount; Conformal structure; Lightweight; Dragless structure; Rigid; Resistant to environmental conditions	
Aircraft communication	Wire; Tail cap; Sleeve; Monopole; Notch; Blade; MSA	Almost same as above; LF/HF/MF bands; VHF/UHF bands	
ETS/V: Portable	2-circular MSA	1.5-GHz band; Placed on lid of T-unit	
Ship	Improved SBF	Low-speed data transmission, compact, high efficiency, wide bandwidth	
Aircraft	16-MSA phased array	Sequential array	
INMARSAT: Ship, Land	Drooping-dipole; Quadrifiler	1.5-GHz band	
PROSAT	5-turn helix; 16-crossed dipole array SBF	1.5 GHz band; Narrow bandwidth	

Note: ETS/V = Engineering Test Satellite V; IFA = inverted-F antenna; ILA = inverted-L antenna; INMARSAT = International Maritime Satellite Organization; LCX = leaky coaxial cable; MSA = microstrip antenna; PROSAT is a European Space Agency program for the promotion of small terminal techniques for satellite systems; SBF = short backfire antenna.

window glass for receiving broadcast and monopole antennas fed by electromagnetic coupling through the window glass (called an *on-glass antenna*) for the automobile telephone system, have been developed and used. Antennas for portable radio phones, cordless telephones, and pagers are required for personal convenience and comfort, and consequently antenna dimensions should be small enough to be mounted either on or built into the equipment. In order to meet these requirements, antennas such as short whip, normal mode helix, and small loop antennas are mainly used. In Japanese portable telephone systems, two elements, a monopole element and a planar IFA, which can be housed inside the equipment, have been used, thus achieving space diversity performance [37].

Since pagers are usually carried in the bearer's vest pocket or attached to his or her belt, small, built-in rectangular loop antennas have been employed. It is usually necessary to eliminate proximity effects, which generally degrade antenna performance but for pager loops, antenna performance is actually enhanced by using both the proximity and body effects as follows: (1) conductors existing near an antenna element are used to generate dipole mode currents, so that nearly omnidirectional receiving patterns can be obtained, and (2) the body effect is used to enhance the receiver sensitivity as a result of superposition of the field produced by the image of the loop [38]. Magnetic-type antennas such as loop, microstrip, and planar IFA are compatible with the above requirments [2]. Radiation patterns of such a radiating system must be taken in three dimensions. Polarizations in this case may become composite; for instance, it may be not only vertical, but also horizontal, even when a vertical monopole element is used, as a result of radiation from the body of the unit. Antenna gain must also be evaluated in three dimensions.

The gain of a portable unit should be as high as possible so that the transmitter power can be made lower and the use of a smaller battery becomes possible. This will result in making the size of the mobile unit smaller and lighter. In turn, it may also allow a longer time between recharging the battery, thus extending the transmitter operation time. When a transmitter is operated very close to the human body, the effect of electronmagnetic energy on the human organs such as brain and eyes must be carefully taken into account in the antenna design.

Environmental conditions must also be seriously considered in the design and practical use of portable terminals. In mobile communication environments, especially in urban areas, the diversity reception technique has been applied to both base station and mobile station antennas in order to overcome multipath fading problems. Diversity antenna systems are required to have branches with small correlation factors and must be economically constructed to be of compact size. In Japanese mobile telephone systems, diversity reception has been adopted in mobile stations as well as in base stations [17]. Very compact diversity antennas, which are composed of two elements but have an appearance of a single element, have been developed for mobile stations [39]. One of them is the vertically spaced sleeve antenna for the vehicle mount and another is the antenna system composed of a whip and an inhouse planar IFA for a portable radio unit.

In digital modulation systems, problems due to delay spread in incident waves may be mitigated by angle or directivity diversity schemes. The development of new adaptive antenna systems having a capability to cope with problems caused by multipath propagation, interference, and so forth are now viable.

Various antennas have been developed for mobile satellite systems, which require either a conical beam or a satellite tracking facility in order to link continuously with a satellite when mounted on a moving vehicle. Mechanical, electronic, and combination tracking systems have been developed [40]. MSA phased arrays, for instance, have been introduced for receiving satellite broadcasts on trains, ships, and buses and for data reception on cars.

2.3.2 Diversity Techniques

2.3.2.1 Diversity Performance

Since signal fading in the mobile radio environment causes severe reception problems, diversity techniques are used to reduce fading effects. Usually the diversity is applied at the receiving site. It is therefore a passive device and does not cause any interference, and it can take the form of space diversity, field component diversity, polarization diversity, frequency diversity, and directivity diversity, each of which will be described later in this section. The diversity performance is dependent on the number of diversity branches and on the correlation coefficient between the received branches. Each diversity scheme can reach the same performance if the branch correlation coefficients are the same. Two diversity signals are received from two antennas if space, polarization, and directivity schemes are used, or from one antenna if frequency and field diversity are used. We also have to consider how to combine two signals which are received from the diversity scheme, and an appropriate combining technique can yield better performance. There are four general combining techniques. Maximal ratio combining (MRC) maximizes the signal-to-noise ratio after the combining. Equal-gain combining (EGC) brings the received signals to a common phase and combines them in voltages. Selective combining (SEC) selects the strongest of the two received signals. Switch combining (SWC) is based on a threshold level below which the signal switches to the other antenna. The MRC, EGC, and SEC require more complex equipment, perform better, and cost more than SWC, which needs only one front end. Among the MRC, EGC, and SEC, the performance of MRC is the best, but the equipment is complicated. The performance of SEC is lower than that of MRC and EGC, but the equipment is simple to build. Although combining techniques are not within the scope of this book, the implications for antenna design are relevant.

This section illustrates the performance of a two-branch SEC diversity signal and its variants. Assume that the average CIR Γ of two-branch signals are the same. Then the probability that the combined signal CIR is γ and is less than the threshold R is [41]:

$$P(\gamma \leq R) = [1 - \exp(R/\Gamma)]^2 \tag{2.25}$$

Equation (2.25) is valid for two uncorrelated-branch signals. Sometimes the two-branch signals received are correlated because the required separation of two base stations cannot be physically wide enough to achieve the uncorrelated condition between two signals [42]. Then the combined signal γ of two correlated signals can be expressed as [41]

$$P(\gamma \leq R) = 1 - \exp(-R/\Gamma)[1 - Q(a, b) + Q(b, a)] \qquad (2.26)$$

where

$$Q(a, b) = \int_b^\infty \exp[-(1/2)(a^2 + x^2)]I_0(ax) \cdot x\,dx$$

$$a = \sqrt{\frac{2R}{\Gamma(1 + |\rho|^2)}} \qquad (2.27)$$

$$b = \sqrt{\frac{2R}{\Gamma(1 - |\rho|^2)}}$$

when the correlation ρ approaches zero, $Q(a, b) = Q(b, a)$, and then (2.26) reduces to (2.25). Equation (2.26) is plotted in Figure 2.15 with different values of correlation coefficient ρ. At a level of 10 dB below the mean CIR level, 10% of the signal is below the level for $\rho = 1$ (no diversity), 2% of the signal is below the level for $\rho = 0.7$, and 1% of the signal is below the level for $\rho = 0$. The best performance of a diversity receiver is trying to make the correlation coefficient ρ of two signals approach zero.

2.3.2.2 Space Diversity

Many cellular design engineers have asked: Why does horizontal antenna separation lead to better diversity performance than vertical separation at a cell site? There is an important reason why. The greater the antenna horizontal separation, the less likely the fades of the two received signals will occur simultaneously. Thus, the diversity gain for reducing the effect of the fades increases as the separation increases and relies on the concept that the signal strength of two signals should be nearly equal. If the two received signal strengths are not equal, as is generally the case for vertical separation, then the diversity gain cannot be achieved, regardless of the requirements of antenna separation.

Designing a diversity antenna scheme is based on the parameter η, which depends [42] on the real antenna height (h) and the antenna separation (D):

$$\eta = h/D \qquad (2.28)$$

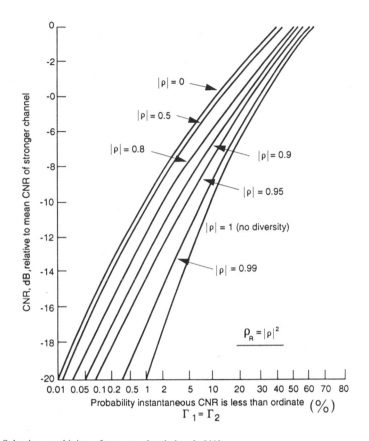

Figure 2.15 Selective combining of two correlated signals [41].

Horizontal Separation

It has been determined experimentally that the optimum value of η is 11 for horizontal antenna separation [42]. For example, if the antenna h is 100 ft, the optimum D is 9 ft. Therefore, the higher the antenna, the more separation will be needed for optimum diversity gain. Incidentally, the h is the antenna height at the cell site when the system is designed.

In a real-world environment, vehicles and portable units are at various ground elevations in all directions from the cell site. Because of this, the effective antenna height (h_e) measured at the cell site varies based on the real-time locations of vehicles and portables. The same actual antenna height may have two different effective antenna heights according to two different vehicle or portable units. For a specific base-to-mobile/portable transmission, when h_e is less than the h, the signal received by the vehicle will be weaker [43]. This is expressed as a gain or loss ΔG thus:

$$\Delta G = 20\log_{10}(h_e/h) \quad \text{dB} \tag{2.29}$$

During system operation, the value of η_e, obtained from $\eta_e = h_e/D$, will vary depending on the present location of the vehicle or portable units. When h_e is less (greater) than the h, the value of η_e becomes less (greater) than 11 and the diversity gain typically varies as in Figure 2.16.

Vertical Separation

Usually the vertical antenna separation is greater than the horizontal antenna separation for achieving a given diversity gain [42]. Let the antenna height of the lower antenna be h_1 and the antenna height of the higher antenna be h_2. The vertical separation is D_v, = $h_2 - h_1$. The difference in reception gain Δg between the two effective antenna heights of the two antennas can be found by using the following equation [43].

$$\Delta g = 20\log_{10}(h_e'/h_e) = 20\log_{10}[1 + (D_v/h_e)] \tag{2.30}$$

where h_e and h_e' = effective antenna heights of two vertically separated antennas (Figure 2.17). For example, when the difference in Δg is 4 dB, $1 + (D_v/h_e) = 10^{4/20}$ or $D_v = 0.58h_e$, in which case the values of D_v and h_e could be 15 and 26 ft, respectively, for a given vehicle location. Although $\eta_e = h_e/D$ is 1.73, there will be no diversity gain, since Δg is too large [44].

Although the horizontal-separation antennas can provide higher diversity gain, there can sometimes be limitations when vehicles are in certain areas with respect to the base station, when no diversity gains are observed [42] because of the relative antenna heights.

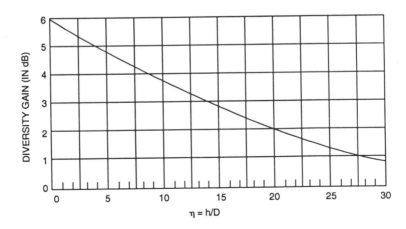

Figure 2.16 Diversity gain versus η [44].

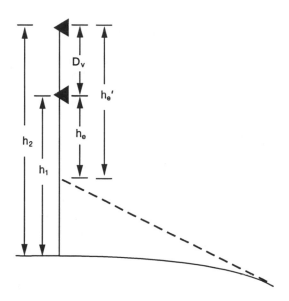

Figure 2.17 Vertical antenna separation [44].

2.3.2.3 Polarization Diversity Schemes

Two polarizations, vertical and horizontal, from two antennas can carry two signals on one radio frequency over satellite-ground links or microwave links. Because there is no coupling effect introduced by the medium, no mutual interference would occur. However, in the mobile radio environment, strong mutual coupling effect occurs. This means that after the signal propagates through the mobile radio medium, the signal energy in the vertical polarization wave can be cross-coupled (leaking) into the horizontal polarization wave and vice versa. Let us define the symbols. Γ_{11} = transmit vertical, receive vertical Γ_{12} = coupling vertical into horizontal (from base to mobile). Γ_{21} = coupling horizontal into vertical (from base to mobile). Γ_{22} = transmit horizontal, receive horizontal. The two differently polarized waves (vertical E_v and horizontal E_h) received by the two polarization antennas at the mobile unit can be expressed as

$$E_v = \Gamma_{11} + \Gamma_{21} \quad E_h = \Gamma_{22} + \Gamma_{12} \tag{2.31}$$

See Figure 2.18. Since the principle of reciprocity is applied to the polarization components, the two differently polarized waves, E_v' and E_h', arriving at the base station can be expressed as

$$E_v' = \Gamma_{11} + \Gamma_{12} \quad E_h' = \Gamma_{22} + \Gamma_{21} \tag{2.32}$$

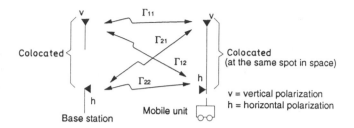

Figure 2.18 Reciprocity of horizontally and vertically polarized waves. (Lee, W. C. Y., "Polarization Diversity for Mobile Radio," *IEEE Trans. on Comm. Tech.*, No. 5, October 1972, p. 921.)

If in some cases $\Gamma_{12} \neq \Gamma_{21}$, then $E_v = E_v'$ and $E_h = E_h'$. However, the leaking energy is very small compared to the main stream; that is, $\Gamma_{11} \gg \Gamma_{12}$ and $\Gamma_{22} \gg \Gamma_{21}$. Therefore, from the measurement we found that $E_v \approx E_h$. Also, it can be found that $E_v \approx E_v'$ and $E_h \approx E_h'$.

For antennas on an infinite ground plane, the array pattern in the elevation angles could be [45]

$$P_{\text{vertical polarization}}(\theta) = \left[1 + \cos\left(\frac{4\pi h}{\lambda}\sin\theta\right) \right]^2 \qquad (2.33)$$

$$P_{\text{horizontal polarization}}(\theta) = \left[1 - \cos\left(\frac{4\pi h}{\lambda}\sin\theta\right) \right]^2 \qquad (2.34)$$

where h is the height of the antenna from the ground plane and the angle θ is the elevation angle from the horizon. Notice that for $h = 1.5\lambda$, the vertical pattern is maximum at $\theta = 0$ deg, but goes into a null at $\theta = 8$ deg and picks up again at 19 deg.

The horizontal pattern is nulling at $\theta = 0$ deg, peaks up to $\theta = 8$ deg, and is nulled again at $\theta = 19$ deg. When they are in a line-of-sight condition, the waves come toward the mobile at one particular elevation angle and the powers of two polarization waves are not equal. In an actual mobile environment, the multipath waves arrive at the mobile terminal from a spread in elevation angles, and the ground plane also works to the advantage of decorrelating the received signals in the sense that the two antennas are aiming at different elevation angles.

Typical estimated mean signal strengths as a function of antenna height are shown in Figure 2.19 for the two polarizations. It was observed that for $h \geq \lambda$, the two branches of power were about the same. However, as the antenna height decreases, the power of the horizontal polarization wave decreases. The rapid Rayleigh fadings received by the two polarization antennas are uncorrelated. The local means (the average signal strength seen as the envelope of the Rayleigh fading signal, sometimes called *long-term fading*)

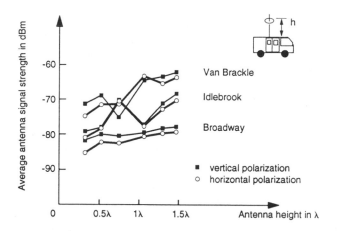

Figure 2.19 Average received signal strength as function of mobile antenna height. (Lee, W. C. Y., "Polarization Diversity for Mobile Radio," *IEEE Trans. on Comm. Tech.,* No. 5, October 1972, p. 921.)

received from two polarization antennas have the same strength. The advantage of polarization diversity in mobile radio is that the two polarization diversity antennas, unlike the two space-diversity antennas, can be placed as close as we wish to achieve full diversity gain. The disadvantage is the 3-dB power reduction at each transmitting antenna due to the split of the transmitter power into the two polarization antennas.

Field Component Diversity [46]

Multiple scattering and fading effects can create cross-polarization such that it is desirable for the antenna to receive both E- and H-field components. This is called an *energy diversity antenna*; an example consisting of two cross semiloops is the basis of many pager antennas. An advantage of this technique is that only one location is needed and the transmit antenna need only have E-field.

Directional Diversity System [47]

Fading signals arriving from different angles at the receiving antenna come via different paths. In directivity diversity, the arriving angles of two fading signals should be separated far enough to avoid correlation among them. Also, it is preferred that the two antennas be in-line with the motion of the vehicle rather than perpendicular to its motion [45]. Then the two antennas can be placed at 0 deg (heading of the vehicle motion) and 180 deg. The effect of antenna directivity has been tested in the mobile radio environment. In an out-of-sight condition, the directivity of a received antenna does not increase the average power, but signal fading is reduced for a narrower beam. This technique is feasible at microwaves, where the directional aperture antennas have a small physical size [46].

2.4 ANTENNA PERFORMANCE EVALUATIONS IN MOBILE ENVIRONMENTS

The performance of a mobile antenna is strongly dependent on its effective gain in a multipath propagation environment, since the gain influences the size of the radio zone, transmitting power of mobile radio equipment, and, in particular, battery capacity in the case of portable radio equipment. Furthermore, in an antenna diversity system, the diversity gain is determined not only by correlation characteristics between signal envelopes received in the antenna elements, but also by the effective gain characteristics of antenna elements. Therefore, for designing a mobile antenna system, it is essential to maximize the effective antenna gain for the anticipated mobile radio environment. However, it is common knowledge that the effective gain of mobile antennas in land mobile propagation environments cannot be evaluated with sufficient accuracy from just antenna directive gain, since the received signals undergo Rayleigh-like fading, and random multipaths exist due to reflection, diffraction, and scattering. Hence, it is common to use the experimental method [48] that measures the mean signal level received over a certain route for evaluating the effective gain. With this method, the mean power levels of the unknown antenna and a reference antenna are obtained by averaging the signal levels received while each antenna moves along the same selected route. The mean effective gain (MEG) of the unknown antenna can, therefore, be related to the reference antenna by comparing the mean power level of the unknown antenna with that of the reference antenna. Half-wavelength dipole antennas have been used as the reference antenna. This method is useful for measuring the MEG of mobile antennas in practical environments, and it has been used for evaluating the MEG of several mobile antennas [16,48,69]. However, measured MEG reflects the interaction between the antenna power gain pattern and the propagation characteristics along the route; it depends on the measurement route. Even though the reference antenna also experiences the same interaction effects, the conventional experimental method is not accurate enough to evaluate the MEG of mobile antennas in general.

On the other hand, antenna diversity reception is an effective technique for mitigating multipath fading in a mobile antenna system. Diversity reception attains effective fading mitigation by the synthesis, selection, and switching of several received signals that are correlated to the smallest possible extent. In the case of antenna diversity, reduction of the correlation coefficient between the receiving antenna branches leads to an enhanced diversity effect. Correlation coefficients have been analyzed mainly by considering only copolarization [50,51]. However, when a diversity antenna is mounted on a portable transceiver, the antenna polarization is varied randomly. As a result, there are situations in which the antenna directivity for the cross-polarization component predominates or the main beam direction of the pattern is oriented in an arbitrary three-dimensional direction.

Hence, it is necessary when designing an antenna diversity system to consider the effect of the cross-polarization component of the arriving wave and the effect of the three-dimensional dispersion of the incident wave direction on the correlation characteristics.

This section introduces a theoretical method for analyzing the MEG of mobile antennas and the correlation characteristics of the antenna diversity branches. The method can treat, in general, the contribution of both vertically polarized (VP) and horizontally polarized (HP) radio waves, the dispersion of incident waves in elevation, and the variations of antenna polarization. Furthermore, the MEG characteristics of half-wavelength dipole antennas are discussed theoretically. The theoretical results are extremely interesting, since dipole antennas are usually used as the reference antennas in MEG measurement. A novel experimental method is described that can accurately evaluate the MEG of any mobile station antenna. We shall also discuss the correlation characteristics of orthogonally crossed dipoles, one of the polarization diversity antenna schemes.

2.4.1 Theoretical Expression of Antenna Performance in the Mobile Environment

2.4.1.1 Mean Effective Gain of Mobile Antennas

First it is necessary to establish a theoretical MEG expression that takes into account the VP and HP incident radio waves in multipath environments. Figure 2.20 illustrates the

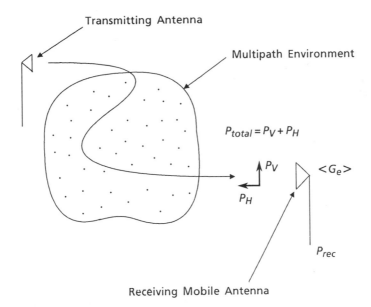

Figure 2.20 Average power arriving at receiving mobile antennas in multipath environments (From [52], © 1990 IEEE.)

notation where the signal transmitted from base station antenna passes through a multipath propagation environment and arrives at a mobile antenna. P_V and P_H are, respectively, the mean incident powers of the VP and HP incident radio waves received while the antenna moves over a random route in the environment. Thus, the total mean incident power arriving at the antenna, averaged over the same route, is $P_V + P_H$. The ratio between the mean received power of antenna over the random route P_{rec} and the total mean incident power $P_V + P_H$ can be considered the MEG of the mobile antenna in the environment. It is assumed that the average over a random route in an environment equals the average over the environment. This ratio is defined as the MEG of the mobile antenna in the environment and is denoted by the symbol G_e:

$$G_e = \frac{P_{rec}}{P_V + P_H} \tag{2.35}$$

The mean incident power ratio P_V/P_H represents the cross-polarization power ratio (XPR):

$$XPR = P_V/P_H \tag{2.36}$$

XPR corresponds to the *cross-polarization coupling* [53] when the polarization of the transmitted radio waves is horizontal, and to the reciprocal of the cross-polarization coupling when the polarization of the transmitted waves is vertical. In a spherical coordinate system, as shown in Figure 2.21, the mean antenna received power P_{rec} is expressed by the following [54, pp. 133–140].

$$P_{rec} = \int_0^{2\pi} \int_0^{\pi} \{P_1 G_\theta(\theta, \phi) P_\theta(\theta, \phi) + P_2 G_\phi(\theta, \phi) P_\phi(\theta, \phi)\} \sin\theta d\theta d\phi \tag{2.37}$$

where $G_\theta(\theta, \phi)$ and $G_\phi(\theta, \phi)$ are the θ and ϕ components of the antenna power gain pattern, respectively, and $P_\theta(\theta, \phi)$ and $P_\phi(\theta, \phi)$ are the θ and ϕ components of the angular density functions of incoming plane waves, respectively. These functions satisfy the following conditions.

$$\int_0^{2\pi} \int_0^{\pi} \{G_\theta(\theta, \phi) + G_\phi(\theta, \phi)\} \sin\theta d\theta d\phi = 4\pi \tag{2.38}$$

$$\int_0^{2\pi} \int_0^{\pi} P_\theta(\theta, \phi) \sin\theta d\theta d\phi = 1 \tag{2.39}$$

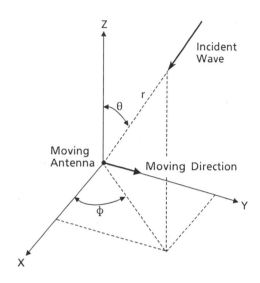

Figure 2.21 Spherical coordinates in mobile radio environments. (From [52], © 1990 IEEE.)

$$\int_0^{2\pi} \int_0^{\pi} P_\phi(\theta, \phi) \sin\theta \, d\theta \, d\phi = 1 \qquad (2.40)$$

where P_1 is the mean power that would be received by an $\mathbf{i}_\theta$-polarized isotropic antenna in the mobile radio environment. Similarly, P_2 is the mean power that would be received by an $\mathbf{i}_\phi$-polarized isotropic antenna. $\mathbf{i}_\theta$ and $\mathbf{i}_\phi$ are unit vectors associated with θ and ϕ, respectively.

In Figure 2.21, since the mobile antenna moves in the X-Y plane, the θ and ϕ components correspond to the VP and HP components. Thus, the terms P_1 and P_2 are, respectively, the mean received power of VP isotropic antennas and that of HP isotropic antennas, and XPR is equal to the ratio P_1/P_2. By using (2.37) and the XPR notation, the expression for MEG can be rearranged to yield the following equation.

$$G_e = \int_0^{2\pi} \int_0^{\pi} \left[\frac{\text{XPR}}{1 + \text{XPR}} G_\theta(\theta, \phi) P_\theta(\theta, \phi) \right. \qquad (2.41)$$

$$\left. + \frac{1}{1 + \text{XPR}} G_\phi(\theta, \phi) P_\phi(\theta, \phi) \right] \cdot \sin\theta \, d\theta \, d\phi$$

When only a VP wave (XPR = ∞) is incoming from a single (θ_s, ϕ_s) direction, which corresponds to line-of-sight propagation with VP wave transmission, the angular density functions in (2.41) are represented as

$$P_\theta(\theta, \phi) = \frac{\delta(\theta - \theta_s) \cdot \delta(\phi - \phi_s)}{\sin \theta_s} \tag{2.42}$$

and

$$P_\phi(\theta, \phi) = 0 \tag{2.43}$$

where $\delta(x)$ is the delta function. It then follows from (2.41) to (2.43) that MEG becomes

$$G_e = G_\theta(\theta_s, \phi_s) \tag{2.44}$$

This means that MEG corresponds to the antenna directive gain in the (θ_s, ϕ_s) direction when incoming signals are centered on the (θ_s, ϕ_s) direction.

If the characteristics of incoming signals in various environments can be represented as statistical distribution functions P_θ, P_ϕ, the MEG given in (2.41) is the mean power gain of the antenna in each environment.

2.4.1.2 Correlation Coefficient of Antenna Diversity Branches

In the spherical coordinate system in Figure 2.21, the incident wave is described by

$$\mathbf{F}(\theta, \phi) = F_\theta(\theta, \phi)\mathbf{i}_\theta + F_\phi(\theta, \phi)\mathbf{i}_\phi \tag{2.45}$$

where F_θ and F_ϕ indicate the random amplitude and phase of the incident electric field in $\mathbf{i}_\theta$ or $\mathbf{i}_\phi$ direction, respectively. Also, the electric field pattern of the antenna $n(n = 1, 2)$ can be given as follows.

$$\mathbf{E}_n(\theta, \phi) = E_{\theta n}(\theta, \phi)\mathbf{i}_\theta + E_{\phi n}(\theta, \phi)\mathbf{i}_\phi \tag{2.46}$$

where $E_{\theta n}$ and $E_{\phi n}$ are the complex expressions of the θ and ϕ components of the electric field pattern. Thus, the received voltages of the two antennas shown in Figure 2.22 are as follows.

$$V_1(t) = C_1 \int_0^{2\pi} \int_0^\pi \mathbf{E}_1(\theta, \phi) \cdot \mathbf{F}(\theta, \phi)e^{-jk\mathbf{u}\cdot\mathbf{r}t} \sin\theta d\theta d\phi \tag{2.47}$$

$$V_2(t) = C_2 \int_0^{2\pi} \int_0^\pi \mathbf{E}_2(\theta, \phi) \cdot \mathbf{F}(\theta, \phi)e^{-jk\mathbf{u}\cdot\mathbf{r}t} e^{jkx} \sin\theta d\theta d\phi \tag{2.48}$$

where C_n $(n = 1, 2)$ is the proportionality constant, $e^{-jk\mathbf{u}\cdot\mathbf{r}t}$ is the Doppler shift caused by the velocity $\mathbf{u}$ of the antenna, k is the wave number, $\mathbf{r}$ is the unit vector in the radiating

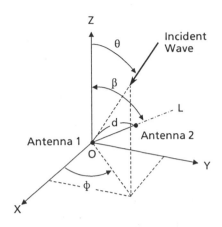

Figure 2.22 Antenna diversity and its coordinate system. (From [55], © 1990 IEICE, Japan.)

direction, and x is the phase difference of the incident waves as seen at the two antennas. As shown in Figure 2.22, $x = d(\sin\theta \sin\phi \sin\beta + \cos\theta \cos\beta)$ if the diversity antenna system lies on the Y-Z plane and the line that connects the antenna branches is inclined by the angle of β from the vertical direction. If it is assumed that there is no correlation between incident waves, then the following equations hold, since the phases of F_θ and F_ϕ are independent for incident waves from different directions $\Omega = (\theta, \phi)$ and $\Omega' = (\theta', \phi')$.

$$\langle F_\theta(\Omega)F_\theta^*(\Omega')\rangle = \langle F_\theta(\Omega)F_\theta^*(\Omega)\rangle\delta(\Omega - \Omega') \tag{2.49}$$

$$\langle F_\phi(\Omega)F_\phi^*(\Omega')\rangle = \langle F_\phi(\Omega)F_\phi^*(\Omega)\rangle\delta(\Omega - \Omega') \tag{2.50}$$

Furthermore, since the phases of F_θ and F_ϕ are independent and uniformly distributed between 0 and 2π, the next relationship is valid.

$$\langle F_\theta(\Omega)F_\phi^*(\Omega')\rangle = 0 \tag{2.51}$$

where the brackets indicate the ensemble average, the asterisk indicates the complex conjugate, and δ is the delta function. $V(t)$ is approximated by the complex Gaussian process with average of zero and satisfies

$$\langle V(t)\rangle = 0 \tag{2.52}$$

Hence, the cross-covariance of the two received voltages can be obtained as follows if (2.45) and (2.46) are substituted into (2.47) and (2.48) and the conditions of (2.49), (2.50), and (2.51) are applied.

$$R_{12} = \langle V_1(t)V_2^*(t) \rangle$$

$$= 2KP_H \int_0^{2\pi} \int_0^{\pi} [\text{XPR} \cdot E_{\theta 1}(\theta, \phi)E_{\theta 2}^*(\theta, \phi)P_\theta(\theta, \phi) \qquad (2.53)$$

$$+ E_{\phi 1}(\theta, \phi)E_{\phi 2}^*(\theta, \phi)P_\phi(\theta, \phi)]e^{-jkx}\sin\theta d\theta d\phi$$

where K is a proportionality constant. Similarly, the standard deviations σ_1 and σ_2 of the complex envelopes of the first and second antennas are

$$\sigma_1^2 = \langle V_1(t)V_1^*(t) \rangle$$

$$= 2KP_H \int_0^{2\pi} \int_0^{\pi} [\text{XPR} \cdot E_{\theta 1}(\theta, \phi)E_{\theta 1}^*(\theta, \phi)P_\theta(\theta, \phi) \qquad (2.54)$$

$$+ E_{\phi 1}(\theta, \phi)E_{\phi 1}^*(\theta, \phi)P_\phi(\theta, \phi)]\sin\theta d\theta d\phi$$

$$\sigma_2^2 = \langle V_2(t)V_2^*(t) \rangle$$

$$= 2KP_H \int_0^{2\pi} \int_0^{\pi} [\text{XPR} \cdot E_{\theta 2}(\theta, \phi)E_{\theta 2}^*(\theta, \phi)P_\theta(\theta, \phi) \qquad (2.55)$$

$$+ E_{\phi 2}(\theta, \phi)E_{\phi 2}^*(\theta, \phi)P_\phi(\theta, \phi)]\sin\theta d\theta d\phi$$

In general, if the complex correlation coefficient is ρ and the correlation coefficient for the observed envelope is ρ_e, then ρ_e is approximately equal to $|\rho|^2$ [50]. Since

$$\rho_e \approx |\rho|^2 = \frac{|R_{12}|^2}{\sigma_1^2 \sigma_2^2} \qquad (2.56)$$

from (2.52), a theoretical expression of the correlation coefficient of antenna diversity can be obtained if (2.53), (2.54), and (2.55) are substituted into (2.56). This expression is a general theoretical equation which takes into account the effect of XPR and the incident wave distribution of the θ and ϕ components. The expression indicates that the correlation coefficient ρ_e becomes zero regardless of XPR or the angular density function of the incident wave if the complex radiation patterns of the θ and ϕ components of the two antennas do not overlap.

2.4.2 Statistical Distribution Model of Incident Waves

2.4.2.1 Theoretical Model

In general, multipath propagation in a land mobile communication environment is caused by the reflection, diffraction, and scattering from topography and buildings. Let us define the secondary wave source generically as the points at which the wave is reflected, diffracted, or scattered immediately before reaching the antenna. The secondary wave sources are widely distributed on the side surfaces and edges of buildings, the ground plane, and objects on the ground (such as trees and vehicles) around the antenna. Their number and location vary widely depending on the style of the city. Hence, if the antenna moves randomly in a metropolitan area with buildings of various heights, shapes, dimensions, and materials, the secondary wave sources observed from the antenna can be assumed to be statistically independent and to be distributed uniformly in the azimuth direction. This assumption corresponds to the scattered ring model [56]. When the antenna moves randomly within a metropolitan area, the building heights and distances from the antenna are considered to be independent variables that are distributed around certain average values. Hence, the secondary wave sources can also be assumed to be distributed around the average elevation direction. If it is assumed that an extremely large number of secondary wave sources are experienced during a random move, this elevation angle distribution can be assumed to be a Gaussian distribution according to the central limit theorem [57].

The preceding assumptions make it quite reasonable to adopt a statistical model in which angular density functions P_θ and P_ϕ are assumed to be Gaussian in elevation and uniform in the azimuth direction, as shown in Figure 2.23. Note that the elevation can take on negative values, since mobile antennas are usually operated above the ground. The distribution functions of incident plane waves are expressed as follows.

$$P_\theta(\theta, \phi) = A_\theta \exp\left\{ -\frac{\left[\theta - \left(\frac{\pi}{2} - m_V \right) \right]^2}{2\sigma_V^2} \right\} \quad (0 \leq \theta \leq \pi) \quad (2.57)$$

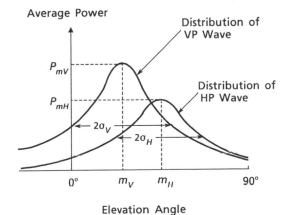

Figure 2.23 Gaussian distribution model of incident waves. (From [52], © 1990 IEEE.)

$$P_\phi(\theta, \phi) = A_\phi \exp\left\{ -\frac{\left[\theta - \left(\frac{\pi}{2} - m_H\right)\right]^2}{2\sigma_H^2} \right\} \quad (0 \leqq \theta \leqq \pi) \tag{2.58}$$

where m_V and m_H are, respectively, the mean elevation angle of each VP and HP wave distribution observed from the horizontal direction, and σ_V and σ_H are, respectively, the standard deviation of each VP and HP wave distribution. A_θ and A_ϕ are constants determined by (2.39) and (2.40). If the mean power strengths of VP and HP waves in the directions of $\theta = \pi/2 - m_V$ and $\theta = \pi/2 - m_H$ are, respectively, P_{m_V} and P_{m_H}, then

$$P_{m_V} = P_1 \cdot A_\theta \tag{2.59}$$

$$P_{m_H} = P_2 \cdot A_\phi \tag{2.60}$$

Therefore, XPR is also determined by P_{m_V} and P_{m_H}, as shown in the following equation.

$$\text{XPR} = \frac{P_V}{P_H} = \frac{P_1}{P_2} = \frac{P_{m_V}}{P_{m_H}} \cdot \frac{A_\phi}{A_\theta} \tag{2.61}$$

2.4.2.2 Validity of the Statistical Model

In order to confirm the validity of the statistical model shown in Figure 2.23, the incident wave distribution had been measured in a Tokyo urban area using a vertically polarized radio signal in the 900-MHz band. The transmitting antenna was a vertically polarized omnidirectional antenna set 87m above the ground. Two measurement routes were selected for the Ningyo-cho and Kabuto-cho areas. Both routes were some 1.2 km from the transmitting antenna, and all receiving points on the routes were out of sight of the transmitting antenna. The receiving antenna was mounted 3.1m above the ground on the roof of a van.

The received power pattern of incident radio waves was measured with a 0.9m-diameter parabolic reflector antenna with a dipole element for the primary radiator. The half-power beamwidth and the first sidelobe level of the antenna were 22 deg and less than −9 dB, respectively. Measurements were taken every 5m for a total of 34 points along the Ningyo-cho route, and every 7m for a total of 30 points along the Kabuto-cho route. At each receiving point, the received power patterns for both VP and HP incident waves were measured by rotating the reflector antenna 360 deg in azimuth, at elevation angles of −10, 0, 20, and 45 deg.

It can be found through the measurements that the number of principal waves is up to five or six at each point, but the incident power and arrival direction vary considerably from point to point even though the measuring points are separated by only 5m. As a result, it can be considered that an extremely large number of incident waves are observed while the antenna moves over a random route, and that the assumption of the model that incident waves arrive from numerous and random azimuthal directions has been confirmed.

Figure 2.24 shows the average power distributions of the incident waves in elevation obtained from the measurement. In this figure, the solid lines express the best approximation of a Gaussian distribution function corresponding to the mean power levels. Figure 2.24 shows, therefore, that the statistical distribution of incident waves can be accurately estimated by the model proposed in this section.

In the distributions of the VP waves, the mean elevation angle m_V is about 20 deg for both routes. This value is consistent with the former experimental results for elevation angles somewhat larger than zero but less than 39 deg [54, p. 149], [2, p. 158]. However, it is found that the dispersion in elevation of the HP waves is larger than that of the VP waves; in fact, the distributions of the HP waves happen to have nearly uniform elevation in this case. This is probably because most buildings are considerably higher than the width of the road along the measurement routes, so that the diffraction over and the multiple reflection from the buildings seem to produce HP waves at relatively high elevation angles. Although further investigation is needed to confirm the statistical distributions for various mobile environments, it is expected that the HP wave distribution will be concentrated around low elevation angles in suburban areas where there are few high buildings. The distribution parameters for the proposed model, m_V, σ_V, m_H, σ_H, can still be obtained empirically, even in the closely uniform distribution case, by approximately the measured values with Gaussian distributions, as shown in Figure 2.24. The distribution parameters for the two routes are shown in Table 2.5. From these parameters, the constants in (2.57) and (2.58), A_θ, A_ϕ, can be determined by (2.39) and (2.40). Thus, XPR can be evaluated by substituting these constants and the maximum mean power levels for both VP and HP waves into (2.61), as shown in Table 2.5. The XPR is evaluated as 5.1 dB in the Ningyo-cho route and as 6.8 dB in the Kabuto-cho route. These XPR values are quite reasonable considering the measured results that show the cross-polarization coupling in urban areas to lie between −9 and −4 dB [45]. This means that XPR in urban areas is larger than about 4 dB but less than 9 dB, because, the VP wave transmission, XPR corresponds to the reciprocal of the cross-polarization coupling.

2.4.3 MEG Characteristics of Dipole Antennas

2.4.3.1 Power Gain Pattern

The half-wavelength dipole antenna and the spherical coordinate system considered here are shown in Figure 2.25. The feeding point of the dipole antenna is situated at the origin

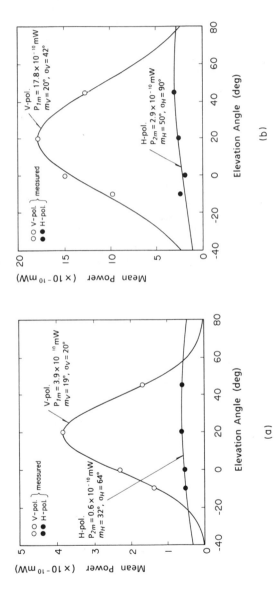

Figure 2.24 Mean power distribution of incident waves versus elevation angles in (a) Ningyo-cho route and (b) Kabuto-cho route. (From [52], © 1990 IEEE.)

Table 2.5
Experimental Results of Statistical Distribution Parameters

Measurement Route	m_V (deg)	σ_V (deg)	m_H (deg)	σ_H (deg)	XPR (dB)
Ningyo-cho	19	20	32	64	5.1
Kabuto-cho	20	42	50	90	6.8

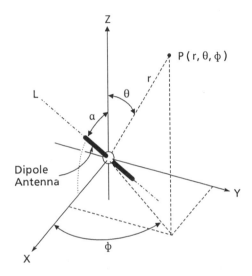

Figure 2.25 Half-wavelength dipole antenna and its coordinate system. (From [52], © 1990 IEEE.)

of the coordinate system, and the antenna elements are on the L-axis inclined at angle α from the Z-axis in the vertical Z-X plane. A thin dipole is assumed and the element radius ignored. The mismatching and ohmic losses in the antenna are also ignored.

The three-dimensional power gain patterns of the dipole antenna with respect to the inclination angle α, from 0 to 90 deg, are shown in Figure 2.26. It is found that when $\alpha = 0$ deg, the antenna is only VP wave sensitive; however, at other inclination angles, it is both VP and HP wave sensitive. In particular, it should be noted that the horizontally oriented half-wavelength dipole antenna is not just HP wave sensitive.

2.4.3.2 MEG Characteristics of Vertical Dipole Antennas

When the inclination angle α is equal to 0 deg the power gain pattern for HP waves G_ϕ is nonexistent, as shown in Figure 2.26. MEG is, therefore, obtained by integrating only (2.41). If, moreover, the incident waves are only VP waves (i.e, XPR is infinitely large),

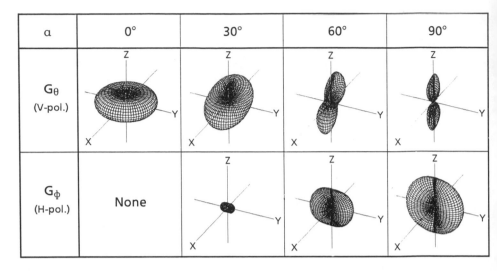

α	0°	30°	60°	90°
G_θ (V-pol.)				
G_ϕ (H-pol.)	None			

Figure 2.26 Power gain patterns of half-wavelength dipole antennas with inclination angle of 0, 30, 60, and 90 deg. (From [52], © 1990 IEEE.)

the coefficient XPR/(1 + XPR) of the first term in the integrand in (2.41) becomes 1, and thus MEG is indpendent of the XPR value. However, when the incident waves have both VP and HP components, MEG is further reduced by the factor XPR/(1 + XPR). This gain degradation due to XPR is represented by the solid line in Figure 2.27. Thus, to clarify the MEG characteristics of vertical dipole antennas, it is sufficient to consider just the MEG characteristics for XPR = ∞.

The dependence of MEG (of the vertical dipole) on the standard deviation σ_V of VP incident waves is shown in Figure 2.28. When $\sigma_V = 0$ deg, MEG equals the directive gain with respect to the incoming direction of incident waves. Furthermore, when $m_V = 0$ deg, and $\sigma_V = 0$ deg (i.e., all incident waves lie completely in the horizontal plane), MEG is equal to the directivity of the half-wavelength dipole antenna at 2.15 dBi, because the incoming direction of incident waves corresponds to the direction with the maximum gain. The more the standard deviation increases, the closer MEG is to the isotropic antenna gain (0 dBi). At $\sigma_V = \infty$ (i.e., when the statistical distribution of the incident waves is completely uniform), MEG is equal to the isotropic antenna gain. As m_V increases from $m_V = 0$ deg, with $\sigma_V = 0$ deg (i.e., incident waves lie in the azimuth plane at the m_V elevation), MEG decreases in proportion to the power gain at the mean elevation angle m_V. However, the more the standard deviation increases, the closer MEG again approaches the isotropic antenna gain (0 dBi). In the actual propagation environments seen in mobile communications, the distribution of the incident waves seems to be spread in elevation. Thus, the effective gain fo the vertical half-wavelength dipole antenna becomes lower han the directivity of 2.15 dBi. MEG is reduced further by a lower XPR, as shown in Figure 2.27.

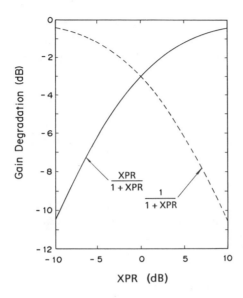

Figure 2.27 Mean effective gain degradation due to the cross-polarization power ratio. (From [52], © 1990 IEEE.)

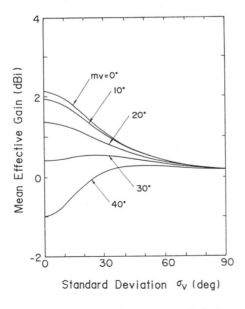

Figure 2.28 Mean effective gain of vertically oriented half-wavelength dipole antennas for vertically polarized incident waves: XPR = ∞. (From [52], © 1990 IEEE.)

2.4.3.3 MEG Characteristics of Inclined Dipole Antennas

Several interesting results are obtained by investigating the MEG characteristics of inclined dipole antennas. Figure 2.29 shows the typical MEG characteristics calculated for several XPR values.

First, it is found that there is a particular inclination angle at which MEG is −3 dBi regardless of the XPR value. This inclination angle is 55 deg and is shown as point A in Figure 2.29. At this angle, the VP radiation power of the antennas is equal to the HP radiation power, as shown in Figure 2.30. MEG variation at point A with the given incident wave distribution parameters (XPR, m_V, m_H, σ_V, σ_H) is less than 0.2 dB. It is proposed that the sum of the incident power of the VP and HP waves, $P_V + P_H$, can be measured by using this property. The power $P_V + P_H$ is twice the average received power measured by a dipole antenna inclined at 55 deg. Therefore, this power level can be used as the reference signal level in the MEG measurement, instead of the mean power level of the vertically oriented dipole antenna, because the latter is strongly affected by the propagation conditions.

Second, in Figure 2.29 the MEG characteristics for XPR = −2 dB shows that there are incident wave parameters that make the MEG constant (−3 dBi) regardless of the antenna inclination angle. According to this theoretical analysis, there are many propagation parameters that yield constant MEG (−3 dBi) characteristics regardless of the antenna

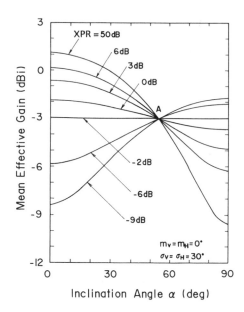

Figure 2.29 Mean effective gain of inclined half-wavelength dipole antennas: $m_V = m_H = 0$ deg. $\sigma_V = \sigma_H = 30$ deg. (From [52], © 1990 IEEE.)

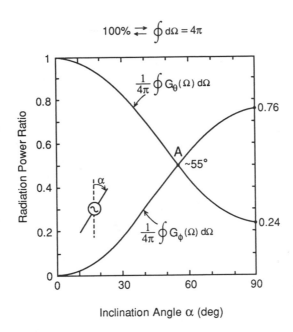

$$100\% \rightleftarrows \oint d\Omega = 4\pi$$

Figure 2.30 Radiation power ratio of vertical or horizontal polarized component of inclined half-wavelength dipole antenna.

orientaton, in addition to the unusual environment of uniform illumination from all directions. In other words, there is a possibility of developing an artificial propagation environment in which the average received signal level of antennas can be made constant regardless of the variation of antenna pattern and polarization. It is expected that these environments could be developed by controlling the polarity of the transmitting antenna for XPR, the height of the transmitting antenna for mean elevation angle, and the beamwidth of the transmitting antenna for standard deviations, but further experimental investigation is required to confirm this.

2.4.3.4 Experimental Results

According to the theoretical consideration described in Section 2.4.3, the reference signal level in the MEG measurement cannot be evaluated definitely by measuring the mean received power level of a vertically oriented half-wavelength dipole antenna; however, it can be evaluated by measuring the mean received power level of a half-wavelength dipole antenna with an inclination angle of 55 deg from the vertical.

To confirm the validity of the theoretical considerations, a 900-MHz band experiment had been performed in the urban area of Tokyo. In this experiment, measurements of the received signal level of a test antenna were carried out over the same routes used in th statistical distribution measurements described in Section 2.4.2. A half-wavelength dipol antenna mounted 3.1m above the ground on the roof of a van was used as the test antenna To evaluate the variation of received power with the direction of the antenna radiatio pattern, the dipole antenna was inclined in the vertical planes with azimuthal angles o 0, 90, +45, and −45 deg from the forward direction of the moving van. The received signal level was digitized by an analog-to-digital (A/D) converter at distances of abou 1 cm, and the average received signal level was calculated by averaging all data sample over the route.

Table 2.6 shows the measurement results for average received signal levels of th dipole antenna with 55-deg inclination and the reference signal levels for the MEG measurement. The average received signal levels were obtained by averaging all measure values digitized over each measurement route. The variation of these values is caused b the mutual influence of the antenna pattern and the lack of the uniformity in azimuth o wave distribution for measurement routes. To evaluate the mean received signal leve without such influences, the average of the measured signal levels for all antenna azimutha orientations must be used as the mean received signal level of the 55-deg inclined dipole Thus, in this case, the average of measured signal levels for four different antenn orientations was adopted as the mean level. The mean was 35.0 dBμV for the Ningyo cho route and 33.5 dBμV for the Kabuto-cho route. The mean received signal leve corresponds to one-half the reference signal level, because the MEG of the 55-deg incline half-wavelength dipole antenna is −3 dBi. Therefore, the reference signal levels wer obtained by adding 3 dB to the mean received signal levels and were evaluated a 38.0 dBμV for the Ningyo-cho route and 36.5 dBμV for the Kabuto-cho route. The MEG value of the test antenna is calculated by normalizing the antenna's average receive signal level with the corresponding reference signal level. Figure 2.31 shows the normalize

Table 2.6
Measured Signal Level of a Dipole Antenna with 55-deg Inclination and
Reference Signal Level for Isotropic Mean Effective Gain

Route	Orientation of Antenna Inclination Plane (deg)	Signal Level (dBμV)		
		Average Received	Mean Received	Reference for $P_V + P_H$
Ningyo-cho	0	35.0	35.0	38.0
	90	34.6		
	+45	35.3		
	−45	35.3		
Kabuto-cho	0	33.5	33.5	36.5
	90	35.2		
	+45	32.4		
	−45	32.2		

Source: [52], © 1990 IEEE.

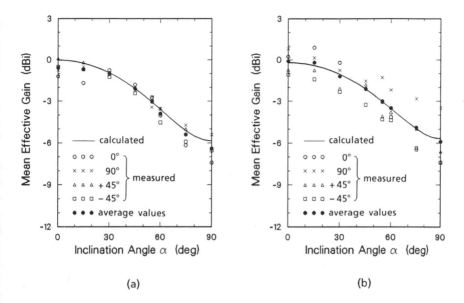

Figure 2.31 Comparison of calculated and measured mean effective gain of an inclined half-wavelength dipole antenna in (a) Ningyo-cho route and (b) Kabuto-cho route. (From [52], © 1990 IEEE.)

measurement results and the theoretical curve of MEG for half-wavelength dipole antennas. The solid line shows the theoretical curve calculated using the empirical distribution parameters shown in Table 2.5. The open circles, crosses, open triangles, and open squares show the measurement results for the antenna orientations of 0, 90, +45, and −45 deg, respectively. The closed circles show the average values of the measured results for each antenna inclination.

The most significant result is that the theoretical curves show excellent agreement, within about 1 dB, with the average values of the measured results. This indicates that the MEG analysis, described in Sections 2.4.1 to 2.4.3, evaluates not only the MEG of mobile antennas (if their movements are sufficiently random so that the statistical distribution of incident waves can be considered completely uniform in the azimuth direction), but also the average of the MEG variation of directional antennas operating in environments whose statistical distribution of incident waves shows a lack of uniformity in the azimuth direction. Furthermore, the experimental results confirm the validity of the presented model and the derived theoretical expression for the MEG.

2.4.4 Correlation Characteristics of Polarization Diversity

It is well known that a cross-dipole antenna made of two orthogonally placed half-wavelength dipole antennas constitutes a polarization diversity branch [45,58,59]. This section describes the analytical derivation of correlation characteristics of this antenna diversity using the theoretical expression described in Section 2.4.1 and the statistical model described in Section 2.4.2.

2.4.4.1 Theoretical Analysis

Figure 2.32 shows the coordinate system for the polarization diversity branch using a cross-dipole antenna. The distance d between the two dipole antennas is zero so that the phase difference x in (2.53) becomes zero. Figure 2.32 exhibits the situation in which the cross-dipole antenna is inclined at an angle α from the vertical (Z) direction in the X-Z plane. The power gain patterns of each dipole antenna for various inclination angles α are shown in Figure 2.33(a) to (d).

The characteristic properties of this polarization diversity antenna are shown in Figures 2.34 through 2.37. The first characteristic is that the correlation coefficient becomes zero (uncorrelated) if one of the dipole antennas is placed vertically and the other horizontally so that $\alpha = 0$ deg. This characteristic can be explained as follows. In the case of $\alpha = 0$ deg, it can be seen from Figure 2.33(a) that the horizontal dipole antenna has radiation patterns of both θ and ϕ components, whereas the vertical dipole has the pattern of only the θ component. Hence, according to (2.53), only the radiation patterns of the θ components are related to the correlation coefficient. The radiation patterns of the θ components are, accordingly, spatially orthogonal. Therefore, the integral of the numerator term in (2.53) is equal to zero so that the diversity branches become theoretically uncorre-

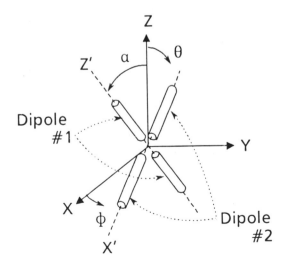

Figure 2.32 Polarization diversity using cross-dipole antennas and their coordinate system. (From [55], © 1990 IEICE, Japan.)

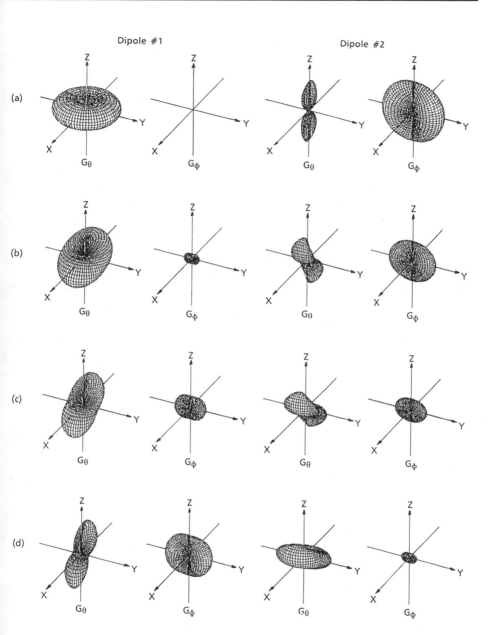

Figure 2.33 Power gain patterns of cross-dipole antennas: (a) $\alpha = 0°$; (b) $\alpha = 30°$; (c) $\alpha = 45°$; (d) $\alpha = 60°$. (From [55], © 1990 IEICE, Japan.)

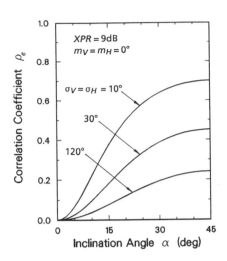

Figure 2.34 Correlation coefficient of polarization diversity using cross-dipole antennas. (From [55], © 1990 IEICE, Japan.)

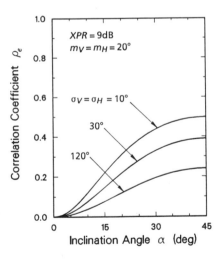

Figure 2.35 Correlation coefficient of polarization diversity using cross-dipole antennas. (From [55], © 1990 IEICE, Japan.)

lated. This lack of correlation does not depend on XPR or the variation in incident wave distribution. According to the experimental results reported to date [45], the correlation characteristics are almost zero for $\alpha = 0$ deg and hence the validity of this analysis appears to be confirmed.

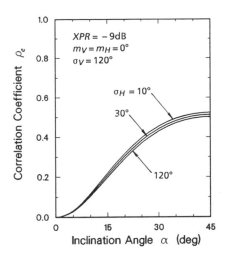

Figure 2.36 Correlation coefficient of polarization diversity using cross-dipole antennas. (From [55], © 1990 IEICE, Japan.)

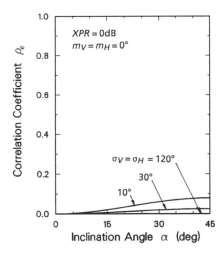

Figure 2.37 Correlation coefficient of polarization diversity using cross-dipole antennas. (From [55], © 1990 IEICE, Japan.)

When the antenna system is inclined, the perfect orthogonality of the antenna radiation patterns described above cannot be maintained, and hence the correlation coefficient increases with the inclination angle α. Such a situation is common in an antenna system installed on a portable transceiver. The second characteristic is the fact that the correlation coefficient becomes maximum for the inclination angle of 45 deg. This is because the radiation patterns of the θ components of the two dipoles in the horizontal (X-Y) plane become identical omnidirectional patterns, while the radiation patterns of the ϕ components are almost identical three-dimensionally. (See Figure 2.33(c).) As a result, if the environment yields an XPR much larger than 0 dB, the correlation increases as the incident wave distribution concentrates more into the horizontal plane, and decreases as it is spread in the elevation angle direction, as shown in Figure 2.34. As shown in Figure 2.35, as the mean elevation angle m_V of the VP waves moves away from the horizontal direction, the correlation is reduced. In an environment where XPR is much smaller than 0 dB, the effect of the radiation pattern of the ϕ component becomes significant. However, since the variation of the ϕ component pattern on the elevation angle is small, as shown in Figure 2.33, the variation of the correlation coefficient for the divergence of the incident wave distribution in the elevation direction is negligible. (See Figure 2.36.) Moreover, since the degree of ϕ component pattern overlap of both antennas is significant, the correlation increases as XPR is reduced.

According to the characteristics clarified above, the integral terms for each polarization component in (2.53) can be reduced, and the correlation becomes extremely small in environments whose XPR values are nearly equal to 0 dB. This low correlation is almost independent of the parameter values of the incident wave distribution and inclination of the diversity antenna. Figure 2.37 shows the correlation characteristics in such cases.

It is clear that the correlation characteristics of a polarization diversity antenna depend on the radiation pattern of the antenna branch, XPR, and incident wave distribution. This dependence is, in fact, also true for all antenna diversity schemes. Hence, to realize optimum antenna diversity, we need to take into account antenna radiation patterns and the propagation environment characteristics. If the dependence of the correlation characteristics on the propagation environment parameters can be used positively, realization of a polarization diversity branch with an extremly low correlation can be expected. According to a theoretical analysis, it is found that this low-correlation diversity branch is realized at XPR = −1.5dB in the diversity branches of a cross-dipole antenna. Furthermore, its realization has been confirmed through an indoor experiment [60].

2.4.4.2 Experiment Results

To confirm the validity of these theoretical considerations, a 900-MHz-band experiment was performed in the urban area of Tokyo. In this experiment, measurements of the correlation characteristics of a cross-dipole antenna were carried out on the Ningyo-cho route passing through all the measurement points described in Section 2.4.2. A cross-dipole antenna was installed on the roof (height: 3.1m above the ground) of a mobile van. Since the radiation pattern of an inclined dipole antenna has an azimuthal deviation as shown in Figure 2.33, it is considered that the measured values might be affected by the azimuthal deviation of the incident wave distribution in the experimental environment. Hence, the direction of the receiving antenna (equal to the direction of the X-axis in

Figure 2.32) took four directions 0, 90, +45, and −45 deg. with respect to the moving direction of the van. For each direction, the inclination angle α of the receiving antenna was varied with respect to the vertical direction and the receiving signal strength between the two branches was measured simultaneously as the van moved at about 20 to 30 km/hr along the route. The measured data were digitized by an A/D converter at intervals of about 1 cm. The correlation coefficient of the envelope of the received signal for one circuit of the measurement route was derived by numerical processing on a computer.

Figure 2.38 shows the correlation coefficient of the cross-dipole antenna calculated from the incident wave distribution parameters in Table 2.5 and the corresponding experimental data. The symbols of open circle, cross, open triangle, and open square are measured values for the antenna azimuth directions of 0, 90, +45, and −45 deg, respectively. The closed circles indicate the average of the values in the four directions, whereas the solid line is the theoretical curve. If the azimuth distribution of the incident wave is uniform, the measured values should not depend on the azimuth direction. Hence, the dependence of the measured values on the antenna azimuth direction indicates that the incident wave distribution on the measurement route is not uniform in the azimuth direction. Also, since the incident wave distribution parameters in Table 2.5 were obtained from the mean level of the measured patterns in the azimuth direction, the theoretical curve in Figure 2.38 represents the average value of the correlation coefficient over various antenna azimuth directions. Hence, to discuss the validity of the analytical results, comparison with the average value of the correlation coefficient measured for many antenna azimuth directions is important. The average values indicated by closed circles are those over only four directions; however, the values agree closely with the theoretical results, and hence the effectiveness of the analysis results is confirmed.

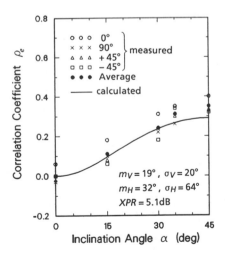

Figure 2.38 Comparison of theoretical correlation curve with measurement value. (From [55], © 1990 IEICE, Japan.)

86

REFERENCES

[1] Kuramoto, M., and H. Tohyama, "General View of Mobile Communication," Ch. 1 of *Basic Technology of Mobile Communications,* Y. Okumura and M. Shinji, eds., IEICE Press (Japanese), 1986, pp. 1–23.

[2] Lee, W. C. Y., *Mobile Communications Engineering,* McGraw-Hill, 1982.

[3] Jakes, W. C., Jr., *Microwave Mobile Communications,* John Wiley, 1974.

[4] Special issue on "Mobile Communications," *IEICE Trans.,* Vol. E74, No. 6, 1991.

[5] Okumura, Y., E. Ohmori, T. Kohno, and K. Fukuda, "Field Strength and Its Variability in VHF and UHF Land-Mobile Radio Service," *Rev. Elec. Comm. Lab.,* Vol. 16, No. 9–10, 1968, pp. 825–873.

[6] "VHF and UHF Propagation Curves for the Frequency Range From 30 MHz and 1000 MHz," CCIR SG-5, Recommendation 370.

[7] Kozono, S., and K. Watanabe, "Influence of Environmental Buildings on UHF Band Mobile Radio Propagation," *IEEE Trans. Com.,* Vol. COM-25, No. 10, 1977, pp. 1133–1143.

[8] Hata, M., "Empirical Formula for Propagation Loss in Land Mobile Radio Service," *IEEE Trans.,* Vol. VT-29, No. 3, 1980, pp. 317–325.

[9] Ikegami, F., S. Yoshida, T. Takeuchi, and M. Umehira, "Propagation Factors Controlling Mean Field Strength on Urban Streets," *IEEE Trans.,* Vol. AP-32, No. 8, 1984, pp. 822–829.

[10] Sakagami, S., "Mobile Propagation Loss Prediction for Arbitrary Urban Environments," *IEICE Trans.,* Vol. J74-B-II, 1991, pp. 17–25.

[11] Fujimoto, K., "Overview of Antenna Systems for Mobile Communications and Prospects for the Future Technology," *IEICE Trans.,* Vol. E74, No. 10, 1991, pp. 3191–3201.

[12] Special issue on "Mobile Radio Communications," *J. IECE Japan,* Vol. 68, No. 11, 1985.

[13] Special issue on "Digital Mobile Telephone System," *J. IEICE,* Vol. 73, No. 8, 1990, pp. 799–844.

[14] Sekiguchi, H., "Switching of Mobile Communications," Ch. 8 of *Mobile Communications,* M. Shinji, ed., Maruzen, 1989.

[15] Nishikawa, T., "RF Front End Circuit Components Miniaturized Using Dielectric Resonators for Cellular Portable Telephone," *IEICE Trans.,* Vol. E74, No. 6, 1991, pp. 1556–1562.

[16] Taga, T., and K. Tsunekawa, "Performance Analysis of a Built-in Planar Inverted F Antenna for 800 MHz Band Portable Units," *IEEE JSAC,* Vol. SAC-5, No. 5, 1987, pp. 921–929.

[17] Yamada, Y., Y. Ebine, and K. Tsunekawa, "Base and Mobile Station Antennas for Land Mobile Radio Systems," *IEICE Trans.,* Vol. E74, No. 6, 1991, pp. 1547–1555.

[18] Ishihara, H., "Mobile Radio Communications for Business Use and Multi-Channel Access System," *J. IECE Japan,* Vol. 68, No. 11, 1985, pp. 1183–1187.

[19] Kanzaki, K., "Current Status and Future Trends on Efficient Spectrum Utilization for Mobile Radio Communications," *J. IECE Japan,* Vol. 68, No. 11, 1985, pp. 1251–1253.

[20] Callendar, M. H., "IMT-2000 (FPLMTS) Standardization" presented at the ITU Malaysia Seminar, Kuala Lumpur, March 1997.

[21] Bullington, K., "Radio Propagation for Vehicular Communications," *IEEE Trans. Vehicular Technology,* Vol. VT-26, No. 4, 1977, pp. 295–308.

[22] Shepherd, N. H., "Radio Wave Loss Deviation and Shadow Loss at 900 MHz," *IEEE Trans. Vehicular Technology,* Vol. VT-26, Nov. 1977, pp. 309–313.

[23] Young, R. W., "Comparison of Mobile Radio Transmission at 150, 450, 900 and 3700 MC/s," *BSTJ,* Nov. 1952, pp. 1068–1085.

[24] Lee, W. C. Y., "Lee's Model," Propagation Ad Hoc Committee of IEEE Vehicular Technology Society appeared in a special issue of *IEEE Transactions on Vehicular Technology,* Feb. 1988, pp. 68–70. Also *IEEE VTS '92 Proc.,* pp. 343–348.

[25] Lee, W. C. Y., "Studies of Base-Station Antenna Height Effects on Mobile Radio," *IEEE Trans. Vehicular Technology,* Vol. VT-29, No. 2, May 1980, pp. 252–260.

[26] Lee, W. C. Y., *Mobile Cellular Telecommunications Systems,* McGraw-Hill, 1989, pp. 102, 126–131.

[27] Lee, W. C. Y., *Mobile Communications Engineering,* McGraw-Hill, 1982, p. 229, p. 46.

[28] Lee, W. C. Y., *Mobile Communications Design Fundamentals,* 2nd edition, John Wiley, 1993, pp. 30, 89–93.

[29] Ibid., p. 30.

[30] Fujimoto, K., "A Loaded Antenna System Applied to VHF Portable Communication Equipment," *IEEE Trans.,* Vol. VT-17, 1968, pp. 6–13.

[31] Hirasawa, K., and K. Fujimoto, "Characteristics of Wire Antenna on a Rectangular Conducting Body," *Trans. IECE Japan,* Vol. J65-B, 1982, pp. 1133–1139.

[32] Sato, K., K. Matsumoto, K. Fujimoto, and K. Hirasawa, "Characteristics of Planar Inverted-F Antenna on a Rectangular Conducting Body," *Trans. IECIE,* Vol. J71-B, 1988, pp. 1237–1243.

[33] Fujimoto, K., A. Henderson, K. Hirasawa, and J. R. James, *Small Antennas,* Research Studies Press, 1986, pp. 131–151.

[34] Fujimoto, K., "Overview of Antenna Systems for Mobile Communications and Prospects for the Future Technology," *IEICE Trans. Comm.,* Vol. E74, No. 10, 1991, p. 3196.

[35] Ibid., p. 3193.

[36] Kagoshima, K., and Y. Yamada, "Mobile Communication Antennas," Ch. 9 in *Mobile Communications,* M. Shinji, ed., Maruzen, 1989.

[37] Tsunekawa, K., "Diversity Antennas for Portable Telephone," *IEEE Proc. VTC '89,* 1989, pp. 50–56.

[38] Fujimoto et al., [33], op cit., pp. 89–112.

[39] Ebine, Y., and Y. Yamada, "A Vehicular-Mounted Vertical Space Diversity Antenna for a Land Mobile Radio," *IEEE Trans.,* Vol. VT-40, No. 2, 1991, pp. 420–425.

[40] Nishikawa, K., K. Sato, and K. Fujimoto, "Phased Array Antenna for Land Mobile Satellite Communications," *IEICE Trans.,* Vol. J72 B-II, No. 7, 1989, pp. 323–329.

[41] Lee, W. C. Y., *Mobile Communication Engineering,* McGraw-Hill, 1982, p. 300.

[42] Lee, W. C. Y., "Mobile Radio Signal Correlation vs. Antenna Height and Spacing," *IEEE Trans. Vehicular Technology,* Aug. 1977, pp. 290–292.

[43] Lee, W. C. Y., *Mobile Communications Design Fundamentals,* Howard W. Sams and Co., 1986, p. 204.

[44] Lee, W. C. Y., "Vertical vs. Horizontal Separations for Diversity Antennas," *Cellular Business,* Dec. 1992, pp. 56–60.

[45] Lee, W. C. Y., and Y. S. Yeh, "Polarization Diversity System for Mobile Radio," *IEEE Trans. Communications,* Vol. COM-20, No. 5, Oct 1972, pp. 912–923.

[46] Lee, W. C. Y., "Statistical Analysis of the Level Crossings and Duration of Fades of the Signal from an Energy Density Mobile Radio Antenna," *Bell Syst. Tech. J.,* Vol. 46, Feb. 1967, pp. 416–440.

[47] Lee, W. C. Y., "Preliminary Investigation of Mobile Radio Signal Fading Using Directional Antennas on the Mobile Unit," *IEEE Trans. Veh. Comm.,* Vol. 15, Oct. 1966, pp. 8–15.

[48] Bach Andersen, J., and F. Hansen, "Antennas for VHF/UHF Personal Radio: A Theoretical and Experimental Study of Characteristics and Performance," *IEEE Trans. Vehicular Tecnology,* Vol. VT-26, No. 4, Nov. 1977, pp. 349–357.

[49] Davidson, A. L., and W. J. Turney, "Mobile Antenna Gain in Multipath Environment at 900 MHz," *IEEE Trans. Vehicular Technology,* Vol. VT-26, No. 4, Nov. 1977, pp. 345–348.

[50] Clarke, R. H., "A Statistical Theory of Mobile-Radio Reception," *Bell Syst. Techn. J.,* Vol. 47, No. 6, July-Aug. 1968, pp. 957–1000.

[51] Awadalla, K. H., "Direction Diversity in Mobile Communications," *IEEE Trans. Vehicular Technology,* Vol. VT-30, No. 3, Aug. 1987, pp. 121–123.

[52] Taga, T., "Analysis for Mean Effective Gain of Mobile Antennas in Land Mobile Radio Environments," *IEEE Trans. Vehicular Technology,* Vol. VT-39, No. 2, May 1990, pp. 117–131.

[53] Cox, D. C., R. R. Murray, H. W. Arnold, A. W. Norris, and M. F. Wazowicz, "Cross-Polarization Coupling Measured for 800 MHz Radio Transmission In and Around Houses and Large Buildings," *IEEE Trans. Ant. Propag.,* Vol. AP-34, No. 1, Jan. 1986, pp. 83–87.

[54] Jakes, W. C., *Microwave Mobile Communications,* Wiley & Sons, 1974.

[55] Taga, T., "Analysis for Correlation Characteristics of Antenna Diversity in Land Mobile Radio Environments," *IEICE Trans. (B-II) of Japan,* Vol. J73-B-II, No. 12, Dec. 1990, pp. 883–895, (in Japanese). (This paper was translated by Scripta Technica, Inc., to *Electronics and Communications in Japan,* Part 1, Vol. 74, No. 8, 1991, pp. 101–115.)

[56] Gans, M. J., "A Power Spectral Theory of Propagation in the Mobile Radio Environment," *IEEE Trans. Vehicular Technology,* Vol. VT-21, No. 1, Feb. 1972, pp. 27–38.

[57] Ikegami, F., and S. Yoshida, "Analysis of Multipath Propagation Structure in Urban Mobile Radio Environments," *IEEE Trans. Ant. Propag.,* Vol. AP-28, No. 4, July 1977, pp. 531–537.

[58] Kozono, S., H. Tsuruhara, and M. Sakamoto, "Base Station Polarization Diversity Reception for Mobile Radio," *IEEE Trans. Vehicular Technology,* Vol. VT-33, No. 4, Nov. 1984, pp. 301–306.

[59] Cox, D. C., "Antenna Diversity Performance in Mitigating the Effects of Portable Radiotelephone Orientation and Multipath Propagation," *IEEE Trans. Communications,* Vol. COM-31, No. 5, May 1983, pp. 620–628.

[60] Taga, T., K. Tsunoda, and H. Imahori, "Correlation Properties of Antenna Diversity in Indoor Mobile Communication Environments," *Proc. 39th IEEE Vehicular Technology Conf.,* San Francisco, 1–3 May 1989, pp. 446–451.

Chapter 3

Advances in Mobile Propagation Prediction Methods

S. R. Saunders

3.1 INTRODUCTION

This chapter surveys recent advances in methods of predicting the coverage and capacity of modern mobile systems. The increasing trend is toward systems with very heterogeneous cell types, including cells of radius from tens of meters up to thousands of kilometers. These cells enable operators to provide an appropriate trade-off between wide-area coverage, using a minimum of equipment, together with high capacity in areas with high traffic densities, tailored to the topography of the local area. The most appropriate propagation prediction technique for a given system is highly dependent on the cell type under investigation. Each of the four major cell types (Figure 3.1) is described in this chapter (macrocells in Section 3.2, microcells in Section 3.3, picocells in Section 3.4, and megacells in Section 3.5) together with a description of state-of-the-art prediction techniques in each case. For each model a basic description is given, frequently with equations and example predictions, together with references to permit the reader to pursue them in more detail. Finally, Section 3.6 provides an exposition of future trends. Much of the contents of this chapter are adapted from [1], to which the reader is referred for further detail.

3.2 MACROCELLS

Macrocells (tens of kilometers) are formed by base stations, having antenna heights that are considerably in excess of the heights of the surrounding clutter, including both trees

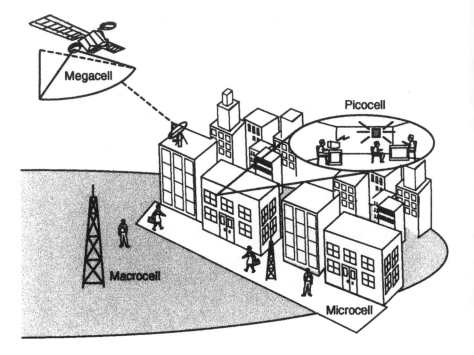

Figure 3.1 Four cell types.

and buildings. Typical installations are shown in Figure 3.2. As a result, the general shape of the coverage area is dominated by the radiation pattern of the antenna, with the environment playing a major role in determining the system outage due to shadowing effects. Although macrocells have been studied empirically for many years, considerable progress has been made more recently in understanding the physical mechanisms to account for the measurements and to create more site-specific techniques. Lee's macrocell model, described in Chapter 2, is one example of an empirical model.

In principle, macrocell propagation could be predicted by deterministic prediction of the loss over every path profile between the base station and every possible mobile location. However, the data describing the terrain and clutter would be very large and the computational effort involved would often be excessive. Even if such resources were available, the important parameter for the macrocell designer is the overall area covered, rather than the specific field strength at particular locations.

The models presented in this chapter treat the path loss associated with a given macrocell as dependent on distance, provided that the environment surrounding the base station is fairly uniform. Consequently, the coverage area predicted by these models for an isolated base station will be approximated as roughly circular. In practice, it is also necessary to consider the variability of path loss at a given distance, usually taken to have a lognormal distribution [1].

Figure 3.2 Typical macrocell installations.

3.2.1 Definition of Parameters

The following terms are used to define the macrocell path loss models in this chapter, and they are illustrated in Figure 3.3:

h_m	mobile station antenna height above local terrain height [m]. often taken as 1.5m;
d_m	distance between the mobile and the nearest building [m];
h_o	typical (usually the mean) height of a building above local terrain height [m];
h_b	base station antenna height above local terrain height [m];
r	great circle distance between base station and mobile [m];
$R = r \times 10^{-3}$	great circle distance between base station and mobile [km];
f	carrier frequency [Hz];
$f_c = f \times 10^{-6}$	carrier frequency [MHz];
λ	free space wavelength [m];

The basic definition of a macrocell is that $h_b > h_o$. Although buildings are not the only obstructions in practice, they are usually by far the most significant at typical

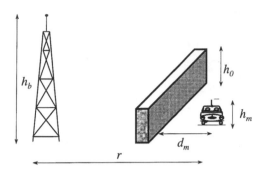

Figure 3.3 Definition of parameters for macrocell propagation models.

macrocellular frequencies. Typical base station heights in practice are around 15m if a mast is used, or around 20m and upward if deployed on a building rooftop. The effective base station height may be increased dramatically by locating it on a hill, overlooking the region to be covered.

3.2.2 Empirical Path Loss Models

The basic propagation models (free-space loss and plane earth or two-wave loss, together with multiple diffraction loss) examined in Chapter 2 account, in principle, for all of the major mechanisms encountered in macrocell prediction. However, to use such models would require detailed knowledge of the location, dimension, and constitutive parameters of every tree, building, and terrain feature in the area to be covered. This is far too complex to be practical, and anyway it would yield an unnecessary amount of detail, because the system designer is not usually interested in the particular locations being covered, but rather in the overall extent of the coverage area. One appropriate way to account for these complex effects is via an empirical model. To create such a model, an extensive set of actual path loss measurements is made, and an appropriate function is fitted to the measurements, with parameters derived for the particular environment, frequency, and antenna heights so as to minimize the error between the model and the measurements.[1] The model can then be used to design systems operated in similar environments to the original measurements. A real example of an empirical model fitted to measurements is shown in Figure 3.4. The very large spread of the measurements at a given distance is the shadowing part, not explicitly predicted by the models to follow [1].

The simplest useful form for an empirical path loss model is as follows:

---

1. Note that each measurement represents an average of a set of samples taken over a small area (around 10–50m), in order to remove the effects of multipath fading (Section 2.2.7). See [2] for further details of measurement procedures.

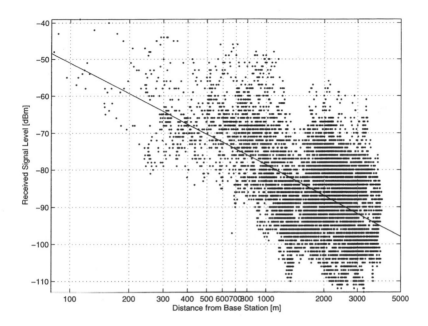

Figure 3.4 Empirical model of macrocell propagation at 900 MHz; the dots are measurements taken in a suburban area, whereas the line represents a best fit empirical model.

$$\frac{P_R}{P_T} = \frac{1}{L} = \frac{k}{r^n}; \text{ or in decibels: } L = 10n\log r + K \qquad (3.1)$$

where P_T and P_R are the effective isotropic transmitted and predicted isotropic received powers, L is the path loss, r is the distance between the base station and the mobile, and $K = 10\log_{10}k$ and n are constants of the model. This will be referred to as a *power law model*. A more convenient form is (in decibels):

$$L = 10n\log\frac{r}{r_{\text{ref}}} + L_{\text{ref}} \qquad (3.2)$$

where L_{ref} is the predicted loss at a reference distance r_{ref}.

Both the free-space loss and the plane earth loss can be expressed in this form. The parameter n is known as the *path loss exponent*. It is found by measurement to depend on the system parameters, such as antenna heights and the environment. The path loss exponent is critical in establishing both the coverage and capacity of a cellular system. Parameter k can be considered as the reciprocal of the propagation loss that would be experienced at 1m range ($r = 1$m).

3.2.2.1 Clutter Factor Models

Measurements taken in urban and suburban areas usually find a path loss exponent close to 4, just as in the plane earth loss, but with a greater absolute loss value [smaller K in (3.1)]. This has led to some models being proposed that consist of the plane earth loss plus (in decibels) an extra loss component, which is referred to as the clutter factor, as shown in Figure 3.5. The various models differ basically in the values which they assign to k and n for different frequencies and environments.

A good example of a clutter factor model is the method of Egli [3], which is based on a large number of measurements taken around American cities. Egli's overall results were originally presented in nomograph form, but [14] has given an approximation to these results for easier computation:

$$L = 40\log R + 20\log f_c - 20\log h_b + L_m \tag{3.3}$$

where

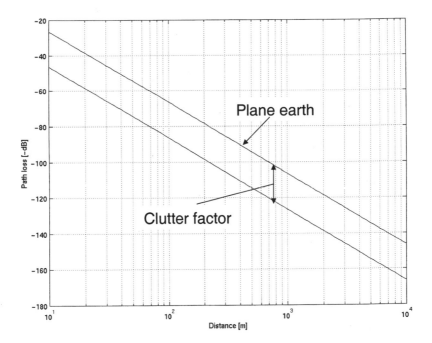

Figure 3.5 Clutter factor model. Note that the y axis in this figure and in several to follow is the negative of the propagation loss in decibels. This serves to make clear the way in which the received power diminishes with distance.

$$L_m = \begin{cases} 76.3 - 10\log h_m \text{ for } h_m < 10 \\ 76.3 - 20\log h_m \text{ for } h_m \geq 10 \end{cases} \tag{3.4}$$

Note that this approximation involves a small discontinuity at $h_m = 10$m. Although plane earth loss is frequency independent, this model introduces an additional f_c^{-2} received power dependence, which is more representative of the results of real measurements. For very large antenna heights, the loss predicted by may be less than the free-space value, in which case the free-space value is used.

The mobile antenna characteristic is found to be linear with height (as with plane earth) for antennas that clear the surrounding terrain features. Elsewhere there is a square root variation for heights in the range of 2–10m, linear above 10m. The transition value (10m) presumably corresponds to the mean obstruction height, although no correction is made for other heights. The average effect of polarization is considered negligible.

3.2.2.2 Okumura-Hata Model

This is a fully empirical prediction method [4], based entirely on an extensive series of measurements made in and around Tokyo city between 200 MHz and 2 GHz. There is no attempt to base the predictions on a physical model such as the plane earth loss. Predictions are made via a series of graphs, the most important of which have since been approximated in a set of formulas by Hata [5]. The thoroughness of these two works taken together has made them the most widely quoted macrocell prediction model, often regarded as a standard against which to judge new approaches. The urban values in the model presented below have been standardized for international use in [6].

The method involves dividing the prediction area into a series of clutter and terrain category definitions, namely *open, suburban,* and *urban,* summarized as follows:

Open area: Open space, no tall trees or buildings in path, plot of land cleared for 300–400m ahead, for example, farm land, rice fields, and open fields.
Suburban area: Village or highway scattered with trees and houses, some obstacles near the mobile but not very congested.
Urban area: Built up city or large town with large buildings and two or more storied houses, or larger villages with close houses and tall, thickly grown trees.

Okumura takes urban areas as a reference and applies correction factors for conversion to the other classifications. This is a sensible choice, since such areas avoid the large variability present in suburban areas and yet include the effects of obstructions better than could be done with open areas. A series of terrain types is also defined for when such information is available. Quasi-smooth terrain is taken as the reference and correction factors are added for the other types.

Okumura's predictions of median path loss are usually calculated using Hata's approximations as follows [5]:

$$L_{dB} = A + B\log R - E; \quad \text{Urban areas}$$
$$L_{dB} = A + B\log R - C; \quad \text{Suburban areas} \quad\quad (3.5)$$
$$L_{dB} = A + B\log R - D; \quad \text{Open areas}$$

where:

$$A = 69.55 + 26.16\log f_c - 13.82\log h_b;$$

$$B = 44.9 - 6.55\log h_b;$$

$$C = 2(\log(f_c/28))^2 + 5.4;$$

$$D = 4.78(\log f_c)^2 + 18.33\log f_c + 40.94; \quad\quad (3.6)$$

$$E = 3.2(\log(11.75h_m))^2 - 4.97; \text{ for large cities, } f_c \geq 300 \text{ MHz}$$

$$E = 8.29(\log(1.54h_m))^2 - 1.1; \text{ for large cities, } f_c < 300 \text{ MHz}$$

$$E = (1.1\log f_c - 0.7)h_m - (1.56\log f_c - 0.8); \text{ for medium-small cities}$$

The model is valid only for 150 MHz $\leq f_c \leq$ 1500 MHz, 30m $\leq h_b \leq$ 200m, 1m $< h_m <$ 10m, and $R > 1$ km. The path loss exponent is given by $B/10$, which is a little less than 4, decreasing with increasing base station antenna height.

Base station antenna height, h_b, is defined as the height above the average ground level in the range 3–10 km from the base station; h_b may therefore vary slightly with the direction of the mobile from the base. The height gain factor varies between 6 and 9 dB/octave as the height is increased from 30m to 1 km. Measurements also suggested that this factor depends on range.

Mobile antenna height gain was found by Okumura to vary between 3 dB/octave up to $h_m = 3$m and 8 dB/octave for greater antenna heights. It depends partially on urban density, apparently as a result of the effect of building heights on the angle-of-arrival of wave energy at the mobile and the consequent shadow loss variation. Urban areas are therefore subdivided into large cities and medium/small cities, where an area having an average building height in excess of 15m is defined as a large city.

Other correction factors are included in Okumura's original work for the effects of street orientation (if an area has a large proportion of streets that are either radial or tangential to the propagation direction) and a fine correction for rolling hilly terrain (used if a large proportion of streets are placed at either the peaks or valleys of the terrain undulations). Application of the method involves first finding the basic median field strength in concentric circles around the base station, then amending these according to the terrain and clutter correction graphs.

Okumura's predictions have been found useful in many cases [7], particularly in suburban areas. However, other measurements have been in disagreement with these predictions, with the reasons for error often being cited as the difference in the characteristics of the area under test with Tokyo. Other authors such as those of [8] have attempted to modify Okumura's method to include a measure of building density, but such approaches have not found common acceptance.

The Okumura-Hata model, together with related corrections, is probably the single most common model used in designing real systems. Several commercial prediction tools essentially rely on variations of this model, optimized for the particular environments they are catering to as the basis of their predictions.

3.2.2.3 COST-231-Hata Model

The Okumura-Hata model for medium-small cities has been extended to cover the band 1500 MHz $< f_c <$ 2000 MHz [9]:

$$L_{dB} = F + B \log R - E + G;$$ (3.7)

where

$$F = 46.3 + 33.9 \log f_c - 13.82 \log h_b$$ (3.8)

E is as defined in (3.6) for medium-small cities and

$$G = \begin{cases} 0 \text{ dB} & \text{medium-sized city and suburban areas} \\ 3 \text{ dB} & \text{metropolitan centers} \end{cases}$$ (3.9)

3.2.2.4 Ibrahim and Parsons Model

This method [10] is based on a series of field trials around London. The method is not intended as a fully general prediction model, but as a first step toward quantifying urban propagation loss. It integrates well with a previous method [11] for predicting terrain diffraction effects since the same 0.5-km^2 database is also used. Each square is assigned three parameters, H, U, and L, which are defined as follows: Terrain height, H, is defined as the actual height of a peak, basin, plateau, or valley found in each square, or the arithmetic mean of the minimum and maximum heights found in the square if it does not contain any such features.

The degree of urbanization factor, U, is defined as the percentage of building site area within the square that is occupied by buildings having four or more floors. For the 24 test squares in inner London that were analyzed, U varied between 2% and 95%, suggesting that this parameter is sensitive enough for the purpose.

Land usage factor, L, is defined as the percentage of the test area actually occupied by any buildings.

These parameters were selected empirically as having good correlation with the data. Two models were proposed. The fully empirical method shows marginally lower prediction errors, but relies on a complex formulation that bears no direct relationship to propagation principles.

The semiempirical method, as with the Egli clutter factor method, is based on the plane earth loss together with a clutter factor β expressed as a function of f_c, L, H, and U. The latter method is examined here, since it has been quoted in later work by the same author [Parsons, 2], and since it forms a better basis for future development. The model is given as:

$$L_T = 40\log r - 20\log(h_m h_b) + \beta$$

$$\text{where } \beta = 20 + \frac{f_c}{40} + 0.18L - 0.34H + K \tag{3.10}$$

$$\text{and } K = 0.094U - 5.9$$

Data are extracted from the databases compiled by local authorities in the United Kingdom. Since information on U is only available in highly urbanized city centers, K is set to zero elsewhere. Considerably lower accuracy may be expected in such areas since $K = 0$ would correspond to $U = 63\%$. RMS errors calculated from the original data on which the model is based vary from 2.0 to 5.8 dB as frequency is increased from 168 to 900 MHz. A comparison is also shown for some independent data, but error statistics are not given.

The model is of limited use in suburban areas since U will normally be zero, giving no measure of building height distribution.

3.2.2.5 Environment Categories

In an empirical model, it is crucial to correctly classify the environment in which the system is operating. The models assume that the characteristics of the environment to be predicted are sufficiently similar to those where the original measurements were taken that the propagation loss at a given distance will be similar. Good results will therefore only be obtained if the correct classification is chosen. The categories of environment should also be sufficiently numerous that the properties of different locations classed within the same category are not too variable. The decision as to which category an environment fits into is usually purely subjective and may vary between individuals and countries. For example, the Okumura-Hata model uses four categories: large cities, medium-small cities, suburban areas, and open areas. Because the original measurements were made in Tokyo, the model relies on other parts of the world having characteristics that are somehow similar to those in Tokyo. While this is an extremely questionable assumption, it is nevertheless true that this model has been applied to many successful system designs.

Many other, more detailed, schemes exist for qualitative classification of land usage (see Table 3.1 for an example). They often correspond to sources of data that are available, such as satellite remote-sensing data that classify land according to the degree of scattering experienced at various wavelengths. This at least avoids the need for ambiguous judgments to be made. Similarly, the Ibrahim and Parsons model uses a clear numerical approach to classification. Nevertheless, there is no guarantee that there is any one-to-one mapping between the propagation characteristics and such measures of land usage. To find more appropriate parameters, the growing tendency in macrocellular propagation is toward models that have a physical basis, and these are examined in the next section.

3.2.3 Physical Models

Although empirical models have been extensively applied with good results, they suffer from a number of disadvantages:

- They can only be used over parameter ranges included in the original measurement set.
- Environments must be classified subjectively according to categories such as "urban," which have different meaning in different countries.
- They provide no physical insight into the mechanisms by which propagation occurs.

The last point is particularly significant, because empirical models are unable to account for factors such as an unusually large building or hill, which may greatly modify propagation in particular locations.

Although the plane earth model has a path loss exponent close to that observed in actual measurements (i.e., 4), the simple physical situation it describes is rarely applicable

Table 3.1
British Telecom Land Usage Categories [12]

Category	Description
0	Rivers, lakes, and seas
1	Open rural areas; e.g., fields and heathlands with few trees
2	Rural areas, similar to the above, but with some wooded areas, e.g., parkland
3	Wooded or forested rural areas
4	Hilly or mountainous rural areas
5	Suburban areas, low-density dwellings, and modern industrial estates
6	Suburban areas, higher density dwellings, e.g., council estates
7	Urban areas with buildings of up to four stories, but with some open space between
8	Higher density urban areas in which some buildings have more than four stories
9	Dense urban areas in which most of the buildings have more than four stories and some can be classed as "skyscrapers." (This category is restricted to the center of a few large cities.)

in practice. The mobile is almost always operated (at least in macrocells) in situations where it does not have a line-of-sight path to either the base station or to the ground reflection point, so the two-ray situation on which the plane earth model relies is hardly ever applicable. To find a more satisfactory physical propagation model, we examine diffraction as a potential mechanism.

3.2.3.1 Allsebrook and Parsons Model

Although this model [13] is based on a series of measurements, it may be regarded as an early attempt to provide a physical basis for urban prediction models.

Measurements were made in three British cities (Bradford, Bath, and Birmingham) at 86, 167, and 441 MHz. These cities cover a wide range of terrain and building classifications. A 40 dB/decade range dependence is again forced, as would be expected for plane earth loss. This results in an Egli-type model with a maximum RMS error of 8.3 dB at 441 MHz. (Note that a least-squares curve fit at this frequency results in a range dependence of only 24 dB/decade). A clutter factor β is introduced to account for excess loss relative to the plane earth calculation.

The frequency dependence of the measured clutter factor is compared with an approximation to the excess loss that would be expected from a 10m absorbing knife edge, placed 30m away from a 2-m-high mobile antenna. The predictions compare reasonably well with the mean values of β at 86 and 167 MHz, but considerably underestimate it at 441 MHz. The knife edge calculation is used as a generalized means of calculating building diffraction adjacent to the mobile, with a UHF correction factor included to force agreement with the measured values above 200 MHz. It is suggested that this deviation is the result of building width being more significant at the higher frequencies, but this is not confirmed by any analysis.

Allsebrook and Parsons' "flat city" model can be expressed as:

$$L_T = L_P + L_B + \gamma \tag{3.11}$$

$$\text{where } L_B = 20\log\left(\frac{h_0 - h_m}{548\sqrt{\dfrac{d_m \times 10^{-3}}{f_c}}}\right)$$

and L_P is the plane earth loss. For ease of computation, the prediction curve for γ can be replaced by the following quadratic approximation:

$$\gamma = -2.03 - 6.67f_c + 8.1 \times 10^{-5}f_c^2 \tag{3.12}$$

Note that, in the calculation of L_P here, the effective antenna heights are those of the base station and of the *building*, resulting in an overall physical model that can be represented by Figure 3.6.

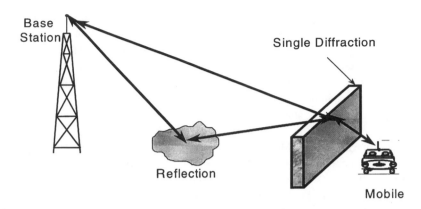

Figure 3.6 Physical interpretation of Allsebrook model.

A published discussion of this model [14] finds that the correction factor γ is necessary in open areas as well as in (sub)urban areas, although the physical cause suggested for γ cannot apply in line-of-sight situations. Additionally, the quoted value of γ is found to be too large in all situations, casting doubt on the model's generality. The model is only physically valid in terms of the final building diffraction; the use of the plane earth calculation suggests the existence of specular ground reflection, which is highly unlikely in a built-up area. Despite this, the model may be considered an improvement over empirical methods since it was the first to make any allowance for the geometry of the specific path being considered.

3.2.3.2 Ikegami Model

This model attempts to produce an entirely deterministic prediction of field strengths at specified points [15]. Using a detailed map of building heights, shapes, and positions, ray paths between the transmitter and receiver are traced, with the restriction that only single reflections from walls are accounted for. Diffraction is calculated using a single edge approximation at the building nearest the mobile, and wall reflection loss is assumed to be fixed at a constant value. The two rays (reflected and diffracted) are power summed, resulting in the following approximate model:

$$L_E = 10\log f_c + 10\log(\sin\phi) + 20\log(h_0 - h_m) \qquad (3.13)$$
$$- 10\log w - 10\log\left(1 + \frac{3}{L_r^2}\right) - 5.8$$

where ϕ is the angle between the street and the direct line from base to mobile and $L_r = 0.25$ is the reflection loss. The analysis assumes that the mobile is in the center of the street. The model therefore represents the situation illustrated in Figure 3.7. It further assumes that the elevation angle of the base station from the top of the knife edge is negligible in comparison to the diffraction angle down to the mobile level.

A comparison of the results of this model with measurements at 200, 400, and 600 MHz shows that the general trend of variations along a street is accounted for successfully. The predictions suggest that field strength is broadly independent of a mobile's position across the street. This is confirmed by the mean values of a large number of measurements, although the spread of values is rather high. Acceptable agreement is also obtained for variations with street angle and width.

Although it accounts reasonably well for "close-in" variations in field strength, the assumption that base station antenna height does not affect propagation is flawed. The same assumption means that the free-space path loss exponent is assumed, so the model tends to underestimate loss at large distances. Similarly, the variation with frequency is underestimated compared with measurements.

3.2.3.3 Rooftop Diffraction

When a macrocell system is operated in a built-up area with reasonably flat terrain, the dominant mode of propagation is multiple diffraction over the building rooftops. Diffraction can occur around the sides of individual buildings, but this tends to become highly attenuated over reasonable distances because many interactions with individual buildings are involved.

The diffraction angle over most of the rooftops is small for typical base station heights and distances, usually less than 1°. In these cases, the diffraction is largely unaffected by the particular shape of the obstacles, so it is appropriate to represent the buildings by equivalent knife edges. The one exception to this is diffraction from the "final building" at which the wave is diffracted from rooftop level down to the street-level antenna of the mobile (Figure 3.8). It is usual to separate these processes into multiple diffraction across the first $(n - 1)$ buildings, treated as knife edges, and a final building.

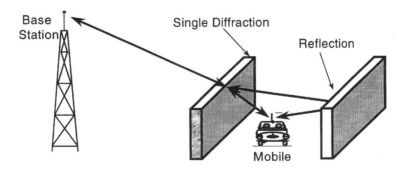

Figure 3.7 Physical interpretation of Ikegami model.

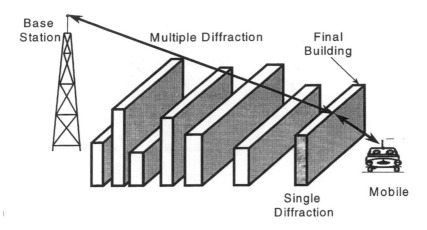

'igure 3.8 Multiple diffraction over building rooftops.

which can be treated either as a knife edge or as some more complex shape for which he diffraction coefficient is known.

The small diffraction angles encountered have two negative consequences for predic- ion of these effects. First, a large number of building rooftops may appear within the 'irst Fresnel zone, all contributing to the propagation loss. Second, the near-grazing ncidence implies that the approximate multiple diffraction models described in Chapter ? will fail, leading to very inaccurate predictions. A full multiple edge diffraction integral must instead be applied [16], which could lead to very long computation times, particularly as it is desired to predict the base station coverage over a wide area, which would require a large number of individual path profiles. Special methods have been developed to enable reasonably rapid calculation of the multiple diffraction integral for cases where accurate results are required and where the necessary data on the building positions and heights are available [17]. Such data are usually too expensive for general use in macrocells. Two simplified solutions with reduced data and computational requirements are therefore examined here.

3.2.3.4 Flat Edge Model

n this model [18], the situation is simplified by assuming all of the buildings to be of qual height and spacing. The values used can be average values for the area under consideration, or can be calculated individually for each direction from the base station f the degree of urbanization varies significantly. The geometry is shown in Figure 3.9, llustrating the following parameters additional to the definitions given in Section 3.2.1: r_1 is the distance from the base station to the first building (m), and α is the elevation angle of the base station antenna from the top of the final building (radians).

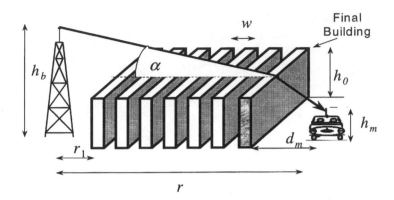

Figure 3.9 Geometry for the flat edge model.

In Figure 3.9, buildings are arranged in the figure normal to the great circle path. Because this will not normally be the case in practice, the value of w used should be an effective one to account for the longer paths between the buildings for oblique incidence.

The excess path loss is then expressed as:

$$L_{ex} = L_{n-1}(t)L_{ke}$$

(3.14)

where L_{ke} accounts for single edge diffraction over the final building and L_{n-1} accounts for multiple diffraction over the remaining $(n-1)$ buildings. It turns out that, provided $r_1 \gg nw$ (i.e., the base station is relatively distant from the first building), the multiple diffraction integral (6.24) can be completely solved in this special case. The result is that L_{n-1} is a function of only a parameter t, given by:

$$t = -\alpha\sqrt{\frac{\pi w}{\lambda}}$$

(3.15)

It is given by the following formula:

$$L_n(t) = \frac{1}{n}\sum_{m=0}^{n-1} L_m(t)F_s(-jt\sqrt{n-m}) \text{ for } n \geq 1; L_0(t) = 1$$

(3.16)

where:

$$F_s(jx) = \frac{e^{-jx^2}}{\sqrt{2j}}\left\{\left[S\left(x\sqrt{\frac{2}{x}}\right) + \frac{1}{2}\right] + j\left[C\left(x\sqrt{\frac{2}{\pi}}\right) + \frac{1}{2}\right]\right\}$$

(3.17)

and $S(.)$ and $C(.)$ are the standard Fresnel sine and cosine integrals defined in [1]. This formulation is extremely quick and simple to compute and it applies for any values of x, even for when the base station antenna height is below the rooftop level. The number of buildings can be increased to extremely high values without difficulty.

The flat edge model can be calculated either directly from (3.16), or the results can be estimated from the prediction curves in Figure 3.10, which show the cases where $h_b \geq h_0$. An alternative approach is to use the approximate formula:

$$L_n(t) = -20\log A_n(t) \tag{3.18}$$
$$= -(c_1 + c_2\log n)\log(-t) - (c_3 + c_4\log n) \text{ [dB]}$$

where

$$c_1 = 3.29; \; c_2 = 9.90; \; c_3 = 0.77; \; c_4 = 0.26$$

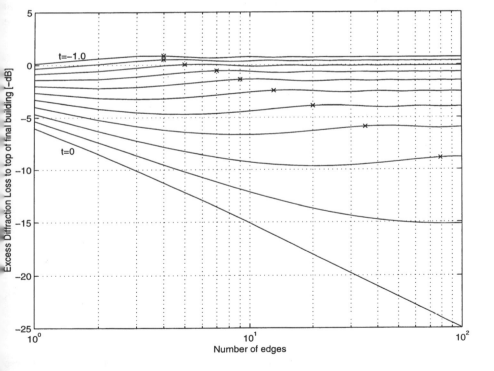

Figure 3.10 Flat edge model prediction curves for elevated base antennas; curves relate to t varying from 0 to −1 in steps of 0.1. The crosses indicate the number of edges required for a settled field according to (3.23).

This approximates the value of (3.16) with an accuracy better than ±1.5 dB for $1 \leq n \leq$ 100 and $-1 \leq t < 0$. It also enables investigate of the behavior of the effective path loss exponent for the flat edge model, since, for fixed n, we can rewrite (3.18), with L being the path loss as a power ratio, as:

$$L \propto (-t)^{\frac{c_2}{10}\log n} = \left(\alpha\sqrt{\frac{\pi w}{\lambda}}\right)^{\frac{c_2}{10}\log n} \approx \left(\frac{h_b - h_0}{r}\sqrt{\frac{\pi w}{\lambda}}\right)^{\frac{c_2}{10}\log n} \tag{3.19}$$

where the approximation holds if $(h_b - h_0) \ll r$. This is the excess field strength, so the overall path loss exponent, including an extra 2 from the free-space part of the loss, is:

$$\text{Path loss exponent} = 2 + \frac{c_2}{10}\log n \tag{3.20}$$

This expression is shown in Figure 3.11, where it is apparent that, for reasonably large numbers of buildings, the path loss exponent for the flat-edge model is close to 4, just as observed in practical measurements. More generally, we can state that *multiple building*

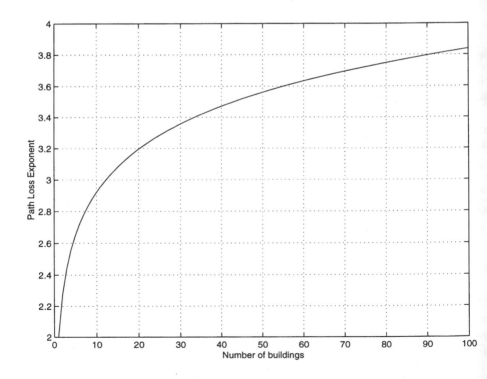

Figure 3.11 Path loss exponent for the flat edge model.

diffraction accounts for the variation of path loss with range that is observed in measurements.

Figure 3.10 shows that, for $(h_b - h_0)$ (i.e., $t < 0$), the field at the top of the final building eventually settles to a constant value as the number of edges increases. This number, n_s, corresponds to the number required to fill the first Fresnel zone around the ray from the base station to the final building. The first Fresnel zone radius r_1 is given approximately by:

$$r_1 = \sqrt{\lambda s} \tag{3.21}$$

where s is the distance along the ray from the field point. Hence, for α small:

$$\alpha = \tan^{-1} \frac{r_1}{n_s w} \approx \frac{\sqrt{\lambda n_s w}}{n_s w} \tag{3.22}$$

So:

$$n_s \approx \frac{\lambda}{\alpha^2 w} = \frac{\pi}{t^2} \tag{3.23}$$

This is marked in Figure 3.10. Note that the number of edges required for settling rises very rapidly with decreasing α. Whenever $\alpha \leq 0$ the field does not settle at all, but decreases monotonically for all n.

The flat edge model is completed by modeling the final building diffraction loss and the reflections from the buildings across the street using the Ikegami model from Section 3.2.3.2. Thus the total path loss is given by:

$$L_T = L_n(t) + L_F + L_E \tag{3.24}$$

where $L_n(t)$ can be found from (3.16), Figure 3.11, or (3.18), L_F is the free-space loss (see Chapter 5), and L_E is given in (3.13).

3.2.3.5 Walfisch-Bertoni Model

This model can be considered as the limiting case of the flat edge model when the number of buildings is sufficient for the field to settle, that is, when $n \geq n_s$. The multiple diffraction process was investigated in [21] using a numerical evaluation of the Kirchhoff-Huygens integral and a power law formula is fitted to the results for the settled field. The Walfisch-Bertoni model was the first to actually demonstrate that multiple building diffraction accounts for the variation of distance with range which is observed in measurements.

The settled field approximation used is as follows:

$$A_{\text{settled}}(t) \approx 0.1 \left(\frac{\alpha}{0.03} \sqrt{\frac{w}{\lambda}} \right)^{0.9} = 0.1 \left(\frac{-t}{0.03} \right)^{0.9} \tag{3.25}$$

This is only valid for $0.03 \leq t \leq 0.4$. For large ranges, we can again put $t \approx -\alpha \sqrt{\frac{\pi w}{\lambda}} \approx -\frac{h_b - h_m}{r} \sqrt{\frac{\pi w}{\lambda}}$. Hence $L_{\text{settled}} \propto r^{-1.8}$. The free-space loss is proportional to r^{-2}, so this model predicts that total propagation loss is proportional to $r^{-3.8}$, which is close to the r^{-4} law which is commonly assumed in empirical models and found in measurements. A single knife edge approximation with a reflection from the building opposite is again used, just as in the Ikegami model, to account for the diffraction from the final building. The complete model is expressed as:

$$L_{ex} = 57.1 + L_A + \log f_c + 18 \log R - 18 \log(h_b - h_0) - 18 \log \left[1 - \frac{R^2}{17(h_b - h_0)} \right] \tag{3.26}$$

where:

$$L_A = 5 \log \left[\left(\frac{w}{2} \right)^2 + (h_0 - h_m)^2 - 9 \log w + 20 \log \left\{ \tan^{-1} \left[\frac{2(h_0 - h_m)}{w} \right] \right\} \right. \tag{3.27}$$

The use of the settled field approximation requires that large numbers of buildings be present, particularly when α is small. Despite this limitation, the Walfisch-Bertoni model is the first to have accounted for observed path loss variation using realistic physical assumptions rather than relying on forcing agreement by means of propagation models of entirely different situations.

3.2.3.6 COST-231-Walfisch-Ikegami Model

The Walfisch-Bertoni model for the settled field has been combined with the Ikegami model for diffraction down to street level and some empirical correction factors to improve agreement with measurements in a single integrated model by the COST-231 project [9]. The total loss is given, for non-line-of-sight conditions, by:

$$L = L_F + L_{msd} + L_{sd} \tag{3.28}$$

where L_F is the free-space loss, L_{msd} accounts for multiple knife edge diffraction to the top of the final building and L_{sd} accounts for the single diffraction and scattering process down to street level. The parameter L is given a minimum value of L_F in case the other terms become negative. The individual terms are:

$$L_{sd} = -16.9 + 10\log f_c + 10\log \frac{(h_0 - h_m)^2}{w_m} + L(\phi) \tag{3.29}$$

where w_m is the distance between the building faces on either side of the street containing the mobile (typically $w_m = w/2$), and the final term accounts for street orientation at an angle ϕ to the great circle path:

$$L(\phi) = \begin{cases} -10 + 0.354\phi & \text{for } 0° < \phi < 35° \\ 2.5 + 0.075(\phi - 35°) & \text{for } 35° \le \phi < 55° \\ 4.0 - 0.114(\phi - 55°) & \text{for } 55° \le \phi \le 90° \end{cases} \tag{3.30}$$

Finally, the rooftop diffraction term is given by:

$$L_{msd} = L_{bsh} + k_a + k_d\log R + k_f\log f_c - 9\log w \tag{3.31}$$

where:

$$L_{bsh} = \begin{cases} -18\log[1 + (h_b - h_0)] & \text{for } h_b > h_0 \\ 0 & \text{for } h_b \le h_0 \end{cases} \tag{3.32}$$

$$k_a = \begin{cases} 54 & \text{for } h_b > h_0 \\ 54 - 0.8(h_b - h_0) & \text{for } R \ge 0.5 \text{ km and } h_b \le h_0 \\ 54 - 0.8\frac{(h_b - h_0)R}{0.5} & \text{for } R < 0.5 \text{ km and } h_b \le h_0 \end{cases} \tag{3.33}$$

$$k_d = \begin{cases} 18 & \text{for } h_b > h_0 \\ 18 - 15\frac{(h_b - h_0)}{h_0} & \text{for } h_b \le h_0 \end{cases} \tag{3.34}$$

$$k_f = -4 + \begin{cases} 0.7\left(\frac{f_c}{925} - 1\right) & \text{for medium-sized city and suburban centers with medium tree density} \\ 1.5\left(\frac{f_c}{925} - 1\right) & \text{for metropolitan centers} \end{cases} \tag{3.35}$$

For approximate work, the following parameter values can be used:

$$h_0 = \begin{cases} 3n_{\text{floors}} & \text{for flat roofs} \\ 3n_{\text{floors}} + 3 & \text{for pitched roofs} \end{cases} \tag{3.36}$$

$$w = 20\text{m to } 50\text{m}; \ d_m = \frac{w}{2}; \ \phi = 90°$$

where n_{floors} is the number of floors in the building. The model is applicable fo 800 MHz $\leq f_c \leq$ 2000 MHz, 4m $\leq h_b \leq$ 50m, 1m $\leq h_m \leq$ 3m, and 0.02 km $\leq R \leq$ 5 km

An alternative approach is to replace the L_{msd} term by $L_n(t)$ from the flat edge model. This would enable the path loss exponent to vary according to the number o buildings and to be uniformly valid for $h_b \leq h_0$. Note, however, that for very lov base station antennas, other propagation mechanisms, such as diffraction around vertica building edges and multiple reflections from building walls, are likely to be significan (see Section 3.3).

3.2.4 Comparison of Models

The path loss predictions of all of the models described in this chapter are compared i Table 3.2, which shows the exponents of power variation predicted by each model. Thu a "−2" in the path loss exponent column means that the model predicts that receivec power is inversely proportional to the square of range. In some cases it is difficult tc express the variation in this form, but otherwise it is useful as a means of comparison.

Table 3.2
Comparison of Macrocell Propagation Models

Model	Path Loss Exponent (−)	h_b	h_m	f_c
Free space (Sec. 2.2.2)	−2	0	0	−2
Plane earth (Sec. 2.2.3)	−4	2	2	0
Egli (Sec. 3.2.2.1)	−4	2	1 ($h_m < 10$) 2 ($h_m > 10$)	−2
Okumura/Hata (Sec. 3.2.2.2)	−4.5 + 0.66 log h_b	1.38 + 0.66 log r	See (3.5)	≈ −2.6
COST231-Hata (Sec. 3.2.2.3)				
Ibrahim (Sec. 3.2.2.4)	−4	2	2	$10^{-f_c/400}$
Allsebrook (Sec. 3.2.3.1)	−4	2	$(h_0 - h_m)^{-2}$	≈ −2
Ikegami (Sec. 3.2.3.2)	−1	0	$(h_0 - h_m)^{-2}$	−1
Flat edge (Sec. 3.2.3.4)	From −2 to −4	≈ 2	$(h_0 - h_m)^{-2}$	≈ −1
Walfisch-Bertoni (Sec. 3.2.3.5)	−3.8	≈ 1.8	≈ 1	−1.05
COST-231 Walfisch-Ikegami (Sec. 3.2.3.6)	−3.8	≈ 1.8	≈ 1	See (3.28)

3.2.5 Computerized Planning Tools

The methods described in this chapter are most often implemented for practical planning within computer software. The development of such software has been motivated and enabled by a number of factors:

- The enormous increase in the need to plan cellular systems accurately and quickly.
- The development of fast, affordable computing resources.
- The development of geographical information systems, with index data on terrain, clutter, and land usage in an easily accessible and manipulable form.

Such techniques have been implemented in a wide range of commercially available and company-specific planning tools. Although most of these are based on combined empirical and simple physical models, it is anticipated that there will be progressive evolution in the future toward more physical or physical-statistical methods as computing resources continue to cheapen, as clutter data improve in resolution and cost, and as research into numerically efficient path loss predication algorithms develops.

3.2.6 Conclusions

Propagation path loss modeling is the fundamental method of predicting the range of a mobile radio system. The accuracy of the path loss predictions is crucial in determining whether a particular system design will be viable. In macrocells, empirical models have been used with great success, but deterministic physical models are being increasingly investigated as a means of improving accuracy, based on the use of multiple rooftop diffraction as the key propagation mechanism. This accuracy comes at the expense of increased input data requirements and computational complexity. Another generation of models is expected to appear that combines sound physical principles with statistical parameters, which can economically be obtained in order to provide the optimum balance between accuracy and complexity.

The path loss for macrocells may be taken, very roughly, to be given by:

$$\frac{P_R}{P_T} = \frac{1}{L} = k\frac{h_m h_b^2}{r^4 f_c} \tag{3.37}$$

It should be emphasized that this expression, and all of the models described in this chapter, account only for the effects of typical clutter on flat or gently rolling terrain. When the terrain variations are sufficient to cause extra obstruction loss, then the models must be supplemented by explicit calculations of terrain diffraction loss [1].

3.3 MICROCELLS

The deployment of microcells is motivated by a desire to reduce cell sizes in areas where large numbers of users require access to the system. Serving these users with limited radio spectrum requires that frequencies be reused over very short distances, with each cell containing only a reduced number of users. This could, in principle, be achieved with base station antennas at the same heights as in macrocells, but this would increase the costs and planning difficulties substantially. In a *microcell,* the base station antenna is typically at about the same height as lamp posts in a street (3–6m above ground level), although the antenna is more often mounted onto the side of a building (Figure 3.12). Coverage, typically over a few hundred meters, is then determined mostly by the specific locations and electrical characteristics of the surrounding buildings, with cell shapes being far from circular.

Pattern shaping of the base station antenna can yield benefits in controlling interference, but is not the dominant factor in determining the cell shape. The dominant propagation mechanisms are then free-space propagation, plus multiple reflection and scattering within the cell's desired coverage area, together with diffraction around the vertical edges of buildings and over rooftops, which becomes significant when determining interference between cochannel cells. Microcells thus make increased use of the potential of the environment surrounding the base station to carefully control the coverage area and hence to manage the interference between sites. More general information on microcell systems is available [22-24].

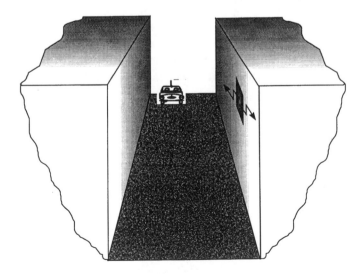

Figure 3.12 A microcell in a built-up area.

3.3.1 Dual-Slope Empirical Models

To model the path loss in microcells, empirical models of the type described earlier, could, in principle, be used. However, measurements (see, e.g., [25]) indicate that a simple power law path loss model cannot usually be used to fit measurements with good accuracy. A better empirical model in this case is a *dual-slope model*. Two separate path loss exponents are used to characterize the propagation, together with a breakpoint distance of a few hundred meters between them where propagation changes from one regime to the other. In this case the path loss is modeled as:

$$\frac{1}{L} = \begin{cases} \dfrac{k}{r^{n_1}} & \text{for } r \leq r_b \\[3mm] \dfrac{k}{\left(\dfrac{r}{r_b}\right)^{n_2} r_b^{n_1}} & \text{for } r > r_b \end{cases} \tag{3.38}$$

or, in decibels:

$$L = \begin{cases} 10 n_1 \log r + L_1 & \text{for } r \leq r_b \\[3mm] 10 n_2 \log \dfrac{r}{r_b} + 10 n_1 \log r_b + L_1 & \text{for } r > r_b \end{cases} \tag{3.39}$$

where L_1 is the reference path loss at $r = 1$m, r_b is the breakpoint distance, n_1 is the path loss exponent for $r \leq r_b$, and n_2 is the path loss exponent for $r > r_b$.

To avoid the sharp transition between the two regions of a dual-slope model, the model can also be formulated as follows, according to an approach proposed by [26]:

$$\frac{1}{L} = \frac{k}{r^{n_1}\left(1 + \dfrac{r}{r_b}\right)^{n_2 - n_1}} \tag{3.40}$$

This can be considered in two regions: for $r \ll r_b$, $\dfrac{1}{L} \approx \dfrac{k}{r^{n_1}}$, while for $r \gg r_b$, $\dfrac{1}{L} \approx \dfrac{k}{\left(\dfrac{r}{r_b}\right)^{n_2}}$. Hence the path loss exponent is again n_1 for short distances and n_2 for larger distances. The model is conveniently expressed in decibels as:

$$L = L_1 + 10n_1 \log r + 10(n_2 - n_1) \log_{10}\left(1 + \frac{r}{r_b}\right) \qquad (3.41)$$

where L_{ref} is a reference value for the loss at 1m. Figure 3.13 compares (3.39) and (3.41).

Typical values for the path loss exponents are found by measurement to be around $n_1 = 2$ and $n_2 = 4$, with breakpoint distances between 200 and 500m, but it should be emphasized that these values vary greatly among individual measurements. (See, for example, [25,27,28].)

To plan the locations of microcells effectively, it is important to ensure the coverage areas of the cochannel cells do not overlap within the breakpoint distance. The rapid reduction of signal level beyond the breakpoint then produces a large carrier-to-interference ratio, which can be exploited to maximize system capacity.

3.3.2 Physical Models

In creating physical models for microcell propagation, it is useful to distinguish line-of-sight (LOS) and non-line-of-sight (NLOS) situations. We will see that it is possible to make some reasonable generalizations about the LOS cases, while the NLOS cases require

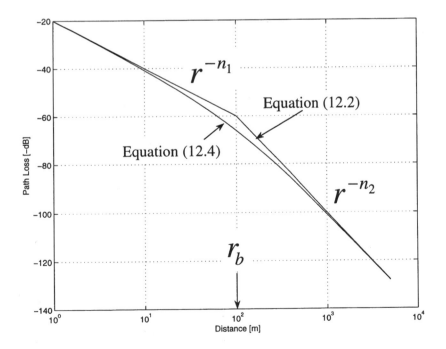

Figure 3.13 Dual-slope empirical loss models. Here $n_1 = 2$, $n_2 = 4$, $r_b = 100$m, and $L_1 = 20$ dB.

more site-specific information. See Figures 3.14 and 3.15 for an example of practical measurements, in which it is clear that the obstructed path suffers far greater variability at a given range than the others. Such effects must be accounted for explicitly in the models.

3.3.2.1 Line-of-Sight Models (Two-Ray Models)

In a line-of-sight situation, at least one direct ray and one reflected ray will usually exist (Figure 3.16). This leads to an approach similar to that followed in the derivation of the plane earth loss (two-wave theory) in Chapter 2, except that it is no longer appropriate to assume that the direct and reflected path lengths are necessarily similar, or that the reflection coefficient necessarily has a magnitude of unity. The loss is then:

$$\frac{1}{L} = \left(\frac{\lambda}{4\pi}\right)^2 \left| \frac{e^{-jkr_1}}{r_1} + R\frac{e^{-jkr_2}}{r_2} \right|^2 \tag{3.42}$$

where R is the Fresnel reflection coefficient for the relevant polarization. In the horizontally polarized case the reflection coefficient is very close to -1, so the path loss exponent

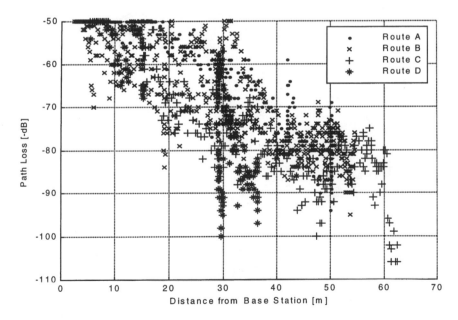

Figure 3.14 Measurements of path loss from a suburban microcell. Routes A, B, and C are radial streets, often with a line-of-sight present, whereas route D is a transverse street, with most locations obstructed. Note how the measurements on route D vary over almost 45 dB, despite the range being almost constant at around 30m. The frequency is 900 MHz.

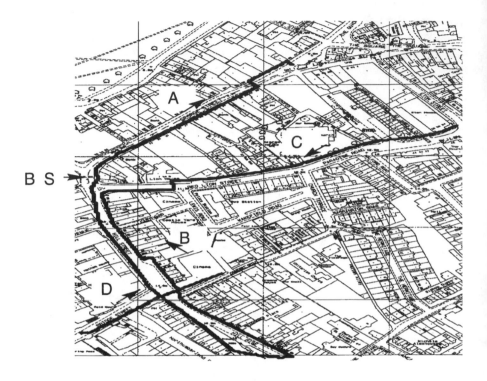

Figure 3.15 Measurement routes corresponding to Figure 3.14.

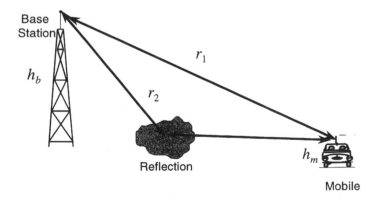

Figure 3.16 Two-ray model of line-of-sight propagation.

tends toward 4 at long distances as in the plane earth loss. For the vertically polarized case, the path loss exponent is essentially 2 at all distances, but the large fluctuations present at short ranges disappear at longer distances. Hence both cases produce two regimes of propagation.

Because the reflection coefficient for vertical polarization is approximately +1 for large distances, the distance at which the rays are in antiphase is closely approximated by the distance at which $r_2 = (r_1 + \lambda/2)$, and this gives the position of the last dip in the vertically polarized signal. This is exactly the definition of the first Fresnel zone. For high frequencies, the distance at which the first Fresnel zone first touches the ground is given approximately by:

$$r_b = \frac{4h_b h_m}{\lambda} \tag{3.43}$$

It has been suggested that this forms a physical method for calculating the breakpoint distance for use in empirical models such as in (3.41) [29].

The two-ray model forms a useful idealization for microcells operated in fairly open, uncluttered situations, such as on long straight motorways where a line-of-sight path is always present and little scattering from other clutter occurs.

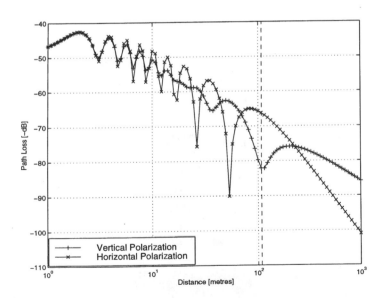

Figure 3.17 Predictions from the two-ray model. Here $h_b = 6$m, $h_m = 1.5$m, $f_c = 900$ MHz, and the constitutive parameters of the ground are $\epsilon_r = 15$ and $\sigma = 0.005$ Sm^{-1}.

3.3.2.2 Street Canyon Models

Although a line-of-sight path frequently exists within microcells, such cells are most usually situated within built-up areas. The buildings around the mobile can all interact with the transmitted signal to modify the simple two-ray regime described earlier. A representative case is illustrated in Figure 3.18. This case assumes that the mobile and base station are both located in a long straight street, lined on both sides by buildings with plane walls. Models that use this canonical geometry are called *street canyon* models.

Six possible ray paths are also illustrated. Many more are possible, but these tend to include reflections from more than two surfaces. These are typically attenuated to a much greater extent, so the main signal contributions are accounted for by those illustrated.

The characteristics of this approach are illustrated by reference to a four-ray model, which considers all three of the singly reflected paths from the walls and the ground. The structure follows, but the reflections from the vertical building walls involve the Fresnel reflection coefficients for the opposite polarization to the ground. A typical result is shown in Figure 3.19. In comparison to Figure 3.17 the multipath fading is more rapid, and there are fewer differences between vertical and horizontal polarization. Eventually the vertically polarized component diminishes with an average path loss exponent of 4, while the horizontally polarized case tends to 2. However, real streets are rarely straight for long enough to observe this distance range.

Figure 3.20 shows the variation of fields predicted by this model as the base station antenna height is varied at a particular range. Neither polarization shows any definite advantage in increasing the antenna height, and the particular positions of the nulls in

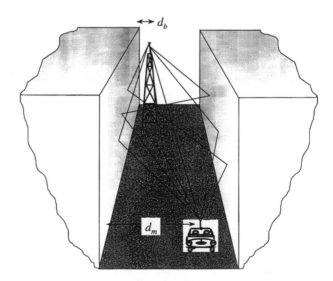

Figure 3.18 Street canyon model of line-of-sight microcellular propagation.

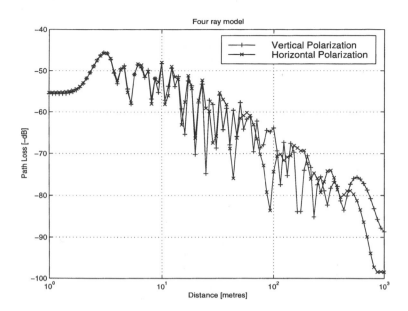

Figure 3.19 Predictions from four-ray model. Here h_b = 6m, h_m = 1.5m, w = 20m, d_m = 10m, d_b = 5m, f_c = 900 MHz, and the constitutive parameters of the ground and buildings are ϵ_r = 15 and σ = 0.005 Sm^{-1}.

this figure are strongly dependent on the range. In general, for line-of-sight microcells, the base station height has only a weak effect on the cell range. There is some effect due to the obstructing effect of clutter (in this case vehicles, street furniture and pedestrians), but we will see in later sections that increasing the base station height does have a significant effect on interference distance. Thus the antenna should be maintained as low as possible, consistent with providing a line-of-sight to locations to be covered.

3.3.3 Non-Line-of-Sight Models

3.3.3.1 Propagation Mechanisms

When the line-of-sight path in a microcell is blocked, signal energy can propagate from the base to the mobile via several alternative mechanisms:

- Diffraction over building rooftops;
- Diffraction around vertical building edges; and
- Reflection and scattering from walls and the ground.

These mechanisms are described further in [30]. At relatively small distances from the base station and low base antenna heights, the angle through which the signal must diffract

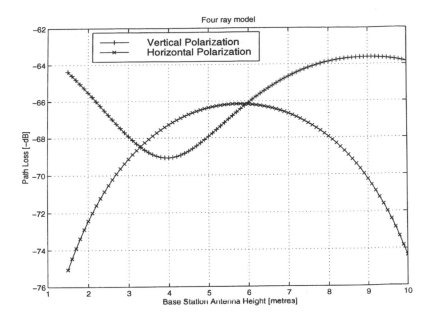

Figure 3.20 Base station antenna height variation according to the four-ray model. Here r = 50m, other parameters are as given in Figure 3.19.

over rooftops in order to propagate is large and the diffraction loss is correspondingly big. At such distances, propagation is dominated by the other two mechanisms described above, where the balance between the diffraction and reflection depends on the specific geometry of the buildings. For instance, Figure 3.21 shows the motion of a mobile across the shadow boundary created by a vertical building edge. Because this building is in an isolated situation, the only possible source of energy in the shadow region is via diffraction and the energy will drop off very rapidly with increasing distance. This contrasts with the case illustrated in Figure 3.22, where the building is now surrounded by others that

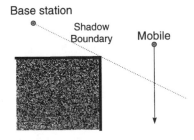

Figure 3.21 Street geometry where diffraction dominates.

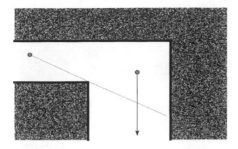

Figure 3.22 Street geometry where reflection dominates.

act as reflecting surfaces. The reflected ray is then likely to be much stronger than the diffracted ray, so that the signal remains strong over much larger distances.

At larger distances still, particularly those involved in interference between cochannel microcells, the rooftop diffracted signal (Figure 3.23) again begins to dominate due to the large number of diffractions and reflections required for propagation over long distances. See, for example, Figure 3.24, which shows the plan view of buildings arranged in a regular "Manhattan grid" structure. In this figure, the short paths A and B involve only a single reflection/diffraction and are likely to be dominant sources of signal energy. By contrast, the long path C is likely to be very weak as four individual reflection losses are involved, and the rooftop-diffracted path D is then likely to dominate. This variation in propagation mechanism with distance is another source of the two slopes in the empirical models of Section 3.3.1.

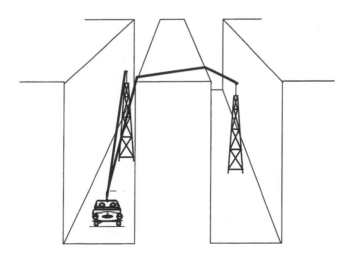

Figure 3.23 Rooftop diffraction as an interference mechanism.

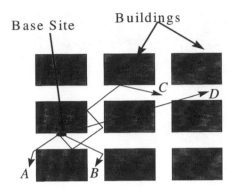

Figure 3.24 Variation of propagation mechanisms with distance for non-line-of-sight microcells.

System range is greatest along the street containing the base site. When the mobile turns a corner off of this street, the signal drops rapidly, often by 20–30 dB. The resultant coverage area is therefore broadly ''diamond'' shaped, as illustrated in Figure 3.25, although the precise shape will depend very much on the building geometry. The curved

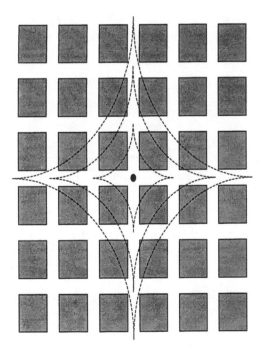

Figure 3.25 Microcellular propagation along city streets; closed circle, position of the base site; dashed line, equal path loss contours.

"sides" of the diamonds in Figure 3.25, which have been confirmed by measurement, have been shown to indicate that the dominant mechanism of propagation into cross streets is diffraction rather than reflection [31].

The variation of the microcell shape with base antenna height in a Manhattan grid structure has been investigated in detail [32] using the multiple diffraction integral in [16] and it is shown that there is a smooth transition from a diamond shape to nearly circular as the antenna height increases. It has also been shown [33] that the characteristic diamond cell shape is obtained even when considering only the vertical corner diffraction plus reflections from building walls. This work also showed that the distance at which the transition between the various mechanisms occurs depends strongly on the distance between the base station and the nearest street corners.

For low antenna heights, the strong scattering present in microcells prevents the efficient use of sectorization since the free-space antenna radiation pattern is destroyed. Efficient frequency reuse can still be provided, however, by taking advantage of the building geometry. In regular street grid structures cochannel microcells should be separated diagonally across the street directions, and with sufficient spacing to ensure that cells do not overlap within their break point distance in order to maintain high signal-to-interference levels.

In more typical environments, where the buildings are not regular in size, more advanced planning techniques must be applied, particularly when frequencies are shared between microcell and macrocell layers. See, for example, [34].

3.3.3.2 Recursive Model

This model is intermediate between an empirical model and a physical model [35]. It uses the concepts of GTD/UTD in that effective sources are introduced for non-line-of-sight propagation at the street intersections where diffraction and reflection points are likely to exist. The model breaks the path between the base station and the mobile down into a number of segments, interconnected by nodes. The nodes may be placed either just at the street intersections or else at regular intervals along the path, allowing streets that are not linear to be handled as a set of piecewise linear segments. An *illusory* distance for each ray path considered is calculated according to the following recursive expressions:

$$k_j = k_{j-1} + d_{j-1} \times q(\theta_{j-1}) \tag{3.44}$$
$$d_j = k_j \times r_{j-1} + d_{j-1}$$

subject to the initial values:

$$k_0 = 1 \tag{3.45}$$
$$d_0 = 0$$

where d_n is the illusory distance calculated for the number of straight line segments along the ray path, n ($n = 3$ in Figure 3.26) and r_j is the physical distance [meters] for the line segment following the jth node. The result is reciprocal.

The angle through which the path turns at node j is θ_j (degrees). As this angle increases, the illusory distance is increased according to the following function:

$$q(\theta_j) = \left(\frac{\theta_j q_{90}}{90}\right)^{\nu}$$ (3.46)

where q_{90} and ν are parameters of the model, with $q_{90} = 0.5$ and $\nu = 1.5$ suggested in [35]. The path loss is then calculated as:

$$L = 20\log\left[\frac{4\pi d_n}{\lambda}D\left(\sum_{j=1}^{n} r_{j-1}\right)\right]$$ (3.47)

where:

$$D(r) = \begin{cases} \dfrac{r}{r_b} & \text{for } r > r_b \\ 1 & \text{for } r \le r_b \end{cases}$$ (3.48)

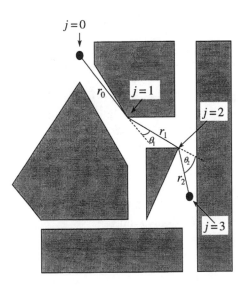

Figure 3.26 Example geometry for recursive microcell model.

This expression, (3.47), creates a dual-slope behavior with a path loss exponent of 2 for distances less than the breakpoint r_b and 4 for greater distances. The overall model is simple to apply and accounts for the key microcell propagation effects, namely, dual-slope path loss exponents and street corner attenuation, with an angle dependence that incorporates effects encountered with real street layouts.

3.3.3.3 Site-Specific Ray Models

Prediction of the detailed characteristics of microcells requires a site-specific prediction based on detailed knowledge of the built geometry. Electromagnetic analysis of such situations is commonly based on use of the geometrical theory of diffraction (GTD) and its extensions, as described in Section 3.5.3. These models are capable of very high accuracy, but their practical application has been limited by the cost of obtaining sufficiently detailed data and the required computation time. More recently progress in satellite remote-sensing has reduced the cost of the necessary data while advanced ray-tracing techniques and cheap computational resources have advanced the art to the point where ray-tracing models are entirely feasible. In practice, however, many operators consider the costs involved with operating such prediction tools to be prohibitive and prefer to deploy their microcells based on the knowledge of experienced planning engineers together with site-specific measurements.

Example predictions from a GTD-based model for a real building geometry are shown in Figure 3.27. This includes contributions from a very large number of multiply reflected and diffracted rays.

3.3.4 Microcell Propagation Models: Discussion

For the practical application of microcell propagation models, there is an important trade-off between the accuracy of the prediction and the speed with which the prediction can be made. Microcells often have to be deployed very quickly, with little engineering effort, often by people who are not necessarily radio experts. Rules of thumb and very rapid statistical planning tools are very important. Also, even with a very high resolution topographic database, propagation may often be dominated by items of street furniture (signs, lampposts, and so on) and by details of the antenna siting and its interaction with the objects on which it is mounted, which no database could hope to have available. These features may also change rapidly with time, as is certainly the case when dealing with the effects of traffic. For example, when a double-decker bus passes close by a microcell antenna, the coverage area of the microcell may change dramatically for a short time. Either these items must be entered by hand or, more likely, systems of the future will have to be capable of adapting their characteristics to suit the environment which they find, by taking measurements from the active network and responding accordingly.

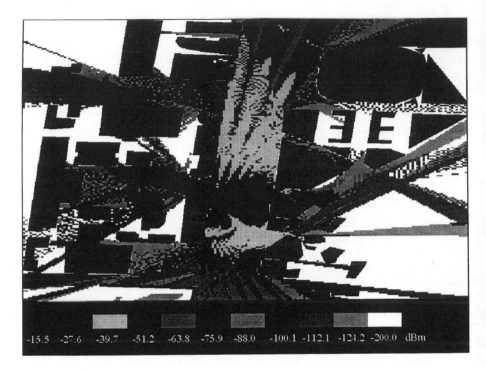

-15.5 -27.6 -39.7 -51.2 -63.8 -75.9 -88.0 -100.1 -112.1 -124.2 -200.0 dBm

Figure 3.27 Predictions for an urban microcell, based on ray tracing and the geometrical theory of diffraction. The shading indicates the received signal strength at a mobile with a quarter-wave monopole antenna (in dBm), with a transmit power of 1W at 900 MHz.

These factors will dramatically change the way in which propagation models are applied, from being processes that are run at the start of a system deployment, and then used to create a fixed set of predictions and recommendations for deployment, to real-time processes that operate within the base station, with assistance from the mobiles, which are optimized on an ongoing basis and are used by the system to assess the likely impact of changes to system parameters such as transmit powers, antenna patterns, and channel assignments.

3.3.5 Microcell Shadowing

The lognormal distribution is applied to shadow fading in microcells, just as for macrocells [1]. Some measurements have suggested that the location variability increases with range, typically in the range of 6–10 dB [36]. To account for microcell shadowing cross-correlation, the shadowing can be separated into two parts, one of which is caused by obstructions very local to the mobile and therefore common to all paths, and another which is specific to the transport of energy from the mobile to a particular base station [37].

3.3.6 Conclusions

Propagation in microcells can be modeled using either empirical or physical models, as was the case for the macrocells in Section 3.2. In either case, however, the clutter surrounding the base station has a significant impact on the cell shape and this must be accounted for to avoid serious prediction errors. In particular, a simple path loss exponent model is inadequate and dual-slope behavior must be accounted for. This clutter also creates difficulties when deploying antennas, because the clutter disrupts the free-space antenna radiation pattern. Nevertheless, the enormous potential offered by microcells in creating high-capacity cellular systems makes them increasingly attractive methods of providing outdoor coverage in areas with high user densities.

3.4 PICOCELLS

When a base station antenna is located inside a building, a *picocell* is formed (Figure 3.28). Such cells are increasingly used in cellular telephony for high-traffic areas such as railway stations, office buildings, and airports. Additionally, the high data rates required by wireless local-area networks restrict cell sizes to picocells and impose a further require-

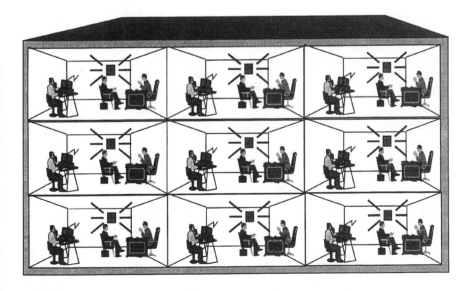

Figure 3.28 Picocells.

ment to predict the wideband nature of the picocell environment. The subject of picocell propagation is also relevant to determining propagation into buildings from both macrocellular and microcellular systems, which could either act as a source of interference to the indoor cells or as a means of providing a greater depth of coverage without capacity enhancement. This section describes both empirical and physical models of picocellular propagation.

3.4.1 Empirical Models of Propagation Within Buildings

3.4.1.1 Wall and Floor Factor Models

Two distinct approaches have been taken here. The first is to model propagation by a path loss law, just as in macrocellular systems, determining the parameters from measurements. This approach, however, tends to lead to large errors in the indoor case because of the large variability in propagation mechanisms among different building types and among different paths within a single building. The same is true if dual-slope models, similar to those used in microcells, are applied.

A more successful approach [38] is to characterize indoor path loss by a fixed path loss exponent of 2, just as in free space, plus additional loss factors relating to the number of floors (n_f) and walls (n_w) intersected by the straight-line distance, r, between the terminals. Thus:

$$L = L_1 + 20\log r + n_f a_f + n_w a_w \qquad (3.49)$$

where a_f and a_w are the attenuation factors (in decibels) per floor and per wall, respectively; L_1 is the loss at $r = 1$m. No values for these factors were reported in [38]. An example prediction using this model is shown in Figure 3.29 for a series of offices leading off a corridor, with the base station inside one of the offices. Contours are marked with the path loss (−dB).

A similar approach is taken by an ITU-R model [39], except that only the floor loss is accounted for explicitly, while the loss between points on the same floor is included implicitly by changing the path loss exponent. The basic variation with frequency is assumed to be the same as in free space, resulting in the following total path loss model (in decibels):

$$L_T = 20\log f_c + 10n\log r + L_f(n_f) - 28 \qquad (3.50)$$

where n is the path loss exponent, given by Table 3.3, and $L_f(n_f)$ is the floor penetration loss, which varies with the number of penetrated floors, n_f according to Table 3.4.

3.4.1.2 COST 231 Multiwall Model

This model of propagation within buildings [9] incorporates a linear component of loss, proportional to the number of walls penetrated, plus a more complex term, which depends

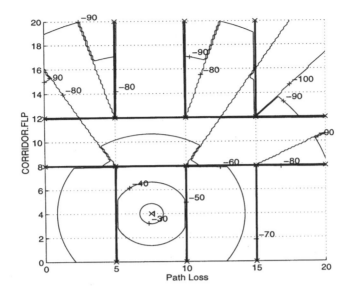

Figure 3.29 Example picocellular path loss prediction.

Table 3.3
Path Loss Exponents, n, for the ITU-R Model (3.50)

		Environment	
Frequency	*Residential*	*Office*	*Commercial*
900 MHz	—	3.3	2.0
1.2–1.3 GHz	—	3.2	2.2
1.8–2.0 GHz	2.8	3.0	2.2
4 GHz	—	2.8	2.2
60 GHz	—	2.2	1.7

The 60 GHz figures apply only within a single room for distances less than around 100m, since no wall transmission loss or gaseous absorption is included.

on the number of floors penetrated, producing a loss that increases more slowly as additional floors after the first are added.

$$L_T = L_F + L_c + \sum_{i=1}^{W} n_{wi} L_{wi} + n_f^{\left[\frac{n_f + 2}{n_f + 1} - b\right]} L_f \qquad (3.51)$$

where L_F is the free-space loss for the straight-line (direct) path between the transmitter and receiver, n_{wi} is the number of walls crossed by the direct path of type i, W is the number of wall types, L_{wi} is the penetration loss for a wall of type i, n_f is the number

Table 3.4
Floor Penetration Factors, $L_f(n_f)$ [dB] for the ITU-R Model (3.50)

Frequency (GHz)	Residential	Environment Office	Commercial
0.9		9 (1 floor)	
		19 (2 floors)	
		24 (3 floors)	
1.8–2.0	$4n_f$	$15 + 4\,(n_f - 1)$	$6 + 3\,(n_f - 1)$

Note that the penetration loss may be overestimated for large numbers of floors, for reasons described in Section 3.4.3.1.

of floors crossed by the path, b and L_c are empirically derived constants, and L_f is the loss per floor. Recommended values for 1800 MHz are $L_w = 3.4$ dB for light walls, 6.9 dB for heavy walls, $L_f = 18.3$ dB, and $b = 0.46$. The floor loss [that is, the last term in (3.51)] is shown in Figure 3.30. Notice that the additional loss per floor decreases with increasing number of floors.

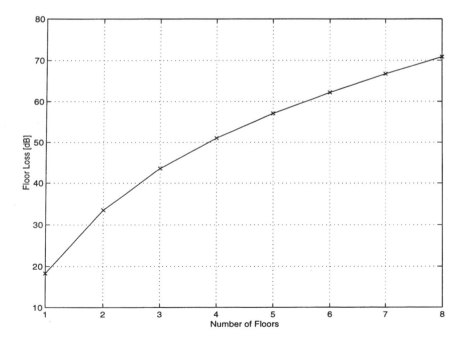

Figure 3.30 Floor loss for COST231 multiwall model.

3.4.1.3 Ericsson Model

In this model, intended for use around 900 MHz, the path loss, including shadowing, is considered to be random, variable, and uniformly distributed between limits that vary with distance as indicated in Table 3.5 [40]. The path loss exponent increases from 2 to 12 as the distance increases, indicating a very rapid decrease of signal strength with distance. A typical prediction from the model is shown in Figure 3.31. The model may be extended for use at 1800 MHz by the addition of an 8.5-dB extra path loss at all distances.

Table 3.5
Ericsson Indoor Propagation Model

Distance (m)	Lower Limit of Path Loss (dB)	Upper Limit of Path Loss (dB)
$1 < r < 10$	$30 + 20 \log r$	$30 + 40 \log r$
$10 \leq r < 20$	$20 + 30 \log r$	$40 + 30 \log r$
$20 \leq r < 40$	$-19 + 60 \log r$	$1 + 60 \log r$
$40 \leq r$	$-115 + 120 \log r$	$-95 + 120 \log r$

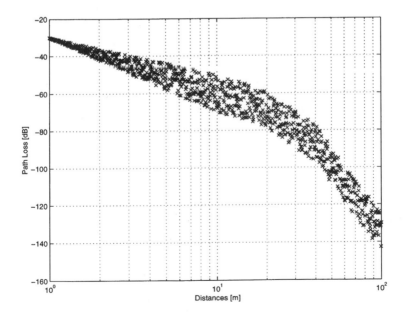

Figure 3.31 Prediction from Ericsson in-building path loss model (900 MHz).

3.4.2 Empirical Models of Propagation Into Buildings

3.4.2.1 Introduction

There are two major motivations for examining signal penetration into buildings. Firstly, because most cellular users spend most of their time inside buildings, the level of service they perceive will depend heavily on the signal strengths provided inside the buildings (the *depth* of coverage). When sufficient capacity exists within the macrocells and the microcells of the network, this indoor coverage is then provided by the degree of penetration into the buildings. When, by contrast, it is necessary to serve very high densities of users within a building (for example, in heavily populated office buildings, railway stations, and airports), the indoor coverage must then be provided by dedicated picocells. It is inefficient to allocate distinct frequencies to these, so it is necessary to reuse frequencies already allocated to macrocells and microcells, via clear knowledge of the extent to which the two cell types will interfere within the building.

3.4.2.2 COST231 Line-of-Sight Model

In cases where a line-of-sight path exists between a building face and the external antenna, the following semiempirical model has been suggested [9], with geometry defined by Figure 3.32. In this figure r_e is the straight path length between the external antenna and a reference point on the building wall; since the model will often be applied at short ranges, it is important to account for the true path length in three dimensions, rather than the path length along the ground. The loss predicted by the model varies significantly as the angle of incidence, $\theta = \cos^{-1}\left(\dfrac{r_p}{r_e}\right)$, is varied.

The total path loss is then predicted using:

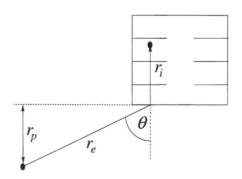

Figure 3.32 Geometry for COST231 line-of-sight building penetration model.

$$L_T = L_F + L_e + L_g(1 - \cos\theta)^2 + \max(L_1, L_2) \tag{3.52}$$

where L_F is the free-space loss for the total path length ($r_i + r_e$), L_e is the path loss through the external wall at normal incidence ($\theta = 0$ deg), L_g is the additional external wall loss incurred at grazing incidence ($\theta = 90$ deg), and:

$$L_1 = n_w L_i; \ L_2 = \alpha(r_i - 2)(1 - \cos\theta)^2 \tag{3.53}$$

where n_w is the number of walls crossed by the internal path r_i, L_i is the loss per internal wall, and α is a specific attenuation (dBm^{-1}), which applies for unobstructed internal paths. All distances are in meters.

The model is valid at distances up to 500m and the parameter values in Table 3.6 are recommended for use in the 900- to 1800-MHz frequency range. These have been found to give good agreement with measurements in real buildings, and implicitly include the effects of typical furniture arrangements.

3.4.2.3 Floor Gain Models

In most macrocell cases, no line-of-sight path exists between the base station and the face of the building. Empirical models of this situation are then most usually based on comparing the path loss encountered in the street outside the building (L_{out} in Figure 3.33) to the path loss within the building at various floor levels. (L_n, where n is the floor number defined in Figure 3.33). It is then possible to define a penetration loss as:

$$L_p = L_n - L_{out} \tag{3.54}$$

Interestingly, the penetration loss has been found to decrease with frequency in some studies; typical values for the ground floor penetration loss L_0 are 14.2, 13.4, and 12.8 dB measured at 900, 1800, and 2300 MHz, respectively [41]. This does not necessarily indicate that the actual wall attenuations follow this trend, since the penetration loss

Table 3.6
Parameters for COST231 Line-of-Sight Model

Parameter	Material	Approximate Value
L_e or L_i (dB m^{-1})	Wooden walls	4
	Concrete with nonmetallized windows	7
	Concrete without windows	10–20
L_g (dB)	Unspecified	20
α (dB m^{-1})	Unspecified	0.6

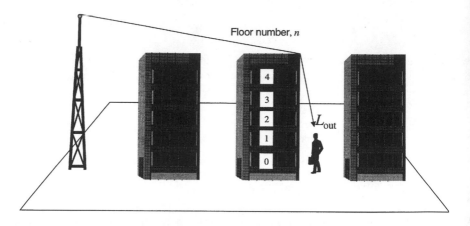

Figure 3.33 Geometry for building penetration in non-line-of-sight conditions.

defined this way makes no attempt to isolate effects due to individual waves. The loss decreases with height from the ground floor upward at a rate of around 2 dB per floor and then starts to increase again with height above around the 9th [42] or 15th [43] floor. The precise variation is likely to be very dependent on the specific geometry of the surrounding buildings.

3.4.2.4 COST231 Non-Line-of-Sight Model

This model relates the loss inside a building from an external transmitter to the loss measured outside the building, on the side nearest to the wall of interest, at 2m above ground level. The loss is given by:

$$L_T = L_{\text{out}} + L_e + L_{ge} + \max(L_1, L_3) - G_{fh} \qquad (3.55)$$

where $L_3 = \alpha r_i$, and r_i, L_e, α, and L_1 are as defined in the COST231 line-of-sight model (Section 3.4.2.2), and the floor height gain, G_{fh}, is given by:

$$G_{fh} = \begin{cases} nG_n \\ hG_h \end{cases} \qquad (3.56)$$

where h is the floor height above the outdoor reference height (m) and n is the floor number, as defined in Figure 3.33. Shadowing is predicted to be lognormal with location variability of 4–6 dB. Other values are as shown in Table 3.7.

Both the line-of-sight and non-line-of-sight models of COST 231 rely on the dominant contribution penetrating through a single external wall. A more accurate estimation may be obtained by summing the power from components through all of the walls.

Table 3.7
Parameters for COST231 Non-Line-of-Sight Model

Parameter	Approximate Value
L_{ge} (dB) at 900 MHz	4
L_{ge} (dB) at 1800 MHz	6
G_n (dB per floor) at 900 or 1800 MHz	1.5–2 normal buildings 4–7 for floor heights above 4m

.4.3 Physical Models of Indoor Propagation

As with microcellular predictions, ray tracing and the geometrical theory of diffraction ave been applied to deterministic prediction of indoor propagation (see, for example, 44]). This can be used for site-specific predictions, provided that sufficient detail of the uilding geometry and materials is available. More advanced electromagnetic prediction echniques, such as the finite-difference time-domain (FDTD) approach are also useful 1 some cases. Such models also yield wideband information and the statistics of multipath ropagation directly. As with physical models of microcellular propagation, however, mitations are associated with using physical models for practical picocell predictions ue to the difficulty of obtaining and using sufficiently accurate physical data. These roblems are particularly significant for picocells, where the influence of furniture and f the movement of people can have a significant (and time-varying) effect on coverage. ome basic physical models can, however, be used to yield insight into the fundamental rocesses affecting building propagation.

4.3.1 Propagation Between Floors

igure 3.34 shows four distinct paths between a transmitter and receiver situated on fferent floors of the same building. Path 0 is the direct path, which encounters attenuation ue to the building floors. Models such as those described in Section 3.4.1 implicitly sume that this path is the dominant source of signal power, although the wall and floor ss factors applied can be modified to account for the average effect of other paths. Paths and 2 encounter diffraction in propagating out of, and back in through, the windows the building, but are unobstructed in propagating between the floors. Finally, path 3 also diffracted through the windows of the building, although this is through a smaller gle than path 2. It is reflected from the wall of a nearby building before diffracting ck into the original building [45].

To analyze the field strength due to paths 2 and 3, the geometry is approximated the double wedge geometry in Figure 3.35, representing the building edges at the ints where the rays enter and leave the building. The propagation is then analyzed using

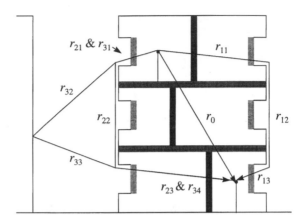

Figure 3.34 Alternative paths for propagation between floors.

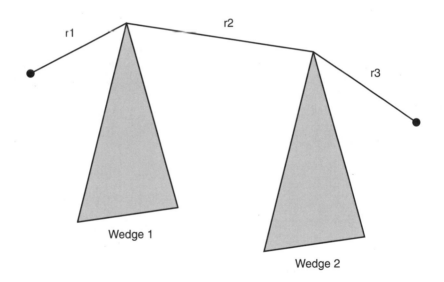

Figure 3.35 Double wedge geometry.

the geometrical theory of diffraction. The source is a point source and therefore radiate
spherical waves. The field incident on wedge 1 is therefore:

$$E_1 = \sqrt{Z_0 \frac{P_T}{4\pi r_1^2}} = \frac{1}{2r_1}\sqrt{\frac{Z_0 P_T}{\pi}} \qquad (3.5$$

where P_T is the effective isotropic radiated power from the source. The diffraction process at wedge 1 then yields a field incident on wedge 2, which is approximated using GTD as:

$$E_2 = E_1 \times D_1 \times \sqrt{\frac{r_1}{r_2(r_1 + r_2 + r_3)}} \qquad (3.58)$$

where the square-root factor is the spreading factor for spherical wave incidence on a straight wedge (see [46], p. 768, for example).

Similarly, the field at the field point is:

$$E_3 = E_2 \times D_2 \times \sqrt{\frac{r_1 + r_2}{r_3(r_1 + r_2 + r_3)}} \qquad (3.59)$$

Hence the power available at an isotropic receive antenna is:

$$P_r = P_T \frac{\lambda^2}{4\pi} \times \frac{E_3^2}{Z_0} = P_T \left(\frac{\lambda}{4\pi}\right)^2 \frac{D_1^2 D_2^2}{r_1 r_2 r_3 (r_1 + r_2 + r_3)} \qquad (3.60)$$

This result can be applied to both paths 1 and 2 by substitution of the appropriate distances. Path 3 also follows in the same way, but (3.60) is multiplied by the reflection coefficient of the nearby building.

The sum of the power from the various contributions is shown in Figure 3.36. It is clear that two regimes are present; for small spacing between the transmitter and receiver, the signal drops rapidly as the multiple floor losses on path 0 accumulate. Eventually the diffracted paths (1 and 2) outside the building dominate, and these diminish far less quickly with distance. When a reflecting adjacent building is present, the diffraction losses associated with this path are less and this provides a significant increase in the field strength for large separations.

3.4.3.2 Propagation on Single Floors

When the transmitter and receiver are mounted on the same floor of a building, the dominant mode of propagation is line-of-sight, as shown in Figure 3.37. However, the floor and ceiling-mounted objects will result in the Fresnel zone around the direct ray becoming obstructed at large distances, and this will give rise to additional loss due to diffraction. The effective path loss exponent will then be increased and the signal strength will fall off very rapidly with distance. The point at which this occurs depends on the specific geometry, with the maximum unobstructed range being obtained when the antennas are mounted at the midpoint of the gap between the highest floor-mounted obstructions and the lowest point of the ceiling-mounted obstructions [47].

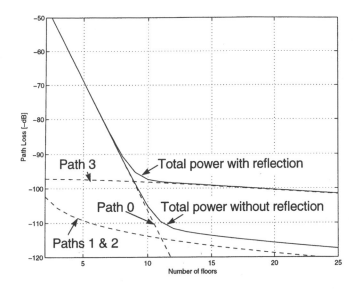

Figure 3.36 Variation of path loss with number of floors; here the floor height is 4m, building width is 30m, distance to the adjacent building is 30m, and the frequency is 900 MHz.

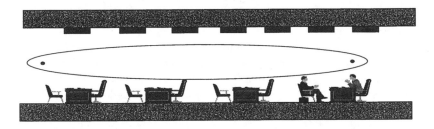

Figure 3.37 Propagation between antennas on a single floor.

3.4.4 Constitutive Parameters for Physical Models

All physical models require both the geometry and constitutive parameters of the building as input. Because walls and floors in buildings are inhomogeneous, predictions should in principle, account for effects such as the reinforcement of concrete walls using steel the periodic layered structures in cavity walls, plus other similar effects for which it i extremely difficult to obtain detailed data. However, useful information on both transmis sion and reflection properties can be obtained using the Fresnel reflection coefficients b assuming that walls and floors are plane and infinite, and by properly accounting fo refraction at both of the interfaces between each wall/floor and free space. Representativ values of the complex permittivity at various frequencies are given in Table 3.8 [39].

Table 3.8
Complex Permittivity of Typical Construction Materials

	1 GHz	*57.5 GHz*	*78.5 GHz*	*95.9 GHz*
Concrete	$7.0-j0.85$	$6.50-j0.43$	—	$6.20-j0.34$
Lightweight concrete	$2.0-j0.50$	—	—	—
Floorboard (synthetic resin)	—	$3.91-j0.33$	$3.64-j0.37$	$3.16-j0.39$
Plaster board	—	$2.25-j0.03$	$2.37-j0.10$	$2.25-j0.06$
Ceiling board (rock wool)	$1.2-j0.01$	$1.59-j0.01$	$1.56-j0.02$	$1.56-j0.04$
Glass	$7.0-j0.10$	$6.81-j0.17$	—	—
Fiberglass	$1.2-j0.10$	—	—	—

3.4.5 Propagation in Picocells: Discussion

Propagation effects in picocells are even more geometry dependent than in microcells, placing even greater burdens on the quality of data and computational requirements if deterministic physical models are to produce useful predictions. In the near future, practical picocell system design is more likely to rely on empirical models and engineering experience. In the longer term, however, combinations of physical models with statistics are expected to yield significant benefits.

3.4.6 Multipath Effects

In macrocells, it is usual to assume that waves arrive with uniform probability from all horizontal angles, leading to the classical Doppler spectrum [1]. By contrast, a more reasonable assumption for the indoor environment, particular when propagation occurs between floors, is that waves arrive with uniform probability from *all* angles. The resulting Doppler spectrum is then relatively uniform so, for simulation purposes, it is reasonable to assume a flat Doppler spectrum, given by:

$$S(f) = \begin{cases} \dfrac{1}{2f_m} & |f| \leq f_m \\ 0 & f > f_m \end{cases} \tag{3.61}$$

where f_m is the maximum Doppler frequency.

With regard to the RMS delay spread of the channel, values encountered in most cases are very much lower than those found in either micro- or macrocells. The variability around the median value is large, however (although there is a strong correlation with the path loss [48]), and there are occasionally cases where the delay spread is very much larger than the median. To provide reasonably realistic simulations both situations must be considered. Tables 3.9 and 3.10 give suitable channels for an indoor office scenario

Table 3.9

Indoor Office Test Environment Wideband Channel Parameters

Median Channel $\tau_{RMS} = 35$ ns		Bad Channel $\tau_{RMS} = 100$ ns	
Relative Delay, τ (ns)	Relative Mean Power (dB)	Relative Delay, τ (ns)	Relative Mean Power (dB)
0	0	0	0
50	−3.0	100	−3.6
110	−10.0	200	−7.2
170	−18.0	300	−10.8
290	−26.0	500	−18.0
310	−32.0	700	−25.2

Doppler spectrum for all taps is flat.

Table 3.10

Outdoor-to-Indoor Test Environment Wideband Channel Parameters

Median Channel $\tau_{RMS} = 45$ ns		Bad Channel $\tau_{RMS} = 750$ ns	
Relative Delay, τ (ns)	Relative Mean Power (dB)	Relative Delay, τ (ns)	Relative Mean Power (dB)
0	0	0	0
110	−9.7	200	−0.9
190	−19.2	800	−4.9
410	−22.8	1200	−8.0
—	—	2300	−7.8
—	—	3700	−23.9

Doppler spectrum for all taps is classical.

and for an outdoor to indoor scenario respectively, intended for evaluation purposes at around 2 GHz [49]. Values for the RMS delay spread for indoor-indoor environment are also shown in Table 3.11; case A represents low, but frequently occurring values; case B represents median values; and case C gives extreme values that occur only rarely [39]. The very high cases can occur particularly if there are strong reflections from buildings situated significant distances from the building under test.

Table 3.11

RMS Delay Spread in Nanoseconds in Indoor-to-Indoor Environments

Environment	Case A	Case B	Case C
Indoor residential	20	70	150
Indoor office	35	100	460
Indoor commercial	55	150	500

More details of the statistics and structure of the indoor wideband channel are available in references such as [50–53]. In particular, [53] proposes that the power-delay profile tends to follow a doubly exponential distribution (Figure 3.38), where the peaks of the individual exponentials can be reasonably accurately predicted from ray-tracing models, but where the associated weaker signals result from rough scattering and fine detail that cannot easily be predicted from a deterministic physical model. It has also been observed that the number of the multipath components follows a Gaussian distribution, with a mean value that increases with antenna separation [54].

3.4.7 Conclusions

Picocell propagation is affected by a wide range of mechanisms, operating on a complex, three-dimensional environment, the details of which are rarely available for propagation predictions. Some simple models can give useful estimates of in-building propagation, however, and further progress in these areas is strongly motivated by the growing importance of in-building communication, particularly for very high data rates. The use of appropriate distributed antenna structures helps considerably in providing controlled coverage around buildings, and it is expected that the provision of intelligence within such units will allow systems to be installed without the need for detailed propagation predictions (see Section 3.6).

3.5 MEGACELLS

Mobile systems designed to provide truly global coverage using constellations of low- and medium-earth orbit satellites are now in operation. These form *megacells*, consisting

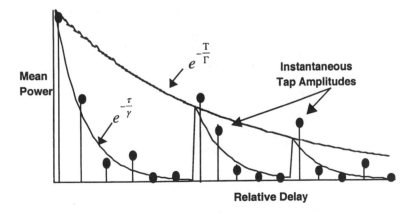

Figure 3.38 Doubly exponential power-delay profile in indoor channels [53].

of the footprint from clusters of spotbeams from each satellite, which move rapidly across the earth's surface. Signals are typically received by the mobile at very high elevation angles, so that only environmental features that are very close to the mobile contribute significantly to the propagation process (Figure 3.39). Atmospheric effects are also signifi- cant in systems operated at SHF and EHF [1]. Megacell propagation prediction techniques must also combine predictions of fast (multipath) fading and of shadowing effects, since these tend to occur on similar distance scales and cannot therefore be easily separated. The predictions tend to be highly statistical in nature, since coverage across very wide areas must be included, while still accounting for the large variations due to the local environment.

Mobile satellite systems are usually classified according to their orbit type: low earth orbits (LEO) involve satellites at an altitude of 500–2,000 km and require a relatively large number of satellites to provide coverage of the whole of the earth (for example, 66 at 780 km in the case of the Iridium system). Medium earth orbits (MEO) involve altitudes of around 5,000–12,000 km and involve fewer, slower moving satellites for whole-earth coverage (for example, 12 at 10,370 km in the case of the Odyssey system). Geostationary satellites (GEO), at the special height of around 36,000 km, have the benefit of requiring

Figure 3.39 Megacell propagation geometry.

only three satellites for whole-earth coverage and require almost no tracking of satellite direction. The large altitudes in GEO systems lead to a very large free-space loss component. For example, the free-space loss for a geostationary satellite at 1.5 GHz would be around 186 dB at zenith. The transmit powers needed at both the satellite and mobile to overcome this loss are excessive, so LEO and MEO systems are far more attractive for mobile communications, while GEO is more usually applied to fixed satellite links. For further details, see [55,56].

In all orbits other than GEO, the satellite position changes relative to a point on the earth, so the free-space loss for a particular mobile position becomes a function of time. In the extreme case of a LEO system, an overhead pass might last only a few minutes, so the path loss will change rapidly between its maximum and minimum values (see, for example, Figure 3.40). Similarly, the motion of the satellite relative to the location of the mobile user will create significant Doppler shift, which will change rapidly from positive to negative as the satellite passes overhead. In the case illustrated in Figure 3.40, the maximum Doppler shift would be around ±37 kHz. This *shift* can be compensated for by retuning either the transmitter or receiver, and should not be confused with the Doppler *spread*, which arises from motion of the mobile relative to sources of multipath, which is of a much smaller size but which cannot be compensated for.

The main local sources of propagation impairments in mobile-satellite propagation are trees, buildings, and terrain. These interact with wave propagation via the following mechanisms: reflection, scattering, diffraction, and multipath.

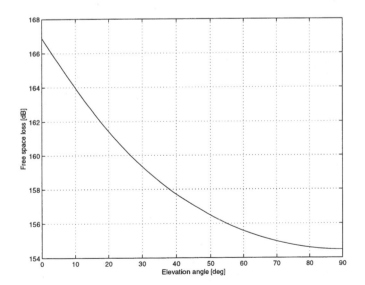

Figure 3.40 Free-space loss for a circular LEO satellite orbit at an altitude of 778 km at 1.625 GHz. This range of values would be encountered over a period of around 7.5 min.

3.5.1 Shadowing and Fast Fading

In mobile satellite systems, the elevation angle from the mobile to the satellite is much larger than for terrestrial systems, with minimum elevation angles in the range of 8–25 deg. Shadowing effects therefore tend to result mainly from the clutter in the immediate vicinity of the mobile. For example, when the mobile is operated in a built-up area, only the building closest to the mobile in the direction of the satellite is usually significant. As the mobile moves along the street, the building contributing to this process changes rapidly, so the shadowing attenuation may also change at a relatively high rate. By contrast, terrestrial macrocellular systems involve elevation angles on the order of 1 deg or less, so a large number of buildings along the path are significant (see Section 3.2).

As a consequence of this effect, rapid and frequent transitions may occur between line-of-sight and non-line-of-sight situations in the satellite-mobile case, causing a variation in the statistics of fast fading that is closely associated with the shadowing process. See Figure 3.41 for a typical measured example of the variation of signal level with location in a suburban area; note that the fade depths are much greater during the obstructed periods.

It is therefore most convenient to treat the shadowing and narrowband fast fading for satellite-mobile systems as a single, closely coupled process, in which the parameters of the fading (such as the Rice k-factor and local mean signal power) are time varying. In subsequent sections we examine the basic mechanisms of channel variation in mobile

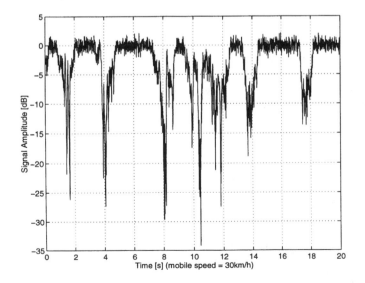

Figure 3.41 Example channel variations measured in a suburban area at 1.5 GHz.

satellite systems, and then describe a number of models which can be used to predict these effects.

3.5.2 Local Shadowing Effects

Roadside buildings are essentially total absorbers at mobile-satellite frequencies, so they can be regarded as diffracting knife edges. They can be considered to block the signal significantly when at least 0.6 of the first Fresnel zone around the direct ray from the satellite to the mobile is blocked (see Figure 3.42). Thus, shadowing may actually be less at higher frequencies due to the narrower Fresnel zones for a given configuration. Once shadowing has occurred, building attenuation may be estimated via the single knife edge diffraction formula given in Section 3.5.2.

Tree shadowing also occurs primarily when the tree is contained within the 0.6 × first Fresnel zone region. In this case, however, the tree is not a complete absorber, so propagation occurs through the tree as well as around it. The attenuation varies strongly with frequency and path length. The simplest approach is to find the path length through the tree and calculate the attenuation based on an exponential attenuation coefficient in dB/m. Single-tree attenuation coefficients have been measured at 869 MHz, and the largest values were found to vary between 1 and 2.3 dB/m, with a mean value of 1.7 dB/m [57].

3.5.2.1 Local Multipath Effects

As well as the existence of a direct path from the satellite to the mobile, reflection and scattering processes lead to other viable wave paths. These multiple paths interfere with each other, leading to rapid fading effects as the mobile's position varies. The multipath may result from adjacent buildings, trees, or the ground. The level of the multipath is usually rather lower than that of the direct path, but it may still lead to significant fading effects, similar to those described in Chapter 2. Note that multiple scattering paths, such as those that reach the mobile via points x and y in Figure 3.43 are attenuated by two reflection coefficients and are therefore unlikely to be significant.

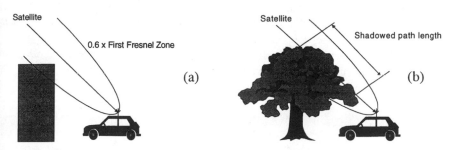

Figure 3.42 Shadowing by (a) buildings and (b) trees.

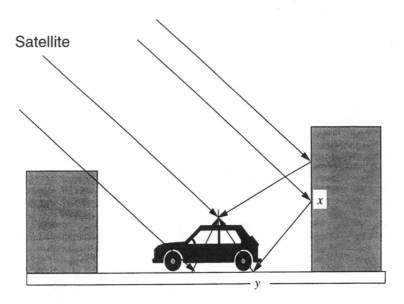

Figure 3.43 Multipath propagation.

The multipath can also lead to wideband fading if the differential path lengths ar
sufficiently large. In practical systems, however, the dominant source of wideband effect
may be the use of multiple satellites to provide path diversity (see Section 3.5.7).

3.5.3 Empirical Narrowband Models

Models for narrowband propagation in mobile satellite systems are different from thos
used in terrestrial systems in two key ways. First, they include the excess path loss an
shadowing effects as dynamic processes, along with the fast fading. Second, they rarel
use direct deterministic calculation of physical effects, since this is not practical fc
predicting satellite coverage of areas exceeding tens of thousands of square kilometer
Instead they use statistical methods, although these may be based on either empirical c
physical descriptions of the channel.

Empirical models, particularly the empirical roadside shadowing (ERS) model, hav
been constructed for mobile satellite systems operated in areas characterized mainly b
roadside trees. The ERS model is expressed as [58]:

$$L(P, \theta) = -(3.44 + 0.0975\theta - 0.002\theta^2)\ln P + (-0.443\theta + 34.76) \quad (3.6.$$

where $L(P, \theta)$ is the fade depth exceeded for P percent of the distance traveled (decibels
at an elevation angle (degrees) to the satellite, where for elevation angles from 7 deg

20 deg, (3.62) is used with $\theta = 20$ deg. If the vehicle travels at constant speed, P is also the percentage of the time for which the fade exceeds L. This model applies only to propagation at L-band (1.5 GHz), elevation angles from 20 deg to 60 deg, and at fade exceedance percentages from 1% to 20%. The result is shown in Figure 3.44.

The ERS model can then be extended up to 20 GHz using the following frequency-scaling function:

$$L(f_2) = L(f_1)\exp\left[1.5\left(\frac{1}{\sqrt{f_1}} - \frac{1}{\sqrt{f_2}}\right)\right] \tag{3.63}$$

where $L(f_2)$ and $L(f_1)$ are the attenuations in decibels at frequencies $f_1 = 1.5$ GHz and f_2, with 0.8 GHz $\leq f_2 \leq 20$ GHz. This function is shown in Figure 3.45.

The ERS model has also been extended to larger time percentages from the original distributions at L-Band. The model extension is given by:

$$L(P, \theta) = \frac{L(20\%, \theta)}{\ln 4} \times \ln\left(\frac{80}{P}\right) \tag{3.64}$$

for $80\% \geq P > 20\%$. Predictions from the extended model at 20 GHz for elevation angles from $20°$ to $60°$ are shown in Figure 3.46.

Foliage effects can also be modeled empirically using a model that was developed from mobile measurements taken in Austin, Texas, in 1995 [59]. The trees had no leaves

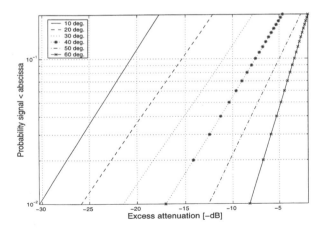

Figure 3.44 Predictions from empirical roadside shadowing model. The curves represent, from left, increasing elevation angle in steps of 10 deg from 10–60 deg.

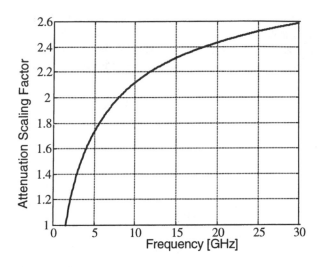

Figure 3.45 Empirical roadside shadowing model frequency-scaling function.

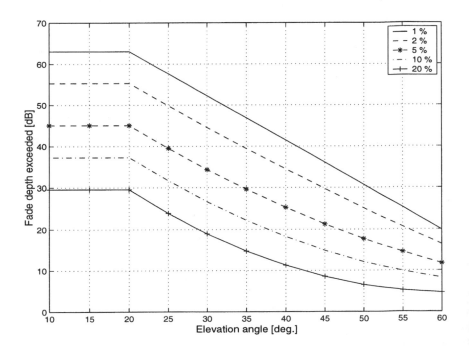

Figure 3.46 Extended empirical roadside shadowing model at 20 GHz.

in February and in May were in full foliage. A least-squares fit to the equal probability levels of the attenuations yielded the following relationship:

$$L_{\text{foliage}} = a + bL_{\text{no\_foliage}}^{c} \qquad (3.65)$$

where the constants are $a = 0.351$, $b = 6.8253$, and $c = 0.5776$ and where the model applies in the range $1 \le L_{\text{no\_foliage}} \le 15$ dB and $8 \le L_{\text{foliage}} \le 32$ dB.

3.5.4 Statistical Models

Statistical models give an explicit representation of the channel statistics in terms of parametric distributions, which are a mixture of Rice, Rayleigh, and lognormal components. Such models use statistical theory to derive a reasonable analytical form for the distribution of the narrowband fading signal and then use measurements to find appropriate values of the parameters in the distribution. These models all have in common an assumption that the total narrowband fading signal in mobile-satellite environments can be decomposed into two parts: a coherent part, usually associated with the direct path between the satellite and mobile, and a diffuse part arising from a large number of multipath components of differing phases. The magnitude of the latter part is assumed to have a Rayleigh distribution. Thus, the multiplicative complex channel, α, corresponding to all such models can be expressed as:

$$\alpha = A_c s_c e^{j\phi} + r s_d e^{j(\theta + \phi)} \qquad (3.66)$$

where A_c is the coherent part, s_c and s_d are the shadowing components associated with the coherent and diffuse parts, respectively, and r has complex Gaussian distribution (that is, its magnitude is Rayleigh distributed).

The simplest model of this form is the Rice distribution [1], which assumes that both components of the signal have constant mean power. More recent work such as that in [60–62] has generalized this model to account for the rapidly changing conditions associated with attenuation and shadowing of both the coherent and diffuse components that arise from mobile motion. The models are summarized in Table 3.12. Note that, as with terrestrial shadowing, the distribution of the mean power arising from s_c and s_d is widely assumed to be lognormal. These models can all be implemented within the structure shown in Figure 3.47.

If the parameters of these models are appropriately chosen, they can provide a good fit to measured distributions over a wide range of environmental and operating conditions, although the Loo model is only really applicable to moderate rural situations. The Hwang model has been shown [62] to include the Rice, Loo, and Corazza models as special cases. Note, however, that Table 3.12 reveals that a further generalization of the Hwang model to allow correlation in the range [0,1] would allow still further generality.

Table 3.12
Parametric Mobile-Satellite Channel Models

Model	Coherent Part	Correlation Between Coherent and Diffuse Part	Diffuse Part
Rice [1]	Constant	Zero	Constant mean power Rayleigh
Loo [60]	Lognormal	Variable	Constant mean power Rayleigh
Corazza and Vatalaro [61]	Lognormal	Unity	Lognormal-Rayleigh
Hwang et al. [62]	Lognormal	Zero	Lognormal-Rayleigh

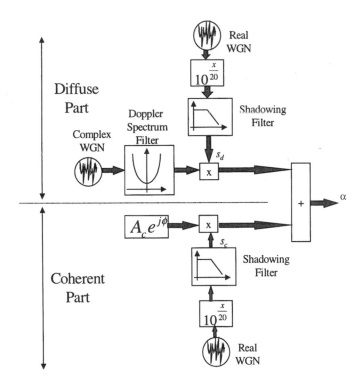

Figure 3.47 Generative structure for analytical-statistical narrowband channel model. WGN is white gaussian noise.

3.5.4.1 Loo Model

This model [60] is specifically designed to account for shadowing due to roadside trees. It is assumed that the total signal is composed of two parts; a line-of-sight component that is lognormally distributed due to the tree attenuation, plus a multipath component that has a Rayleigh distribution. Thus the total complex fading signal α is given by:

$$\alpha = d e^{j\phi_0} + s e^{j\phi} \tag{3.67}$$

where d is the lognormally distributed line-of-sight amplitude, s is the Rayleigh distributed multipath amplitude, and ϕ_0 and ϕ_0 are uniformly distributed phases.

The pdf of the fading amplitude, $r = |\alpha|$, is too complicated to evaluate analytically, but it can be approximated by the Rayleigh distribution for small values and by the lognormal distribution for large values:

$$
p(r) =
\begin{cases}
\dfrac{r}{\sigma_m^2} \exp\left(-\dfrac{r^2}{2\sigma_m^2}\right) & \text{for } r \ll \sigma_m \\[3mm]
\dfrac{1}{20\log r \sqrt{2\pi\sigma_0}} \exp\left[-\dfrac{(20\log r - \mu)^2}{2\sigma_0}\right] & \text{for } r \gg \sigma_m
\end{cases}
\tag{3.68}
$$

where σ_m is the standard deviation of either the real or imaginary component of the multipath part, σ_0 is the standard deviation of $20 \log d$ (dB), and is the mean of $20 \log d$ (dB). An example prediction is shown in Figure 3.48 where the parameters of the Loo model have been chosen to fit the results from the ERS model at an elevation angle of 45 deg. Expressions for the level crossing rate and average fade duration are also given in [60] and it is found that these depend on the correlation between d and s, with the highest crossing rates occurring for low values of the correlation.

3.5.4.2 Corazza Model

This model [61] can be seen as a development of the Loo approach, where both the direct path and multipath are lognormally shadowed, so that the channel amplitude is given by:

$$\alpha = S(e^{j\phi_0} + s e^{j\phi}) \tag{3.69}$$

where S is the lognormal shadowing and other parameters follow the definitions of (3.67). The parameters of the model are then the Rice factor, k, and the lognormal mean μ decibels and standard deviation σ_L decibels. A wide range of environments can be modeled by appropriate choice of these parameters. The following empirical formulations for the parameters were extracted from measurements at the L-band in a rural tree-shadowed environment:

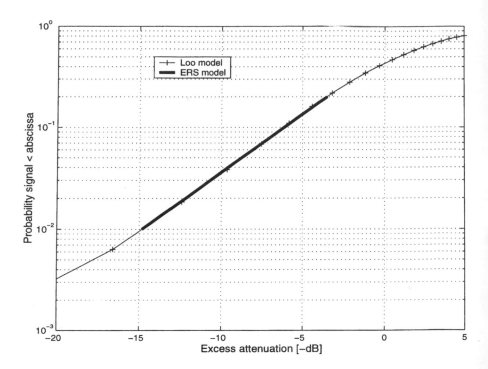

Figure 3.48 Comparison of Loo model (dotted line) and ERS model (solid line) at 1.5 GHz and 45-deg elevation. Parameters are $\sigma_m = 0.3$, $\sigma_0 = 5$ dB, and $\mu = 0.1$ dB.

$$k = 2.731 - 0.1074\theta + 2.774 \times 10^{-3}\theta^2$$
$$\mu = -2.331 + 0.1142\theta - 1.939 \times 10^{-3}\theta^2 + 1.094 \times 10^{-3}\theta^3 \qquad (3.70)$$
$$\sigma_L = 4.5 - 0.05\theta$$

3.5.4.3 Lutz Model

In this model [63], the statistics of line-of-sight and non-line-of-sight states are modeled by two distinct *states*, which is particularly appropriate for modeling in urban or suburban areas where there is a large difference between shadowed and unshadowed statistics. The parameters associated with each state and the transition probabilities for evolution between states are empirically derived. These models permit time-series behavior to be examined and may be generalized to more states to permit a smoother representation of the transitions between LOS and NLOS conditions [64], or to characterize multiple satellite propagation [65]. For example, modeling in an environment that includes building blockage, tree shadowing, and line-of-sight conditions can be modeled via a three-state approach.

In [63], the line-of-sight condition is represented by a "good" state, and the non-line-of-sight condition by a "bad" state (Figure 3.49). In the good state, the signal amplitude is assumed to be Rice distributed, with a k-factor that depends on the satellite elevation angle and the carrier frequency, so that the pdf of the signal amplitude is given by $p_{good}(r) = p_{Rice}(r)$, where p_{Rice} is a given in (10.30). The noncoherent (multipath) contribution has a classical Doppler spectrum.

In the bad state, the fading statistics of the signal amplitude, $r = |\alpha|$, are assumed to be Rayleigh, but with a mean power $S_0 = \sigma^2$, which varies with time, so the pdf of r is specified as the conditional distribution $p_{Rayl}(r|S_0)$, and S_0 varies slowly with a lognormal distribution, $p_{LN}(S_0)$, with mean μ decibels and standard deviation σ_L decibels, representing the varying effects of shadowing within the non-line-of-sight situation. The overall pdf in the bad state is then found by integrating the Rayleigh distribution over all possible values of S_0, so that:

$$p_{bad}(r) = \int_0^\infty p_{Rayl}(r|S_0)p_{LN}(S_0)dS_0 \qquad (3.71)$$

The proportion of time for which the channel is in the bad state is the *time-share of shadowing, A,* so that the overall pdf of the signal amplitude is:

$$p_r(r) = (1 - A)p_{good}(r) + Ap_{bad}(r) \qquad (3.72)$$

Transitions between states are described by a *first-order Markov chain.* This is a state-transition system, in which the transition from one state to another depends only on

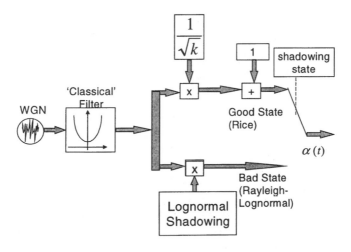

Figure 3.49 Two-state Lutz model of narrowband mobile-satellite fading. WGN is white gaussian noise.

the current state, rather than on any more distant history of the system. These transitions are represented by the state transition diagram in Figure 3.50, which is characterized by a set of state-transition probabilities.

The transition probabilities are:

- Probability of transition from good state to good state: p_{gg};
- Probability of transition from good state to bad state: p_{gb};
- Probability of transition from bad state to bad state: p_{bb}; and
- Probability of transition from bad state to good state: p_{bg}.

For a digital communication system, each state transition is taken to represent the transmission of one symbol. The transition probabilities can then be found in terms of the mean number of symbol durations spent in each state:

- $p_{gb} = 1/D_g$, where D_g is the mean number of symbol durations in the good state; and
- $p_{bg} = 1/D_b$, where D_b is the mean number of symbol durations in the bad state.

The sum of the probabilities leading from any state must sum to 1, so:

$$p_{gg} = 1 - p_{gb} \text{ and } p_{bb} = 1 - p_{bg} \tag{3.73}$$

Finally, the time-share of shadowing (the proportion of symbols in the bad state) is:

$$A = \frac{D_b}{D_b + D_g} \tag{3.74}$$

The model can be used to calculate the probability of staying in one state for more than n symbols by combining the relevant transition probabilities. Thus:

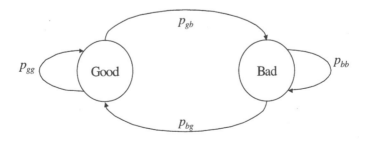

Figure 3.50 Markov model of channel state.

Probability of staying in "good" state for more than n symbols

$$= p_g(>n) = p_{gg}^n \tag{3.75}$$

Probability of staying in "bad" state for more than n symbols

$$= p_b(>n) = p_{bb}^n$$

An example of the signal variations produced using the Lutz model is shown in Figure 3.51, where the two states are clearly evident in the signal variations. Typical parameters, taken from measurements at 1.5 GHz, are shown in Figure 3.52 (highway environment) and Figure 3.53 (city environment).

This two-state model of the satellite-mobile channel is very useful for analyzing and simulating the performance of satellite-mobile systems. It is inaccurate in representing the second-order statistics of shadowing, however, since it assumes that the transition from LOS to NLOS situations is instantaneous. In practice this is not true, due to the smooth transitions introduced by diffraction and reflection effects, particularly at lower frequencies.

One way of overcoming this is to introduce extra states, which represent intermediate levels of shadowing with smaller Rice k-factors than the LOS "good" state. It has been

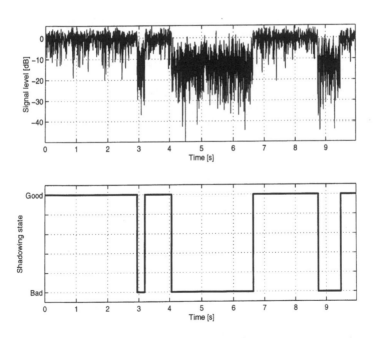

Figure 3.51 Example time-series output (signal level and shadowing state) from Lutz model. Parameters used are $D_g = 24$m, $D_b = 33$m, $\mu = -10.6$ dB, $\sigma_L = 2.6$ dB, $k = 6$ dB, $f_c = 1500$ MHz, and $v = 50$ km h^{-1}.

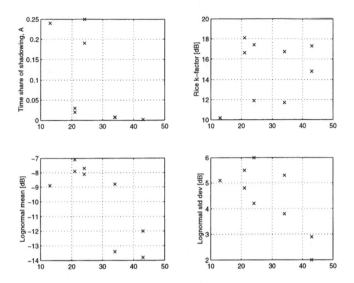

Figure 3.52 Highway parameters for Lutz model; x axis is elevation angle.

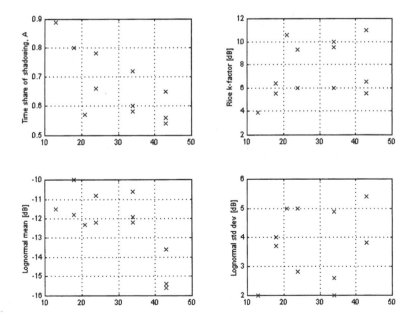

Figure 3.53 City parameters for Lutz model; x axis is elevation angle.

shown [64] that at least three states are needed in order to accurately model these transitions even at 20 GHz and the expectation is that even more would be needed for accurate modeling at L-band. Alternatively, the sharp transitions between states can be smoothed by filtering. In other work [66], the statistics of the fading process vary according to the environment, with several environment types (urban, open, suburban) distinguished as model states.

3.5.5 Physical-Statistical Models for Built-Up Areas

Deterministic physical models of mobile-satellite propagation have been created and have produced good agreement with measurements over limited areas [67]. For practical predictions, however, an element of statistics has to be introduced. One approach is to use physical models to examine typical signal variations in environments of various categories and use these to generalize over wider areas. A more appropriate approach is to use *physical-statistical models,* which derive fading distributions directly from distributions of physical parameters using simple electromagnetic theory. This class of models is described in some detail here, as the modeling approach has been examined only slightly in the open literature.

In modeling any propagation parameter, the aims of modeling are broadly similar. The key point is to predict a particular parameter with maximum accuracy, consistent with minimum cost in terms of the quantity and expense of the input data and in terms of the computational effort required to produce the prediction.

For empirical models, the input knowledge consists almost entirely of previous measurements that have been made in environments judged to be representative of practical systems. An approximation to these data, usually consisting of a curve fit to the measurements, is used for predictions. The input data are then fairly simple, consisting primarily of operating frequency, elevation angle, range, and a qualitative description of the environment (for example, rural or urban). Such models are simple to compute and have good accuracy within the parameter ranges spanned by the original measurements. However, because the models lack a physical basis, they are usually very poor at extrapolating outside of these parameter ranges. There is a classification problem involved in describing the environment, since an environment judged to be urban in some countries may be little more than a small town elsewhere. Additionally, the use of a curve-fitting approach implies that the real data will generally be considerably scattered around the predicted values and this represents a lower limit on the prediction accuracy. For example, predictions of loss are subject to an error resulting from the effects of shadowing and fading.

The input knowledge used in physical models, by contrast, consists of electromagnetic theory combined with engineering expertise that is used to make reasonable assumptions about which propagation modes are significant in a given situation. Provided that the correct modes are identified, the theoretical approach is capable of making very accurate predictions of a wide range of parameters in a deterministic manner. The output that can be given is point-by-point rather than an average value, so the model can apply to very wide ranges of system and environment parameters, certainly well beyond the range within which measurements have been made. To make such predictions, however, the models may require very precise and detailed input data concerning the geometrical and electrical properties of the environment. This may be expensive or even impossible to obtain with sufficient accuracy. Also, the computations required for a full theoretical calculation may be very complex, so extra assumptions often have to be made for simplification, leading to compromised accuracy.

Physical-statistical modeling is a hybrid approach that builds on the advantages of both empirical and physical models while avoiding many of their disadvantages. As in the physical model case, the input knowledge consists of electromagnetic theory and sound physical understanding. However, this knowledge is then used to analyze a *statistical* input data set, yielding a *distribution* of the output predictions. The outputs can still effectively be point by point, although the predictions are no longer linked to specific locations. For example, a physical-statistical model can predict the distribution of shadowing, avoiding the errors inherent in the empirical approach, although it does not predict what the shadowing value will be at a particular location. This information is usually adequate for the system designer. Physical-statistical models therefore require only simple input data such as input distribution parameters (such as mean building height, building height variance). The environment description is entirely objective, avoiding problems of subjective classification, and is capable of high statistical accuracy. The models are based on sound physical principles, so they are applicable over very wide parameter ranges. Finally, by precalculating the effect of specific input distributions, the required computational effort can be very small.

One example of a physical-statistical model has been used to predict the attenuation statistics of roadside tree shadowing, using only physical parameters as input [68]. This modeled the trees as consisting of a uniform slab whose height and width were uniformly distributed random variables. For a given direction from mobile to satellite, the mean and standard deviation of the path length through the block was calculated and used with an empirical model of specific attenuation through trees to calculate the mean and standard deviation of the tree attenuation. These values are then taken as the mean and standard deviation of the lognormal distribution in the Lutz structure (Figure 3.49).

In Sections 3.5.5.2 and 3.5.5.3, two physical-statistical models for megacells operated in built-up areas are described [69,70]. First, the basic physical parameters used in both these models are introduced.

3.5.5.1 Building Height Distribution

The geometry of the situation to be analyzed is illustrated in Figure 3.54. It describes a situation in which a mobile is situated on a long straight street with the direct ray from the satellite impinging on the mobile from an arbitrary direction. The street is lined on both sides with buildings having heights that vary randomly.

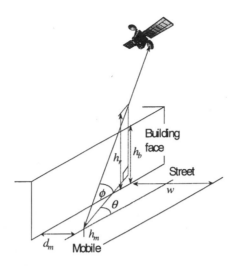

Figure 3.54 Geometry for mobile-satellite propagation in built-up areas.

In the models to be presented, the statistics of building height in typical built-up areas will be used as input data. A suitable form was sought by comparing with geographical data for the city of Westminster and for the city of Guildford, United Kingdom (Figure 3.55). The probability density functions that were selected to fit the data are the lognormal and Rayleigh distributions with parameters the mean value μ and the standard deviation σ_b. The pdf for the lognormal distribution is:

$$p_b(h_b) = \frac{1}{h_b \sqrt{2\pi}\sigma_b} \cdot e^{\dfrac{\ln^2(h_b/\mu)}{2\sigma_b^2}} \tag{3.76}$$

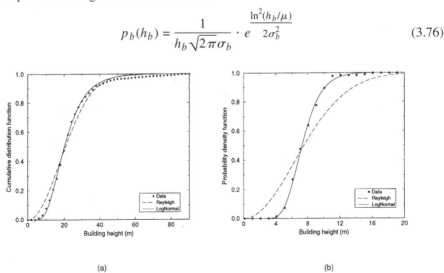

(a) (b)

Figure 3.55 Building height distribution for (a) city of Westminster and (b) city of Guildford.

and for the Rayleigh distribution is:

$$p_b(h_b) = \frac{h_b}{\sigma_b^2} \cdot e^{\frac{h_b^2}{2\sigma_b^2}} \tag{3.77}$$

The best fit parameters for the distributions are shown in Table 3.13. The lognormal distribution is clearly a better fit to the data, but the Rayleigh distribution has the benefit of greater analytical simplicity.

3.5.5.2 Time-Share of Shadowing

This model [69] estimates the time-share of shadowing [A in (3.72)] for the Lutz model [63] using physical-statistical principles. The direct ray is judged to be shadowed when the building height h_b exceeds some threshold height h_T relative to the direct ray height h_s at that point. A can then be expressed in terms of the probability density function of the building height, $p_b(h_b)$ as:

$$P_s = \Pr(h_b > h_T) = \int_{h_T}^{\infty} p_b(h_b)dh_b \tag{3.78}$$

Assuming that the building heights follow the Rayleigh distribution, (3.77), this yields:

$$A = \int_{h_T}^{\infty} \frac{h_b}{\sigma_b^2} \exp\left(-\frac{h_b^2}{2\sigma_b^2}\right) dh_b = \exp\left(-\frac{h_T^2}{2\sigma_b^2}\right) \tag{3.79}$$

The simplest definition of h_T is obtained by considering shadowing to occur exactly when the direct ray is geometrically blocked by the building face (a more sophisticated approach would account for the size of the Fresnel zone at that point). Simple trigonometry applied to this yields the following expression for h_T:

Table 3.13
Best Fit Parameters for the Theoretical pdfs

| City | Log-Normal pdf | | Rayleigh pdf |
	Mean (μ)	Standard Deviation (σ_b)	Standard Deviation (σ_b)
Westminster	20.6	0.44	17.6
Guildford	7.1	0.27	6.4

$$h_T = h_r = \begin{cases} h_m + \dfrac{d_m \tan\phi}{\sin\theta} & \text{for } 0 < \theta \le \pi \\[2ex] h_m + \dfrac{(w - d_m)\tan\phi}{\sin\theta} & \text{for } -\pi < \theta \le 0 \end{cases} \qquad (3.80)$$

Figure 3.56 compares this model with measurements of A versus elevation angle in city and suburban environments at L-band, taken from [63,71,72]. The model parameters are $\sigma_b = 15$, $w = 35$, $d_m = w/2$, $h_m = 1.5$, and $\theta = 90$ deg.

3.5.5.3 Time-Series Model

Although the approach in the last section allowed one of the parameters in the Lutz model to be predicted using physical-statistical methods, the remainder of the parameters still has to be determined empirically. All of the statistical models described earlier assume that the lognormal distribution was valid for predicting the shadowing distribution. However, Chapter 9 shows that this distribution comes as a result of a large number of individual

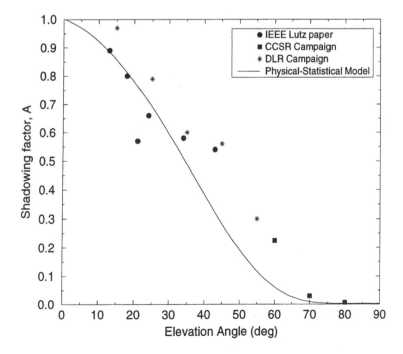

Figure 3.56 Comparison of theoretical and empirical results for time-share of shadowing. Measurements include L and S band examples.

effects acting together on the signal. In the case being treated here, only a single building is involved, so the lognormal approximation is questionable. The model described here avoids this assumption and directly predicts the statistics of attenuation [70]. The power received in this case is predicted as a continuous quantity, avoiding the unrealistic discretization of state-based models such as [63].

The total received power for a mobile in a built-up area consists of the direct diffracted field associated with the diffraction of the direct path around a series of roadside buildings, plus a multipath component whose power is set by computing reflections from the buildings on the opposite sides of the street and from the ground. The direct field is given by:

$$u_o(P) = u_o(P)\left\{1 - \frac{j}{2}\sum_{m=1}^{N}\left[\left(\int_{-\infty}^{v_{x2m}} e^{-\frac{i\pi}{2}v_x^2}dv_x\right)\cdot\left(\int_{v_{y1m}}^{v_{y2m}} e^{-\frac{i\pi}{2}v_y^2}dv_y\right)\right]\right\} \quad (3.81)$$

$$\begin{cases} v_{x1} = -\infty, \ v_{x2} = \sqrt{\dfrac{2(d_1 + d_2)}{\lambda d_1 d_2}}(x_{22} - x_m) \\[3mm] v_{y1} = \sqrt{\dfrac{k(d_1 + d_2)}{\pi d_1 d_2}}(y_{2i} - y_m), \ i = 1, \ 2 \end{cases} \quad (3.82)$$

where x_{22} defines the building height, y_{21} defines the position of the left building edge, y_{22} defines the position of the right building edge, d is the distance from the satellite to the building, and d_2 is the distance from the building to the mobile. This result is similar to the Fresnel integral formulation of single knife edge diffraction used earlier, but accounts for diffraction from around the vertical edges of buildings as well as over the rooftops. For the direct-diffracted field, $d_2 \equiv d_m$ and $x_3 = h_m$. For the wall reflected path, (3.81) is again used, but applied to the image of the source, so $d_2 = 2w - d_m$. For the ground-reflected field $x_3 = -h_m$. The satellite elevation angle and azimuth angles and, respectively, are introduced using the following geometrical relationships, with R being the direct distance from mobile-satellite:

$$y_{sat} = \frac{R\cos\phi}{\sqrt{1 + \tan^2\theta}}$$

$$x_{sat} = R\sin\phi + h_m \quad (3.83)$$

$$d_1 = \sqrt{R^2\cos^2\phi - y_{sat}^2} - d_2$$

The formulation above is then used with a series of roadside buildings, randomly generated according to the lognormal distribution of (3.76), with parameters applicable to the environment under study, including gaps between the buildings to represent some open areas. Figure 3.57 illustrates measured and simulated time-series data for a suburban environment at 18.6 GHz with the same sampling interval and with a 90-deg azimuth angle and a 35-deg elevation angle. The other model parameters were w = 16m, d_m = 9.5m, open area = 35% of total distance, μ = 7.3m, and σ_b = 0.26. For the building and ground reflections, the conductivity was set to 0.2 and 1.7 Sm^{-1} and the relative permittivity was set to 4.1 and 12, respectively. Qualitatively, the characteristics of the signal variation are similar, although the statistical nature of the prediction implies that the model should not be expected to match the predictions at any particular locations.

Figures 3.58 and 3.59 illustrate comparison results for the first-order and second-order statistics for the same satellite measurements as in Figure 3.57 and also for helicopter measurements [64]. The frequency for the helicopter measurements was 1.2 GHz, with a 90-deg azimuth angle and a 60-deg elevation angle.

3.5.6 Wideband Models

Wideband measurement and modeling has received relatively little attention for LMS systems, partly because the delay spreads encountered are far smaller than in most terrestrial systems, rendering the channel essentially narrowband for most first-generation LMS

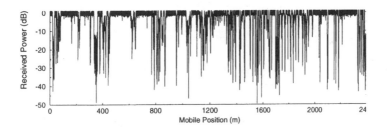

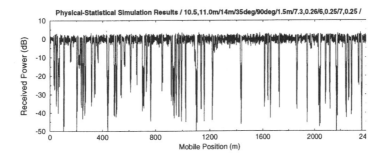

Figure 3.57 Comparison of theoretical output with real outdoor measurement data from a satellite.

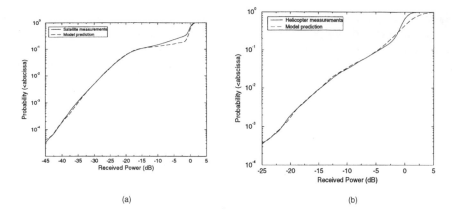

(a)   (b)

Figure 3.58 Comparison of first-order statistics with real outdoor (a) satellite and (b) helicopter measurements.

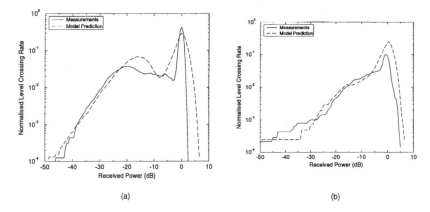

(a)   (b)

Figure 3.59 Comparison of second-order statistics with real outdoor (a) satellite and (b) helicopter measurements.

systems. However, future systems will offer multimedia services requiring very large channel bandwidths and will use spread spectrum techniques to provide high reliability and capacity, increasing the significance of wideband effects.

We saw previously that shadowing effects in mobile-satellite systems are dominated by the clutter in the immediate vicinity of the mobile. This is also true when considering wideband scattering, where the significant scatterers tend to contribute only relatively small excess path delays. The resulting wideband channel has been found to have an RMS delay spread of around 200 ns. The wideband channel model is then very similar to the terrestrial case, except that, in general, the tap gain processes may each be represented as instances of narrowband models such as those described in Sections 3.5.3, 3.5.4, and 3.5.5. Additionally, the first arriving tap will usually have a significant coherent component and hence a relatively high Rice k-factor compared to the later echoes. Several measurements have been conducted [72,73]. The latter work has led to results for the variation of delay spread at the L- and S-bands in five different environmental categories at elevation angles from 15 deg to 18 deg. A tapped delay-line model has been created based on the analysis of a representative fraction of the data [74].

A more sophisticated structure has also been proposed [72] in which the earliest arriving echoes, which arise from scatterers within 200m excess delay (600 ns) of the mobile, are assumed to have exponentially decreasing power with increasing delay. In this region the number of echoes is assumed to be Poisson distributed and the delays are exponentially distributed. Occasional echoes also occur with longer delays, with the delays being uniformly distributed up to a maximum of around 15 μs.

3.5.7 Multisatellite Correlations

When considering the effects of multiple-satellite diversity (or for more accuracy in the single-satellite case) it is crucial to consider the effects of the correlation between the fading encountered for satellites at different elevation *and* azimuth angles. Two satellites at the same elevation angle may exhibit very different outage probabilities, when, for example, one is viewed perpendicular to the direction of a building-lined street, while the other is viewed down the street and therefore much more likely to be unshadowed.

One method of accounting for these effects is to generalize the Lutz model (Section 3.5.4.3) to four states, representing all the "good" and "bad" combinations of the two mobile-satellite channels, as shown in Figure 3.60 [65]. The transition probabilities can again be found using information from measurements and models of time-share of shadowing; it is also necessary, however, to have knowledge of the correlation between the shadowing states of the two channels.

When two satellites are present, their locations are defined by two pairs of elevation and azimuth angles (θ_1, ϕ_1) and (θ_2, ϕ_2). The shadowing state of satellite i is defined as:

$$S_i = \begin{cases} 0 & \text{if } h_b \geq h_r \text{ (bad channel state)} \\ 1 & \text{if } h_b < h_r \text{ (good channel state)} \end{cases} \quad (3.84)$$

where (θ_i, ϕ_i) is used to find the corresponding value for h_r. The correlation between these shadowing states is then defined by:

$$\rho = \frac{E[(S_1 - \overline{S}_1) \cdot (S_2 - \overline{S}_2)]}{\sigma_1 \cdot \sigma_2} = \frac{1}{\sigma_1 \cdot \sigma_2} \cdot [E(S_1 S_2) - \overline{S}_1 \cdot \overline{S}_2] \quad (3.85)$$

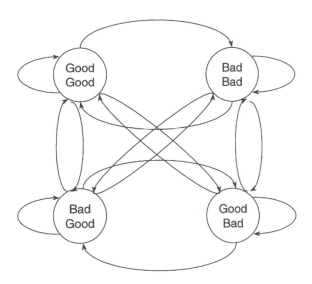

Figure 3.60 Four-state model of satellite diversity.

where $E[S] = \overline{S}$. If ρ is close to unity, the satellites suffer simultaneous shadowing always, and satellite diversity does not increase system availability significantly. If ρ is close to zero, however, the satellites suffer simultaneous shadowing only rarely, and the system is available far more of the time than would be the case with a single satellite. Still better is the situation where ρ is close to -1, and the availability of the satellites complements each other perfectly.

Measurements of the correlation are available [76] that have used the results of circular flight-path measurements to derive a correlation function, which is then used to derive appropriate transition probabilities for the model. Other approaches are to derive the correlations using a fish-eye lens photographs [77] or physical expressions [75]. In all cases, the correlation encountered in built-up areas diminishes rapidly with increasing azimuth angle between the satellites, being sufficiently small for appreciable diversity gain above around 30, difference (see Figure 3.61). Negative correlations are also possible when the environment has a particular geometrical structure. Parameters such as the time-share of shadowing can also be extracted from fish-eye lens photographs and a very close correspondence between measurements and predicted parameters has been found [78].

3.5.8 Overall Mobile-Satellite Channel Model

The overall mobile-satellite channel model is shown in Figure 3.62. There are two parts to the model: a *satellite process,* which includes effects between the satellite and the earth's surface, and a *terrestrial process,* which accounts for all the effects in the vicinity of the mobile.

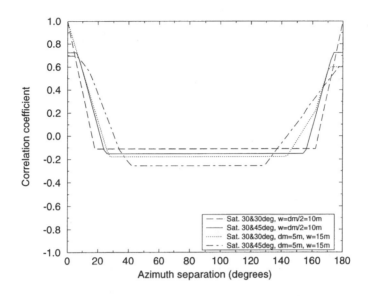

Figure 3.61 Example prediction of correlation coefficient from [75].

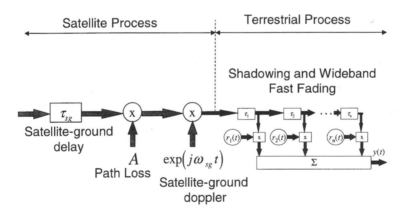

Figure 3.62 Complete mobile-satellite channel model.

The satellite process includes a delay of τ_{sg}, arising from the total propagation path length, a total path loss (excluding shadowing) A, which includes a free-space and atmospheric loss component, and a Doppler shift through a frequency of $f_{sg} = \omega_{sg}/2\pi$, which arises from the relative speed of the satellite and a point on the ground adjacent to the mobile.

The terrestrial process is modeled as a time-variant transversal filter (tapped delay line) representation of the wideband channel, with tap-gain processes $r_1(t)$, $r_2(t)$, ...,

$r_n(t)$. Each of these processes may be modeled using any of the narrowband models described earlier in this chapter, so essentially the same parametric representation applies. Note that there are two quite distinct sources of Doppler in the megacell channel: One is a Doppler *shift* arising from satellite motion relative to the ground, whereas the other is a Doppler *spread* arising from motion of the mobile relative to the scatterers in the immediate vicinity. It is assumed in this structure that all waves arriving at the mobile are subject to the same Doppler shift, which is a good approximation for scatterers in the near vicinity of the mobile.

3.6 THE FUTURE

This brief chapter provides a speculative and highly personal view of the likely developments in future understanding and treatment of the wireless communication channel. Although the fundamental properties of radio waves that arise from Maxwell's equations have been well understood for almost a century, the application of these properties to practical wireless systems is an evolving field. The points raised in this chapter should indicate some of the key future challenges.

3.6.1 Intelligent Antennas

To meet the changing needs of various wireless systems, antenna structures developed from the generic families of antennas described in the rest of this book are being developed on an ongoing basis. A more radical shift, however, is emerging from the possibilities of merging the functionality of digital signal processing with multiple antenna elements. This creates *intelligent antennas,* whose characteristics are varied over time to optimize the antenna characteristics with respect to specific system goals, such as coverage, capacity, or quality. Intelligent antennas include the diversity and adaptive antenna concepts as special cases, but they can also merge the concepts of equalizers to overcome wideband effects, adaptive matching to improve power delivery, and interference cancellation to enhance system capacity. Initially these concepts are being applied primarily at base stations, where the extra processing and relatively large numbers of antenna elements are relatively easily accommodated, but developments in compact multielement antenna structures and available signal processing power will enable intelligent antennas to be a standard feature of future mobile terminals as well. Ultimately, an understanding of how these structures operate in conjunction with the characteristics of the propagation channel will enable increases in user densities and channel data rates that are many orders of magnitude beyond those possible with today's technology.

3.6.2 Multidimensional Channel Models

For the most part, this chapter examines the variations of the channel in space, time, and polarization as separate topics. However, future systems will increasingly require knowledge of the *joint* behavior of channels with respect to these variables in order to optimize their performance, particularly to support large numbers of users with high data rates. A clear example is the joint angle-of-arrival/time-of-arrival scattering maps described in [1]. This will require the creation of new models that account for the correlations between various channel parameters in detail, and which permit the design, characterization, and verification of advanced processing systems such as time-space beamforming or multiple-use detectors.

3.6.3 High-Resolution Data

The demand for high-resolution data for use as input to propagation predictions has increased substantially in recent years. This has coincided with the expansion in the use of small macrocells and microcells, which require a greater level of detail for useful propagation predictions. The data available include terrain, building height data, vector building data, land usage categories, and meteorological information to increasingly fine resolution and high accuracy. Nevertheless, the costs of such data are still prohibitively high for many applications when compared with making measurements. These costs are expected to fall over time as the demand increases further and the technology for acquiring such data becomes cheaper.

3.6.4 Analytical Formulations

There is an ongoing need for the solution of basic electromagnetic problems in order to permit propagation models to be based on sound physical principles. These are essential in order to gain value from the high-resolution input data described above. Particular needs are in the consideration of three-dimensional problems involving diffraction and rough scattering from multiple obstacles. Such formulations will permit rapid evaluation compared to brute-force numerical techniques and allow predictions to make the best possible use of any available data.

3.6.5 Physical-Statistical Channel Modeling

Section 3.5.5 described a physical-statistical propagation prediction methodology within the context of mobile satellite systems. These methods are also attractive in terrestrial applications, particularly when full deterministic input data may be too expensive for a given application, or when the channel is randomly time-varying due to the motion of vehicles, people, or other scatterers. It is expected that physical-statistical models will increasingly be seen as an appropriate compromise between the accuracy and applicability of deterministic physical models and the coarse but rapid results produced by empirical models.

3.6.6 Real-Time Channel Predictions

In the past, system designers made predictions of propagation parameters and system performance as part of the initial design process for a new system. However, this approach requires that considerable fade margins be included, leading to system operation that is far from optimum at any particular point in time. Significant performance gains are possible if the parameters of the wireless system (such as power level, modulation and coding rates, antenna patterns, channel reuse) are permitted to evolve over time to meet changing constraints resulting from the behavior of the users and the propagation channel. This can best be done by allowing the system to maintain an evolving model of the propagation characteristics with which it can test the likely best choice of parameters. Such models are likely to take a very different form from those currently used in non real-time situations, because they will necessarily need to run much faster and also to integrate both theoretical predictions and measurements.

3.6.7 Overall

The fast-moving developments in the field of wireless communications imply that the state of the art may not be described by this chapter by the time it is read. It is hoped, however, that by focusing on fundamental physical mechanisms and a wide range of wireless system types, the book has been able to give an indication of methodologies for analyzing and understanding antennas and propagation that will be useful for many years to come. For more information on the latest developments in the areas listed above the reader is invited and encouraged to visit the following web site: http://www.ee.surrey.ac.uk/CCSR/AP/.

REFERENCES

[1] Saunders, S. R., *Antennas and Propagation for Wireless Communication Systems*, John Wiley, Chichester, 1999.

[2] Parsons, J. D., *The Mobile Radio Propagation Channel*, John Wiley, Chichester, 1992.

[3] Egli, J. J., "Radio Propagation above 40MC over Irregular Terrain," *Proc. IRE*, pp. 1383–1391, 1957.

[4] Okumura, Y., et al., "Field Strength and Its Variability in VHF and UHF Land Mobile Radio Service," *Rev. Electr. Commun. Lab.*, Vol. 16, pp. 825–873, 1968.

[5] Hata, M., "Empirical Formula for Propagation Loss in Land Mobile Radio Services," *IEEE Trans. Vehicular Technology*, Vol. VT-29, pp. 317–325, 1980.

[6] International Telecommunication Union, "ITU-R Recommendation P. 529-2: Prediction Methods for the Terrestrial Land Mobile Service in the VHF and UFH Bands," Geneva, 1997.

[7] COST-207 Management Committee, *Digital land mobile radio communications*, Final Report, Commission of the European Communities, Luxembourg, 1989.

[8] Kozono, S., and K. Watanabe, "Influence of Environmental Buildings on UHF Land Mobile Radio Propagation," *IEEE Trans. Comm.*, Vol. 25, No. 10, pp. 1133–1143, 1977.

[9] COST 231 Committee, "Digital Mobile Radio: COST 231 View on the Evolution Towards 3rd Generation Systems," Final Report, European Commission-COST Telecommunications, Brussels, Belgium, 1998.

[10] Ibrahim, M. F., and J. D. Parsons, "Signal Strength Prediction in Built-Up Areas," *Proc. IEE*, Vol. 130F, Vol. 5, pp. 377–384, 1983.

[11] Edwards, R., and J. Durkin, "Computer Prediction of Service Areas for VHF Mobile Radio Metworks," *Proc. IEE*, Vol. 116, No. 9, pp. 1493–1500, 1969.

[12] Huish, P. W., and E. Gurdenli, "Radio Channel Measurements and Prediction for Future Mobile Radio Systems," *British Telecom Tech. J.*, Vol. 6, No. 1, pp. 43–53, 1988.

[13] Allsebrook, K., and J. D. Parsons, "Mobile Radio Propagation in British Cities at Frequencies in the VHF and UHF Bands," *IEEE Trans. Vehicular Technology*, Vol. VT-26, No. 4, pp. 95–102, 1977.

[14] Delisle, G. Y., et al., "Propagation Loss Prediction: A Comparative Study with Application to the Mobile Radio Channel," *IEEE Trans. Vehicular Technology*, Vol. 26, No. 4, pp. 295–308, 1985.

[15] Ikegami, F., T. Takeuchi, and S. Yoshida, "Theoretical Prediction of Mean Field Strength for Urban Mobile Radio," *IEEE Trans. Ant. Prop.*, Vol. 39, No. 3, pp. 299–302.

[16] Vogler, L. E., "An Attenuation Function for Multiple Knife-Edge Diffraction," *Radio Science*, Vol. 17, No. 6, pp. 1541–1546, 1982.

[17] Saunders, S. R., and F. R. Bonar, "Prediction of Mobile Radio Wave Propagation over Buildings of Irregular Heights and Spacings," *IEEE Trans. Antennas and Propagation*, Vol. 42, No. 2, pp. 137–144, February 1994.

[18] Saunders, S. R., and F. R. Bonar, "Explicit Multiple Building Diffraction Attenuation Function for Mobile Radio Wave Propagation," *Electron. Lett.*, Vol. 27, No. 14, pp. 1276–1277, July 1991.

[21] Walfisch, J., and H. L. Bertoni, "A Theoretical Model of UHF Propagation in Urban Environments," *IEEE Trans. Antennas & Propagation*, Vol. AP-36, No. 12, pp. 1788–1796, Dec. 1988.

[22] Greenstein, L. J., et al., "Microcells in Personal Communication Systems," *IEEE Communications Magazine*, Vol. 30, No. 12, pp. 76–88, 1992.

[23] Sarnecki, J., et al., "Microcell Design Principles," *IEEE Communications Magazine*, Vol. 31, No. 4, pp. 76–82, 1993.

[24] Madfors, M., et al., "High Capacity with Limited Spectrum in Cellular Systems," *IEEE Communications Magazine*, Vol. 35, No. 8, pp. 38–46, 1997.

[25] Green, E., "Radio Link Design for Microcellular Systems," *British Telecom Tech. J.*, Vol. 8, No. 1, pp. 85–96, 1990.

[26] Harley, P., "Short Distance Attenuation Measurements at 900 MHz and 1. 8 GHz Using Low Antenna Heights for Microcells," *IEEE J. Selected Areas in Communinications*, Vol. 7, No. 1, pp. 5–11, 1989.

[27] Chia, S. T. S., "Radiowave Propagation and Handover Criteria for Microcells," *British Telecom Tech. J.*, Vol. 8, No. 4, pp. 50–61, 1990.

[28] Bultitude, R. J. C., and D. A. Hughes, "Propagation Loss at 1. 8GHz on Microcellular Mobile Radio Channels," *Proc. 7th IEEE Int. Symp. Personal, Indoor and Mobile Radio Communications, PIMRC '96*, Taipei, Taiwan, Oct. 1996, pp. 786–790.

[29] Xia, H. H., et al., "Radio Propagation Characteristics for Line-of-Sight Microcellular and Personal Communications," *IEEE Trans. Antennas and Propagation*, Vol. 41, No. 10, pp. 1439–1447, 1993.

[30] Dersch, U., and E. Zollinger, "Propagation Mechanisms in Microcell and Indoor Environments," *IEEE Trans. Vehicular Technology*, Vol. 43, No. 4, pp. 1058–1066, 1994.

[31] Goldsmith, A. J., and L. J. Goldsmith, "A Measurement-Based Model for Predicting Coverage Areas of Urban Microcells," *IEEE J. Selected Areas in Communications*, Vol. 11, No. 7, pp. 1013–1023, 1993.

[32] Maciel, L. R., and H. L. Bertoni, "Cell Shape for Microcellular Systems in Residential and Commercial Environments," *IEEE Trans. Vehicular Technology*, Vol. 43, No. 2, pp. 270–278, 1994.

[33] Erceg, V., A. J. Rustako, and P. S. Roman, "Diffraction Around Corners and Its Effects on the Microcell Coverage Area in Urban and Suburban Environments at 900-MHz, 2-GHz, and 6-GHz," *IEEE Trans. Vehicular Technology*, Vol. 43, No. 3, Pt. 2, pp. 762–766, 1994.

[34] Dehghan, S., and R. Steele, "Small Cell City," *IEEE Communications Magazine*, Vol. 35, No. 8, pp. 52–59, Aug. 1997.

[35] Berg, J. E., "A Recursive Method for Street Microcell Path Loss Calculations," *PIMRC '95*, Vol. 1, pp. 140–143, 1995.

[36] Feuerstein, M. J., et al., "Path Loss, Delay Spread and Outage Models as Functions of Antenna Height for Microcellular System Design," *IEEE Trans. Vehicular Technology*, Vol. 43, No. 3, pp. 487–498, 1994.

[37] Arnold, H. W., D. C. Cox, and R. R. Murray, "Macroscopic Diversity Performance Measured in the 800 MHz Portable Radio Communications Environment," *IEEE Trans. Antennas and Propagation*, Vol. 36, No. 2, pp. 277–280, 1988.

[38] Keenan, J. M., and A. J. Motley, "Radio Coverage in Buildings," *British Telecom Tech. J.*, Vol. 8, No. 1, pp. 19–24, Jan. 1990.

[39] International Telecommunication Union, "ITU-R Recommendation P. 1238: Propagation Data and Prediction Models for the Planning of Indoor Radiocommunication Systems and Radio Local Area Networks in the Frequency Range 900 MHz to 100 GHz," Geneva, 1997.

[40] Akerberg, D., "Properties of a TDMA Picocellular Office Communication System," IEEE Global Telecommunications Conference Globecom '88, Hollywood, pp. 1343–1349, 1988.

[41] Turkmani, A. M. D., and A. F. Toledo, "Propagation Into and Within Buildings at 900, 1800 and 2300 MHz," *IEEE Vehicular Technology Conference*, 1992.

[42] Turkmani, A. M. D., J. D. Parsons, and D. G. Lewis, "Radio Propagation into Buildings at 441, 900 and 1400 MHz," *Proc. 4th Int. Conf. on Land Mobile Radio*, 1987.

[43] Walker, E. H., "Penetration of Radio Signals into Buildings in Cellular Radio Environments," *IEEE Vehicular Technology Society Conference*, 1992.

[44] Catedra, M. F., et al., "Efficient Ray-Tracing Techniques for Three Dimensional Analyses of Propagation in Mobile Communications: Application to Picocell and Microcell Scenarios," *IEEE Antennas and Propagation Magazine*, Vol. 40, No. 2, pp. 15–28,1998.

[46] Balanis, C. A., *Advanced Engineering Electromagnetics*, John Wiley & Sons, New York, 1989.

[47] Honcharenko, W., et al., "Mechanisms Governing UHF Propagation on Single Floors in Modern Office Buildings," *IEEE Trans. Vehicular Technology*, Vol. 41, No. 4, pp. 496–504, 1992.

[48] Hashemi, H., "Impulse Response Modeling of Indoor Radio Propagation Channels," *IEEE J. Selected Areas in Communications*, Vol. 11, No. 7, 1993.

[49] European Telecommunication Standards Institute, "Selection Procedures for the Choice of Radio Transmission Technologies of the Universal Mobile Telecommunications System (UMTS)," DTR/SMG-50402, 1997.

[50] Hashemi, H., "The Indoor Radio Propagation Channel," *Proc. IEEE*, Vol. 81, No. 7, pp. 943–967, 1993.

[51] Hashemi, H., and D. Tholl, "Statistical Modeling and Simulation of the RMS Delay Spread of the Indoor Radio Propagation Channel," *IEEE Trans. Vehicular Technology*, Vol. 43, No. 1, pp. 110–120, 1994.

[52] Rappaport, T. S., S. Y. Seidel, and K. Takamizawa, "Statistical Channel Impulse Response Models for Factory and Open Plan Building Radio Communication System Design," *IEEE Trans. Communications*, Vol. 39, No. 5, pp. 794–806, 1991.

[53] Saleh, A. A. M., et al., "Distributed Antennas for Indoor Communication," *IEEE Trans. Communications*, Vol. COM-35, No. 11, pp. 1245–1251, 1987.

[54] Hashemi, H., "Impulse Response Modeling of Indoor Radio Propagation Channels," *IEEE J. Selected Areas in Communications*, Vol. 11, No. 7, 1993.

[55] Evans, B. G., ed., *Satellite Communication Systems*, 3rd ed., IEE, London, 1999.

[56] Pattan, B., *Satellite-Based Cellular Communications*, McGraw-Hill, New York, 1998.

[57] Vogel, W. J., and J. Goldhirsh, "Tree Attenuation at 869 MHz Derived from Remotely Piloted Aircraft Measurements," *IEEE Trans. Antennas and Propagation*, Vol. AP-34, No. 12, pp. 1460–1464, 1986.

[58] International Telecommunication Union, "ITU-R Recommendation P. 681-3: Propagation Data Required for the Design of Earth-Space Land Mobile Telecommunication Systems," Geneva, 1997.

[59] Goldhirsh, J., and W. J. Vogel, "Propagation Effects for Land Mobile Satellite Systems: Overview of Experimental and Modeling Results," NASA Reference Publication 1274, Feb. 1992.

[60] Loo, C., "A Statistical Model for a Land Mobile Satellite Link," *IEEE Trans. Vehicular Technology*, Vol. VT-34, No. 3 , pp. 122–127, 1985.

[61] Corazza, G. E., and F. Vatalaro, "A Statistical-Model for Land Mobile Satellite Channels and Its Application to Nongeostationary Orbit Systems," *IEEE Trans. Vehicular Technology*, Vol. 43, No. 3, Pt. 2, pp. 738–742, 1994.

[62] Hwang, S., et al., "A Channel Model for Nongeostationary Orbiting Satellite System," *47th IEEE International Vehicular Technology Conference*, Phoenix, Arizona, May 5–7, 1997.

[63] Lutz, E., et al., "The Land Mobile Satellite Communication Channel-Recording, Statistics and Channel Model," *IEEE Trans. Vehicular Technology*, Vol. 40, No. 2, pp. 375–385, May 1991.

[64] Ahmed, B., et al., "Simulation of 20 GHz Narrowband Mobile Propagation Data Using N-state Markov Channel Modeling Approach," *10th International Conference on Antennas and Propagation*, Edinburgh, Apr. 14–17, 1997, pp. 2.48–2.53.

[65] Lutz, E., "A Markov Model for Correlated Land Mobile Satellite Channels," *Int. J. Satellite Communications*, Vol. 14, 1996.

[66] Vucetic, B., and J. Du, "Channel Model and Simulation in Satellite Mobile Communication Systems," *IEEE Trans. Vehicular Technology*, Vol. 10, pp. 1209–1218, 1992.

[67] van Dooren, G. A. J., et al., "Electromagnetic Field Strength Prediction in an Urban Environment: A Useful Tool for the Planning of LMSS," *IMSC '93*, June 16–18, 1993, pp. 343–348.

[58] Barts, R. M., and W. L. Stutzman, "Modeling and Measurement of Mobile Satellite Propagation," *IEEE Trans. Antennas and Propagation*, Vol. 40, No. 4, pp. 375–382, April 1992.

[59] Saunders, S. R., and B. G. Evans, "A Physical Model of Shadowing Probability for Land Mobile Satellite Propagation," *Electron. Lett.*, Vol. 32, No. 17, pp. 1548–1589, Aug. 15, 1996.

[70] Tzaras, C., B. G. Evans, and S. R. Saunders, "A Physical-Statistical Analysis of the Land Mobile Satellite Channel," *Electron. Lett.*, Vol. 34, No. 13, p. 1355–1357, 1998.

[71] Parks, M. A. N., B. G. Evans, and G. Butt, "High Elevation Angle Propagation Results Applied to a Statistical Model and an Enhanced Empirical Model," *Electron. Lett.*, Vol. 29, No. 19, pp. 1723–1725, Sep. 1993.

[72] Jahn, A., H. Bischl, and G. Heiss, "Channel Characterization for Spread-Spectrum Satellite-Communications," *IEEE Fourth Int. Symp. Spread Spectrum Techniques & Applications*, Vols. 1–3, Chap. 261, pp. 1221–1226, 1996.

[73] Parks, M. A. N., et al., "Simultaneous Wideband Propagation Measurements for Mobile Satellite Communications Systems at L- and S-bands," *Proc. 16th Int. Communications Systems Conference (ICSSC96)*, Washington, D.C., Feb. 25–29, 1996, pp. 929–936.

[74] Parks, M. A. N., S. R. Saunders, and B. G. Evans, "Wideband Characterisation and Modeling of the Mobile Satellite Propagation Channel at L and S Bands," *10th Int. Conference on Antennas and Propagation*, Edinburgh, Apr. 14–17, 1997, pp. 2.39–2.43.

[75] Tzaras, C., S. R. Saunders, and B. G. Evans, "A Physical-Statistical Propagation Model for Diversity in Mobile Satellite PCN," *48th IEEE Int. Vehicular Technology Conference*, Ottawa, Canada, May 18–21, 1998, pp. 525–529.

[76] Bischl, H., M. Werner, and E. Lutz, "Elevation-Dependent Channel Model and Satellite Diversity for NGSO S-PCNS," *Proc. IEEE 46th Vehicular Technology Conference*, Vols. 1–3, Chap. 385, pp. 1038–1042, 1996.

[77] Meenan, C., et al., "Availability of First Generation Satellite Personal Communication Network Service in Urban Environments," *Proc. IEEE Int. Vehicular Technology Conference*, 1998, pp. 1471–1475.

[78] Lin, H. P., R. Akturan, and W. J. Vogel, "Photogrammetric Prediction of Mobile Satellite Fading in Roadside Tree-Shadowed Environment," *Electron. Lett.*, Vol. 34, No. 15, 1998, pp. 1524–1525.

Chapter 4

Land Mobile Antenna Systems I: Basic Techniques and Applications

K. Kagoshima and T. Taga

4.1 ANTENNAS

Land mobile communication systems employ base stations and mobile stations. These stations employ different types of antennas, and the design criteria also differ. Figure 4.1 shows important items to be considered in designing a base station antenna. Although antenna design in the narrow sense means electrical design, in reality it includes a wider area, and it is important to derive the antenna hardware specifications from system requirements. In order to determine the hardware specifications, it is necessary to perform an evaluation that compares electrical and mechanical characteristics and the trade-off between performance and cost, as shown in Figure 4.2. Performance and cost considerations are sometimes the first step, while determination of the electrical and mechanical design is the second step [1].

In designing the practical antenna, it is important to assess how the antenna hardware will be installed after manufacturing. For base station and mobile station antennas (e.g., vehicular antennas), installation fees may be larger than the cost of the antennas themselves. For this reason, it is important to consider not only the reduction of construction costs, but also the creation of an antenna design which makes installation easy.

Figure 4.3 shows some items to be considered in designing a mobile station antenna. Mobile station antennas are classified into two categories: antennas for mobile mounting, such as on vehicles, and antennas for mounting on portable radio equipment. Mobile

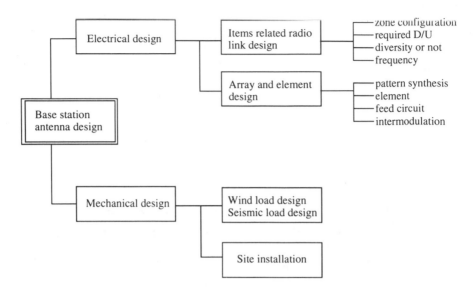

Figure 4.1 Key items in designing base station antennas.

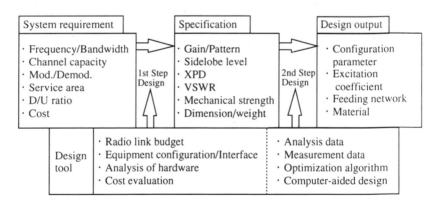

Figure 4.2 Steps of antenna design.

station antennas are more independent of system parameters than are base station antennas, and should be designed for easy handling and customer convenience.

Because mobile communication antennas are not used in free space, but within a multipath environment, antenna gain or radiation patterns specified for the system design should be designed for their environment. Therefore, it is not necessarily meaningful for an antenna to have superior performance in free space if its performance seriously degrades in a multipath environment. Although it has been pointed out that the prediction of antenna

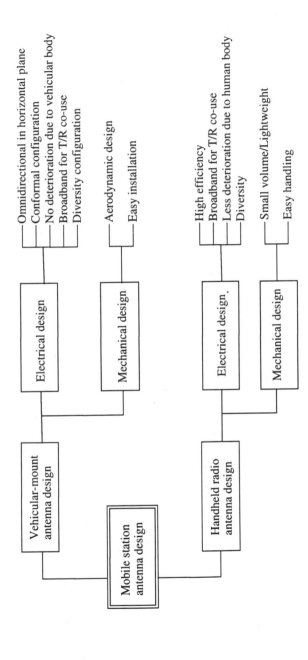

Figure 4.3 Key items in designing mobile station antennas.

performance should be carried out under the multipath environment, a method to predict antenna gain and radiation patterns has only recently been proposed. Taga and Ebine [2,3] developed the prediction method by assuming the amplitude distribution of the incoming wave is a Gaussian distribution, while the plane wave is incident in free space. Since this was discussed in more detail in Chapter 2, further description is abbreviated here.

Analysis and measurement technologies for land mobile communication antennas are also very important, as is true for other kinds of antennas. Since mobile antennas are usually made of wires or metal plates, the method of moments is highly effective for analyzing these antennas. An antenna, consisting of metal plates or box(es), can be analyzed by simulating the box(es) with a multiple-wire grid [4], or by dividing the surface of the box(es) into small triangular patches [5]. The method of moments has been modified and extended to analyze antennas made of dielectric materials, and computer simulations can be effectively carried out in lieu of direct experiments. Moreover, the progress of computer technology has allowed remarkable advances in time-domain analysis, such as transmission line matrix method (TLM) and finite difference time-domain (FDTD). This has allowed the development of more complex antennas and models, which more closely emulate real antennas. Thus, many valuable results have been reported [6-8]. Several tutorial books on antenna analysis have been published, so interested readers can refer to these for more details [9,10]. As for measurement, mobile communication antennas should be measured in a real propagation environment. Therefore, evaluation of antenna performance under such circumstances is more difficult and laborious than is usual for communication antennas used in line of sight.

4.2 PROPAGATION PROBLEMS

Propagation in mobile communication occurs within a diffraction region and is not free-space propagation. Propagation path loss in free space is simple and is proportional to the square of the distance, but in mobile communications it depends on various factors, such as the propagation environment, antenna height, and frequency. It is important to define such terms clearly in order to discuss propagation problems.

The height of a base station antenna, the h_{te}, is defined as $h_{te} = h_{ts} - h_{ga}$, as shown in Figure 4.4, where h_{ts} is the sea level height of the base station antenna, and h_{ga} is the averaged sea level height of the geometrical profile within 3–15 km around the base station [11]. The height of a mobile station antenna is defined as the height measured from the ground level of the mobile station.

Mobile propagation environments are very complex, but are roughly categorized into three kinds of environment [11].

- Open area: There are few obstacles, such as high trees or buildings, in the propagation path. Roughly speaking, free spaces of about 300m to 400m in length lie between the base and mobile stations.

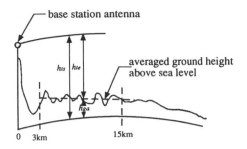

Figure 4.4 Definition of effective height of a base station antenna. (After [11].)

- Suburban area: There are some obstacles around the mobile stations, but they are not dense. Roughly speaking, it is an area of trees and low houses.
- Urban area: There are many buildings or other high structures. Roughly speaking, it is an area with high, close buildings, or a densely mixed area of buildings and high trees.

Propagation-Path Loss Characteristics

Although there are many aspects of mobile propagation, path loss characteristics are most closely related to the antenna design. Path loss in a mobile propagation environment is larger than that in free space due to the existence of various obstacles. Figure 4.5 shows measured field strength versus distance from the antenna with the parameter of base station antenna height. These measurements were recorded at a frequency of 453 MHz with vertical polarization and a mobile station height of 3m [11].

4.3 BASE STATION ANTENNA TECHNIQUES

4.3.1 Antenna System Requirements

The role of antennas in mobile communication systems is to establish a radio transmission line between radio stations, at least one of which is moving. There are two types of mobile communication systems: one where a transmitter and receiver communicate directly, and the other where they communicate through a base station. It is the latter type that has advanced around the world in recent years. Examples include automobile telephone systems, portable telephone systems and MCA, which is a multichannel access system for private use. Automobile telephone and portable telephone systems adopt a cellular structure, and the relation between system requirements and the necessary antenna technology is illustrated in Figure 4.6. In order for the base station to communicate with the mobile stations located in the service area, radio wave energy must be radiated uniformly inside

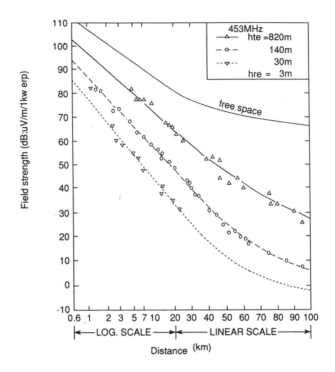

Figure 4.5 Measured mean value of field strength versus distance from a base station (f = 453 MHz, urban area). (After [11].)

the area. Moreover, antenna gain should be as high as possible. Since the width of the service area is already specified, antenna gain cannot be increased by narrowing the beam in the horizontal plane. Therefore, it is necessary to narrow the antenna beam in the vertical plane to increase gain; a vertically arrayed linear array antenna is effective to achieve this. Normal cellular systems use antennas with a gain from 7 to 15 dBd as the base station antennas.

In order for the base station antenna to communicate with many mobile stations simultaneously, multiple channels must be handled. This requires wide-frequency characteristics and a function for branching and/or combining the channels. For example, the base stations of a Japanese cellular system using the 800-MHz band use one antenna for both transmitting and receiving. The required bandwidth of the antenna is more than 7% where the specified voltage standing wave ratio (VSWR) is less than 1.5. Moreover, if the antenna is shared by several systems (e.g., an analog land mobile telephone and a digital land mobile telephone), a wider antenna frequency bandwidth is required. To comply with Radio Regulation allocations, the frequency bandwidth for land mobile around 800 MHz runs from 810 to 960 MHz. Therefore, in order to cover this bandwidth with one antenna, a 17% bandwidth is necessary. When the antenna both transmits and

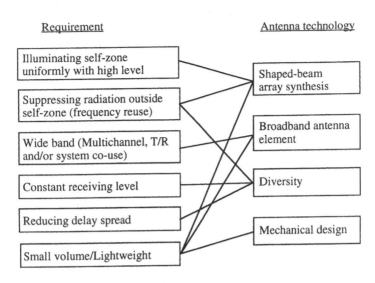

Figure 4.6 System requirement and antenna technology.

receives, passive intermodulation arises and this increases interference. This problem is discussed in Section 4.3.5.

Due to the rapid growth of demand, the lack of communication channels has become a serious problem for metropolitan areas in the United States, Europe, and Japan, and technologies for effective frequency reuse are strongly needed. Although the cellular system has an advantage in terms of reusing frequency, its efficiency significantly depends on the radiation pattern of the base station antenna. Technologies for main-beam tilting and beam shaping have been developed and effectively contribute to frequency reuse. These technologies are described in detail in Section 4.3.3.

One of the most common features of mobile communication is that the base station and the mobile station do not fall within line of sight of each other. Moreover, the mobile station moves within a complex propagation environment. As a result, fading occurs constantly at the base and mobile stations, and the receiving levels may fluctuate by 10 dB or more. If system design takes the minimum receiving level into account, the load on the devices is excessive and system cost becomes too high.

One technology for overcoming fading is diversity reception, which has been studied since the 1960s. Its effectiveness has been confirmed both experimentally and theoretically. Reception diversity was first used commercially in the Advanced Mobile Phone System (AMPS) of the United States in 1982. It is also used in the Large-Capacity System of Japan. This technology is also described in detail in Section 4.3.4.

Various base station antenna technologies have been developed to enhance the system through the achievement of high performance and new functions. Figure 4.7 shows the historical trends of base station antenna technology.

182

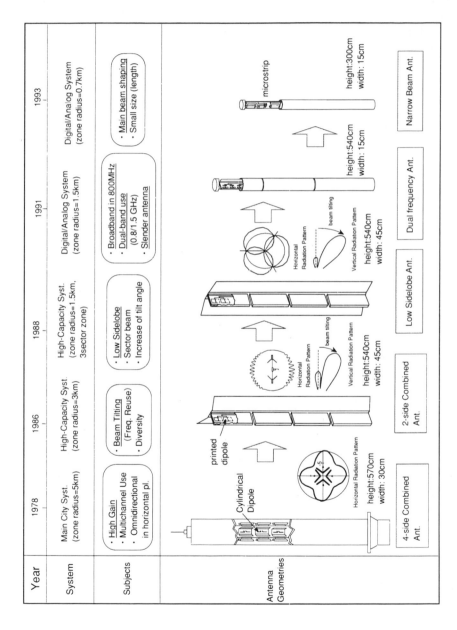

Figure 4.7 Historical trends in base station antenna technology.

4.3.2 Types of Antennas

Base station antenna configurations depend on the size and shape of the service area and the number of cells and channels. In a private mobile communication system whose service area is small, the base station antenna is as small as the vehicular-mount antenna used in automobile telephone systems. If the service area is limited within a restricted angle in the horizontal plane, a corner reflector antenna is often used. When the service area is wide, as in a pager system, maritime telephone system, or aeronautical telephone system, a linear array antenna, which has large directivity in the vertical plane, is used. It is generally uniformly excited, and examples of this type are shown in Figure 4.8. The linear array antenna is often used in cellular systems, too. Since the base station antenna of a cellular system must simultaneously handle 30 to 60 channels, it is important to realize a feeding circuit with low loss. NTT used an omnidirectional pattern antenna in its early commercial system. It consisted of four sector beam antennas combined with 3-dB hybrid circuits [12]. A multibeam antenna in which a Butler matrix is used for the feeding circuit was also introduced, but was not used commercially.

In the early stage of cellular system development, the length of the base station antenna was determined by the required gain. To achieve higher gain, an array antenna was usually excited uniformly. However, in order to reuse frequency more effectively, cells should be subdivided. Given this situation, it is more important for the base station antenna to have a larger ratio of desired-to-undesired signal strength (D/U ratio) than to have high antenna gain. Therefore, main-beam tilting, either mechanically or electrically, has been adopted throughout the world. Experiments have determined that cochannel interference can be reduced by about 10 dB, as shown in Figure 4.9, and it is recognized that beam tilting is essential for enhancing frequency reuse. Moreover, as shown in Figure 4.10, sidelobe suppression adjacent to the main beam, achieved by synthesizing appropriate array antenna patterns, is also effective to decrease the frequency reuse distance.

As for diversity antennas, space diversity, in which two antennas are separated by 5 to 10 wavelengths, is commonly used. A special diversity antenna, such as the pattern

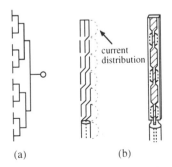

(a)   (b)

Figure 4.8 Uniformly excited array antennas: (a) parallel feed; (b) series feed.

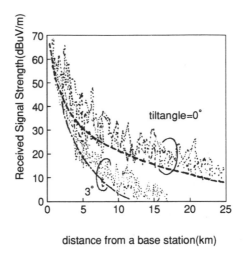

distance from a base station(km)

Figure 4.9 Effect of beam tilting for frequency reuse: — · — and ------- calculated value under the assumption of free-space pattern. (After [13].)

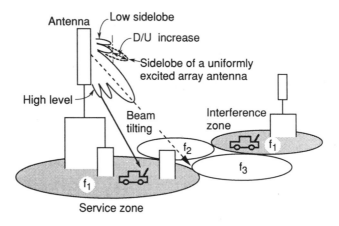

Figure 4.10 Effect of sidelobe reduction for frequency reuse.

diversity antenna [13] or the polarization diversity antenna [13], have also been developed and are being used in commercial systems as base station antennas. Figure 4.11 categorizes the types of base station antennas from the viewpoints of functions and antenna characteristics.

4.3.3 Design of Shaped-Beam Antennas

The shaped-beam technique, which enhances spatial frequency reuse, is described in this section. A base station antenna in a cellular system is required to radiate energy at as low

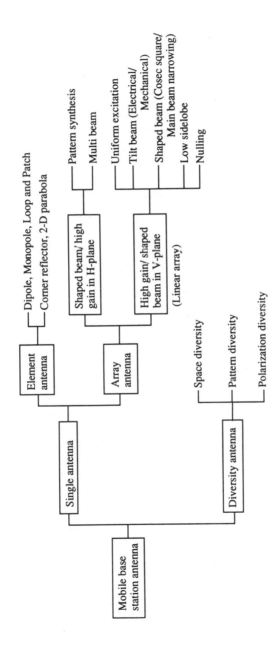

Figure 4.11 Classification of base station antennas.

a level as possible toward the cell where the same frequency is used; conversely, it is required to illuminate the service area at as high a level as possible. There are two types of shaped-beam antennas. One shapes its radiation pattern in the horizontal plane so that a sector beam is needed, and the other shapes the pattern in the vertical plane so that a cosecant beam is desired.

Strictly speaking, main-beam tilt is not really a shaped-beam technique, but since the purpose is the same, the simplified configuration of a beam-tilting antenna and its measured performance are presented. First, the relation between frequency reuse distance and antenna radiation pattern is explained. Next, the design of a sector beam antenna is described, and finally, the shaped-beam antenna design in the vertical plane for a linear array antenna is described. This section limits discussion on shaped-beam design to the specific application of cellular mobile systems. Readers interested in array antenna pattern synthesis or the numerical technique to obtain the excitation coefficient of the array, should refer to the publications listed in the reference section.

Frequency Reuse Distance

From the cellular system shown in Figure 4.12 [14], the following formula can be derived.

$$N = 1/3(D/R)^2 \tag{4.1}$$

where N is the number of cells, R is the radius of a cell, and D is the distance between the centers of adjacent cells [15]. The worst value of the CIR, expressed in decibels, appears at the edge of the cell and is given by:

$$\text{CIR} = -\alpha \cdot 10\log\{R/(D-R)\} \tag{4.2}$$

where α is the attenuation constant of the path loss characteristic curve. In (4.2), the antenna pattern difference between the desired wave direction (θ_d, ψ_d) and the interference wave direction (θ_i, ψ_i), $(C/I)_{\text{ANT}}$, expressed in decibels, is not included.

Therefore, the total CIR of antenna pattern $P(\theta, \psi)$, expressed in decibels, is obtained by:

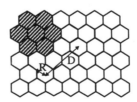

Figure 4.12 Zone configuration in cellular system (cluster number = 7). (From [14]. © 1991 IEICE.)

$$\text{CIR} = -\alpha \cdot 10\log\{R/(D-R)\} + (C/I)_{\text{ANT}} \tag{4.3}$$

$$(C/I)_{\text{ANT}} = P(\theta_d, \psi_d)/P(\theta_i, \psi_i) \tag{4.4}$$

Figure 4.13 shows the relation between D/R and CIR when $(C/I)_{\text{ANT}}$ is varied and $\alpha = 3$ [14]. From this figure, the significance of $(C/I)_{\text{ANT}}$ can be understood.

Figure 4.14 shows a typical radiation pattern of a base station antenna in the vertical plane and radiation levels at the edge of the cell and toward the interference direction. If the two directions are separated enough compared to 50% of the main beamwidth of the antenna, interference points exist in the sidelobe region of the radiation pattern, and the sidelobe level must be decreased to increase $(C/I)_{\text{ANT}}$.

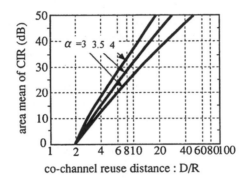

Figure 4.13 Relation between D/R and CIR. (From [14]. © 1991 IEICE.)

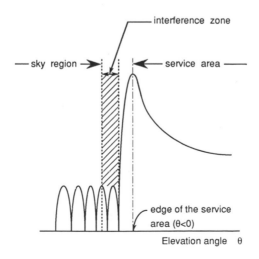

Figure 4.14 Typical radiation pattern of a base station antenna that is effective in reducing interference.

If the interference direction approaches the desired direction (N becomes small), interference lines may exist in the main beam as well as the desired direction. To increase $(C/I)_{ANT}$ in this case, it is necessary to increase the length of the antenna or to narrow the main beamwidth with beam-shaping techniques without increasing the length of the antenna.

When sector beams with the angle of θ_s, are used in place of omnidirectional (circular) beams, the interference distance $(D - R)$ in (4.2) lengthens to $(D^2 + R^2 - 2DR \cos \theta_s)^{1/2}$, as shown in Figure 4.15. These are more advantageous in frequency reuse than omnidirectional beams.

Sector Beams

The horizontal pattern of a base station antenna is usually omnidirectional. However, a sector beam may effectively cover the service area if the service area is not a circle, but rather a hemicircle or a sector. Moreover, in a cellular system, the frequency reuse distance with a sector zone arrangement is shorter than that with a circular zone arrangement, as described in the previous section. Consequently, the sector zone arrangement is used in several recent automobile telephone systems. A typical sector-beam antenna is of the corner reflector type, while a two-dimensional parabolic reflector antenna fed by two primary radiators has also been developed [16].

A corner reflector antenna has the advantage that the beamwidth can be adjusted by controlling the aperture angle of the reflector. Figure 4.16 shows the fundamental configuration of the corner reflector antenna. In fact, a base station antenna is a vertically arrayed antenna consisting of the elements shown in Figure 4.16. It achieves high directivity

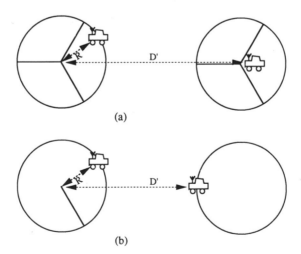

(a)

(b)

Figure 4.15 Interference distance of (a) sector zones and (b) omni zones.

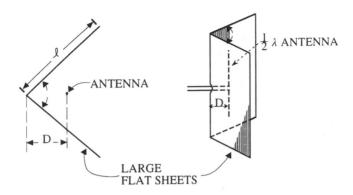

Figure 4.16 Fundamental geometry of a corner reflector antenna.

by narrowing the main beam in the vertical plane and can radiate a shaped radiation beam by controlling the excitation coefficient, as described in the next section. In this section, discussion is confined to a corner reflector antenna with one element for easier understanding.

Figure 4.17 shows the relationship between the aperture angle of the corner reflector and a half-power beamwidth in the horizontal plane, as well as the relationship between the aperture angle and the directivity when the primary radiator is a half-wavelength dipole. Sector beams with beamwidths from 60 to 180 deg can be obtained by setting the aperture angle from 60 to 270 deg [17]. The results shown in Figure 4.17 are somewhat different from those presented by Kraus [18], who assumes an infinite corner width. When the aperture angle is 180 deg, corresponding to a flat plate, the beamwidth becomes

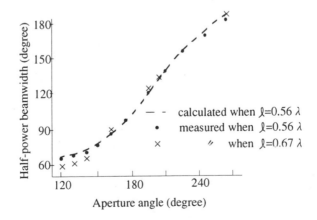

Figure 4.17 Relation between aperture angle and half-power beamwidth of a corner reflector antenna ($f = 900$ MHz, $D = 0.28\lambda$). (After [17].)

approximately 120 deg. If one would like to obtain a sector beam with a beamwidth more than 120, a corner reflector with an aperture angle of more than 180 deg (i.e., the so-called *superior angle corner reflector*) is needed.

If two sector beam antennas of 180 deg beamwidth are combined with moderate spacing (approximately larger than 6λ), an omnidirectional pattern is realized, as shown in Figure 4.18. In this case, large ripples appear in the direction of ±90 deg due to the interference of the two patterns. However, measurements confirm that these ripples disappear in a multipath environment, as shown in Figure 4.18.

Concerning a corner reflector beam antenna, a dual-frequency design method has been reported [19]. To realize a dual-frequency corner reflector antenna, it is necessary to prepare a dual-frequency primary radiator. This is possible with a dipole with a closely spaced parasitic element as shown in Figure 4.19. Such an antenna has the advantage of being very compact. When the primary radiator is shown in Figure 4.19 is used, sector beams of equal beamwidth with frequencies f_1 and f_2 can be obtained by determining the aperture angle α, distance d between reflector and primary radiator, and the width of the corner l. Typical parameter curves are shown in Figure 4.20 for frequencies in the 900- and 1,500-MHz band. An effective parameter for controlling beamwidth is not aperture angle α, but corner width l, and a dual-frequency sector beam with angles from 60 to 150 deg can be realized by adjusting l.

Shaped Beams in the Vertical Plane

When a limited horizontal area is to be illuminated with equal received signal level from an antenna fixed at a certain height, as shown in Figure 4.21, it is known that this can be achieved by a cosecant squared shaped-beam power pattern in the vertical plane. If the path loss is larger than that in free space ($\propto r^{-2}$), as in the case of mobile communication systems, the pth power order of a cosecant shaped-beam power pattern is necessary to

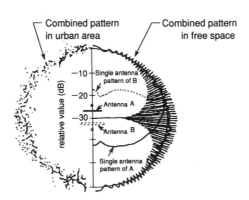

Figure 4.18 Omnidirectional pattern by combining two 180-deg sector beam antennas. (After [13].)

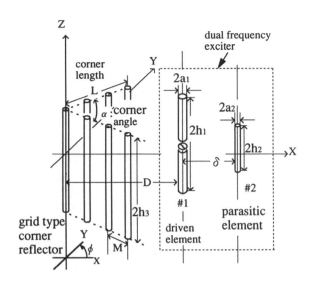

Figure 4.19 Geometry of a dual-frequency corner reflector antenna. (After [19].)

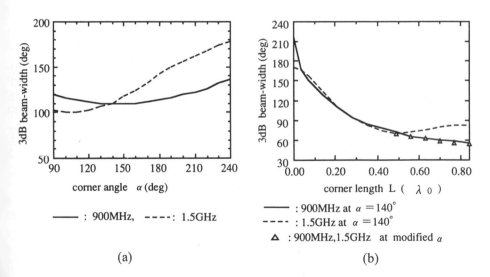

(a)

(b)

Figure 4.20 Beamwidth characteristics of a dual-frequency corner reflector antenna: (a) beamwidth versus corner angle; (b) beamwidth versus corner length. (After [19].)

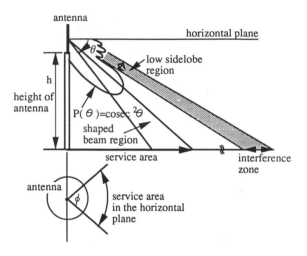

Figure 4.21 Shaped beam for illuminating the service area uniformly with low interference.

achieve equal received signal level at all points in the area. However, the significance of a shaped beam in mobile communication systems, especially cellular systems, may be more in suppressing the radiation toward the cell where the same frequency is reused than for illuminating the self-zone uniformly, as illustrated in Figure 4.22. If frequencies are reused, and the cells are closely packed, a part of the main beam illuminates the reuse cell. Therefore, it may be effective to tilt the main beam down to suppress the inteference, even if the received signal level within the self-zone weakens. By reducing the main beamwidth while maintaining the length of the antenna, it is possible to increase $(C/I)_{\text{ANT}}$.

The remainder of this section describes major techniques and results obtained from shaped beams, with particular emphasis on practical use.

Beam Tilt

The principal idea of the beam tilt-down technique is to tilt the main beam in order to suppress the direction level toward the reuse cell and to increase $(C/I)_{\text{ANT}}$. In this case, the carrier level also decreases in the zone edge. However, the interference level decreases more than the carrier level, so the total $(C/I)_{\text{ANT}}$ increases. This is an advantage from the viewpoint of system design, and this technique is used in most cellular systems in the world. Figure 4.23 shows the comparison between antennas with and without beam tilt, which verify the effectiveness of the beam tilt technique [20]. It can be easily understood from this figure that the distance from the base station inside which the interference level exceeds the threshold level of the system can be significantly reduced. In Japan, beam tilt is achieved electrically by adjusting the excitation coefficient of the array, while in Europe it is mainly achieved mechanically.

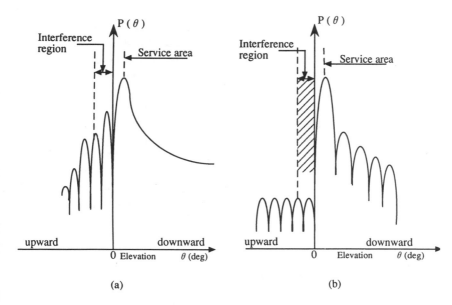

Figure 4.22 Comparison of synthesized pattern: (a) smooth cosecant beam with high sidelobe level; (b) ripple constant beam with low sidelobe level.

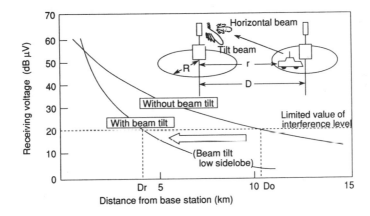

Figure 4.23 Beam tilt effect to reduce the frequency reuse distance.

Rippled Radiation Pattern with Low-Sidelobe Level

As shown in Figure 4.18, the sharp null in the radiation pattern is nearly filled in the urban area due to multipath wave incidence. Noticing this point, Kijima derived a synthesized pattern expression with finite null depth by using a modified Schelknoff's unit circle

[21] and obtained the excitation coefficients of the linear array [22]. Pattern synthesis to achieve the desired pattern shown in Figure 4.24(a) was carried out, and the results are shown in Figure 4.24(b). From this figure, it is understood that the excitation coefficients of the rippled pattern are almost symmetrical, both in amplitude and in phase, and that the difference between the maximum value and minimum value of the excitation amplitude is decreased. This is advantageous in realizing the feeding network.

If a smoothed p th power order cosecant pattern is required to illuminate the self-zone uniformly, the precise difference between the maximum value and minimum excitation coefficients becomes large, and it may be difficult to realize the desired excitation coefficients. As a result, the realized pattern is worse than the correct pattern. On the other hand, when a rippled radiation pattern, whose envelope is $\csc^p \theta$, is synthesized, the excitation coefficient distribution is smoother than the smoothed radiation pattern. The above result is a typical example taking propagation characteristic into account for antenna design.

Shaped Beam with a Locally Suppressed Sidelobe Level

The necessary directions for suppressing sidelobe level to reduce the interference do not exist in all sidelobe regions, but only in limited angle regions. A pattern synthesized with locally suppressed sidelobe levels was first studied in the field of radar [23]. However, since this is located closely to the main beam direction, and sidelobe level is closely related to the main beamwidth, sidelobe suppression must be carried out carefully. As shown in Figure 4.25, Kijima et al. obtained a main beam 30% narrower than that of a uniformly excited array by suppressing only several sidelobes near the main beam and setting the other sidelobes at a comparatively high level [24]. An antenna with this radiation

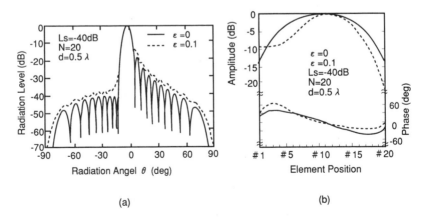

(a)

(b)

Figure 4.24 Synthesized excitation coefficient depending on the pattern shape with or without nulls: (a) synthesized pattern; (b) excitation coefficients. (After [22].)

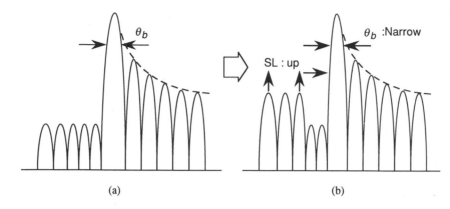

Figure 4.25 Main-beam narrowing technique by controlling sidelobe levels: (a) uniformly suppressed sidelobe pattern; (b) restricted angle low-sidelobe pattern.

pattern can increase the level at the zone edge by approximately 1.5 dB if the interference level is constant, a great advantage in system design.

Dual-Frequency Shaped Beams in a Vertical Plane

The previous section described a dual-frequency shaped beam in the horizontal plane, and the feasibility of creating a dual-frequency shaped beam in the vertical plane with a linear array antenna was presented by Kijima [25]. The difficulty of designing a shaped beam with dual-frequency operation results in a compromise of the performance at each frequency, and Kijima found that the optimum frequency for dual-frequency design tends to occur at lower frequencies.

4.3.4 Diversity Antenna Systems

Effect of Reception Diversity

The effect of reception diversity of the base station was first reported in 1965 by [26]. They showed that fading reduction could be achieved by placing two antennas approximately ten wavelengths apart in the horizontal plane. Figure 4.26 shows the cumulative probability of received level using either one isolated antenna or two antennas with a correlation coefficient of 0, 0.5, and 0.8, respectively [27]. From this it can be understood that the received level at a probability of 1% with the diversity antenna is larger by 8 dB than that with a single antenna.

Although two or more ports are necessary to carry out reception diversity, it significantly reduces fading. As a result, the transmitting power of the mobile station is reduced,

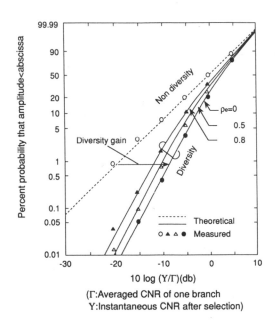

Figure 4.26 Diversity effect versus correlation coefficient: ρ_e is the correlation coefficient assumed in the theoretical prediction. (From [27], © 1991 IEICE.)

and the quality of the transmission is enhanced. This is a great advantage from the total-system point of view. Reception diversity in the base station has been in commercial use in AMPS since 1982 in the United States and in NTT's Large-Capacity System since 1985 in Japan. Since the theory of diversity reception is discussed in detail in Chapter 2 of this book as well as in other publications listed in the reference section, only those items pertaining to the design of base station antennas are described in this section.

Configuration of Base Station Diversity Antennas

Many versions of base station antennas are in use in commercial systems in the United States, England, and Japan [14]. In each base station antenna, sector beam antennas with a 3-dB beamwidth of 120 deg are used. Diversity antennas in the United States and Japan are arranged at angular increments of 120 deg, antennas in England are composed of six sector beam antennas with a 3-dB beamwidth of 60 deg, and at the border line of the sector zone, the port with the higher receiving level is chosen.

There are three types of diversity antenna configurations: space diversity, pattern diversity, and polarization diversity. In these three configurations, space diversity is most commonly used. Since the design method and the characteristics of the space diversity antenna are described in detail in the next section, examples of a pattern diversity antenna and a polarization diversity antenna are briefly introduced in the following.

Figure 4.27 depicts the pattern diversity antenna for an omnizone. This configuration places the two antennas shown in Figure 4.18 at 90 deg to each other [13]. Whereas 180-deg sector beam antennas for synthesizing an omnidirectional pattern are placed apart in space, the centers of the two omnidirectional antennas coincide with the center of the platform, and the spacing between the two antennas is regarded as zero. Therefore, the difference of the received power for each antenna is considered to be the cause of the difference in the radiation pattern. When the 180-deg sector beam antenna spacing is six wavelengths, as in Figure 4.27, measurements found that the correlation coefficient between two antennas is less than 0.2 in urban areas.

Figure 4.28 presents an example of a polarization diversity antenna developed by NTT [28]. It appears as a single antenna and has the advantage of small volume. The element of this antenna is a circular disk microstrip antenna with two feeding ports for diversity reception which are orthogonal to each other. The correlation coefficient is sufficiently low at 0.2 in urban areas. However, because the incident wave is mostly vertically polarized and the average received power in each port differs considerably, improvement in the received power is not as large as that achieved by other diversity schemes. Therefore, this type of diversity antenna is used only where the installation space is limited.

Design of the Base Station Diversity Antenna

When the received power level at the receiving terminals of the antennas take a Rayleigh distribution, the relationship between the correlation coefficient of the diversity terminals

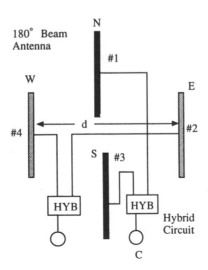

Figure 4.27 Configuration of pattern diversity antenna with omnidirectional pattern ($d = 6\lambda$ typical).

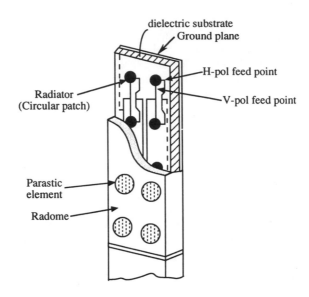

Figure 4.28 Configuration of polarization diversity antenna (60-deg sector beam). (After [28].)

and carrier-to-noise (CNR) level at the cumulative probability of 1% is as shown in Figure 4.29. From this figure, it can be understood that the ideal improvement of CNR is 9.5 dB (when the correlation coefficient is 0), and that the improvement of CNR remains at 8 dB even if the correlation coefficient rises to 0.6. It is usual, therefore, to design the diversity antenna using such antenna spacings or antenna radiation patterns in order to achieve a correlation coefficient of less than 0.6. To provide design data for the diversity antenna, the relationship between the configuration of the diversity antenna and its correlation coefficient is described below.

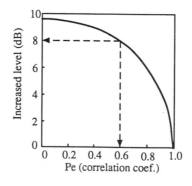

Figure 4.29 Increased level due to diversity reception with selection scheme.

A Horizontally Spaced Diversity Antenna

Figures 4.30 (a,b) shows the relationship between antenna spacing and the correlation coefficient in urban and suburban areas. Antenna heights are 120m (□) and 45m (•) in (a) and 65m in (b). It can be understood from this figure that the antenna spacing should be larger than five wavelengths in urban areas to achieve a correlation coefficient of less than 0.6, while more than 20 wavelengths are required in suburban areas. It is also understood that the correlation coefficient increases with the antenna height. In this figure, the calculated value of the correlation coefficient, derived from the equation given in Chapter 2 is also shown, where S is the standard deviation of the amplitude distribution of the incoming wave. The value of S in the horizontal plane has not yet been obtained, while its value in the vertical plane was determined from a comparison of measured antenna gain dependent on antenna height, and calculated antenna gain using an incoming wave of Gaussian distribution with standard deviation S, as shown in the next section.

A Vertically Spaced Diversity Antenna

No vertically spaced diversity antenna has been commercially used as a base station yet, although this type is used for mobile station antennas. However, a sufficiently low correlation coefficient can be obtained when the antenna height is just above the average height of surrounding buildings. This may be practical in an urban area. Figure 4.31 shows the correlation coefficient between antennas composed of 24 element arrays as a function of their center spacing [29]. In this figure, the solid line shows the calculated correlation coefficient under the assumption that the incident wave is a Gaussian distribution with a standard deviation S. The measured value and calculated value agree well with each other

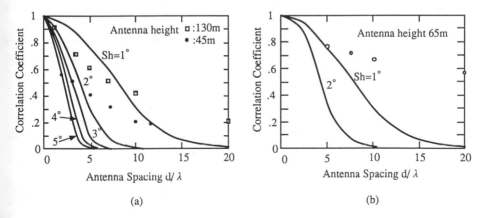

(a)   (b)

igure 4.30 Relation between correlation coefficient and antenna spacing: (a) urban area; (b) suburban area. Sh is the standard deviation of the incoming move profile in the horizontal plane.

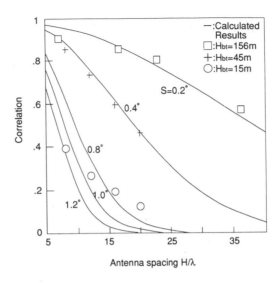

Figure 4.31 Relation between antenna spacing and correlation coefficient in vertically spaced diversity antenna. (After [29].)

when an adequate value of S is assumed. Ebine found from this and the antenna gain characteristic of the antenna height obtained by the Okumura Curve that standard deviation S is determined by antenna height, as shown in Figure 4.32 [29]. The value of S gained from this curve can be used to estimate the correlation coefficient of a vertically spaced diversity antenna.

4.3.5 Intermodulation Problems in Antennas

Passive intermodulation (PIM) must be considered when an antenna is used for both transmitting and receiving. Multiple transmitting channels are intermodulated due to the nonlinear effect of the metal heterojunctions that exists between the antenna radiating elements and the feed line; an inteference wave with the same frequency as that of receiving wave may occur in the receiving circuit. Therefore, in order to serve simultaneously for transmitting and receiving, the antenna should be designed and manufactured in such a way that the intermodulation power is less than the specified value (usually $-10\text{dB}\mu\text{W}$ in the case of an automobile telephone system) [14].

Principle of Passive Intermodulation

Since the theory of the occurrence mechanism for PIM has already been published [30] only minimum material need be presented in this section. Assuming that the frequencies

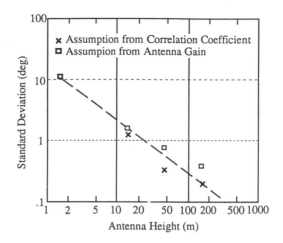

Figure 4.32 Prediction curve of standard deviation of Gaussian distribution incident wave. *S* is the standard deviation of incoming move profile in the vertical plane. (After [29].)

of the two transmitting waves are f_i and f_j, frequency f_{IM} of the intermodulated wave of the $(m + n)$th order is given by the following equation.

$$f_{IM} = mf_i \pm nf_j \qquad (4.5)$$

where m, n are positive odd integers. The possibility of f_{IM} being an interference wave on the receiving frequency band depends on the frequency distance between transmitting and receiving and the value of $(m + n)$. For example, in frequencies used in the Japanese analog automobile telephone system, the lowest order intermodulation that can be considered as interference is the seventh order, as shown in Figure 4.33, and the interference

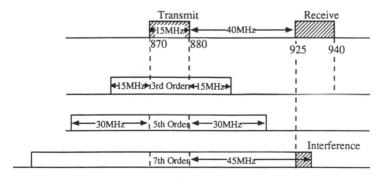

Figure 4.33 Higher object intermodulation and interference.

power is limited to $-10dB\mu W$, which is below the threshold level for transmitting and receiving antenna co-use, even if the total input power to the antenna is 200W.

Studies of the relationship between the order of PIM and the generated power have found that power is approximately given by 10 dB/$(m + n)$. Therefore, if the frequency separation between the transmitting wave and the receiving wave is narrow, fifth or third order PIM may become interference with levels 20 or 40 dB higher than that of the seventh order, and may exceed the threshold level permitted for antenna co-use.

Occurrence Points and the Suppressing Technique of PIM

Generated power of PIM depends on the kinds of metal in contact and their configuration. Table 4.1 shows techniques used to suppress PIM. Although the fundamental theory of PIM has been established, it cannot be said that the relationship between contact point structure and PIM generation has been investigated quantitatively. On the other hand, due to the rapid growth of demand in mobile communications, the number of base stations will continuously increase. Co-use antennas offer great advantage to the system, especially from the standpoint of system economy, and it is expected that co-use antennas will be used more often. Thus, it is becoming more urgent to investigate PIM, especially the suppression techniques.

4.4 MOBILE STATION ANTENNA TECHNIQUES

Mobile antennas should be designed to reduce the required transmitting power of mobile radio equipment while ensuring that the required service qualities (i.e., speech quality, outage probability within a whole cell (service coverage), and so on) are satisfied. The optimum design realizes small but economical mobile telephone units. This section outlines the vehicular antenna characteristics of cellular mobile systems and the antenna technolo-

Table 4.1
Suppression Methods of Passive Intermodulation

Items	Methods
Antenna:	
Radiator	Printed instead of wire element
Connection between radiator and flange	Increase contact area between flange and printed element; tightly fixed with silver material
Welding	Put paste to surrounding contact-point area
Cable:	
Heterojunction of metal contact	Prohibit using contact of aluminum and nickel
Rust	Use grooves in case of manufacturing and integration to prevent oxide generation

gies used in their design. The requirements of and technologies for portable telephone equipment are also noted.

Vehicular Antennas

The first generation of cellular mobile radio systems were analog mobile radio systems that used the 400-MHz band or 800- to 900-MHz band. Some are still in commercial use in the United States, the United Kingdom, the Nordic nations, Japan, and other countries, but the number of subscribers using second-generation equipment has been rapidly increasing. The requirements for vehicular antennas include the operating frequency, bandwidth, directivity, pattern characteristics, polarization, and the support of diversity reception. The requirements of operating frequency, bandwidth, and directivity of antennas are different for the system of each country.

Since the typical mobile station moves randomly in a radio zone, an omnidirectional azimuth pattern is required for mobile antennas. Particularly in suburban areas, the base station and the mobile station are in line of sight, and, hence, if the antenna pattern were not omnidirectional, the received signal level would vary considerably. Therefore, omnidirectional antennas are usually required in a mobile radio system. Experimental results from suburban and urban environments indicate that angular distributions of arriving waves extending from 0 to 50 deg in elevation are quite common. Since the mean elevation angle of their angular distribution depends on the environment, a mobile antenna whose elevation angle of maximum radiation could be aligned to match the mean elevation angle of the incident wave distribution seems to be effective for ensuring the maximum received power. At present, however, practical concerns require the radiation pattern to be maximum in the horizontal direction. Vertical polarization is usually used in most mobile systems because it makes broadband omnidirectional antennas very easy to develop (e.g., whip and dipole antennas).

It is commonly known that diversity reception is very effective for mitigating multipath fading. However, in most cellular mobile systems, diversity reception is an optional technology in the system specifications. NTT's system adopts diversity reception for both the base and mobile stations, and postdetection selection diversity is used. To develop the diversity branches, space diversity has been adopted, and the correlation between the antenna branches has been held to less than 0.6.

Antennas for Portable Telephones

Most portable telephone services are provided by using the radio channels of the vehicular telephone system; hence, the requirements for operating frequency and bandwidth are the same as those of vehicular antennas. However, because of limited battery capacity, the transmitting power of portable telephone units must be less than that of vehicular telephone units. Furthermore, antenna gain is generally less than that possible with vehicular antennas, because only small antennas can be used and there is gain degradation due to the proximity of the human body. Under these conditions, the antenna's requirements for portable telephone units are to develop the highest possible gain over the required bandwidth. The effective gain means the mean effective gain in a multipath mobile radio environment, and is the same as the mean effective gain described in Chapter 2. Improving the antenna effective gain is very effective in reducing the size and weight of portable telephone units.

There are some distinctive features of antennas to be mounted on portable telephone units. The first feature is that the polarization direction and radiation pattern of the antennas are not fixed, since the portable telephone unit is randomly directed when used. The second feature is that their radiation pattern and radiation efficiency vary considerably when close to the human body. This means that it is impractical to require the antennas of portable units to have vertical polarization and omnidirectional radiation pattern. In the design of portable telephone antennas, it is important to try to optimize the effective gain. Theoretical analyses must consider variation of the radiation pattern due to the effect of the unit's housing, the degradation of the antenna's radiation efficiency due to the proximity effect of the human body, and the variation of pattern and polarization due to human operation. It is necessary to determine the following characteristics when designing an antenna: (1) the effective gain in the multipath propagation environment; (2) the effective gain when the portable unit is operated in the speaking, carrying, and dialing positions.

It should be noted that an evaluation based on the free-space radiation pattern is not accurate enough to assess the practical performance of an antenna because the effect of the human body is not considered. Finally, it also should be noted that the influence of the RF power radiated from the antenna on the human body is very important for designing antennas for portable telephone units.

REFERENCES

[1] Kagoshima, K., "Key Factor and Technique for Antenna Design," *J. IEICE,* Vol. 71, No. 6, 1988 pp. 607–609.

[2] Taga, T., "Analysis for Mean Effective Gain of Mobile Antennas and in Land Mobile Radio Environments," *IEEE Trans. VT,* Vol. 39, No. 2, 1990, pp. 117–131.

[3] Ebine, Y., "Antenna Characteristics in Vertically Gaussian Distributed Mobile Radio Propagation," *IEE AP-S'90,* Dallas, Texas, 7–11 May 1990, pp. 1800–1803.

[4] Sato, K., K. Matsumoto, K. Fujimoto, and K. Hirasawa, "Characteristics of a Planar Inverted-F Antenna on a Rectangular Conducting Body," *IEICE Trans.,* Vol. J71-B, No. 11, 1988, pp. 1237–1243.

[5] Analoui, M., and Y. Kagawa, "Surface Patch Analysis of a Built-in Planar Inverted-F Antenna," *Proc. 1992 Int. Symp. on Antennas and Propagation,* Sapporo, Sep. 22–25, 1992, pp. 661–664.

[6] Reineix, A., and B. Jecko, "Analysis of Microstrip Patch Antennas Using Finite Difference Time Domain Method," *IEEE Trans. AP,* Vol. 37, No. 11, 1989, pp. 1361–1368.

[7] Wu, C., K. L. Wu, Z. Q. Bi, and J. Litva, "Accurate Characterization of Planar Printed Antennas Using Finite-Difference Time-Domain Method," *IEEE Trans. AP,* Vol. 40, No. 5, 1992, pp. 526–534.

[8] Kagoshima, K., K. Tsunekawa, and A. Ando, "Analysis of a Planar Inverted-F Antenna Fed by Electromagnetic Coupling," *IEEE AP-S'92,* Chicago, July 18–25, 1992, pp. 1702–1705.

205

[9] Mittra, R., ed., *Computer Techniques for Electromagnetics,* Pergamon Press, 1973.

[10] Stutzaman, W. L., and G. A. Thiele, "Antenna Theory and Design," John Wiley, 1981.

[11] Okumura, Y., and A. Akeyama, "Radio Wave Propagation in Mobile Communications," Ch. 2 of *Basic Technology of Mobile Communications,* Y. Okumura, and M. Shinji, eds., IEICE Press (Japanese), 1986, pp. 24–59.

[12] Mishima, H., Y. Ebine, and K. Watanabe, "Base Station Antennas and Multiplexer System for Land Mobile Telephone System," *Elect. Com. Lab. Tech. J.,* Vol. 26, No. 7, 1977, pp. 2011–2036.

[13] Nakajima, N., H. Mishima, and Y. Yamada, "Mobile Communications Antennas," Ch. 10 of *Basic Technology of Mobile Communications,* Y. Okumura, and M. Shinji, eds., IEICE Press (Japanese), 1986, pp. 239–260.

[14] Yamada, Y., Y. Ebine, and K. Tsunekawa, "Base and Mobile Station Antenna for Land Mobile Radio Systems," *IEICE Trans.,* Vol. E74, No. 6, 1991, pp. 1547–1555.

[15] Sakamoto, M., "System Configuration and Control," Ch. 8 of *Basic Technology of Mobile Communications,* Y. Okumura, and M. Shinji, eds., IEICE Press (Japanese), 1986, pp. 188–217.

[16] Mishima, H., Y. Ebine, and K. Watanabe, "Base Station Antennas and Multiplexer System for Land Mobile Telephone System," *Elect. Com. Lab. Tech. J.,* Vol. 26, No. 7, 1977, pp. 2011–2036.

[17] Yamada, Y., T. Nara, S. Kameo, Y. Chatani, and H. Abe, "A Variable Beamwidth Corner Reflector Antenna," *IECE Nat. Conv. Record,* No. 694, 1986.

[18] Kraus, J. D., *Antennas,* 2nd ed., McGraw-Hill, 1988, pp. 549–558.

[19] Suzuki, T., and K. Kagoshima, "Corner Reflector Antenna With the Same Beamwidth in Two Frequency Bands," *IEICE Trans.,* Vol. J75-B-II, No. 12, 1992, pp. 950–956.

[20] Nara, T., Y. Ebine, and N. Nakajima, "Beam Tilting Effect of Base Station Antenna," *IECE Nat. Conv. Record,* No. S5-15, 1985.

[21] Elliott, R. S., Ch. 4, Sec. 4.4 in *Antenna Theory and Design,* Prentice-Hall, 1981, p. 128.

[22] Kijima, M., and Y. Yamada, "Relationship Between Array Excitation Distribution and Radiation Pattern Ripple Depth," *IEICE Trans.,* Vol. J73-B-II, No. 12, 1990, pp. 860–868.

[23] Elliott, R. S., Ch. 5 in *Antenna Theory and Design,* Prentice-Hall, 1981.

[24] Kijima, M., and Y. Yamada, "Beam Narrowing Method for Radiation Pattern With Suppressing Some Sidelobes," *IEICE Tech. Rep. on Antennas and Propagation,* A·P91–125, 1992.

[25] Kijima, M., and Y. Yamada, "Determining Excitation Coefficients of Dual-Frequency Shaped Beam Linear Array Antenna for Mobile Base Station," *IEEE AP-S'91,* London, Ontario, 24–28 June 1991, pp. 932–935.

[26] Clark, R. H., "A Statistical Theory of Mobile-Radio Receptions," *B.S.T.J.,* Vol. 47, No. 6, 1968, pp. 957–1000.

[27] Yamada, Y., K. Kagoshima, and K. Tsunekawa, "Diversity Antennas for Base and Mobile Stations in Land Mobile Communication Systems," *IEICE Trans.,* Vol. E74, No. 10, 1991, pp. 3202–3209.

[28] Nara, T., Y. Ebine, and Y. Yamada, "Characteristics of Polarization Diversity Base Station Antenna," *IECE Nat. Conv. Rec.,* No. 2362, 1986.

[29] Ebine, Y., T. Takahashi, and Y. Yamada, "A Study of Vertical Space Diversity for a Land Mobile Radio," *IEICE Trans.,* Vol. J73-B-II, No. 6, 1990, pp. 286–292.

[30] Gardiner, J. G., and R. E. Fudge, "Aerials and Base Station Design," Ch. 4 of *Land Mobile Radio Systems,* R. J. Holbeche, ed., IEE Telecommunications Series 14, Peter Peregrinus, 1985, pp. 45–71.

Chapter 5

Advances in Base Station Antennas

Y. Yamada and Y. Karasawa

5.1 RECENT BASE STATION ANTENNAS

The fundamental technical subjects needed for designing antennas were discussed in Chapter 4. In this section, recent trends of base station antennas, practical antenna configurations, and antenna performance are presented.

In Japan, PDC (personal digital communication) systems started in 1994. The number of subscribers has been increasing at a rate of two times every year. To meet this rapid increase, small cell radii of less than 1 km are selected for highly dense areas such as Tokyo metropolitan area. A large number of base station antennas are planned for installation on the civil buildings. Severe requirements such as saving of installation space, light weight, and low wind load should be satisfied. To meet these requirements, dual-frequency antennas, dual-beam antennas, and polarization diversity antennas were developed.

PHS (personal handy phone systems) started in 1995. A very small cell radius of 100m to 500m is utilized. Base station antennas are installed on structures in streets or nearby buildings. Low cost and very lightweight base stations and antennas were developed. Recently, adaptive antennas were introduced to increase antenna gain and suppress interference.

Moreover, PHS is employed to achieve a WLL (wireless local loop) system. Twelve base station antennas are used. Multibeam and adaptive antenna techniques are employed to increase system capacity.

A satellite phone system began operation in 1996. The geostational N-STAR satellite is utilized. To conquer the large distance, moderate-gain planar array antennas with satellite tracking mechanisms were developed.

5.1.1 Base Station Antennas for Cellular Systems

Typical radio cell parameters for Japanese mobile communication systems are shown in Table 5.1 [1]. Frequency reuse of four-cell repetition gives nearly the maximum frequency spectrum utilization within most mobile systems. To achieve this repetition pattern, interference reduction is important. For this purpose, the antenna beam is tilted downward as the first step, and sidelobes facing interference cell directions should be suppressed to very low levels. The three-sector cell configuration is also effective for achieving the four-cell repetition pattern by reducing interference cell numbers. For two-branch reception diversity, two antennas are employed. One of these antennas is co-used as a transmission antenna. As for the sector cell configuration, a six-sector configuration is preferable in order to increase the system capacity.

Antenna specifications are shown in Table 5.2. The antenna gain in the 900-MHz band is 17 dBi for a 120-deg beamwidth in the horizontal plane. The antenna gain in the 1,500-MHz band is determined based on the 900-MHz value by considering the loss increase in high-frequency bands. The bandwidth of the 900-MHz digital system is

Table 5.1
Outline of the Objective System

System Parameters	Specifications	
Frequency bands	Uplink	Downlink
800-MHz analog	915–940	860–885
800-MHz digital	940–956	810–826
1,500-MHz digital	1429–1441	1477–1489
	1453–1465	1501–1513
Cell radius	0.5–3 km	
Frequency reuse	Four-cell repeat pattern	
Sector configuration	Three-sector cell	
Diversity scheme	Two-branch switching	

Table 5.2
Antenna Specifications

Item	Specifications
Antenna height	5.4m
Antenna gain	17 dBi (810–956 MHz)
	17.5 dBi (1429–1503 MHz)
Sidelobe level below the horizontal direction	Less than −20 dB
Electrical beam tilt	Type 1: 0–5 deg
	Type 2: 3–11 deg
Beamwidth in a horizontal plane	120 deg

146 MHz, which corresponds to a 17% bandwidth ratio. The bandwidth ratio of the 1,500-MHz band is 6%. The sidelobe levels are required to be less than −20 dB. Electrical beam tilt expands the antenna's applicability to a variety of cell radii and building heights. For building height (H), beam tilt angle θ is determined by the relation of $\tan\theta = H/R$, where R is the cell radius. By taking into account building heights and cell radii in the Tokyo metropolitan area, the maximum value of tilt angle is determined. The tilt angle coverage from 0 to 11 deg is too wide to maintain low sidelobe characteristics. As a consequence, two antenna types having shallow and deep tilt usage were prepared. As for antennas having six-sector cell usage, a compact antenna configuration is required in order to avoid increasing the numbers of antennas and to ease the installation conditions.

In this section, three types of antennas, a dual-frequency antenna, a dual-beam antenna, and a polarization diversity antenna, are described. The dual-frequency antenna is used in the three-sector cell configurations. The dual-beam antenna is used in the six-sector cell configuration. Antenna design outlines and achieved characteristics are discussed.

5.1.1.1 Dual-Frequency Antenna

The exterior view of a dual-frequency antenna is shown in Figure 5.1. This antenna is composed of three cylindrical blocks. Beside the antenna is a beam tilt box. The box has two terminals with 900-MHz and 1,500-MHz bands. The antenna is connected to the beam tilt box with six coaxial cables whose lengths are 5m, respectively. The antenna and coaxial cables are used at the dual-frequency bands. The configuration of the antenna feed network is shown in Figure 5.2. All antenna components such as radiators, transmission lines, and power dividers are mounted on the same dielectric substrate. The beam tilt panel (contained in the beam tilt box) is separated from the antenna body and placed on the antenna base. The design concept is for all components shown in Figure 5.2 to be commonly used in dual-frequency bands. Array elements are excited through the feed network composed of transmission lines and power dividers. Each element of the feed network has wide frequency band characteristics. Array elements are composed to have dual-frequency operation. Twenty-six elements are selected to achieve the required antenna gain. Because the suppression of the grating lobe in the higher frequency is important in determining radiator spacing, an array element spacing of λ (one wavelength) in the 1,500-MHz band was chosen. At 900 MHz, the array element spacing becomes 0.6 wavelength. The objective excitation coefficients (amplitude and phase) of array elements are achieved by designing the feed network. Excitation amplitudes are determined by setting power ratios at power dividers. Excitation phases are determined by setting the length of transmission lines. Here, because the line length is frequency dependent, changes of excitation phases occur in dual-frequency operation. Accordingly, low sidelobe characteristics will be degraded in dual-frequency operation.

The electrical beam tilt is achieved by giving a specific phase difference to subarrays (composed of M array elements) [2]. This phase difference is achieved at a beam tilt

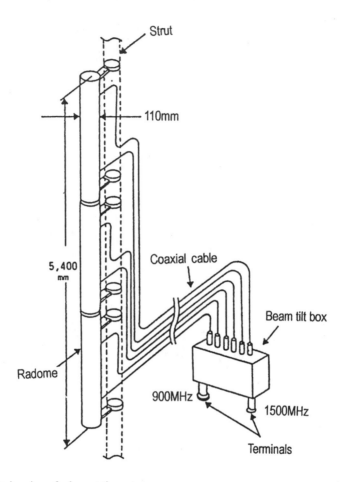

Figure 5.1 Exterior view of a base station antenna.

panel. The wavefront of the subarrays is tilted by θ_{t0} as shown by solid lines. The antenna's electrical beam tilt angle is set to θ_t as shown by the broken line. Change in electrical beam tilt is achieved by replacing the beam tilt panel. When θ_t does not equal θ_{t0}, abrupt phase jumps between the edges of subarrays occur. This phase jump produces many grating lobes in the low sidelobe area. Here, the number of subarray elements (M) is the key factor in determining the grating lobe positions. In this case, six subarrays are employed whose numbers of element are 5, 4, 4, 4, 4, and 5. Namely, M numbers are 4 and 5.

An example of a base station antenna is shown in Figure 5.3. As dual-band array elements, the dual radiator configuration is employed. The 900- and 1,500-MHz elements are printed dipole configurations. The 1,500-MHz element is placed in front of the

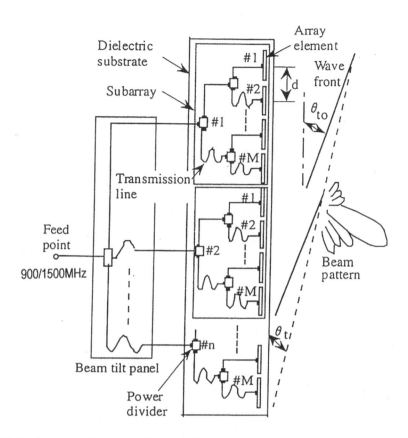

Figure 5.2 Configuration of an antenna feed network.

900-MHz element. The small element spacing of 0.6 wavelength would introduce mutual coupling. To suppress mutual coupling, stubs are inserted between array elements. A reflector is placed behind these radiators to achieve a 120-deg beam in the horizontal plane. The radome diameter is 110 mm.

The radiation characteristics of the base station antenna are expressed by the following equation:

$$G(\theta) = g_o(\theta) \sum_{n=1}^{N/M} \sum_{m=1}^{M} I_{nm} \times \exp(j\alpha\phi_{nm}) \times \exp(jk_o \alpha d_{nm} \sin\theta) \times \exp(-jk_o \alpha d\phi_r)$$

$$(5.1)$$

Here I_{nm} and ϕ_{nm} represent the excitation amplitude and phase of the nm array element (radiator), respectively, n is the subarray number, m is the array element number in a subarray, the suffix nm represents element number of $m + (n - 1)M$; α indicates the

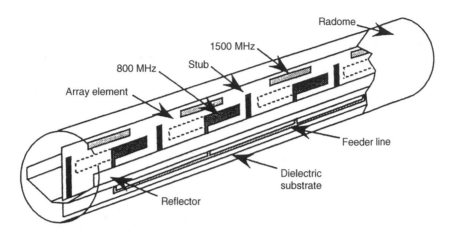

Figure 5.3 Actual antenna configurations.

frequency ratio where $\alpha = 1$ corresponds to a frequency of 900 MHz; $g_o(\theta)$ expresses a radiation pattern of an array element.

The first exponential function in (5.1) expresses the radiation pattern dependence on excitation phase component, the second exponential function expresses the radiation pattern dependence on the component of path lengths of all the array elements, and the third exponential function expresses the radiation pattern dependence on added phases in order to achieve electrical beam tilt. In (5.1), d_{nm} means $d(m + (n - 1)M)$, and ϕ_T is given by the next expression:

$$\phi_T = m \sin \theta_{t0} + (n - 1)M \sin \theta_t \tag{5.2}$$

The exponent of the second exponential expression of (5.1) has the periodicity of $k_0 \alpha d \sin \theta = 2\pi$. Therefore, those angles θ_{t0} and θ_G determined by the next formula have the same beam pattern:

$$k_0 \alpha d (\sin \theta_{t0} - \sin \theta_G) = 2\pi p \tag{5.3}$$

where p is an integer. The radiation peak at θ_{t0} is repeated at θ_G, which is called a grating lobe. In the case of Figure 5.2, abrupt phase jumps occur at intervals of Md, so the remarkable lobes at angles θ_{Gt} are given by replacing d in (5.3) with Md. Moreover, by taking into account the beam tilt angle θ_t, we can express θ_{Gt} as follows:

$$\sin \theta_t - \sin \theta_{Gt} = \lambda_0 / (\alpha Md) \tag{5.4}$$

Radiation characteristics of $\alpha = 1(900 \text{ MHz})$ and $\alpha = 1.67(1,500 \text{ MHz})$ are shown in Figure 5.4, where $g_0(\theta)$ is assumed to be constant over all angles. This is the case for

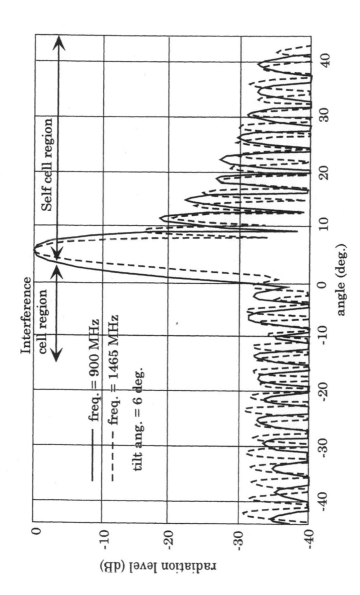

Figure 5.4 Designed antenna radiation patterns.

type 2 antennas. A beam tilt angle of 6 deg is selected as the design object. The feature of these sidelobes is their asymmetry. Sidelobes in the interference cell region are suppressed to SL_0. The interference cell region is considered to be 15 deg from the edge of the main lobe in the case of 11-deg beam tilt. Low sidelobe levels in this region is the main design objective. Sidelobes on the right side of the main lobe illuminate the self-cell. Electrical beam tilt (main beam shift) is performed toward the self-cell direction.

To achieve sidelobe asymmetry, the excitation coefficients of array elements can be designed by the LMS (least mean square) method [3] or by Elliot's perturbation method [4]. The LMS method achieves sidelobe levels such that the errors of sidelobe levels between the objective and designed are minimized. Information needed in designing is objective sidelobe levels. On the other hand, Elliot's method has the feature of achieving accurate sidelobe levels if accurate null positions can be designated for all sidelobes.

In the LMS method, the excitation coefficients for objective radiation pattern are calcuated as follows:

$$[\mathbf{V}] = [[\mathbf{B}^t]*[\mathbf{B}]]^{-1}[\mathbf{B}^t]*[\mathbf{A}] \qquad (5.5)$$

Here, [$\mathbf{V}$] is a column matrix of the excitation coefficient to be determined. [$\mathbf{B}$] is a matrix that incorporates information about the radiation element's spatial phases concerning radiation directions. [$\mathbf{A}$] is a column matrix of the objective radiation pattern, which is numerically given. [$\mathbf{B}^t$] denotes transpose and [$\mathbf{B}$]* denotes complex conjugate matrices. The excitation coefficients designed by using the LMS method are shown in Figure 5.5. This distribution corresponds to the first exponential component of (5.1). Transmission

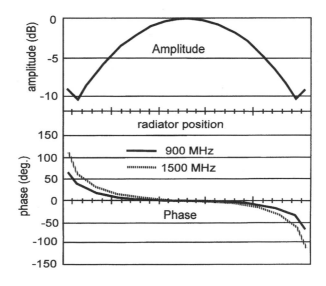

Figure 5.5 Excitation coefficients of array elements.

line lengths were determined according to the characteristics of $\alpha = 1$. Power divider ratios were determined according to the amplitude distribution. Power divider ratios were considered to be frequency invariant. However, the path lengths of the transmission lines change with the frequency. The path lengths of the transmission lines in $\alpha = 1.67$ become large compared to the $\alpha = 1$ case. Accordingly, the phase distribution in $\alpha = 1.67$ is increased. This phase shifts of Figure 5.5 are proportional to the α value.

Measured and designed radiation patterns in the vertical plane for different beam tilt angles are shown in Figures 5.6 through 5.8. At two frequencies, the same beam tilt angles are achieved. Sidelobes in the interference region are suppressed to under −20 dB. Main beam tilt angles in two frequencies coincide well. Measured sidelobes of Figure 5.7 increase remarkably from the designed levels. The actual excitation phases are deformed from designed values of Figure 5.5 due to manufacturing errors, mutual coupling between array elements, and refraction from the radome. In Figures 5.6 and 5.8, many grating lobes have appeared. Grating lobe positions can be calculated by inserting $M = 4$, $\alpha = 1(900$ MHz$)$ and $d/\lambda_0 = 0.6$ in (5.4). An angle between main lobe and grating lobes $(\theta_{Gt} + \theta_t)$ of 24.6 deg, is obtained. For the 1,500-MHz case, $\theta_{Gt} + \theta_t = 14.5$ deg is obtained by using $\alpha = 1.67$. These values agree well with the radiation patterns of Figure 5.6 and 5.8. Important characteristics of grating lobes are (1) angles between main lobe and grating lobes are kept constant for different tilt angles and (2) grating lobe peaks are proportional to tilt angle differences from the center tilt angle of 6 deg.

The cross-sectional configuration of a developed antenna is shown in Figure 5.9. The feature of this configuration is the sidewalls attached to both sides of the reflector. Adding sidewalls is effective to achieve a 120-deg beamwidth with a small reflector size. Next, the design chart of radiator position is shown in Figure 5.10. By selecting an adequate radiator position, a 120-degree beamwidth can be achieved. For the 800-MHz band (810–960 MHz), a radiator position of 0.05m is adequate. For the 1,500-MHz band (1,429–1,501 MHz), a radiation position of 0.07m is adequate. Measured radiation patterns are shown in Figure 5.11. In the 800- and 1,500-MHz bands, a 3-dB beamwidth of 120 deg was achieved. The bandwidths of the 800- and 1,500-MHz bands are 17% and 5%, respectively. That is, greater beamwidth change is observed in the 800-MHz band.

Measured antenna gains for the type 2 antenna are shown in Figure 5.12. At tilt angle of 6 deg and antenna gain of 17.1 dBi is achieved in the 900-MHz band. Here, losses of the tilt panel and coaxial cables (length of 5m) are included. Loss of a tilt panel is 1.2 dB and that of a coaxial cable is 0.9 dB. For 1,465 MHz, and antenna gain of 17.7 dBi is achieved. Loss of a tilt panel is 1.8 dB and that of a coaxial cable is 1.0 dB. In tilt angles of 4–10 deg, the specified antenna gains shown in Table 5.2 are nearly achieved. At tilt angles of 3 and 11 deg, gain cannot reach the specified values. In these tilt angles, gain reduction becomes large because of the great increase of grating lobe peak levels.

5.1.1.2 Dual-Beam Antenna

In cellular systems, six-sector cell configurations are employed to increase the system capacity. To achieve the diversity reception capability, 12 base station antennas are needed.

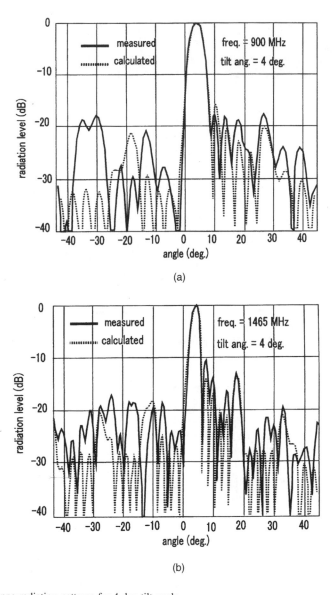

Figure 5.6 Antenna radiation patterns for 4-deg tilt angle.

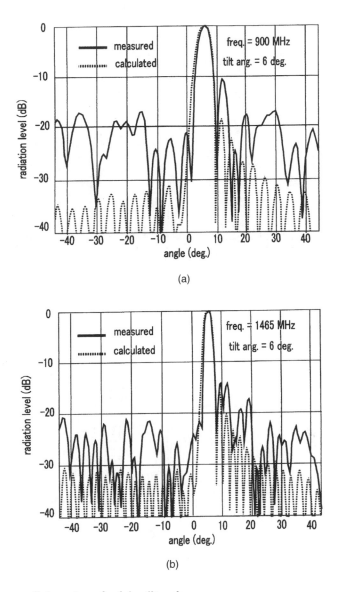

Figure 5.7 Antenna radiation patterns for 6-deg tilt angle.

Reduction of the number of antennas has been requested to ease the antenna installation on the tower.

In this section the novel 60-deg dual-beam antenna is explained. The fundamental antenna configuration is shown in Figure 5.13 [5]. Two radiators are combined by a

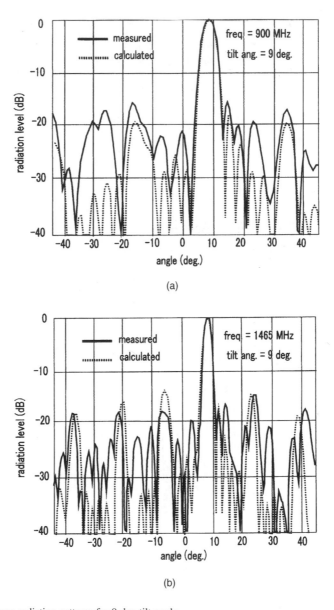

Figure 5.8 Antenna radiation patterns for 9-deg tilt angle.

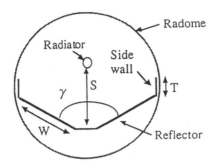

Figure 5.9 Antenna cross-sectional configuration.

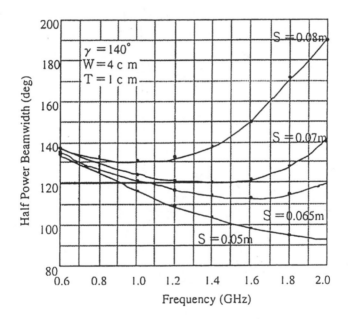

Figure 5.10 Beamwidth change due to radiator position.

90-deg hybrid circuit. By excitation of the A and B ports, respectively, beams of A and B are generated. By setting the radiator separation (*D*) to be half a wavelength, the beamwidth of each beam becomes 60 deg centering at $\theta_T = 30$ deg direction. Two sidewalls are attached in order to refine beam shapes. An antenna configuration for an actual base station is shown in Figure 5.14. Two vertical array antennas are arranged side by side. Subarrays are combined with a hybrid circuit. By connecting each port of the hybrid circuits to a beam tilt panel, electrical beam tilt is achieved. Radiation pattern designing

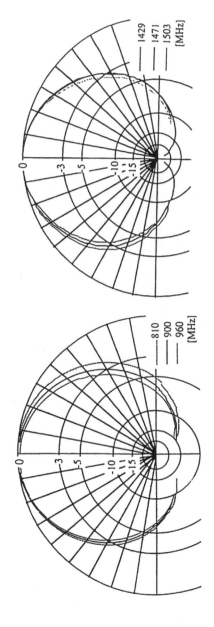

Figure 5.11 Radiation characteristics in the horizontal plane.

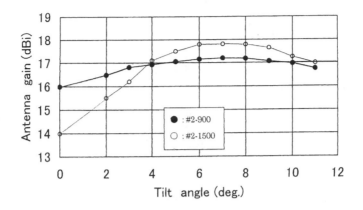

Figure 5.12 Antenna gain.

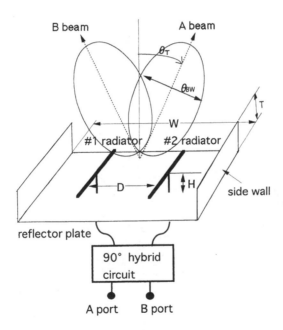

Figure 5.13 Fundamental configuration of a dual-beam antenna.

of this antenna is shown in Figure 5.15. The main parameter of designing low sidelobe characteristic is length (T) of a sidewall. This result is one sample of showing the effectiveness of a sidewall. By increasing the height, sidelobes near 70 deg can be effectively reduced. Measured and designed characteristics are shown in Figure 5.16. To achieve an excellent 60-deg beam and low sidelobe characteristics, a parasitic element is

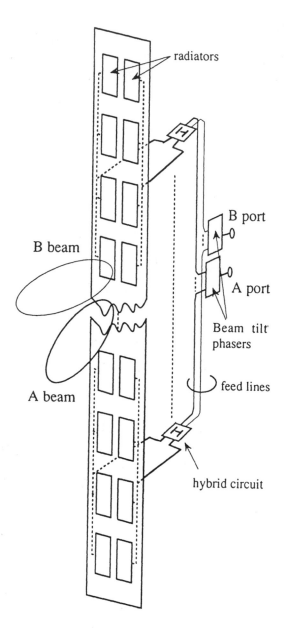

Figure 5.14 Dual-beam base station antenna configuration.

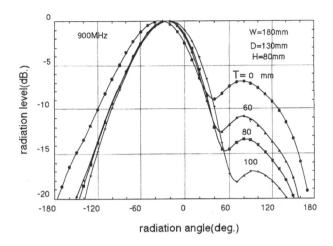

Figure 5.15 Sidelobe reduction by a sidewall.

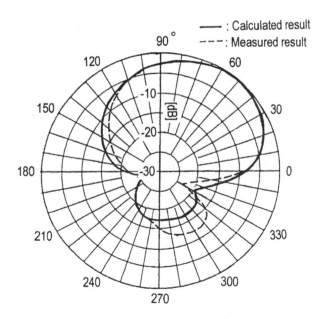

Figure 5.16 Measured and designed radiation patterns, $D = 0.32\lambda$, $W = 0.68\lambda$, $T = 0.08\lambda$, and $H = 0.14\lambda$.

placed in front of the radiator [6]. The most suitable parameters are also shown. Although a low sidelobe characteristic of less than −17 dB was expected, the achieved level was −12.5 dB.

To introduce this dual beam antenna into a six-sector system, observe the layout shown in Figures 5.17(a) and (b). First of all, two dual-beam antennas are combined to achieve the four-beam antenna as shown in Figure 5.17(a). Next, three four-beam antennas are mounted on an antenna tower. In each sector direction two beams are always installed. Diversity reception becomes available in all sectors. As a result, a six-sector system is achieved by employing 6 dual-beam antennas. This is very simple compared with conventional antenna use, which requires 12 antennas.

5.1.1.3 Polarization Diversity Antenna

In most mobile communication systems, vertical polarization has been used because of easy implementation of vertical antennas at mobiles and base stations. For the reverse link (the link of mobile to base station), diversity reception at the base station is the prerequisite for enhancing the receive performance to compensate for the inferior mobile

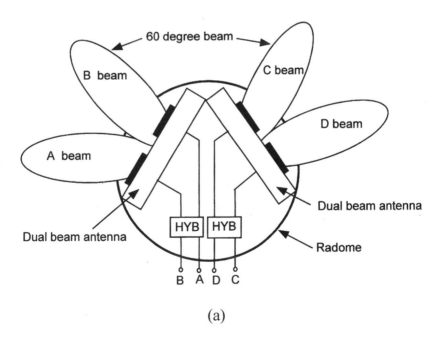

(a)

Figure 5.17 Dual-beam antenna layout for a six-sector cell configuration: (a) four-beam antenna configuration of dual-beam antennas; (b) diversity antenna layout for a six-sector cell configuration.

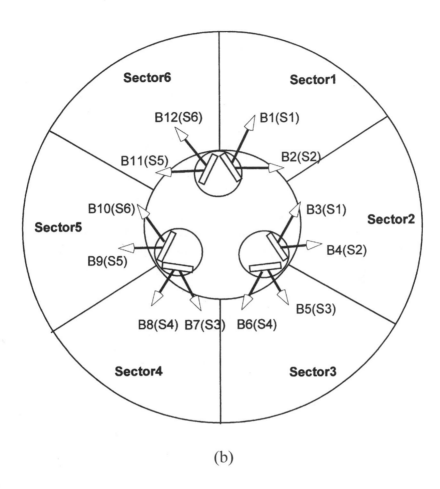

(b)

Figure 5.17 (continued).

transmit capability. Among various types of diversity schemes, space diversity has traditionally been used.

Space Diversity

Space diversity needs at least a pair of independent receive antennas for each sector at a base station for receiving signals via different paths from a mobile. The antennas have to be separated by at least 10 wavelengths, about 3m for the 800-MHz system, from each other for securing low correlation between the signals to obtain sufficient diversity gain. For a base station having a three-sector configuration, at least six receive antennas and three transmit antennas are required. However, it is getting very difficult to acquire cell sites in large cities like Tokyo for installing such a large number of antennas on the tops of buildings.

Shared use of an antenna element for reception and transmission is a mandatory necessity for reducing the number of antennas. The approach of accommodating two sector elements inside a cylindrical radome is being applied to most base station antennas in order to reduce the apparent number of antennas for multisector cell sites. An example of this scheme is illustrated in Figure 5.18, which is a top view of a three-sector antenna system consisting of three poles which contain two radiating elements each.

Pole 1 contains 2 vertical arrays, $A_{1,1}$ and $A_{3,2}$, facing azimuth angles of 0 deg for sector 1, and 240 deg for sector 3, respectively. The array $A_{1,1}$ is to receive the signal $Rx_{1,1}$ to be combined with $Rx_{1,2}$ received by array $A_{1,2}$ in pole 2 for diversity of sector 1. Array $A_{1,1}$ is commonly used for transmitting the signal Tx_1 for sector 1. Signal Tx_2 is the signal transmitted by array $A_{2,1}$, which is shared for receiving the signal $Rx_{2,1}$ from the 120-deg direction. Signal $Rx_{2,2}$ is received by array $A_{2,2}$ in pole 3 and combined with $Rx_{2,1}$ for diversity of sector 2. Transmission and diversity reception of sector 3 are accomplished in the same manner. This configuration needs three antenna poles separated by 10 wavelengths each for three-sector space diversity system.

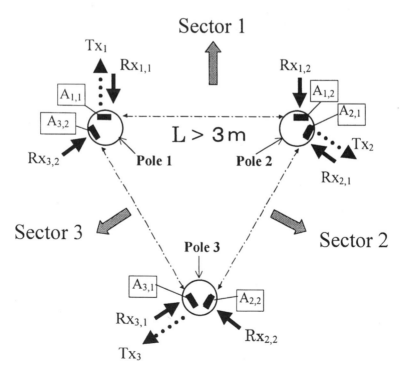

Figure 5.18 Typical configuration of three-sector space diversity system.

Polarization Diversity

Polarization diversity, on the other hand, does not require two spatially separated antennas. Multiple dipole elements with orthogonal polarization can be alternately mounted on a piece of dielectric substrate in a vertical radome. Elements for the orthogonal polarization may be of vertical/horizontal or of ±45-deg cross dipoles depending on the particular design. Polarization diversity is a well-known diversity technique, but has not traditionally been used in cellular phone systems, because it was not necessarily effective in the vertical polarization environment where mobiles with vertical trunk-lid antennas on board cars dominated.

The number of cellular phone subscribers exceeded 50 million in August 1999 in Japan. Now, more than 99.9% of phones are of the handheld type. Users hold their phones in a tilted position around 60 deg from zenith when they engage in conversation. Under such a circumstance, it is clear that a strong horizontal component is dominant and uplink polarization diversity should be effective.

Figure 5.19 shows a top view of the newly developed polarization diversity antenna with compact structure, which should be compared with Figure 5.18. This is an antenna consisting of a single piece of vertical pole containing three-sector transmit and receive arrays with diversity capability inside. Three vertical arrays, A1, A2, and A3, are bound

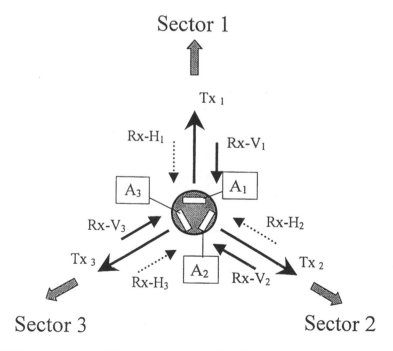

Figure 5.19 New configuration of three-sector polarization diversity system.

together in a triangular pillar shape and are accommodated in a cylindrical radome. Array A1, for example, contains a set of vertical and horizontal dipoles stacked alternately on a dielectric substrate to receive vertically polarized signal Rx-V1 and horizontally polarized signal Rx-H1 for sector 1. Either vertical or horizontal dipole elements of array A1, or both, can be shared for transmission of the signal Tx1. The other two arrays, A2 and A3, have configurations identical to that of A1.

Results of Propagation Tests

Based on the concept above, the applicability of polarization diversity in the cellular phone system was reexamined. A number of tests were conducted [7–10] to confirm the effect of polarization diversity, typical results of which are described here.

A handy phone was held by a test person up to his ear at 60 deg from zenith to transmit signals in the field. He was instructed to walk around a circle for a while at each location to vary the antenna orientation. For measuring the received signals the random field measurement method [11] was employed. The received signals from both ports at the base station diversity antennas were switched at a speed of 300 Hz and fed to a receiver to record the fading signals virtually at the same time.

Figure 5.20(a) shows the cumulative distribution of received level at a base station using space diversity, while the handheld phone transmitted signals from a line-of-sight location in the base station coverage. The two solid lines on the left-hand side show the received levels at the two separate vertical ports and the third solid line designated as "Diversity" shows the cumulative distribution of selection-combine diversity. Figure 5.20(b) shows the case for a line-of-sight location with polarization diversity reception. The two solid lines show the distribution of output levels of vertical and horizontal ports and the third solid line designated as "Diversity" again shows that of selection-combine diversity. As can be seen, the polarization diversity gain is 10.6 dB at the 1% level of cumulative distribution, whereas that of space diversity is 3.3 dB, suggesting the effectiveness of polarization diversity. Similar results were obtained for the case where the mobile transmitted signals from an out-of-sight area. The results are shown in Figures 5.20(c) and (d). Although the effect is not as significant as that for the line-of-sight case, diversity gain for polarization diversity is 10.3 dB which is greater than 9.4 dB for space diversity.

Development of New Polarization Diversity Antenna

Based on positive test results as described in the previous section, new polarization diversity antennas were developed with the goal of achieving superior receive performance in the handheld mobile environment and, at the same time, reducing the size and weight to cope with the stringent environmental requirement. Now, this type of polarization diversity antenna is being used by all CDMA base stations and by some PDC base stations at IDO Corporation of Japan [12,13]. This has resulted in easy cell site selection and

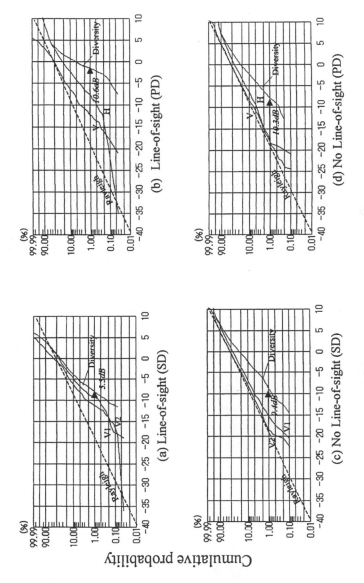

Figure 5.20 Cumulative distribution of received level: (a,b) line of sight; (c,d) out of sight.

reduction of construction costs because of the simple structure of the antenna in addition to superior performance.

Figure 5.21 shows an example of actual installation of a 16-dBi gain antenna in Tokyo. The excellent capability of this type of antenna was also demonstrated in Hong Kong in their CDMA network by installing its 12-dBi version for commercial use as seen in Figure 5.22. The large structure seen on the right-hand side is a conventional transmit-only antenna to be compared with this single piece of antenna covering three sectors.

The tower on the right-hand side in Figure 5.23 is for a conventional three-sector PDC cell site with space diversity, operated by another carrier. This is to be compared

Figure 5.21 New polarization diversity antenna with a gain of 16 dBi in Tokyo. (Courtesy of IDO Corporation.)

Figure 5.22 New polarization diversity antenna with a gain of 12 dBi in Hong Kong. (Courtesy of IDO Corporation.)

with the simple structure of the adjacent pole seen on the left-hand side. Two three-sector polarization diversity antennas are installed on top for the PDC and CDMA systems, respectively. If conventional antennas were used, at least six antenna poles would have been needed.

Figure 5.24 shows a simplified inner structure for this type of antenna. Arrays for three sectors are accommodated in the shape of a triangular tube in a cylindrical radome of 23 cm in diameter. In this example, three subarrays are stacked for achieving high gain. Downtilting is performed by changing the phase angles of the top and bottom subarrays in an opposite sense relative to the central subarray. The tilting is given indepen-

Figure 5.23 A three-sector space diversity antenna system contained in three poles for PDC (right) and two polarization diversity antennas for CDMA and PDC (left). (Courtesy of IDO Corporation.)

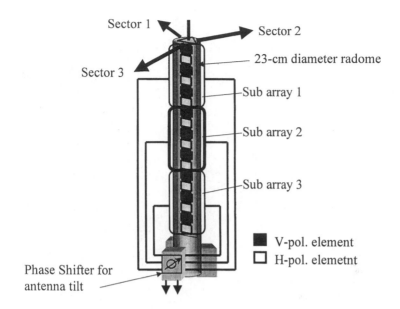

Figure 5.24 Inner structure of a new polarization diversity antenna.

dently to each transmit/receive and vertical/horizontal polarization combination. Each subarray consists of three vertically polarized dipoles and horizontally polarized dipoles stacked alternately [14].

Figure 5.25 shows a simplified structure of the dipoles consisting of the vertical array in vertical polarization [15]. A pair of dipoles is arrayed to make beamwidths in the H-plane smaller to match that of the E-plane. Similar dipoles are used for the horizontal polarization array. The protruded portion of the parasitic element is effective to widen the frequency range in which the beamwidth in the magnetic plane is kept constant. Table 5.3 lists the specifications for the newly developed antennas.

Care has to be taken to reduce the coupling between dipoles of different polarization elements as well as that between transmit and receive ports. Also important is the intermodulation product that would be generated elsewhere around the antenna when multiple carriers are fed to the antenna.

One of the potential features of this antenna is transmission diversity using different polarizations. This can be readily applied to third-generation CDMA mobile communications system.

Performance of the New Polarization Diversity Antenna

Radiation patterns in the horizontal plane of this antenna for vertical and horizontal polarizations are shown in Figure 5.26. Great care has been taken to keep the beamwidths

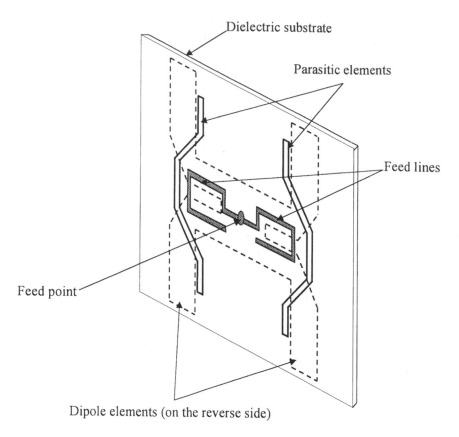

Dielectric substrate

Parasitic elements

Feed lines

Feed point

Dipole elements (on the reverse side)

Figure 5.25 Example structure of the dipole consisting of the vertical array.

in both polarizations for transmit and receive bands almost identical. Cross-polarization characteristics are rather important for such antennas. Figure 5.27 shows those of the new polarization diversity antenna.

To verify the effectiveness of the new polarization diversity antenna, a series of measurements was taken in a suburban area before the IDO's CDMA system was put into commercial service. Comparisons were made between polarization diversity and space diversity for the vertical dipole and tilted dipole at the transmitting mobile. The received level at a base station without diversity was used as a reference for evaluation. In the CDMA system, when the received signal performance in the uplink at the base station is better than specified, the mobile is instructed to reduce its transmit power. Therefore, if the transmit power of the test mobile in a diversity system is lower than that for a nondiversity system, it is judged that the system has some diversity gain.

In the test environment, there was no other mobile and no other base station, thus no interference problem was observed. In such a case, the reduced amount of transmit

Table 5.3
Specifications of Polarization Diversity Antennas

(a) Electrical Specifications		
Type	AN-951-1	AN-951-3
Antenna element	Array of two printed dipoles	
Frequency range	818–958 MHz	
Gain	16 dBi	12 dBi
VSWR	Less than 1.5	
−3-dB beamwidth (Vertical plane)	Approx. 18 deg	Approx. 5.5 deg
−3-dB beamwidth (Horizontal plane)	Approx. 70 deg	
Cross-polarization ratio	Greater than 30 dB	
Isolation between V-H ports	Greater than 40 dB	
Isolation between sector beams	Greater than 40 dB	

(b) Mechanical Specifications		
Type	AN-951-1	AN-951-3
Radome height	Approx. 3,200 mm	Approx. 1,200 mm
Radome diameter	239 mm	
Rated wind velocity	60 m/sec (survival)	
Connector	BFX-20D	
Weight (with electrical tilt box)	Approx. 124 kg	Approx. 94 kg

power could be used as an indicator of CDMA quality instead of monitoring E_b/N_0, which can be used when many mobiles exist [16].

Figure 5.28(a) shows the reduced amount of mobile transmit power relative to the nondiversity condition, when a vertical dipole was used at a test mobile. Measurements were taken with a test mobile in the field traveling in the longitudinal and transversal directions to the base station receive antenna at a driving speed of 40 and 6 km/h, respectively. Measurements were taken both in the boresight area of the main beam and in the 3-dB-down beam edge area of the base station receive antenna in order to take care of base station antenna performance effect. As seen in the figure, slightly better performance is observed in the case of space diversity than polarization diversity.

Similar measurements were taken in the case of the mobile antenna tilted 60 deg from zenith simulating the average cellular user's manner when using a phone. As is clearly seen from Figure 5.28(b), the mobile transmit power was significantly reduced, that is, ranging from 3 to 8 dB, for polarization diversity versus space diversity, indicating better receive signal performance.

.1.2 Antennas for Personal Handy Phone System

.1.2.1 Collinear Array and Diversity Antennas

In the Japanese personal handy phone system (PHS), TDD (time-division duplex) transmission scheme is employed. This system provides service in the 1,900-MHz band. A block

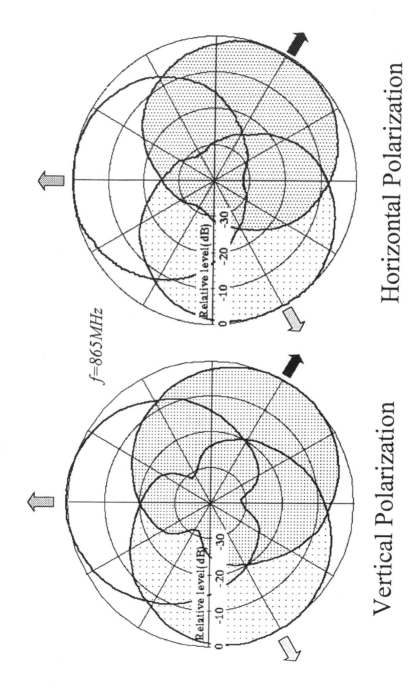

Figure 5.26 Horizontal patterns of three-sector antenna in V- and H-polarization.

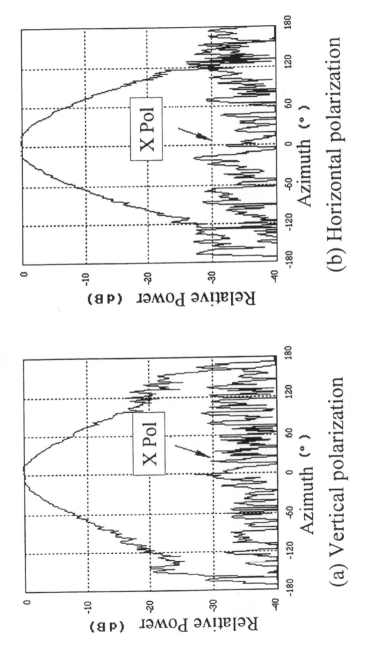

Figure 5.27 Cross-polarization characteristics of polarization diversity antenna.

238

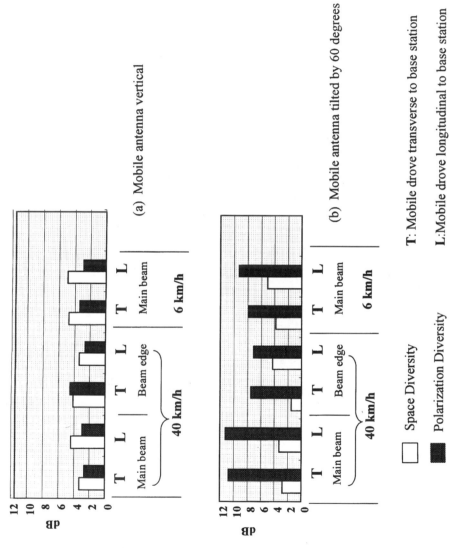

(a) Mobile antenna vertical

(b) Mobile antenna tilted by 60 degrees

T: Mobile drove transverse to base station

L:Mobile drove longitudinal to base station

☐ Space Diversity

■ Polarization Diversity

diagram of the transmission systems with transmission diversity is shown in Figure 5.29 [17]. The base station has two antennas and receivers. The portable station has only one antenna and receiver. In a TDD system, a single carrier frequency is used to provide two-way communication (upward channel, PS to BS; downward channel, BS to PS). The BS is able to predict the received signal strength at the PS because of the reciprocity between upward and downward channels. The BS receives an upward link signal from the PS using the reception diversity method, and measures the received signal strength during receiving period. In addition, it predicts which antenna gives the highest received signal strength at the PS. Then, the selected antenna is used for transmission. The signal frame configuration is shown in Figure 5.30. CH1 is for control channel, and CH2 to CH4 are user channels. At CH2, transmission and reception are repeated every 2.5 ms. In this short time interval, propagation conditions are supposed to be stable. So, the same propagation condition in an upward channel is supposed to exist in a downward channel.

Practical base station antennas are shown in Figures 5.31(a) to (d). Antennas shown in Figures 5.31(a) to (c) are omnidirectional collinear antennas. As a diversity antenna strip, a space diversity system is employed. Antennas are set up in a vertical and a horizontal plane. In Figure 5.31(a), where a vertical diversity antenna setup is employed, two antennas are contained in a single pole. The base station of the Figure 5.31(b) type antenna is most convenient. In the case of Figure 5.31(c), four antennas are employed and four-branch combination diversity is achieved. The radio cell radius of this type is some 100m. Figure 5.31(d) shows the indoor-type base station antennas, which are conventional circular patches that meet the low profile condition. Diversity antenna setups are shown in Figure 5.32. The antenna setup orthogonal to a road has superior diversity gain. For antenna separation(s) larger than 50 cm, sufficient diversity gains are achieved.

A typical omnidirectional collinear antenna configuration is shown in Figure 5.33 [18]. The inner microstrip patch antennas are excited. The cylindrical parasitic elements are used as omnidirectional radiators. The element separation is 0.7 wavelength. Five radiation elements are employed for the upper and lower antennas, respectively. Antennas are covered with a radome whose diameter is 17 mm. Antenna beamwidth in the vertical plane is 16 deg. Antenna gain of 7.5 dBi is achieved.

5.1.2.2 Adaptive Array Antenna

Adaptive array antennas have the ability to increase antenna gain at the desired signal direction and suppress interference at undesired directions. Therefore, it is expected to save transmission power and increase the system capacity. DDI Corp. introduced this antenna in 1998. The block diagram of an adaptive base station antenna is shown in Figure 5.34 [19]. Four antennas are used for transmission and reception. For the output signal of each antenna, suitable weight is determined in the CPU module to achieve the best bit error rate (BER) value by means of CMA algorithm. In this case, a maximum antenna radiation pattern is achieved in the desired signal direction and pattern nulls are

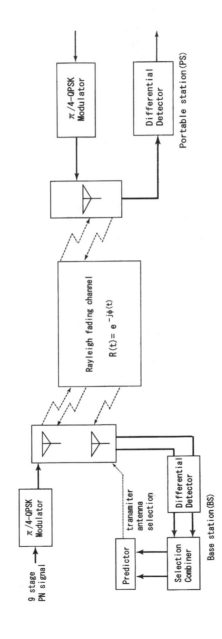

Figure 5.29 Linear predictive transmission diversity system block diagram.

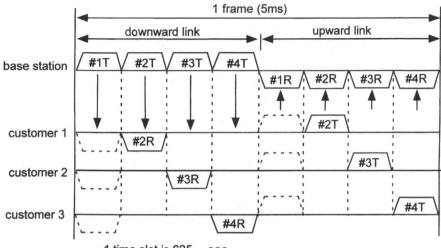

· 1 time slot is 625 μ sec
· #1 is control channel, #2~#4 are customer channels

Figure 5.30 Signal frame configuration.

achieved in the undesired directions. The TDD system has the ability to provide the same propagation conditions for an upward and a downward channel. So, the determined weights are also effective in the case of transmission to suppress interference in the undesired directions and increase antenna gain in the desired direction.

The base station antenna configuration is shown in Figure 5.35 [19]. Each antenna has antenna gain of 10 dBi. Antenna spacing is five wavelengths. A radio unit is installed at the foot of the antenna.

A study to determine antenna spacing was conducted [20]. The antenna setup is shown in Figure 5.36. In this case, four half-wavelength dipole antennas are used for the base station antenna. Antenna spacing is denoted as d. The measured SINR (signal-to-interference-noise ratio) for various antenna spacings is shown in Figure 5.37. Here, ϕ_d denotes the direction of desired signal and ϕ_i denotes the direction of interference, respectively. In every interference angle, SINR increases for large antenna spacing. SINR saturates near the antenna spacing of five wavelengths. So, the antenna spacing of a base station is determined to be five wavelengths.

Antenna radiation pattern changes in the case of interference suppression are shown in Figure 5.38 [21]. A deep pattern null is achieved in the direction of interference. Adaptive antenna performance of interference suppression is ensured.

A field test was conducted with a conventional space diversity antenna having a transmission power of 500 mW and an adaptive antenna with a total transmission power of 125 mW [22]. Nearly the same communication quality in the same cell size was

(a)

Figure 5.31 Practical base station antennas with diversity: (a) top of a telephone box; (b) attached to a telegraph pole; (c) top of a building; (d) ceiling of a subway station.

confirmed. As a result, a 6-dB increase in antenna gain was obtained. Received power is also compared with a conventional diversity antenna. The average received power of an adaptive antenna was 3 dB higher than that of a conventional diversity antenna.

5.1.3 Adaptive Array Antenna for Wireless Local Loop

An adaptive array antenna system (AAAS) applied to a WLL system is introduced [23]. The system is called Super PHS-WLL, because the PHS standards are applied to the

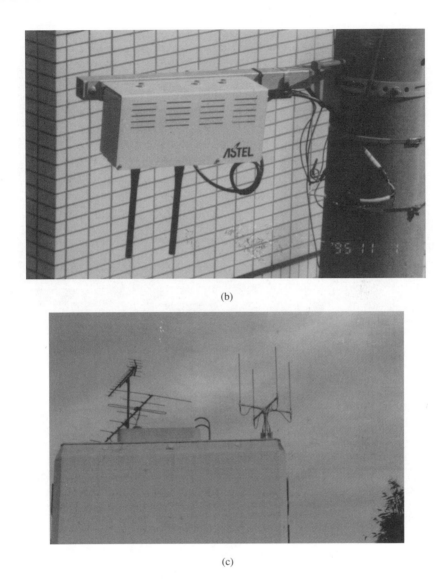

(b)

(c)

Figure 5.31 (continued).

WLL, and wideband radios and spatial channel processing are employed to enhance system performance. By means of adaptive beamforming, the channel capacity is increased, the number of multipath signals is reduced, the coverage is expanded, and the flexible configuration of the coverage for each base station to match the local propagation environment is made feasible.

The system concept is briefly illustrated in Figure 5.39. The figure shows an example of a system configuration that is composed of a BS (base station), a single and four-line

(d)

Figure 5.31 (continued).

SU (subscriber unit), and connection to the local exchange. The operating frequency of the system is the same as that of the PHS, 1,880–1,930 MHz, and for the TDMA-SDMA system for the channel access and the duplex system, TDD, are employed.

The AAAS, which produces multibeam by DBF (digital beamforming) technology, performs SDMA functions. The beams are automatically directed to a plural number of SU, wherever they are located. The received signal at the BS in the multipath environment is processed adaptively to enhance the processing gain, thereby improving the quality of the link. It also contributes to rejecting interferences by directing a null against it and enhancing the reuse of the same frequency, thereby increasing the number of channels.

The system performance that can dynamically adapt to the changeable propagation environments is achieved by DBF technology. The DBF software processes in real time the phase and time differences among the incoming signals in the multipath environment so that the signal transmission can be directed to each of the desired SU. By this means a stable and robust link can be established.

Other factors that contribute to enhancing the channel capacity are the multiple-frequency operation that is made possible by using a plural number of wide-band radios in addition to the AAAS and use of the PHS standard TDD, which has four time slots in one data frame that can be used for both transmitting and receiving of four channels at the same time in one frequency.

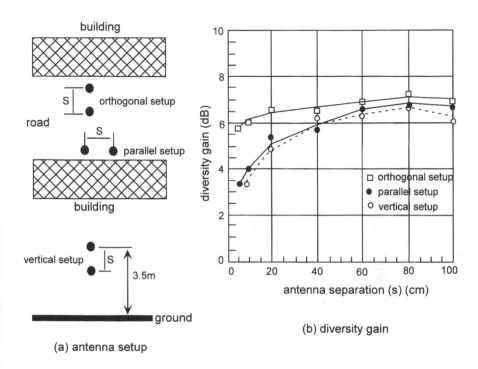

(a) antenna setup

(b) diversity gain

Figure 5.32 Diversity antenna: (a) setup configurations; (b) diversity gain.

A base station of the Super PHS-WLL system, which has been operated practically, is composed of 12 antennas and 12 radios, and uses 16 radio frequencies for the 16-multiplex operation. By combining the spatial channel processing with the AAAS, a spatial channel efficiency of about 2.5 is achieved. Consequently, the voice channel capacity achieved is up to 155; that is, 4 (time slot) × 16 (frequency) × 2.5 (spatial channel efficiency) − 5 (control channel). It can serve 2,730 subscribers as the total traffic capacity, and is capable of covering up to about 15 km.

Figure 5.40 shows a pole-mounted base station antenna, which is comprised of 12 elements. With the number of antennas and radiators at 8 each, the channel capacity (voice) may be changed to 120. The system can claim remarkable features such as enhanced capacity, coverage, and quality; in addition to these, the increased channel capacity contributes to a significant reduction in the cost per subscribers.

5.1.4 Antennas for N-STAR Systems

In March 1996, NTT Mobile Communication Network Inc. (NTT DoCoMo) initiated the N-STAR mobile satellite communication services, the goal of which is to complement the terrestrial cellular system service areas and expand the service areas for maritime

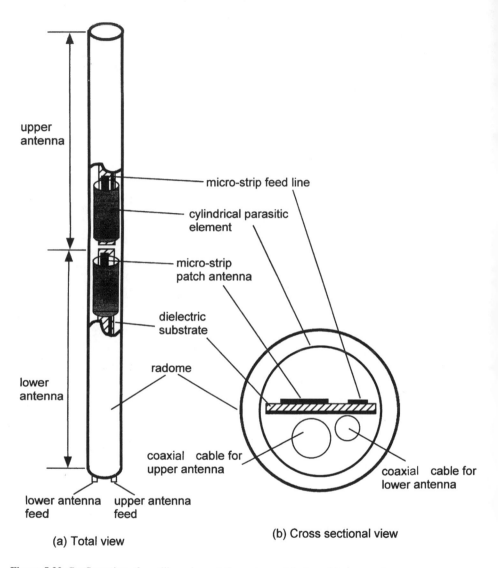

upper antenna

micro-strip feed line

cylindrical parasitic element

micro-strip patch antenna

dielectric substrate

radome

lower antenna

coaxial cable for upper antenna

coaxial cable for lower antenna

lower antenna feed

upper antenna feed

(a) Total view

(b) Cross sectional view

Figure 5.33 Configuration of a collinear base station antenna: (a) view; (b) cross-section.

communications systems. An overview of this system is shown in Figure 5.41 [24]. Frequency bands are S-band (2.6/2.5 GHz) for the service link and the C-band (6/4 GHz) for the feeder link. The service areas are covered by four radio beams, the radii of which are approximately 600 km.

Three kinds of terminals have been developed: car mounted, maritime, and portable [25,26]. The car-type terminal is shown in Figure 5.42. This terminal is used as a dual-

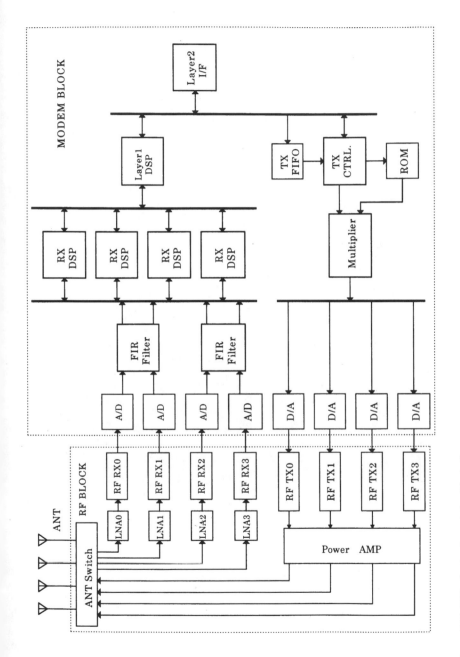

Figure 5.34 PHS adaptive base station block diagram.

Figure 5.35 PHS adaptive base station antenna configuration. (Courtesy of Kyocera Corporation.)

mode (satellite and cellular) terminal. The terminal is composed of an antenna unit and a radio unit. The antenna unit is mounted on the roof of a car. The size of the antenna unit is 30×34 cm and the weight is about 3 kg. The interior of the antenna unit is shown in Figure 5.43 [27]. A nine-element planar array antenna is employed and a satellite tracking mechanism in the azimuth plane is adopted. The antenna radius is 27 cm. As a radiation element, a circular patch antenna equipped with a parasitic element is employed. This antenna is fed in two points in order to produce circular polarization.

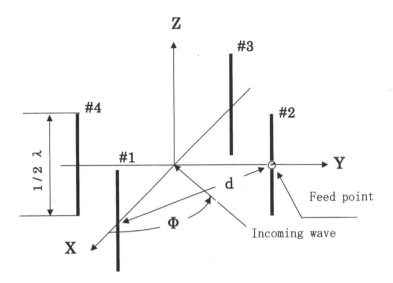

Figure 5.36 Base station antenna setup.

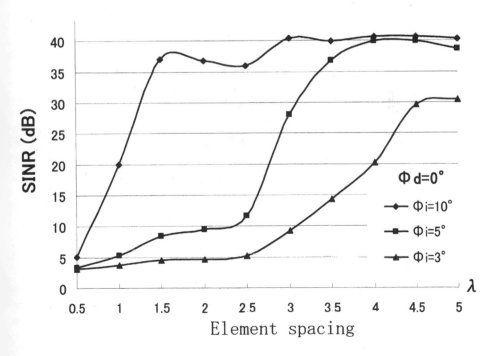

Figure 5.37 Measured SINR for various antenna spacing.

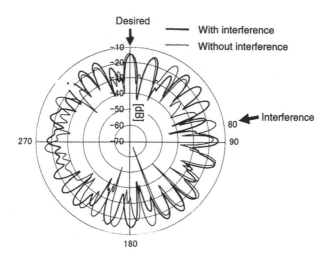

Figure 5.38 Antenna radiation pattern of interference suppression.

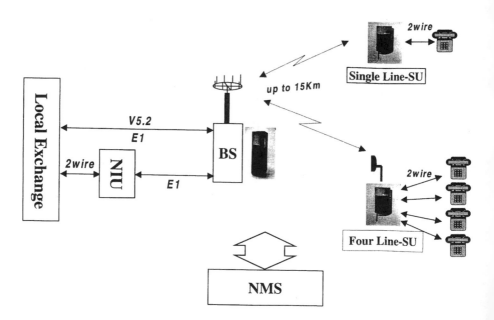

Figure 5.39 System concept of PHS-WLL.

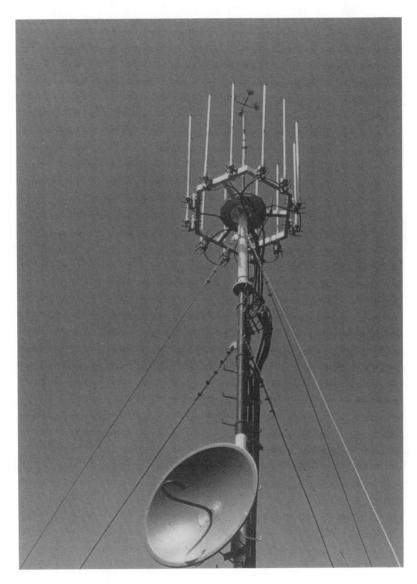

Figure 5.40 Pole-mounted base station antenna. (Courtesy of Kyocera Corporation.)

The antenna radiation pattern in the elevation plane is shown in Figure 5.44. The N-STAR direction is about 50 deg. This beam is designed to operate in the angular range of 40–60 deg. In this angular range, antenna gain of 10 dBi is achieved and axial ratio is less than 2 dB.

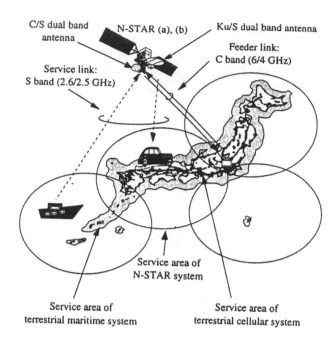

Figure 5.41 Overview of the N-STAR mobile communications system.

The maritime-type terminal is shown in Figure 5.45, which is exclusively used for the satellite system. The diameter of the antenna unit is 30 cm and the weight is about 5 kg. The antenna unit is usually installed on a mast. The maritime-type terminal uses the same antenna. However, both the azimuth and elevation of the antenna plane are controlled, thus this type can be used in an undulating environment. The portable-type terminal is shown in Figure 5.46. The portable-type terminals are provided as both dual-mode use and exclusively for use with a satellite system. The radio unit and the antenna unit comprise one unit. The total weight is 2.7 kg. Subscribers should turn the antenna unit toward the satellite when using this type. A planar array antenna of 18.5 cm^2 is used. The same circular patch elements used in the car-mounted antenna are used. Four antenna elements are employed and a 3-dB beamwidth of 40 deg in azimuth and elevation direction is achieved.

5.2 DIGITAL BEAMFORMING ANTENNAS IN MOBILE COMMUNICATIONS

5.2.1 What Do We Mean by "Digital Signal Processing Antennas"?

Conventionally, adaptive signal processing antennas are referred to as *adaptive antennas* (or adaptive arrays) [28–30]. Digital beamforming (DBF) antennas realize such antenna

Figure 5.42 Car-type terminal. (Courtesy of NEC.)

Figure 5.43 Car-type antenna unit. (Courtesy of NEC.)

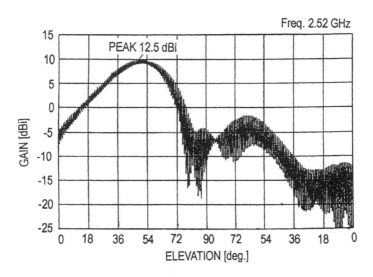

Figure 5.44 Radiation pattern of a car-type antenna in the vertical plane.

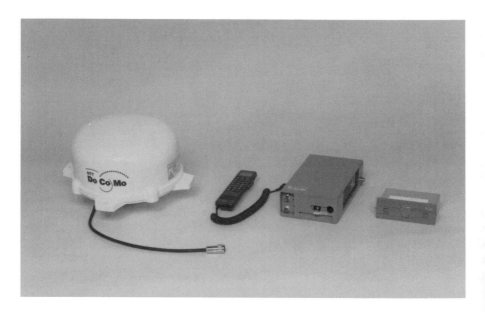

Figure 5.45 Marine-type terminal. (Courtesy of NEC.)

Figure 5.46 Portable-type terminal. (Courtesy of NEC.)

characteristics by digital signal processing. DBF antennas also use a technique for forming fixed patterns such as multibeam patterns and low sidelobe patterns, and in this respect they differ from adaptive antennas slightly [31]. The term *adaptive antenna* refers to its functional aspects, whereas the term *DBF antenna* refers to the configuration aspects. In recent years, the application of adaptive antennas to mobile communications has become widespread, and adaptive antennas developed for this purpose are referred to as *smart antennas* [32]. However, these do not have a strict definition and the term is not necessarily used properly. For example, regarding the smart antenna, there is an idea that "the smart antenna is another name for the adaptive antenna" [33], and there is another idea that "the smart antenna is a comparatively advanced adaptive antenna based on space-time signal processing" [32]. Furthermore, the concept of a "software antenna" has been proposed that can reconfigure its adaptive algorithm by means of programmable logic circuits [34]. The software antenna can be made to more dynamically and adaptively track environmental changes. In [35], the term *software antenna* has a more general meaning closer to that of "smart antenna." Here, the term *digital beamforming antenna* will be used commonly to refer to digital signal processing antennas.

5.2.2 Basic Configuration of DBF Antennas

The DBF antenna receives signals by an antenna array comprising many antenna elements, following which analog-to-digital converters (A/D) convert them without any combining and dividing. It thus realizes desired functions such as multibeam formation, low sidelobe characteristics, and adaptive behavior by digital signal processing. Although the main use of a DBF antenna is as a receiving antenna from the viewpoint of adaptive signal processing, transmitting signals can also be generated based on the information of a receiving DBF antenna (result of adaptive reception), and an intelligent transmitting antenna may be constructed in the same way. Here, however, we focus on receiving functions.

The structure and characteristic features of DBF antennas are easier to understand in comparison with phased-array antennas. Figure 5.47 shows the basic structures of both types. As for phased-array antennas, by controlling the phase of the feed circuit to/from each antenna element, the desired antenna pattern can be produced, and the pattern can be controlled adaptively according to the variation of the surrounding radio wave environment. To achieve this function, a number of phase shifters, which are composed of analog devices, are attached to every antenna element. Nevertheless, the number of faces (= beam patterns) of the antenna is always one at every instant. On the other hand, the DBF antenna converts the signals from each antenna element into digital signals, and assigns weights to each to synthesize the resulting signal. Because different combinations can be generated without limitation by parallel signal processing, the antenna may simultaneously have plural "faces" varying with time. It is theoretically possible to achieve the

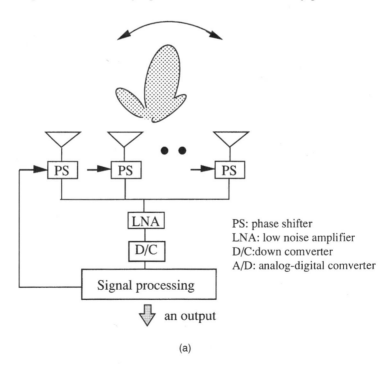

PS: phase shifter
LNA: low noise amplifier
D/C:down comverter
A/D: analog-digital comverter

(a)

Figure 5.47 Basic configuration of a phased-array antenna and a DBF antenna.

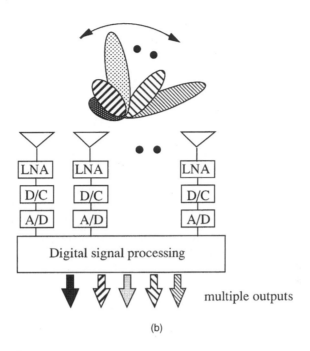

(b)

Figure 5.47 (continued).

same function using analog circuits (Butler-Matrix circuit, etc.), however the feed network circuit configures a complicated matrix, the antenna becomes more complex the greater the number of elements, and there is no flexibility in beam management (beam configuration, etc.). In the case of DBF, the signal processing is in the last stage, so low-noise amplifiers are indispensable for avoiding the effect of noise after reception, and the antenna has a structure wherein antenna elements and active devices are integrated in one piece.

A DBF antenna can generate a desired pattern in parallel by signal processing, and it is therefore suitable for mobile communications where multibeam formation and adaptive operation in a time-varying radio wave environment are required.

5.2.3 Analog-to-Digital Conversion

In extracting the functions of the DBF antenna to the maximum, the analog-to-digital (A/D) conversion unit (= sampling unit), which acts as a bridge between the analog unit (RF signal unit) and digital signal processing unit, plays a major role. In particular, due to the progress of DSP device technology together with higher speeds for A/D conversion and digital signal processing, sampling at higher frequencies than conventional baseband sampling has become possible, and the number of choices has increased. As a result, a highly functional antenna system can be designed. In this section, we focus on this A/D conversion unit (= sampling unit), and describe the structure and characteristic features of various sampling schemes.

5.2.3.1 A/D Conversion Stage

The signals from each antenna (analog signals) are A/D converted at a certain stage and input to the signal processing circuit (DSP unit). Figure 5.48 shows an example of A/D conversion in a baseband unit and an intermediate frequency (IF) unit. In the case of the baseband unit [Figure 5.48(a)], as a large amount of processing is performed in the analog unit, the signal processing function of DSP can be demonstrated to the maximum extent. Specifically, this method is suitable for the processing of broadband signals, and in this case, twice as many A/D converters as the number of antenna elements is needed. In the case of the A/D conversion in the intermediate frequency unit [Figure 5.48(b)], for the DSP unit to perform IQ detection, the number of A/D converters should be equal to the number of antenna elements. Although circuit construction of the analog unit is

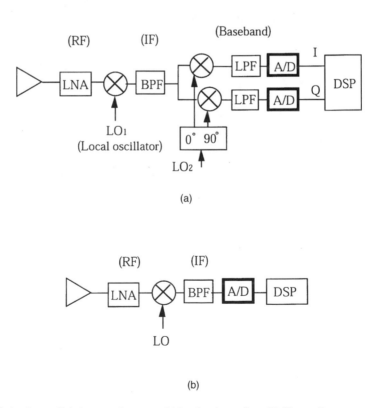

(a)

(b)

Figure 5.48 Analog-to-digital conversion stage. (a) baseband sampling; (b) IF sampling.

simplified when A/D conversion is performed in the IF band, the computational load on the digital signal processing unit is that much larger. Due to the limitation of sampling or processing speed, the intermediate frequency input to the A/D converter is often set from one to several times the signal bandwidth. In principle, a scheme is also possible wherein RF band signals are sampled directly, but as in the case of the GHz band, a GHz response is required for the A/D converter and DSP; this is still far from being practically feasible.

5.2.3.2 Sampling Schemes

Nyquist Sampling and Oversampling

The classification mentioned above was that of a circuit arrangment that focused attention on the A/D conversion stage, but another classification is that of the relation between signal bandwidth and sampling rate, that is, that of digital signal processing. In this case, oversampling is sampling at a higher rate than that of Nyquist sampling, and undersampling is sampling at a lower rate than that of Nyquist sampling, where the sampling rate of Nyquist sampling is twice that of the maximum frequency component of an analog signal.

In Nyquist sampling, the sampling is performed at twice the maximum frequency of the baseband signal, or twice the maximum frequency of modulated signal over the intermediate frequency. Signal components appear within the range $DC - f_s/2$ (f_s = sampling frequency). This is the minimum sampling rate so that the aliasing component does not appear in this frequency range. Steep low-pass filtering characteristics are required as an antialiasing filter, but as the sampling rate is the minimum at which information in the analog signal is not lost. In oversampling, the sampling is performed at a higher rate than Nyquist sampling (e.g., two or four times), and the requirements for the antialiasing filter are less stringent. There is also an advantage in the signal-to-noise ratio (SNR) relative to quantization error because the A/D conversion improves with increasing sampling rate.

Undersampling (Bandpass Sampling)

Undersampling, applied to the sampling of bandpass signals in the RF band or IF band, is also referred to as bandpass sampling. In this method, sampling is performed at a rate below the Nyquist sampling, and downconversion and sampling are performed simulataneously by using the fact that input signals are converted at frequencies of less than one-half the sampling rate. This is known as *undersampling,* but it is understood that a rate higher than the Nyquist sampling rate (= twice the signal bandwidth) is required for signal bandwidth as a matter of course.

The principle of extracting the desired information by undersampling is described in Figure 5.49. Let the center frequency of the IF band signal be f_0, and the signal

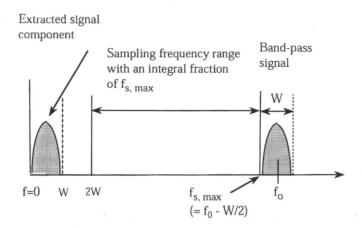

Figure 5.49 Determination of a sampling frequency for the undersampling scheme.

bandwidth (in this case, 100% bandwidth, which includes the whole signal component) be W. If a temporary sampling frequency is set to the lower band edge $(f_0 - W/2)$ as $f_{s,\max}$, a desired signal component will be reproduced in the low-frequency part $(0 - W)$. If the sampling frequency f_s is an integral fraction of $f_{s,\max}$ with a condition of $f_s > 2W$, a desired signal component is reproduced in the same low-frequency part $(0 - W)$. Undersampling can be understood to be a method where bandpass signals having the center frequency f_0 are directly downconverted to a lower center frequency $W/2$ (however, $f_{s,\max}$ are not necessarily limited to the band edge, and may not always be $W/2$). As an example, let us consider the case where the IF is 450 MHz and the signal bandwidth W is 20 MHz. The lower band edge frequency $f_{s,\max}$ then becomes 440 MHz, so candidate sampling frequencies will be integral fractions of this above 40 MHz, for example, 220 MHz (1/2), 147 MHz (1/3), 110 MHz (1/4), 55 MHz (1/8), and so on. If a lower rate is chosen, $f_s = 55$ or 44 MHz. The circuit configuration of undersampling is fundamentally the same as that of the A/D conversion scheme in the IF band mentioned in Figure 5.48(a).

Undersampling is a technique wherein IF signals (and RF signals in future) are directly downconverted, and input to a DSP unit. It is fully compatible with DBF antennas or software radio in a wider sense, and will no doubt find extensive use. Problems to be solved in its implementation are (1) broadband characteristics are required of the A/D converter to sample the phase of IF signals correctly even if the sampling rate is low, and (2) steep bandpass characteristics equal to those of a low-pass filter (LPF) for the baseband signal are required for the band filter (antialiasing filter), leading to stringent filter design specifications in the front stage of the A/D converter.

Quantization Error and Dynamic Range Accompanying A/D Conversion

An A/D converter determines dynamic range according to the number of bits, and the level fluctuation below the minimum bit is the quantization error. In an ordinary R bit

A/D converter that has a fixed quantization step width, when the input signal is a sine wave, the SNR ratio for the quantization error relative to the signal in Nyquist sampling is $6.02R + 1.76$ dB. This SNR specifies a noise floor or dynamic range for a time sequence signal after A/D conversion. By performing oversampling, the SNR is improved by $10\log(f_s/f_{sN})$ (dB) compared with Nyquist sampling (sampling frequency: f_{sN}). In an N-element antenna array, the SNR is improved by a further $10\log N$ (dB) compared to Nyquist sampling in the case of a single element.

From the above computation, the dynamic range is at most 70 dB for a 12-bit A/D converter. In the radio wave propagation environment of mobile communications, a fluctuation range that exceeds this level including the part due to distance dependence occur frequently, so some expansion of dynamic range is desirable. This is usually provided by an automatic gain control (AGC) in the IF or RF stage. In the case of an antenna array, in order to avoid degradation of the antenna directivity, a common AGC that makes all branch gains even is needed. However, regarding the AGC control method, the required conditions are different depending on adaptive algorithm adopted in the system. This has not yet been studied sufficiently, and is a topic for future research.

5.2.4 Element-Space and Beam-Space Beamforming

In performing digital signal processing and extracting a signal with a desired quality, it is common to synthesize the signal by giving an optimum weighting to each antenna element. On the other hand, it is also possible to confer directivity, form multiple beams which easily receive radio waves in certain directions, and synthesize the signal by weighting these outputs. The former processing is referred to as *element-space* and the latter is referred to as *beam-space*.

Figure 5.50 shows the basic construction of a beam-space array. For systems based on orthogonal conversion such as multibeam forming by spatial FFT described later, there is no essential difference between techniques because the overall amount of information has not changed. However, DBF antennas often appear in the beam-space scheme (wherein a multibeam is first formed preprocessing an incoming signal). Whereas signal processing using all element signals is always required in element-space, signal processing using a couple of beam-space outputs that contain higher power signals is often acceptable [36,37]. This leads to a considerable decrease of circuit scale and computational load.

Multibeam forming, which corresponds to a preprocessing to signal processing, is well suited to the DBF antenna. As a way of forming this multibeam with a small amount of computation, the FFT algorithm is often adopted. It is used to form N beams from an antenna array comprising N elements arranged at equal intervals (typically 1/2 wavelength intervals). In ordinary beamforming, N^2 computations are required for both addition and multiplication. If the FFT algorithm is used for this computation, additions can be reduced to $N\log_2 N$ times and multiplication can be reduced to $2N\log_2 N$ times; therefore, the effect appears at $N > 8$ and this effect becomes conspicuous the larger the number of antennas.

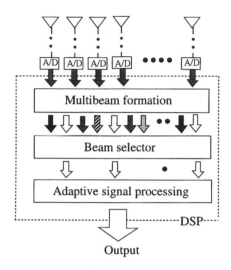

Figure 5.50 Beam-space-type DBF antenna.

Reduction of computation time by such a technique is very important in order to extract the maximum performance from a DBF antenna in the context of limited processing capacity.

5.2.5 Adaptive Algorithm

5.2.5.1 Overview

In discussing the operating principle of adaptive antennas, it is convenient to recognize a criterion (= a measure of the optimization) and an adaptive algorithm (= optimization method to realize the criterion). As examples of such a criterion, minimum mean square error (MMSE) and maximum likelihood sequence estimation (MLSE) are typical. MMSE is a method in which the difference between a reference signal (= replica of transmitted signal) and the received signal synthesized by weighting is taken as an evaluation parameter, and the optimum weighting is determined by means of an adaptive algorithm where the difference (= error) is minimized. It is widely used for spatial signal processing, which performs antenna pattern control, and time signal processing, which performs waveform equalization. MLSE estimates a signal pattern that produces a result most like the received waveform obtained based on transmission path estimation, and is often used for signal processing of time domain.

The adaptive algorithm that tries to achieve this by controlling weights based on the criterion is the optimizing method. For MMSE, the least mean square (LMS) algorithm based on the steepest descent method and the recursive least square (RLS) algorithm that aims for optimization using previous sample values are often used. LMS has a small-scale computational load, but longer time is required for convergence; on the other hand, RLS involves complex computations, but convergence is rapid.

For modulation schemes with a constant envelope such as FSK, GMSK, and PSK, an algorithm that minimizes envelope fluctuations using the fact that the envelope level is constant as prior knowledge is the constant modulus algorithm (CMA). The criterion is close to MMSE, and for the optimizing method, the steepest descent method can be used. As for the MLSE, the Viterbi algorithm that aims to make computation more efficient is utilized frequently. Figure 5.51 is a functional block diagram of an adaptive array based on the MMSE criterion as an example.

5.2.5.2 MMSE Adaptive Array

It may be said that the most traditional adaptive algorithm based on the MMSE criterion was the LMS method proposed by Widrow [28]. In the MMSE criterion, a reference signal with which comparison should be made is known in advance, and operations are

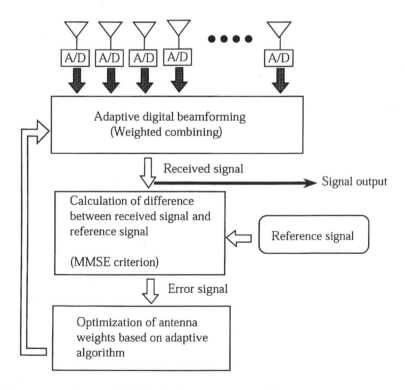

Figure 5.51 Functional diagram of MMSE adaptive array.

performed to minimize the difference between the actual received signal after combining with weighting and the reference signal. This is a system suited to communications where a pilot signal or predetermined signal is available.

An error signal $e(t)$ in a received signal $y(t)$ given by the difference between $y(t)$ and a reference signal $d(t)$ can be expressed by the following equations:

$$e(t) = d(t) - y(t) = d(t) - \mathbf{W}^H \mathbf{X}(t) \tag{5.6}$$

where $\mathbf{X}$ is input signal vector, and $\mathbf{W}$ is a weight vector, respectively, given by

$$\mathbf{X} = [x_1 \, x_2 \ldots x_N]^T \tag{5.7}$$

$$\mathbf{W} = [w_1 \, w_2 \ldots w_N]^T \tag{5.8}$$

where N is the number of antenna elements, and superscripts T and H are a transpose and an Hermitian (or complex conjugate) transpose.

The expected value of the error power is as follows:

$$E[|e(t)|^2] = E[|d(t)|^2] - \mathbf{W}^T \mathbf{r}_{xd}^* - \mathbf{W}^H \mathbf{r}_{xd} + \mathbf{W}^H R_{xx} \mathbf{W} \tag{5.9}$$

Here, $\mathbf{r}_{xd}$ is defined as the correlation vector of the reference signal and input signal vector and R_{xx} is the correlation matrix of $\mathbf{X}$. They are respectively given by:

$$\mathbf{r}_{xd} \equiv E[\mathbf{X}(t) \cdot d^*(t)] \tag{5.10a}$$

$$R_{xx} \equiv E[X(t)X^H(t)] \tag{5.10b}$$

where * means a complex conjugate. The weight vector that minimizes the expected value of (5.9) satisfies the following formula:

$$\nabla_w E[|e(t)|^2] = -2\mathbf{r}_{xd} + 2R_{xx}\mathbf{W} = 0 \tag{5.11}$$

where ∇_w is the derivative by weight vector. Therefore, the optimum weight $\mathbf{W}_{\text{opt}}$ is given by the following equation:

$$\mathbf{W}_{\text{opt}} = R_{xx}^{-1} \mathbf{r}_{xd} \tag{5.12}$$

Some typical methods of obtaining the optimum weight are the (1) LMS algorithm, (2) RLS algorithm, and (3) SMI (sample matrix inversion) method. We briefly describe the features of these methods.

LMS Algorithm

In this method, the weighting is moved sequentially toward an optimum value so that error is minimized [28]. In calculating the $(m + 1)$th weight $\mathbf{W}(m + 1)$ from the weight $\mathbf{W}(m)$ found on the mth occasion by the steepest descent method, the basic equation is:

$$\mathbf{W}(m + 1) = \mathbf{W}(m) - \frac{\mu}{2} \nabla_W E[|e(m)|^2] \tag{5.13}$$

where μ is a real positive value which gives the step size.

Now, introducing (5.11) and substituting the time average value by an instantaneous one, the following equation is obtained:

$$\mathbf{W}(m + 1) = \mathbf{W}(m) + \mu \mathbf{X}(m) e^*(m) \tag{5.14}$$

The LMS method finds the optimum weight by repeating this computation. It has the advantage that computation load is very light. However, there are disadvantages in that there is still some remaining uncertainty in the selection of step size, and the number of iterations necessary for convergence is uncertain.

RLS Algorithm

This method is used for performing sequential optimization while giving exponential weighting to the error [30]. A quantity $T(m)$ minimized on the mth occasion is expressed by the following equations:

$$T(m) = \sum_{i=1}^{m} \beta^{m-i} |e(i)|^2 \tag{5.15}$$

$$e(i) = d(i) - \mathbf{W}^H \mathbf{X}(i) \tag{5.16}$$

Parameter β is a constant such that $0 < \beta < 1$. The optimum weight is expressed in the following form:

$$\mathbf{W}(m + 1) = \mathbf{W}(m) + \delta R_{xx}^{-1}(m) \mathbf{X}(m + 1) e^*(m + 1) \tag{5.17}$$

$$e(m + 1) = d(m + 1) - \mathbf{W}^H(m) \mathbf{X}(m + 1) \tag{5.18}$$

$$\delta = \frac{1}{\beta + \mathbf{X}^H(m + 1) R_{xx}^{-1}(m) \mathbf{X}(m + 1)} \tag{5.19}$$

The inverse matrix $R_{xx}^{-1}(m)$ contained in this formula can also be found by the following updating formula:

$$R_{xx}^{-1}(0) = \eta^{-1}I \tag{5.20a}$$

$$R_{xx}^{-1}(m) = \frac{1}{\beta}R_{xx}^{-1}(m-1) - \frac{R_{xx}^{-1}(m-1)\mathbf{X}(m)\mathbf{X}^H(m)R_{xx}^{-1}(m-1)}{\beta^2 + \beta\mathbf{X}^H(m)R_{xx}^{-1}(m-1)\mathbf{X}(m)} \tag{5.20b}$$
$$(m = 1, 2, 3 \ldots)$$

where η is a positive constant.

SMI Method

This method is a method of solving the inverse matrix of (5.12) directly, and if the signal received at each antenna can be extracted, the optimum weight can be calculated by one computation [29]. However, it has the disadvantage that when the number of antenna elements increases, the computational load becomes larger.

Summary of Three Methods

Summarizing the overall advantages and disadvantages according to the application, when rapid convergence is required (such as when the signal transmission rate is high), the order of priority of these methods is SMI > RLS > LMS; and when there are a large number of antenna elements and it is desired to reduce the computation load in one operation (i.e, when rapid optimization is not requested), the order will be LMS > RLS > SMI.

The adaptive array based on MMSE criterion assumes that a reference signal is available, and it is thus more effective in communications systems because they often have preassigned signals such as a pilot signal, preamble signal, and so on. For example a training signal that is inserted in bursts in communication systems using the TDMA scheme can be used as a reference signal, and its application to mobile communications is now attracting attention.

5.2.5.3 CMA Adaptive Array

For constant-envelope-type modulation schemes such as FSK, GMSK and PSKs, a constant modulus algorithm (CMA) adaptive array is applicable [38,39]. In the method, each weight is controlled so that the envelope of received signal is constant. By doing so, interfering signals or delayed signals that cause envelope fluctuations can be eliminated. The main feature of this method is not to use any reference signals, namely, a blind algorithm.

The principle of CMA adaptive processing is now briefly described. CMA is an algorithm that controls a weight so that the envelope of the combined output ($|y|$) becomes constant. Therefore, an optimum weight is obtained by minimizing the evaluation function expressed by the following formula:

$$J = \frac{1}{4}E[\{|y|^2 - \sigma^2\}^2] \qquad (5.21)$$

where σ is the desired envelope value and $E[\,]$ is an expected value.

Because the solution of **W** that minimizes the evaluation function J cannot be found analytically, iterative computation by a nonlinear optimization technique is required. When the steepest descent method is used as computing algorithm, the update of the weight follows this equation:

$$\mathbf{W}(m+1) = \mathbf{W}(m) - \mu\mathbf{X}(m)y^*(m)\{|y(m)|^2 - \sigma^2\} \qquad (5.22)$$

where μ is a constant that specifies a feedback step size. Because CMA allows blind processing, it is suited to mobile communications where the reference signal is not available.

5.2.6 Spatial Signal Processing Antennas

It is thought that one of the biggest reasons why adaptive antennas based on spatial signal processing are now attracting attention to communications application is because they have broken the limitation of channel capacity due to removal of cochannel interference by antenna pattern control in cellular mobile communications systems. However, due to the inevitability of increased hardware and higher cost compared to a conventional antenna, it is more realistic to apply these antennas to a base station rather than a user terminal. They also have the advantage that if they are applied to a base station, the coverage of the base station can be expanded due to the high directivity of the antenna. In fact, due to the rapid spread of cellular phones, there is considered to be a serious lack of frequencies, and intensive research is being carried out worldwide on the commercialization of adaptive signal processing antennas to cellular base stations. The following are actual examples of this.

5.2.6.1 CMA Adaptive Array with DSP Implementation

The feasibility in communication applications of adaptive antenna arrays has been studied since around 1990, when experimental research began on a prototype. In particular, the research carried out at CRL (Japan) was probably the first in the world to prove their effectiveness in real fading environments [40,41]. The prototype antenna was a DBF antenna consisting of four omnidirectional monopole antennas in a horizontal plane at the corners of a square with a 0.444-wavelength interval, and was developed as a vehicle-mounted mobile communications antenna using 1,431 MHz. The weight updating algorithm used was CMA, which is a type of blind algorithm, the radio transmitting rate was 256 kbps, and the multiplexing method was 24-channel (8 kbps per channel) TDM (time-division multiplexing). Figure 5.52 shows the construction of the adaptive array, and Table 5.4 shows parameters of the experimental system. The DSP used for array adaptive signal processing and demodulation was a 16-bit fixed point processor having a performance of about 5 MIPS per chip. Sufficient performance was obtained with two DSP chips, but 4 DSPs were used with one DSP per antenna element. A/D conversion was four samples (sampling rate 1 MHz) per symbol with a 10-bit resolution. The prototype antenna was mounted on an experimental vehicle which traveled around the laboratory test area, and into suburban areas and urban areas to verify its performance. Figure 5.53 shows results of BER characteristics in running tests in urban areas. It might be expected that the results would show a large amount of frequency selective fading, but an improvement of 20 dB or more was obtained at a BER = 10^{-2} compared to the case when an adaptive array was not used.

5.2.6.2 Beam-Space Adaptive Array with FPGA Implementation

In ATR (Japan), a beam-space adaptive array with FPGA (field programmable gate array) implementation was developed [42,43]. Utilizing the flexibility of FPGA, reconfiguration of two adaptive algorithms was realized in the antenna. One is CMA, which is capable of interference cancellation for constant envelope signals such as FSK and PSK modulation signals. The other is a maximal ratio combining (MRC) algorithm based on a self-beam steering (SBS) scheme that is characterized by maximal signal-to-noise reception. Because both algorithms are implemented after multibeam formation, they are called *BSCMA (beam-space CMA)* and *BSMRC (beam-space MRC)*, respectively. The antenna was developed for a proof-of-concept demonstration of the DBF functions, and was not limited to base station antenna applications.

A block diagram of the antenna system developed in ATR is shown in Figure 5.54 [42,43]. The DBF antenna consists of 16 patch antennas arranged in a 4 × 4 square, and the element spacing is a half wavelength. The carrier frequency of the system is the L-band (1.54 GHz). In the DSP section, 16 multibeams are synthesized. To realize sufficient speed of calculation convergence, signal processing at the final stage (i.e., CMA or MRC) is carried out using a limited number of beam outputs (four beams are selected at the beam selector in this system). The vector rotation in the figure plays a role in removing phase discontinuity at each beam crossover. The DBF system employs $\pi/4$-QPSK as its modulation method, and the data rate of the system is 16 kbps (maximum 64 kbps).

The DSP section of the developed system employs general-purpose rewritable FPGAs. The advantage of such an FPGA-based system is that, once functions are verified

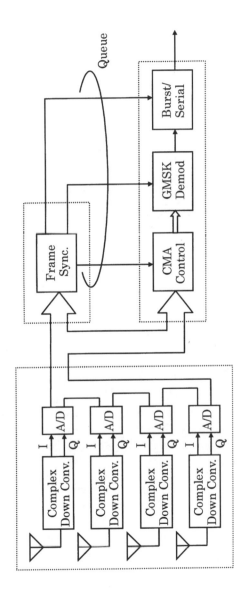

Figure 5.52 Configuration of CMA adaptive array developed in CRL. (From [40], © 1993 IEEE.)

Table 5.4
Parameters of the Experimental System

Radio-frequency	1,431.5 MHz
Modulation	GMSK
Transmission rate	256 kbps
Max TX power	37 dBm
Multiplexing	TDM
Total channel number	24
Array element	4
Adaptive algorithm	CMA
Sampling frequency	1 MHz

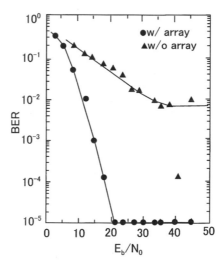

Figure 5.53 Results of field experiment on BER characteristics with and without adaptive array. (From [41], © 1993 IEEE.)

their design can be transferred as is on custom LSIs such as mask-type gate arrays called ASIC (application specific IC). Ten FPGAs, each equivalent to a nominal 25,000 gates mounted on a printed circuit board (23.3 × 34.0 cm), make it possible to construct a circuit equivalent to a 250,000-gate ASIC. Figure 5.55 shows a view of the DSP section. Table 5.5 lists the number of gates on a hardware scale required by each component of the DSP section. Calculations in each DSP block are performed at 8- and 12-bit fixed point precision. The operation clock of the DSP section runs at 7.04 MHz.

A number of experiments took place in a large radio anechoic chamber [44]. The desired wave was radiated from a scan azimuth angle of 0 deg and the interference wave from a scan azimuth angle of −23 deg. This angular difference corresponded to the antenna

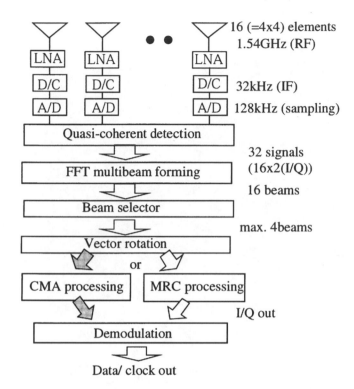

Figure 5.54 Functional block diagram of the DBF antenna developed in ATR realizing BSCMA and BSMRC.

beamwidth. For BSCMA, tracking characteristics for radiated interference waves with one-symbol delay are shown in Figure 5.56. Excellent tracking and interference suppression characteristics were confirmed and no bit errors were detected during the experiment.

Figure 5.57 shows the scene of an ETS-V satellite experiment using a van equipped with the DBF array antenna [45]. Figure 5.58 shows a BSMRC array output and the largest multibeam output measured when the van was driven in a circle on flat ground under the line-of-sight condition at an angle speed of 13 deg/s (10 km/h). The signal level degradation observed in the multibeam reception is compensated and the level variation is reduced based on SBS processing.

5.2.6.3 Smart Antennas

ERA (United Kingdom) introduced a smart antenna concept into AMPS (Advanced Mobile Phone Service), and performed field tests in the Washington, D.C., area in the United

Figure 5.55 A DSP board composed of 10 FPGAs [42].

States to improve C/I in areas with a high user density distribution [46]. Advantages of introducing the smart antenna into a cellular system are improvement of user capacity and increased coverage from the base station. The former is particularly important in high-traffic regions such as urban areas, in which case the vital factor is not steering by a high-gain antenna but improvement of C/I at the base station and user station. This suggests that to improve user capacity, gain suppression (null) in the interfering wave direction is more important than increasing the gain in the desired wave direction. Also, in considering an antenna array with a limited number of elements, it is for example

Table 5.5
Estimation of Required Logic Gates Realizing the BSCMA Array

Functional Section	Number of Gates	
	BSCMA	BSMRC
Multibeamforming	106,125	
Beam selection	2,889	
Vector rotation	6,107	
Adaptive processing	8,546	9,449
Others	3,605	
Total	127,272	128,175

simpler to reduce the gain by 10 dB in the interfering wave direction than to increase the gain by 10 dB in the desired wave direction. In the prototype antenna used in this test, an adaptive signal processing antenna was used on the receiving side, CMA was used as the adaptive algorithm, and C/I was maximized by beam steering to the desired wave direction, null steering in the interfering wave direction, and maximum ratio combining of the multipath wave. On the transmitting side, switching by a section antenna divided every 120 deg in the horizontal plane was used. In this system, as shown in Figure 5.59, the signal, which has been combined by digital signal processing, can be reconverted to an analog signal by a D/A converter, and after returning to RF, it is input to a conventional demodulator. In this way, performance is improved simply by adding antennas and circuits and making as few modifications to the conventional receivers as possible.

In the European Community (EC), as part of the RACE program (currently ACTS), a research group known as the TSUNAMI consortium was organized, and devised a two-year plan for research activities and testbed development starting in 1994 with a proof-of-concept demonstration of SDMA in radio communications systems as its final goal [47]. In TSUNAMI, adaptive array antennas were used as a core technology to implement SDMA in next-generation mobile communications networks. This is because it is possible to spatially separate multiuser signals by performing adaptive array signal processing in several different ways.

In TSUNAMI, a number of technical problems can be mentioned which are required to implement SDMA. First, one example of a beamforming technique is digital beamforming (DBF) in the baseband. In DBF, much more precise pattern control can be achieved by 16-bit digital resolution than beamforming by an analog phase shifter. The testbed hardware consisted of an eight-channel system employing a patch antenna array. The block diagram is shown in Figure 5.60 [47]. The baseband system provides two independent bidirectional wide-band beamformer channels to the DECT (Digital European Cordless Telephone) radio system. The DBF cards were based on two DBF1108 chips developed by ERA Technology. The two DECT radio subsystems were designed such that each operated only on a single fixed frequency and time slot. Channel multiplexing/demultiplex-

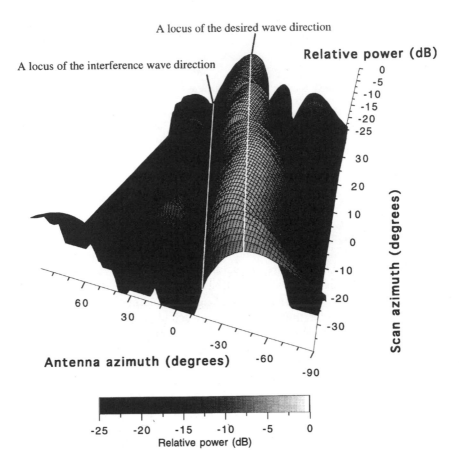

A locus of the desired wave direction

A locus of the interference wave direction

Figure 5.56 Tracking characteristics of the desired and interference waves for BSCMA (π/4-QPSK, interference signal: −3 dB with one-symbol delay; "antenna azimuth" means the angle measured from the antenna center direction; "scan azimuth" means the scan angle of the antenna center direction) [44].

ing in FDMA (frequency-division multiple access) can be integrated with DBF by filtering with DSP, and not only beamforming but also functions such as calibration or user position identification can be realized by spatial signal processing. However, problems such as system complexity or cost, frequency characteristics, dynamic range, and processing speed must be resolved. In TSUNAMI, DBF is used not only for receiving but also for transmitting.

5.2.7 Space-Time Signal Processing

Space-time signal processing, which combines the time region with a spatial region, offers the advantages of both [48–50]. Although the configuration becomes complex, it is suite

ETS-V
(1.5GHz)

Figure 5.57 Satellite experiment using a van equipped with the DBF array antenna.

to the case where adaptive operation with respect to broadband signals is emphasized. In spatial signal processing, multipath waves that did not coincide with the timing of the reference signal (i.e., replica) were removed as an intersymbol interference component, but it can be incorporated as a wanted signal based on time-region signal processing. Space-time signal processing can be understood as a matched receiving function for multipath waves dispersed in space and time, and more sophisticated research on an optimum receiver which performs this as preprocessing and MLSE estimation at a later stage is now in progress.

5.2.7.1 Multipath Separation and Combining Using a Multibeam Adaptive Array

A desired wave (direct wave in the case of line-of-sight communication, for example) and delayed wave are separated and received by a multibeam adaptive array, and combined after adjusting the delay of the separated signals as shown in [51]. Figure 5.61 shows the basic configuration. The above path separation processing using a multibeam has a disadvantage that cochannel interference (CCI) cannot be fully suppressed when the

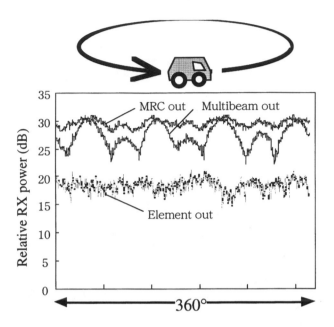

Figure 5.58 Receiving power with BSMRC while the van was circling on flat ground [45].

desired wave or delayed wave to be incorporated is close to CCI in direction. The signal processing algorithm itself, however, is relatively simple.

5.2.7.2 Space-Time Processing Based on Cascaded Adaptive Array and Equalizer

In this arrangement, where an equalizer is connected in cascade with the output of an adaptive array, CCI is basically suppressed in the adaptive array part, the delayed wave responsible for ISI (intersymbol interference) is incorporated, and ISI is compensated in the equalizer [52,53].

In [53] not only the direct wave but also short-delayed waves (e.g., one-symbol delay or so) are captured together with the adaptive array, and the equalizer equalizes waveform distortion caused by short-delayed waves. This can be done by suitably designing the training sequence. The configuration and the training sequence are shown in Figure 5.62. A one-symbol delayed wave can be incorporated by the adaptive array in addition to the desired wave by making two identical symbols follow each other in the training sequence. Multipath waves with long delay and CCI are suppressed by the adaptive array. Equalization is performed in a maximum likelihood sequence estimation (LMSE) unit following the adaptive array part, and a path diversity gain due to the desired wave and a delayed wave with a short delay is obtained. The signal input to the equalizer consists

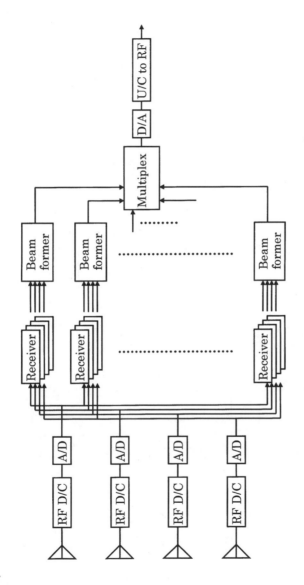

Figure 5.59 DBF architecture for AMPS applications. (From [46], © 1995 IEEE.)

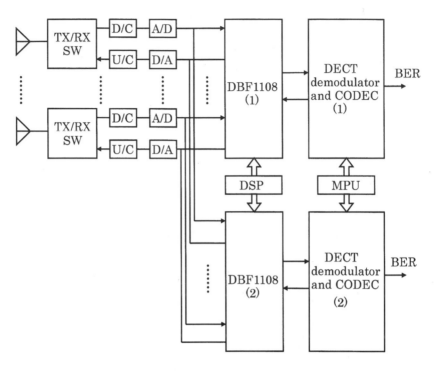

Figure 5.60 RACE TSUNAMI testbed architecture. (From [47], © 1996 IEEE.)

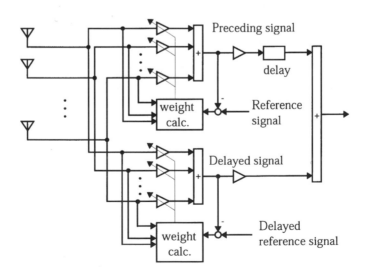

Figure 5.61 Multipath separation and combining using a multibeam adaptive array [51].

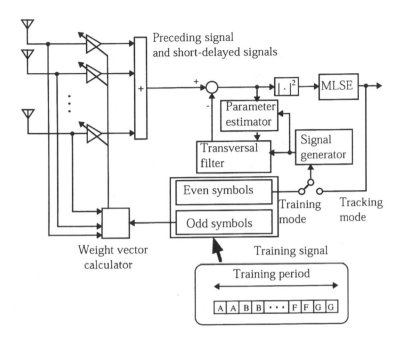

Figure 5.62 Cascade connection of an adaptive array and an equalizer [53].

only of the desired wave and delayed wave with short delay, so the amount of computation of the Viterbi algorithm is relatively small.

5.2.7.3 Space-Time Equalizer by TDL Array

Multipath waves generally arrive from different directions, and a space-time equalizer performs equalization using information about the incoming wave direction and delay. Figure 5.63 is a schematic diagram of an adaptive antenna array comprising a tapped delay line (TDL) adaptive array, and z^{-1} introduces a delay of one sampling interval. The TDL array may be viewed also as a matched filter, which takes account not only of delay but also of incidence angle information.

A receiver that performs MLSE after the TDL adaptive array is known as an optimal receiver [49,54]. The TDL antenna is considered as a whitening matched filter in space and time, but usually it is insufficient to show perfect performance due to limited scale of the antenna size and TDL length. This imperfection is compensated by MLSE. Figure 5.64 shows the configuration. This may also be referred to as an extension of adaptive MLSE to time and space. Performance of optimum receiver in terms of BER is summarized in [49].

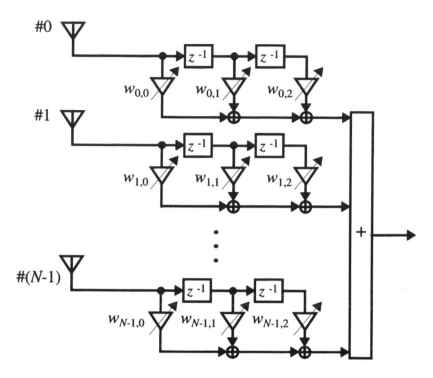

Figure 5.63 Space-time equalizer by TDL array.

5.2.7.4 Application of Space-Time Adaptive Array to CDMA Communications

Studies on adaptive array for CDMA systems are being carried out energetically with a view to mobile communications applications (i.e., IS-95 and IMT-2000) [55–60]. Two-dimensional (2D) RAKE, which assimilates multipath waves dispersed in space and time by means of maximal ratio combining is a promising countermeasure for improving signal-to-interference power ratio [56,57,60]. This is the expansion of conventional RAKE reception in time domain to space and time domains.

Figure 5.65 shows the basic configuration of an element-space-type 2D-RAKE scheme. Figure 5.66 shows an image of antenna patterns in which each pattern receives each delayed signal optimally. In the case of CDMA where the signal-to-interference ratio of each branch is fairly low due to a large number of users, combining of wanted signals dispersed in space and time becomes quite difficult. For mitigating this problem, a beam-space-type 2D-RAKE scheme is being investigated [60]. Figure 5.67 shows the configuration. The signal received by the N-element antenna array is sent to a multibeam forming circuit where at least one of the multibeam outputs contains a higher signal-to-interference ratio signal. In the kth user channel ($k = 1, 2, \ldots, K$), each beamforming circuit output

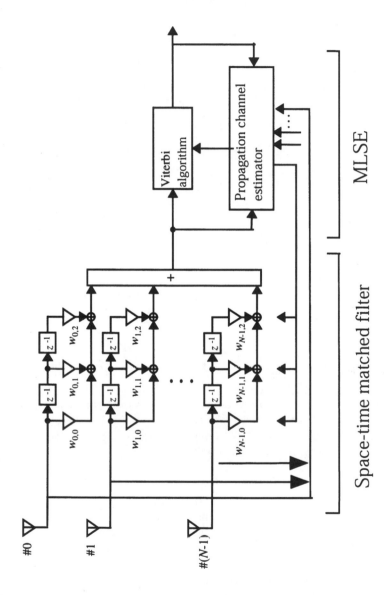

Figure 5.64 Optimal receiver composed of TDL array and MLSE equalizer. (From [54], © 1994 IEEE.)

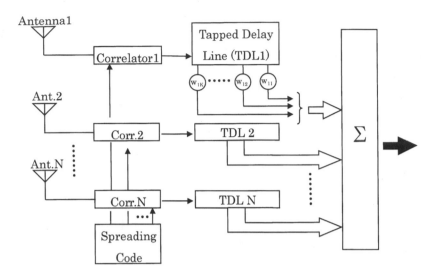

Figure 5.65 Basic configuration of an element-space-type 2D-RAKE scheme. (From [56], © 1994 IEEE.)

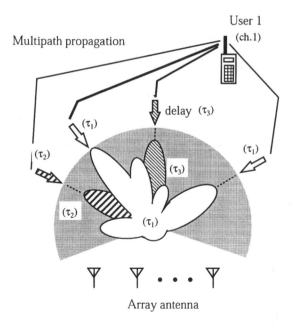

Figure 5.66 An image of antenna patterns in which each pattern receives each delayed signal optimally.

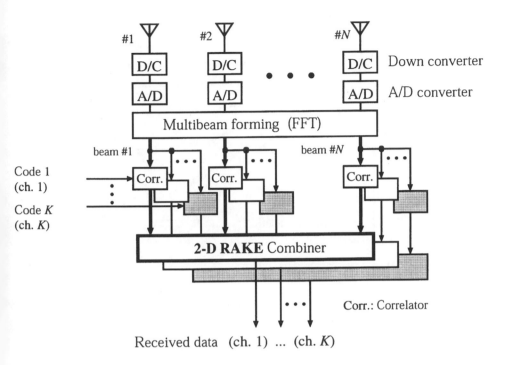

Figure 5.67 Beam-space-type 2D-RAKE for DS-CDMA systems [60].

signal is individually subjected to despreading by N correlators, and is sent to the kth 2D-RAKE combiner. Signals from other users act as interfering waves, but when the number of users is large, the total amount of interference is averaged out in space and there is less dependence on the antenna pattern for multiuser interference. The improvement of channel capacity is due to an increase in desired signal power.

Figure 5.68 shows the average BER performance of the 1D-RAKE (conventional RAKE) and 2D-RAKE schemes as a parameter of the number of multiusers [60]. Solid curves are results by means of a simplified BER estimation, while the square dots with broken lines are those by a direct BER simulation. In the direct BER simulation, the average BER performance for a multiuser system is calculated by adding multiuser interference one by one in each multipath environment based on a statistical propagation model, consequently, this is a strict evaluation method that takes enormous calculation time. In the simple BER estimation, on the other hand, the average BER performance is estimated by assuming that averaged total multiuser interference power is equivalent to thermal noise power. In the figure, the calculation condition is that the process gain is 31, and delay spread of the multipath environment is 1 chip period for (a) and (b), the number of antenna elements is 16, and angular spread of each incoming signal is 6 deg for (b). From the figure, (1) the simple method agrees fairly well with the direct simulation result, and (2) the merit of 2D-RAKE, represented by channel capacity improvement, can be seen.

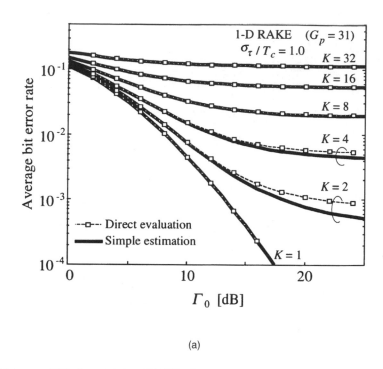

(a)

Figure 5.68 Average BER characteristics of RAKE scheme under multiuser environments (Γ_0 is the average SNR at an antenna element; K is the number of users in a cell): (a) 1D-RAKE; (b) 2D-RAKE. [60].

One of the objectives of the RACE TSUNAMI project was to investigate the potential deployment of adaptive antennas in third-generation mobile systems, and analysis of a microcellular CDMA (DS-CDMA) system employing adaptive antennas was also included [47]. The raytracing tool that was developed in [61] was used to obtain the spatial and temporal information necessary for a site-specific capacity analysis study. From these data the optimized radiation pattern was calculated in order to obtain the highest output signal-to-interference-plus-noise ratio over an eight-element array using various types of MMSE algorithms to control the beamformer by applying temporal reference optimization via the embedded user code synchronizer. Figure 5.69 shows an example of the optimized radiation pattern with the square root recursive least squares (SQRLS) algorithm for a typical microcellular scenario with 15 interfering users, where the effect of mutual coupling between the antenna elements has also been considered. This process was repeated many times with a random deployment of users and interfering sources. The results indicate that at least a fivefold increase in the overall spectrum efficiency of the DS-CDMA network for microcellular operation can be achieved with adaptive antennas.

In W-CDMA mobile radio, an adaptive array can be combined with the coherent RAKE receiver. This is called the coherent adaptive antenna array receiver (CAAAR)

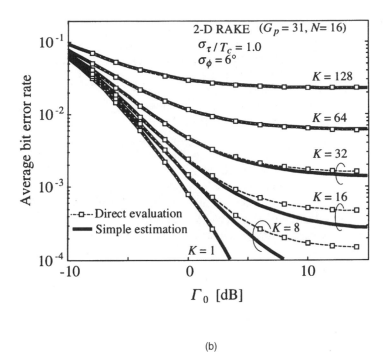

(b)

Figure 5.68 (continued).

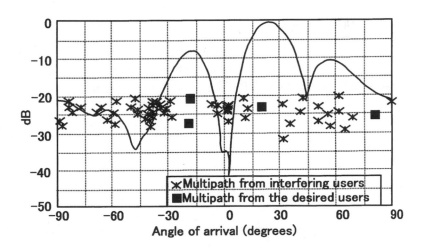

Figure 5.69 Optimized radiation pattern for a microcellular application with DE-CDMA. (From [47], © 1997 IEEE.)

[58,62]. The CAAAR receiver is practically useful for multimedia communications in which different users are transmitting with different data rates. At the beginning of communication, the CAAAR receiver starts from the omnibeam pattern and forms the optimum beam pattern quickly by the aid of an LMS adaptive algorithm. The receiver can also be designed to direct the beam toward each resolved path of each user and to realize coherent RAKE combination even though their angles are quite different. Both the known pilot symbols, received periodically, and data symbols can be exploited for this purpose. The block diagram of a decision-directed coherent CAAAR receiver is shown in Figure 5.70. For a performance comparison of CAAAR and two-antenna diversity reception, the measured average BERs of a simple two-element CAAAR receiver are plotted in Figure 5.71 as a function of received power ratio of interfering user to desired user [62]. The data rate and the chip rate are 64k symbols/s and 4.096M chips/s, respectively. The average BER can be reduced by about one order of magnitude compared to the antenna diversity case even if the interfering user's power is 10 dB higher.

5.2.8 Reconfigurable Adaptive Array: Software Antenna

All of adaptive algorithms mentioned above have their respective advantages and disadvantages, and there is no universal algorithm suitable for any environments. Indeed, even if there were such an algorithm, the signal processing involved would require an extremely

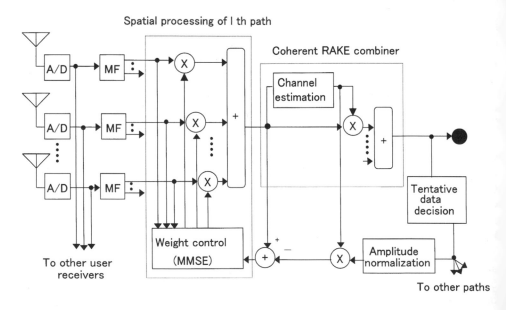

Figure 5.70 Block diagram of CAAAR receiver. (From [62], © 1998 IEEE.)

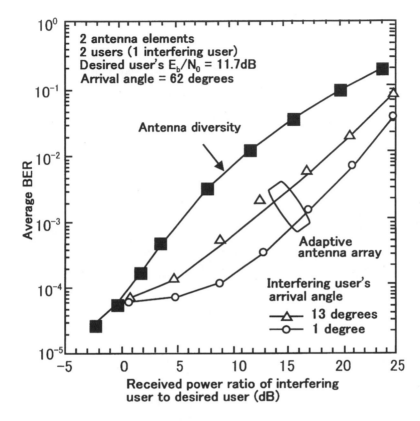

Figure 5.71 Measured BER characteristics under multiuser interference. (From [62], © 1998 IEEE.)

complex construction in terms of space (large number of array elements) and time (long TDL).

To enhance the adaptation to a radio environment more dynamically, the concept of a software antenna has been proposed [34,63]. It enables a finer selection of the most appropriate algorithm through reconfiguration of both software (algorithms) and hardware (logic circuits). Using an array antenna, and by performing signal processing which overcomes the limitations of each individual algorithm, it adapts to the existing environment at any time in the manner of a chameleon. It therefore might be described as a kaleidoscopic intelligent antenna.

Figure 5.72 is a schematic illustration of a software antenna that would be used in a mobile communication base station. It comprises (a) a function for precisely evaluating changes in the environment and finding the optimum algorithm at any time, (b) a number of algorithms for a variety of possible communication environments, and (c) DBF by a reconfiguring logic circuit (FPGA, DSP, etc.), which implements the selected algorithm

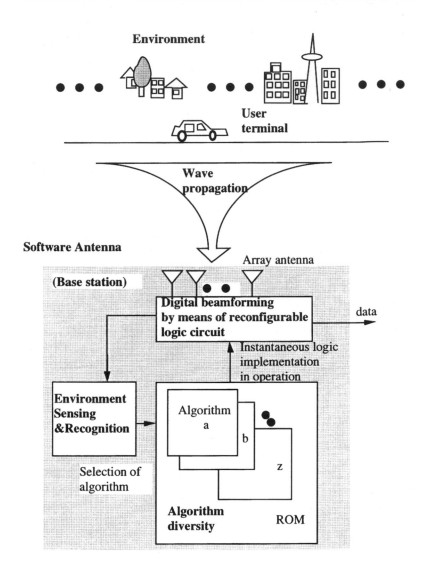

Figure 5.72 A basic configuration of the software antenna with application to a base station antenna for mobile radio systems [34].

instantaneously. Although the software antenna is still in the conceptual phase, the antenna will play an important role in parallel with software radio technology development in the near future.

REFERENCES

[1] Kijima, M., Y. Ebine, and Y. Yamada, "Development of a Dual-Frequency Base Station Antenna for Cellular Mobile Radios," *IEICE Trans. Communication*, Vol. E82-B, No. 4, April 1999.

[2] Yamada, Y., and M. Kijima, "Low Sidelobe and Tilted Beam Base-Station Antennas for Smaller Cell Systems," *IEEE Antennas and Propagation Society Int. Symp.*, June 1989, pp. 138–141.

[3] Hirasawa, H., and B. J. Strait, "On a Method for Array Design by Matrix Inversion," *IEEE Antennas and Propagation Society Int. Symp.*, May 1971, pp. 446–447.

[4] Elliot, R. S., *Antenna Theory and Design*, Ch. 5, Prentice-Hall, 1981, pp. 165–180.

[5] Yamada, Y., and M. Kijima, "A Slender Two Beam Base Station Antenna for Mobile Radio," *IEEE Antennas and Propagation Society Int. Symp.*, June 1994, pp. 352–355.

[6] Ebine, Y., and M. Ito, "Design of a Dual Beam Antenna Used for Base Station of Cellular Mobile Radios," *IEICE Trans. Communication*, Vol. J79-B-II, No. 11, Nov. 1996, pp. 909–915.

[7] Arai, H., and M. Nakano, "Up-link Polarization Diversity and Antenna Gain Measurement of Hand Held Terminal at 900 MHz," *MDMC'94*, Nov. 1994.

[8] Nakano, M., T. Satoh, and H. Arai, "Up-link Polarization Diversity Measurement of Hand-Held Terminal at 900 MHz," Technical Report of IEICE, AP95-13, May 1995, pp. 13–19 (in Japanese).

[9] Nakano, M., T. Satoh, and H. Arai, "Up-link Polarization Diversity and Antenna Gain Measurment of Hand-Held Terminal," *IEEE AP-S Digest*, July 1995, pp. 1940–1943.

[10] Nakano, M., T. Satoh, and H. Arai, "Up-link Polarization Diversity and Antenna Gain Measurement of Hand-Held Terminal," *IEICE National Convention*, B-11, 1995 (in Japanese).

[11] Anderson, J., and F. Hansen, "Antennas for VHF/UHF Personal Radio: A Theoretical and Experimental Study of Characteristics and Performance," *IEEE Trans. Vehicular Technology*, Vol. VT-26, No. 4, Nov. 1977, pp. 349–357.

[12] Nakano, M., et al., "Up-link Polarization Diversity Measurement for Cellular Communication Systems Using Hand-Held Terminal," *IEEE Antenna and Propagation Society Int. Symp.*, 66.8, July 1997, pp. 1360–1363.

[13] Nakano, M., et al., "Up-link Polarization Diversity for Cellular Communication System," Technical Report of IEICE, A-P98-15, May 1998, pp. 13–20 (in Japanese).

[14] Nakano, M., et al., "Small-Sized Polarization Diversity Antenna for Cellular Base Stations," *National Society Convention of IEICE*, B-1-42, 1997 (in Japanese).

[15] Nakano, M., and T. Satoh, "Polarization Diversity for Base Stations on cdmaOne Cellular System," *National Society Convention of IEICE*, B-1-14, 1998 (in Japanese).

[16] Moriyasu, S., et al., "Printed Dipole Pair with Modified Parasitic Elements and Reflector," *National Society Convention of IEICE*, B-1-150, 1999 (in Japanese).

[17] Kondo, Y., and K. Suwa, "Linear Predictive Transmission Diversity for TDMA/TDD Personal Communication Systems," *IEICE Trans. Communication*, Vol. E79-B, No. 10, Oct. 1996, pp. 1586–1591.

[18] Cho, K., et al., "Bidirectional Collinear Antenna with Arc Parasitic Plates," *Proc. IEEE Antennas and Propagation Society Int. Symp.*, June 1995, pp. 1414–1417.

[19] Yokota, T., et al., "The Development of PHS Base Station with Adaptive Array Antenna," *National Society Convention of IEICE*, B-5-74, 1998 (in Japanese).

[20] Yuan, H., et al., "A Study of the Interelement Spacing on Adaptive Array Antenna for PHS Base Station," *National Society Convention of IEICE*, B-5-75, 1998 (in Japanese).

[21] Uchibori, J., et al., "The Radiation Characteristics of Adaptive Array Antenna for PHS Base Station," *National Society Convention of IEICE*, B-5-76, 1998 (in Japanese).

[22] Fujitsuka, Y., et al., "A Field Test for the PHS Base Station with Adaptive Array Antenna," *National Society Convention of IEICE*, B-5-77, 1998 (in Japanese).

[23] Kimura, S., "Kyocera Super PHS-WLL Technology," Technical Seminars, The 6th Vietnam International Exhibition on Communication, Computer & Office System & the 4th Vietnam International Exhibition on Broadcasting Technology, December 1997.

[24] Furukawa, K., et al., "N-STAR Mobile Communications System," *IEEE GLOBECOM*, 1996, pp. 390–395.

[25] Mizoguchi, S., et al., "Portable Terminals for N-STAR Mobile Satellite Communication System," *National Convention of IEICE*, B-3-36, 1997 (in Japanese).

[26] Ueda, T., et al., "Special Issue of Mobile Satellite Communication System, Mobile Station," *NTT DoCoMo Tech. J.*, Vol. 4, No. 2, 1996, pp. 24–28 (in Japanese).

[27] Kuramoto, A., et al., "The 9th Element Patch Array Antenna for Satellite Communications of the Vehicle," *National Convention of IEICE*, B-1-65, 1997 (in Japanese).

[28] Widrow, B., "Adaptive Antenna System," *Proc. IEEE*, Vol. 55, No. 12, Dec. 1967, pp. 2143–2159.

[29] Hudson, J. E., *Adaptive Array Principles*, Peter Peregrinus Ltd., 1981.

[30] Compton, Jr., R. T., *Adaptive Antennas—Concepts and Performance*, Prentice-Hall, 1988.

[31] Steyskal, H., "Digital Beamforming Antennas—An Introduction," *Microwave J.*, Vol. 30, No. 1, 1987, pp. 107–124.

[32] Goldsmith, A. J., "Smart Antennas," *IEEE Personal Commun.*, Vol. 5, No. 1 (special issue on Smart Antennas), Feb. 1998, p. 9.

[33] Litva, J., and T. K. Lo, *Digital Beamforming in Wireless Communications*, Artech House, 1996.

[34] Karasawa, Y., T. Sekiguchi, and T. Inoue, "The Software Antenna: A New Concept of Kaleidoscopic Antenna in Multimedia Radio and Mobile Computing Era," *IEICE Trans. Commun.*, Vol. E80-B, No. 8, 1997, pp. 1214–1217.

[35] Kohno, R., "Spatial and Temporal Communication Theory Using Software Antennas for Wireless Communications," in *Wireless Communications: TDMA vs CDMA*, S. G. Glisic and P. A. Leppanen, eds., Kluwer Academic Pub., 1997, pp. 293–321.

[36] Takao, K., and K. Uchida, "Beamspace Partially Adaptive Antenna," *IEE Proc.*, Pt. H, No. 6, 1989, pp. 439–444.

[37] Chiba, I., W. Chujo, and M. Fujise, "Beam Space CMA Adaptive Array Antennas," *Trans. IEICE*, Vol. J77-B-II, No. 3, 1994, pp. 130–138.

[38] Treichler, J. R., and B. G. Agee, "A New Approach to Multipath Correction of Constant Modulus Signals," *IEEE Trans. Acoustics, Speech, and Sig. Process.*, Vol. ASSP-31, No. 2, 1983, pp. 459–472.

[39] Gooch, R., and J. Lundell, "The CMA Array: An Adaptive Beamformer for Constant Modulus Signals," *Proc. ICASSP*, Vol. 4, 1986, pp. 2523–2526.

[40] Ohgane, T., et al., "An Implementation of a CMA Adaptive Array for High Speed GMSK Transmission in Mobile Communications," *IEEE Trans. Vehicular Technology*, Vol. 42, No. 3, Aug. 1993, pp. 282–288.

[41] Ohgane, T., et al., "BER Performance of CMA Adaptive Array for High-Speed GMSK Mobile Communication—A Description of Measurements in Central Tokyo," *IEEE Trans. Vehicular Technology*, Vol. 42, No. 4, Nov. 1993, pp. 484–490.

[42] Tanaka, T., et al., "An ASIC Implementation Scheme to Realize a Beam Space CMA Adaptive Array Antenna," *IEICE Trans. Commun.*, Vol. E78-B, No. 11, Nov. 1995, pp. 1467–1473.

[43] Tanaka, T., R. Miura, and Y. Karasawa, "Implementation of a Digital Signal Processor in a DBF Self-Beam-Steering Array Antenna," *IEICE Trans. Commun.*, Vol. E80-B, No. 1, Jan. 1997, pp. 166–175.

[44] Tanaka, T., et al., "Interference Cancellation Characteristics of a BSCMA Adaptive Array Antenna with a DBF Configuration," *IEICE Trans. Commun.*, Vol. E80-B, No. 9, 1997, pp. 1363–1371.

[45] Miura, R., et al., "A Land-Mobile Satellite Experiment on a DBF Self-Beam Steering Array Antenna," *1996 Int. Symp. Antennas and Propagat. (ISAP'96)*, Chiba, Japan, 1996, Vol. 3, pp. 1189–1192.

[46] Kennedy, J., and M. C. Sullivan, "Direction Finding and Smart Antennas Using Software Radio Architectures," *IEEE Commun. Mag.*, Vol. 33, No. 5, May 1995, pp. 62–68.

[47] Tsoulos, G., M. Beach, and J. McGeehan, "Wireless Personal Communications for the 21st Century: European Technological Advances in Adaptive Antennas," *IEEE Commun. Mag.*, Vol. 35, No. 9, Sep. 1997.

[48] Paulraj, A. J., and C. B. Papadias, "Space-Time Processing for Wireless Communications," *IEEE Signal Process. Mag.*, Nov. 1997, pp. 49–83.

[49] Kohno, R., "Spatial and Temporal Communication Theory Using Adaptive Antenna Array," *IEEE Personal Commun.*, Feb. 1998, pp. 28–35.

[50] Sekiguchi, T., "Space-Time Adaptive Signal Processing Technologies," TRICEPS Inc., White Series No. 189, Ch. 6 of *Software Antenna Technologies for Wideband Wireless Communication Systems*, Y. Karasawa, ed., March 1999, pp. 115–153 (in Japanese).

[51] Ogawa, Y., Y. Nagashima, and K. Itoh, "An Adaptive Antenna System for High-Speed Digital Mobile Communications," *IEICE Trans. Commun.*, Vol. E75-B, No. 5, May 1992, pp. 413–421.

[52] Ishii, N., and R. Kohno, "Spatial and Temporal Equalization Based on an Adaptive Tapped-Delay-Line Array Antenna," *IEICE Trans. Commun.*, Vol. E78-B, No. 8, Aug. 1995, pp. 1162–1169.

[53] Doi, Y., T. Ohgane, and Y. Karasawa, "ISI and CCI Canceller with Preselecting Adaptive Array and Cascaded Equalizer in Digital Mobile Radio," *IEICE Trans. Commun.*, Vol. E81-B, No. 3, Mar. 1998, pp. 674–682.

[54] Nagatsuka, M., R. Kohno, and H. Imai, "Optimal Receiver in Spatial & Temporal Domains," *Proc. IEEE Int. Symp. Inform. Theory & Its Applications*, 1994, pp. 893–898.

[55] Nagatsuka, M., and R. Kohno, "A Spatially and Temporally Optimal Multi-User Receiver Using an Array Antenna for DS/CDMA," *IEICE Trans. Commun.*, Vol. E78-B, No. 11, 1995, pp. 1489–1497.

[56] Iwai, H., T. Shiokawa, and Y. Karasawa, "An Investigation of Space-Path Hybrid Diversity Scheme for Base Station Reception in CDMA Mobile Radio," *IEEE J. Selected Areas in Communication*, Vol. 12, No. 5, June 1994, pp. 962–969.

[57] Naguib, A., and A. Paulraj, "Performance of CDMA Cellular Networks with Base-Station Antenna Arrays," *Proc. Int. Zurich Seminar on Digital Commun.*, Zurich, Switzerland, March 1994, pp. 87–100.

[58] Tanaka, S., M. Sawahashi, and F. Adachi, "Pilot Symbol-Assisted Decision-Directed Coherent Adaptive Array Diversity for DS-CDMA Mobile Radio Reverse Link," *IEICE Trans. Fundamentals*, Vol. E80-A, Dec. 1997, pp. 2445–2454.

[59] Dohi, T., Y. Okumura, and F. Adachi, "Further Results on Field Experiments of Coherent Wideband DS-CDMA Mobile Radio," *IEICE Trans. Commun.*, Vol. E81-B, No. 6, 1998, pp. 1239–1247.

[60] Inoue, T., and Y. Karasawa, "Two-Dimensional RAKE Reception Scheme for DS/CDMA System in Beam Space Digital Beam Forming Antenna Configuration," *IEICE Trans. Commun.*, Vol. E81-B, No. 7, 1998, pp. 1374–1383.

[61] Athanasiadou, G. E., A. R. Nix, and J. P. McGeehan, "A Ray Tracing Algorithm for Microcellular Wideband Modelling," *Proc. 45th VTC*, Chicago, IL, 1995, pp. 261–265.

[62] Adachi, F., M. Sawahashi, and H. Suda, "Wideband DS-CDMA for Next-Generation Mobile Communications Systems," *IEEE Comm. Mag.*, Vol. 36, No. 9, 1998, pp. 56–69.

[63] Karasawa, Y., Y. Kamiya, T. Inoue, and S. Denno, "Algorithm Diversity in a Software Antenna," *IEICE Trans. Comm.*, Vol. 83-B, No. 6, 2000, pp. 1229–1236.

Chapter 6

Land Mobile Antenna Systems II: Pagers, Portable Phones, and Safety

R. Mumford, Q. Balzano, T. Taga, and C. K. Chou

6.1 PRACTICAL REQUIREMENTS OF AND CONSTRAINTS ON PAGER ANTENNA DESIGN

Radio paging is a cost-effective solution for locating staff, alerting personnel, and transmitting one-way messages or data. In Europe, the industry has developed from single onsite systems, the first of these being an inductive loop system installed in St. Thomas's Hospital in England by Multitone Electronics in 1956. Transmissions were in the 30- to 50-kHz range, and each receiver was tuned to its own unique frequency. Nationwide area systems in Europe started by the Netherlands PTT in 1964 were followed up by public operator licenses being granted to private consortia. A pan-European paging system with access from any country is now planned and specified.

Three main types of pagers are currently available: the audible-tone alerter, the numeric message displaying up to 20 digits, and the alphanumeric displaying about 100 characters and storing up to 5,000-character messages. Figure 6.1 shows a typical VHF tone alert pager with a strip antenna. Paging receivers have been substantially reduced in size and greatly increased in features in the last 30 years. This has had the effect of reducing possible integral antenna size and bringing RF noisy digital circuitry (e.g., gate arrays and microprocessors) much closer to the antenna and the RF front-end circuits.

Paging receivers are used in private onsite situations, usually with one low-power transmitter radiating 1W to 5W of RF power, and in large wide-area systems where there are multiple transmitters of 100W to 200W, each covering a 5- to 20-mi radius. It is

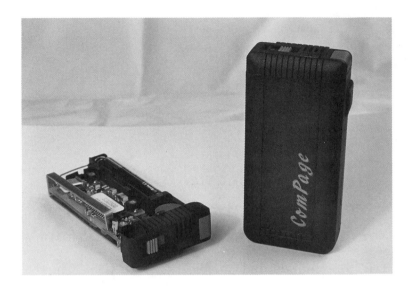

Figure 6.1 VHF tone alert pager. (Courtesy Multitone Electronics, Ltd.)

important in the latter case for the paging receiver to be as sensitive as possible in order to economize on the number of transmitters in a large system. The onsite pager, which may be operating, for example, on a warehouse site, a brewery, hotel, or supermarket, must be sensitive enough to operate in antenna shadow regions and between buildings. The pager's own internal antenna is arguably the most important factor in defining sensitivity, since it is much smaller than a portable two-way radio and is worn on the belt or in the shirt or jacket top pocket.

Frequency bands allocated to paging vary according to country across the world and are in the total range of 25 to 910 MHz. Traditionally in Europe and the United States, wide-area systems, such as British Telecom's service (covering nearly all the populated areas of the United Kingdom) and large North American radio common carrier subscriber networks, operated on frequencies in the VHF and 147 to 174 MHz. In Japan, frequencies ranged up to 280 MHz. However, the attraction of greater penetration and propagation in buildings has been one factor contributing to UHF frequencies in the 450- to 512-MHz band being used for citywide networks such as the Cityruf system in Germany. Various small integral antenna designs are therefore required for this very wide frequency range.

The designer's aim is to optimize the antenna performance in the presence of a wide variety of problems such as the following.

- The small size of the unit;
- The proximity of the circuit components;
- The RF noise generated by the rest of the receiver circuitry;

- The ease of receiver alignment and the maintenance of the alignment when the pager is worn on the body;
- The effect of the proximity of the human body on antenna performance;
- The unit is frequently transported at widely varying speeds;
- Use inside of a building or vehicle is commonplace.

It is evident that in order to deal with the first two of these problems, the aerial development must include the pager chassis as an inseparable part of the antenna. If possible, its metal parts should be used to aid the performance. To avoid loss of calls when the transmitter direction is in a pager antenna pattern null, any such nulls must be made as shallow as possible. A somewhat obvious solution to this problem would be to use two or more antennas arranged so that their nulls do not coincide.

The latter three problems in the above list all stem from the nature of an electromagnetic field near a reflective or absorptive surface. Fades caused by absorption cannot be counteracted as the power is dissipated within the surface of the object. With the reflective nulls, however, the situation is different. We will now discuss some of these issues in more detail.

6.1.1 Effect of the Human Body on Antennas

Over the frequency range under consideration (20 to 1,000 MHz), the human body has a conductivity and relative permittivity on the order of 1.5 S/m and 75, respectively, and acts predominantly as a reflector [1] with varying degrees of efficiency. The low impedance, presented to the wave by the body, reduces the electric field close by, while increasing the magnetic field. At a distance of a quarter wavelength away from the body towards the transmitter, a high impedance is presented to the approaching wavefront which enhances the electric field while reducing the magnetic field. This effect is periodic in nature.

On the far side of the body from the transmitter, there is a deep null caused by absorption of power by the body. This, of course, affects both electric and magnetic fields. It is evident from the above that for good onbody performance, a magnetic antenna must be used. It is also evident that because of the conductive nature of the body any magnetic field close to the body will be tangential to the body. This is useful because it means that if a magnetic loop antenna is used and positioned such that the axis of the loop is tangential to the body, the maximum magnetic field available will always be intercepted, whatever the orientation of the body [2]. Figure 6.2 shows a plan view of the magnetic field around a cylindrical lossy dielectric object representing the human body.

Magnetic Antennas

The magnetic field enhancement of approximately 6 dB (at 150 MHz, 3 dB at 450 MHz, and 0 dB at 900 MHz) close to the body has led to the use of predominantly magnetic

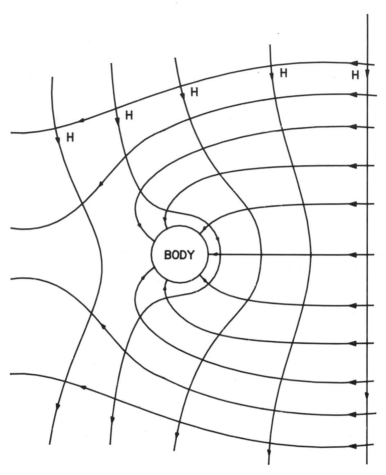

Figure 6.2 Plan view of magnetic field around the human body.

antennas, which naturally perform best on the body. These antennas are based mainly on the electrically small loop [3,4], and the performance of the latter when the loop is body-mounted, compared to the free-space behavior, is shown in the azimuth radiation pattern plots of Figure 6.3. The free-space pattern has two finite nulls in the 90- and 270-deg positions. When body-mounted, the resulting pattern shape depends on the body location and loop orientation. Simple small-loop theory tends to assume constant current around the loop with no phase change. This in turn would predict infinitely deep nulls. In practice, the loop size is large enough to produce a nonuniform current distribution around the loop, which has the effect of filling in the two nulls. For electrically larger loops, the notch depth is decreased, and if the circumference approaches half a wavelength, the null disappears altogether and the antenna exhibits electric rather than magnetic characteristics,

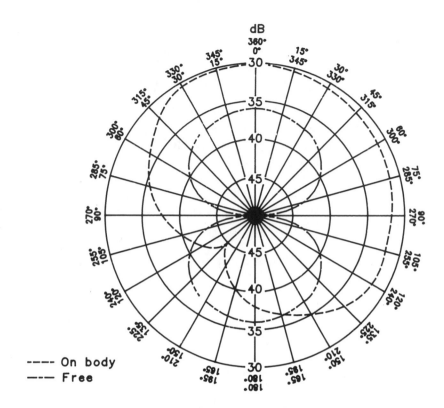

Figure 6.3 Radiation pattern of typical magnetic antenna expressed as ratio of electric field (μV/m) to antenna output voltage (μV).

with the associated poorer onbody response. A typical one-turn air loop mounted on a pager chassis has a null depth of about 15 to 20 dB (depending on frequency) while maintaining a predominantly magnetic effect, and is thus very effective on the body.

An improvement in performance may be obtained by integrating metallic parts of the receiver into the antenna. However, the antenna cannot be reoriented such that it is wound around the maximum pager cross-sectional area, since this has the effect of rotating the radiation pattern by 90 deg, and creating a null in the forward direction, which is clearly not acceptable.

Electric Antennas in Free Space

Electric field antennas have not been used in pagers in the VHF frequency band, principally because of the deep null in the electric field close to the human body. However, in a free-space situation, there are several advantages to be gained by using an electric field antenna, especially if it is used in conjunction with a magnetic antenna.

Experiments have proved that the most efficient electric field antenna that is compatible with a pager chassis is the top-loaded monopole consisting of a short length of wire loaded with a widely spaced coil. Typically, four turns, wound conformally to the extreme dimensions of the chassis, are found to be resonant in the pager bands. The azimuth pattern performance of this antenna is compared with magnetic antenna performance in Figure 6.4. These patterns show a significantly superior free-space performance on the part of the electric field antenna, as well as a total absence of nulls in this azimuth plane. The loaded monopole radiation resistance is typically five times that of the magnetic antennas for typical pager dimensions and accounts for the superior performance of the latter of some 2 to 6 dB. For some applications, a pager will be required to function when not body-mounted. It is found that the top loading coil acts as an antenna in its own right when the chassis is placed on its side or face, which is a useful design compromise when the pager is lying on a desk or table.

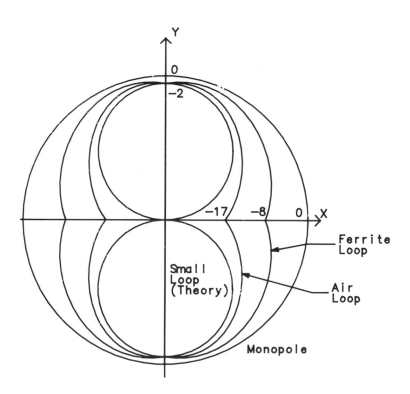

Figure 6.4 Top-loaded monopole performance compared with magnetic antennas.

6.1.2 Aspects of Manufacture

Noise

Thermal noise is unavoidable, but can be minimized by conventional design procedures; however, noise generated within the receiver circuitry requires careful consideration. The noise may originate from local oscillator-derived signals, limiter signals, data signals, and any combination of the latter. This may be minimized by judicious layout, screening, and interfering signal level control. It is also evident that antenna immunity to this problem will depend very much on the geometry and near-field response of the antenna itself. Experiments have established that the loop antenna is much more susceptible to internally generated interference than the ferrite loop antenna, due to the fact that the near fields of the latter are condensed closer to the ferrite core.

Antenna Tuning and Matching

Antenna tuning and matching can be one of the most critical processes in manufacturing a radio pager. The main requirements are:

- The physical movements required to accurately tune and match the antenna should not be too small.
- When the antenna is correctly tuned and matched, the presence of nearby objects and particularly the human body must not be significant.

Both the above constraints require the antenna Q factor to be designed to less than a given value. A design guideline based on the angle of rotation of a commercially available trimmer capacitor is

$$Q < \frac{9f}{f_2 - f_1} \tag{6.1}$$

where f is the nominal frequency of the pager and f_2 and f_1 the upper and lower limits of the band. For instance, Q should be less than 67 for a 20-MHz band centered on 150 MHz to ensure sufficient angular physical movement while tuning up. As regards body proximity effects, induced losses have the effect of reducing sensitivity, but the detuning effect can be reduced by limiting Q so that

$$Q < C/\Delta C \tag{6.2}$$

where C is the tuning capacitance and ΔC is the change of capacitance induced into the antenna by the proximity of the human body. For the air-cored loop antennas of Figure

6.5, typical values are $C = 16$ pf, $\Delta C = 0.2$ pf; hence, $Q < 80$, while for the ferrite-cored loops $C = 8$ pf and $Q < 40$. For more complicated antenna circuits (i.e., having capacitive or inductive matching taps), these capacity values will differ in magnitude.

6.1.3 Measurement of Pager Antenna Performance

The accurate measurement of pager antenna performance requires a certain amount of care in decoupling the device under test from any conductors in the immediate vicinity. Such conductors (e.g., cables) could have better antenna performance than the pager under test.

The method often used in testing pager performance is to detect incoming signals at the threshold of the pager sensitivity. If this parameter is known by bench measurements, then the antenna equivalent area or its gain can be determined if the incident electric and magnetic fields are also known. This methodology has the clear advantage of using the pager acoustic or mechanical altering devices to indicate the presence of a signal strong enough to be detected by the receiver. In this condition, no metal wires need to be connected to the pager, whose antenna performance can be severely affected by the presence of such conductors. Another method consists of packing a low-power (few milliwatts) transmitter in the form factor of the paging receiver and using it to test antenna

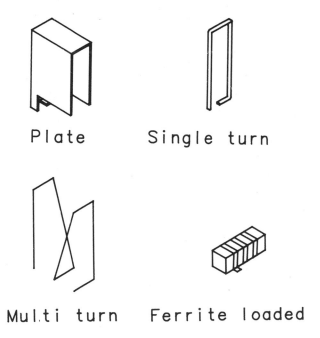

Plate   Single turn

Multi turn   Ferrite loaded

Figure 6.5 Loop antennas.

performance. Although this procedure can be used to establish the effects of the human body on the angular coverage of the pager, the real performance of the antenna cannot be measured until the radio receiver with all its printed circuit (PC) boards and components is placed in the package. In order to collect meaningful results, the testing fixture for the pager antenna should resemble the final package of the product in the geometry of the conducting surfaces.

Let us now quantify the considerations of the method to detect pager antenna performance. From basic theory [5], the following relationship exists between the gain above isotopic and the equivalent area of an antenna with no match loss:

$$G = 4\pi A / \lambda^2 \tag{6.3}$$

where λ is the wavelength of the incident electromagnetic energy. The power density of the incident wave is simply given by

$$P_d = |E|^2 / 120\pi \tag{6.4}$$

in watts per square meter or watts per square centimeter, depending on the units of $|E|$, the electric field. The power received by an antenna of equivalent area A is then

$$P_r = \frac{|E|^2 \lambda^2 G}{480\pi^2} \quad \text{watts} \tag{6.5}$$

The power received by the antenna is delivered to a load, representing the input impedance of the receiver. For the convenience of the design effort, the receiver input impedance is set very often at $(50 + j0)\Omega$. Using the voltage delivered by the antenna to a 50Ω load, V is

$$V = |E|\lambda \sqrt{\frac{G}{94.7}} \quad \text{volts} \tag{6.6}$$

Equation (6.6) simply states the obvious result that a constant cross-section antenna maintains its performance as the frequency decreases. It is often convenient to measure the incident electric field in microvolts/meter (μV/m) and express the pager sensitivity in the same units; in the process it is also convenient to measure the received voltage in terms of dBμV (decibels relative to microvolts across a 50Ω load). Equation (6.6) expressed in decibels becomes

$$V_{\text{dB}\mu\text{V}} = E_{\text{dB}\mu\text{V/m}} + 20\log\lambda + G_{\text{dBi}} - 19.8 \tag{6.7}$$

In (6.7), the wavelength λ is measured in meters and the gain G in decibels above isotropic. Equation (6.7) relates all the important characteristics of the paging receiver.

In a good antenna range or an anechoic chamber, the incident E field can be calibrated to within ± 0.25 dB; the received RF power necessary to trigger a pager can be measured within 0.1 dB with commercially available instruments. Using (6.7), the absolute gain of the pager antenna can be measured with an uncertainty of ± 0.35 dB, which is acceptable for most commercial applications.

Let us now briefly discuss the characterization of pager antenna performance. Clearly, from (6.7) it is possible to measure the gain $G(\theta, \phi)$ for any given angular direction. For directional antennas pointing to specific transmitters or targets, the performance is normally evaluated in terms of maximum gain and sidelobe level below the peak of the beam. This criterion is strictly a way to evaluate the energy transfer from one antenna to another or from one antenna to itself by means of an echo from a target. For pager antennas, the use of the maximum gain criterion only may not be the best for the application, because if there are wide gaps in the coverage of pager antenna, the user may not receive some messages incident from the blind directions. A better evaluation method than the maximum gain criterion is to average the gain of the pager antenna over the 360-deg angular sector at the horizon of the user. This hypothesis is correct in most cases. Given the slow angular variation of the gain of small antennas, the receive characteristics remain practically constant in the elevation sector ± 10 deg around a direction at the horizon. Using (6.7), the performance parameter can be obtained directly as a gain average or by averaging the square of the received voltages in microvolts.

Table 6.1 summarizes the measured performance of pagers from a variety of manufacturers at frequency bands allocated to paging. There is some spread of values in each frequency band, depending on antenna size and packaging of the receiver. The values given in the second column of Table 6.1 represent an eight-point average of the gain of the antenna when the pager is worn at the belt level of a human-equivalent phantom [6]. In the third column, the maximum measured value of the gain is tabulated. In the last column, the maximum spread of gain value over the eight-point sample is shown at a given frequency.

At the very low frequencies (30 to 80 MHz), the performance of the antennas is uniformly poor. The paging device is probably just coupling into the currents excited on the simulated human, which is the true antenna at these long wavelengths. The only

Table 6.1
Measured Pager Antenna Performance (Tone-Only Signals)

Frequency Band	Average Gain (dBi)	Maximum Gain (dBi)	Gain Spread (dB)	Receiver Sensitivity (dBm)
30–80 MHz	−32	−30	5	−129
VHF	−23	−20	10	−129
UHF	−19	−14	18	−128
900 MHz	−9.0	−2.5	30	−127

positive quality of these antennas is that they provide almost uniform coverage in function of an azimuthal angular reference. The gain values of pagers show a substantial improvement in the 150- to 280-MHz band. The average values given in the table reflect a somewhat uniform coverage in the azimuthal plane, with variations in gain at the horizon of no more than 5 dB. At 450 MHz, the peak value of pager gain approximates 0 dBi, but the coverage ceases to be uniform. Deep nulls, on the order of −10 dB below the maximum, appear in the direction opposite to where the pager is located on the phantom. The human equivalent is a column of salt water 30 cm in diameter and 180 cm high. At 450 MHz, the phantom casts a penumbra at the backside of the incident wave. Finally, at 800 to 900 MHz, the shadow effect from the phantom becomes very clear. Deep nulls (20 to 30 dB) below the peak gain value are found in an angular region 20 to 30 deg opposite the location of the antenna.

The values given in Table 6.1 are representative of the expected performance on an average human, if such a person exists. The designer should expect some variation in average and maximum value if a specific individual is used in the measurement. Experiments have shown that belt-level results were so dependent on the individuals wearing the pager that the human phantom, however crude, was a welcome standard setter [6].

6.1.4 Pocket Pager Size Constraints

The antenna size (it must fit in a small pocket) constitutes the essential problem for this class of radio receivers and is often the only obstacle to good performance of the equipment. All the comments relative to loss versus antenna size presented in the previous sections can be repeated here with stronger emphasis. In addition, the orientation of the pager with respect to the incident field is no longer predictable as in the case of the belt-worn receivers. With these constraints in mind, let us review the few options available to the designer.

To minimize matching losses, the frequency of operation of the pager should be as close as possible to the frequency of resonance of the antenna. The use of self-resonant antennas is strongly recommended. It is the designer's good fortune that most pocket-sized equipment is requested for the newer available paging channels, which are at higher frequencies than older ones.

Finally, there are obvious exceptions to the considerations of the section. A pen-sized pager has one dimension (length), which gives the designer an obvious advantage. The design of pocket-sized antennas becomes problematic if the form factor of the pager is similar to that of a credit card, a small cigarette lighter, a watch, a lipstick container, or other items normally found in the pocket or the purse, and the frequency of operation is relatively low.

6.1.5 Concluding Remarks

This section has been descriptive and qualitative rather than analytical and quantitative, due to the impossibility of clearly defining the antenna for a very small, pocket-sized pager. As already pointed out, the components or the PC board of the device will be practically part of the antenna because of the tight packing requirements. In addition, the presence of other objects in the pocket or in the purse can radically affect the performance of the paging device. While from a purely experimental point of view there seems to be little relief for the design effort, analytical methods are very much on the verge of providing the tools to solve some of the problems presented in this section. Moment methods, finite element methods, or other analytical techniques that can solve Maxwell's equations for very complicated boundary conditions represent practically the only chance to properly design these antennas. The use of a modern high-speed computer to simulate deterministically the pager package performance as an antenna and, statistically, the influence of nearby external objects is the only tool on the horizon that can rationalize the design of these devices. Certainly, the availability of high-frequency, low-loss ferrites, low-loss dielectric materials, and high-temperature superconductors will help the performance level of pocket-sized pagers. However, the predictability of the performance of these devices will be based essentially on analytical techniques and early product definition, which, properly coupled, are the basis of any successful engineering design effort.

6.2 PAGER TYPES AND PERFORMANCE

6.2.1 Design Considerations

The salient design factors for a pager antenna are as follows. First, the paging service in Japan uses the VHF band, and since the pager must be less than 10 cm long, the antenna is electrically small for the service frequency wavelength. The pager's antenna, which is built into the body, must thus be an electrically small antenna. The radiation pattern resistance of such an antenna is smaller than the ohmic loss of the antenna element [7–9]; therefore, the efficiency of the pager's antenna is low. Second, when the antenna is built into the pager, it must be in close proximity to metal parts or a battery; this alters the impedance and the gain of the antenna [10]. Third, pagers are carried in shirt pockets or hung on the belt [11–15]. For good pager sensitivity, the antenna must have a high efficiency that is not unduly reduced in the vicinity of a human body or metal parts. Small-loop antennas are known to retain their gain near the human body, and are one of the most promising antennas for pagers. The characteristics of a small-loop antenna and basic design formulas are now summarized.

Antenna Characteristics Near the Human Body

The influence of the human body on the electric and magnetic components of an electromagnetic wave is a substantial effect [11]. The electric component is reduced significantly near the human body, whereas the magnetic component increases. Small-loop antenna are strongly affected by the magnetic component. A loop antenna can be regarded as magnetic dipole normal to the loop area [16], as shown in Figure 6.6. Therefore, if the loop antenna is built into a pager so that it is perpendicular to the human body, the magnetic dipole is parallel to the human body and the sensitivity of the pager is not reduced. The magnetic component in the high-frequency band can be captured very efficiently if the antenna is reinforced with ferrite material [17,18]. Unfortunately, the loss of ferrite is excessive in the VHF band [19].

Efficiency

The efficiency of η of a small-loop antenna is [2,4]:

$$\eta = \frac{R_r}{(R_r + R_{\text{loss}})} \tag{6.8}$$

where R_r = radiation resistance and R_{loss} = ohmic loss of the loop element. There are two methods of increasing the efficiency: one is to increase R_r and the other is to decrease R_{loss}. A small-rectangular-loop antenna, whose longer side is l_a and whose shorter side is l_b, has a radiation resistance R_r and an ohmic loss R_{loss} [2] given by

$$R_r = 320\pi^4 \left(\frac{l_a^2 l_b^2}{\lambda^4} \right) \tag{6.9}$$

$$R_{\text{loss}} = \frac{1}{\pi a}(l_a + l_b)\sqrt{\frac{\pi \mu f}{\sigma}}$$

where a is the wire radius of the loop element, μ is the permeability, f is the frequency, λ is the wavelength, and σ is the conductivity of the element.

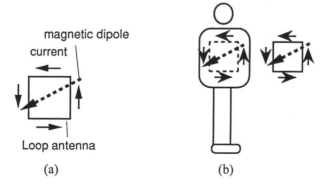

(a)　　　　　　　　(b)

Figure 6.6 The response of a small-loop antenna in the vicinity of a human body: (a) action of small-loop antenna; (b) response in the vicinity of a human body.

R_r can be effectively increased by increasing the loop area or the number of loop turns [7]. The ohmic loss can be reduced by thickening the diameter of the element, by using platelike elements to increase the element surface, or by using low-loss material for the antenna element. In commercial pager antennas, the ohmic loss is reduced by using plate-shaped elements.

Matching Circuit

The input resistance of a small-loop antenna is very low (less than 1Ω), and a matching circuit is required to connect the antenna to the receiving circuit. Thus, the loss of the matching circuit affects pager sensitivity significantly. Pagers frequently employ L-shaped matching circuits with two capacitors. The principle of the matching circuit is shown in Figure 6.7(a) [20], and in general the circuit consists of an inductor and a capacitor. The admittance from a-a' toward the right-hand side in Figure 6.7(a) is expressed as

$$Y = \left(\frac{R + \dfrac{X_1}{Q_L}}{\left(R + \dfrac{X_1}{Q_L}\right)^2 + X_1^2} + \frac{Q_C}{X_2(1 + Q_C^2)} \right) + j\left(\frac{Q_C^2}{X_2(1 + Q_C^2)} - \frac{X_1}{\left(R + \dfrac{X_1}{Q_L}\right)^2 + X_1^2} \right)$$

$$X_1 = \omega L = 2\pi f L \tag{6.10}$$

$$X_2 = \frac{1}{\omega C} = \frac{1}{2\pi f C}$$

where R is the load resistance, L is the inductance, and C is the capacitance shown in Figure 6.7(a). Q_L and Q_C are Q factors of the inductance L and the capacitance C respectively. X_1/Q_L and X_2/Q_C are the loss of the inductance and the capacitance respectively. Generally, the loss of an inductance is larger than that of a capacitor

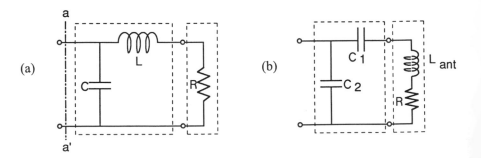

Figure 6.7 Matching circuit for: (a) load R; (b) small-loop antenna. (From [20], © 1964 Ohm Co., Ltd.)

Impedance matching is realized if inductance L tunes with C, and then the real part of the right-hand side of (6.10) is the required input resistance of the receiving circuit.

For impedance matching of a small-loop antenna to a receiving circuit, the low-loss matching circuit can be constructed with just capacitors C_1 and C_2 because of the large inductive reactance L_{ant} of the loop antenna [see Figure 6.7(b)]; ωL in Figure 6.7(a) is then equivalent to the series combination of ωL_{ant} and $1/\omega C_1$ in Figure 6.7(b).

Another matching arrangement with low loss can be designed by unifying the antenna and the matching circuit [21]. This circuit is shown in Figure 6.8(a). Since there is only one capacitor in the circuit, the loss of the matching section is much lower than that in the first example. In this case, impedance matching between the antenna and the circuit is carried out by capacitance tuning at the paging system frequency and by selecting a tapping point on the loop element so that the input impedance of the antenna equals the input impedance of the circuit. The equivalent circuit of this circuit is shown in Figure 6.8(b). The input impedance Y_{in} is expressed as (6.11), given the approximation of a small-loop antenna, $\omega L, 1/\omega C \gg R_r + R_{loss}$.

$$Y_{in} = \frac{\omega^2 (R_r + R_{loss})(L_1 + M)^2}{\left[\omega^2 (L_1 + M)^2 - \omega L_1\left(\omega L_T - \dfrac{1}{\omega C}\right)\right]^2} + j\,\frac{\left(\omega L_T - \dfrac{1}{\omega C}\right)}{\omega^2 (L_1 + M)^2 - \omega L_1\left(\omega L_T - \dfrac{1}{\omega C}\right)}$$

$$L_T = L_1 + L_2 + 2M$$

$$(6.11)$$

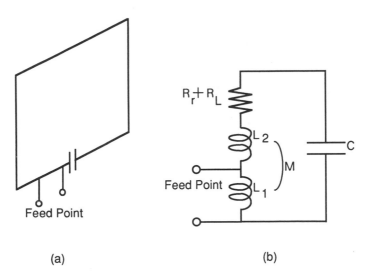

(a)   (b)

Figure 6.8 Matching circuit unifying the antenna: (a) construction figure; (b) equivalent circuit. (From [11], © 1973 Matsushita Electric Industrial Co., Ltd.)

where L_1 and L_2 are the inductances and C is the loading capacitance shown in Figure 6.8(b). M is the mutual inductance existing between the inductances L_1 and L_2. At resonance, the input admittance $Y_{\text{in/res}}$ is expressed as (6.12), where $\omega L_T - 1/\omega C = 0$.

$$Y_{\text{in/res}} = \frac{(R_r + R_{\text{loss}})}{\omega^2(L_1 + M)^2} \tag{6.12}$$

L_1 and M are decided by the tap position.

Directivity and Polarization

When a pager is carried, it is turned in all directions. Thus, the pager antenna must be omnidirectional, and the vertical and horizontal polarization characteristics of the antenna must be carefully designed. A commercial pager can have a built-in antenna that is both vertically and horizontally polarized. The antenna design specifies unbalanced current flow on the loop element and circuit boards and that the circuit board functions as part of the antenna [2]. Figure 6.9 illustrates balanced and unbalanced systems of the loop antenna, which is placed on the side of the pager. The behavior of the antenna in this

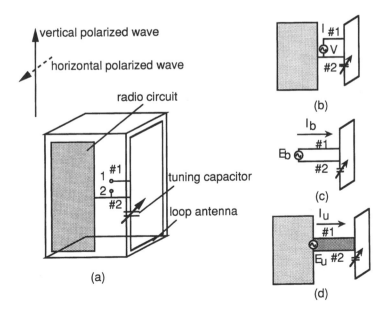

Figure 6.9 The behavior of a small built-in loop antenna: (a) small-loop antenna built into the pager; (b) original; (c) balanced mode; (d) unbalanced mode. (From [11], © 1973 Matsushita Electric Industrial Co., Ltd. Reprinted with permission.)

location is partly balanced and partly unbalanced, as shown in Figure 6.9. In the balanced mode, a balanced current flows on the loop element and the antenna receives the vertical polarized wave [shown in Figure 6.9(a,c)]. In the unbalanced mode, the loop antenna and the ground of the circuit board function as an unsynthesized dipole antenna, and the antenna receives the horizontal polarized wave [shown in Figure 6.9(a,d)]. Figure 6.10 illustrates the receiving pattern of this combination antenna in free space. The polarization of the loop antenna is the same as the polarization of the incident wave in case A. The polarization of the loop antenna is perpendicular to the incident wave in case C; the receiving level in case C is not unduly reduced compared to that in case A (the receiving level is 5 dB less than in case A). This is why this pager antenna responds as a combination antenna. On the other hand, when the loop antenna is built into the pager so that it is perpendicular to the polarization of the incident wave in case B, the receiving level has two nulls that are 15 dB less than they would be if the loop antenna were parallel to the polarization of the incident wave. Figure 6.11 illustrates the receiving pattern of the combination antenna in the vicinity of a human body. The directional gain increases about 5 dB from the free-space value.

6.2.2 Card-Sized Equipment

A card-type pager has a thickness of only 5 mm, which is half that of the tone-only type. Thus, it is difficult to achieve an antenna aperture that yields satisfactory pager sensitivity. To solve this problem, the antenna is constructed as shown in Figure 6.12(a) (i.e., the pager body is made of two large metal plates). One edge of both side plates is connected

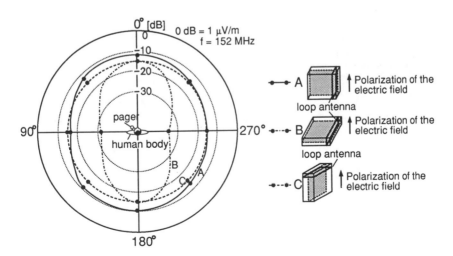

Figure 6.10 Receiving pattern of a pager in free space. (From [11], © 1973 Matsushita Electric Industrial Co., Ltd.)

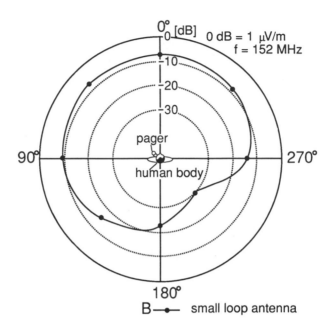

Figure 6.11 Horizontal receiving pattern of the pager in the vicinity of the human body. (From [11], © 1973 Matsushita Electric Industrial Co., Ltd.)

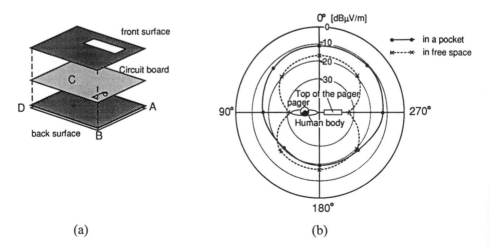

(a)   (b)

Figure 6.12 Antenna of the card type of pager: (a) construction of the antenna; (b) receiving pattern. (b) is courtesy Kokusai Electric Co., Ltd.)

to the input circuit and some of the other edges are shorted, so the pager becomes a loop antenna. Using the two large side plates allows the loop aperture to be maximized while reducing the element loss. The receiving pattern of this antenna is shown in Figure 6.12(b). The polarization direction of this antenna runs from the feed point to the shorted point [22]. If the shorted point is moved along the edge of the pager, the polarization direction is changed because the currents on the antenna (metal plate) are changed. Figure 6.13 illustrates the polarization characteristic when the shorted point is moved. This characteristic can be used to increase pager sensitivity by matching the polarization of the antenna to that of the incident wave.

Another method of controlling the polarization direction is to use electronic circuits [22]. This is carried out by impressing voltages on points A and B and shorting the points C and D (see Figure 6.13). When the voltages have the same amplitudes and phases, the receiving pattern is maximum at $\phi = 90$ and 270 deg [shown in Figure 6.14(a)]. On the other hand, if the phases are opposite, the receiving pattern is changed and the maximum directions are 0 and 180 deg [shown in Figure 6.14(b)]. The reason is as follows. When the phases are the same, the current on the antenna flows mainly between A and D and between B and C. Few currents flow between A and B or D and C. When the phases are in opposition, the current flows mainly between A and B, and few currents flow between A and D or B and C.

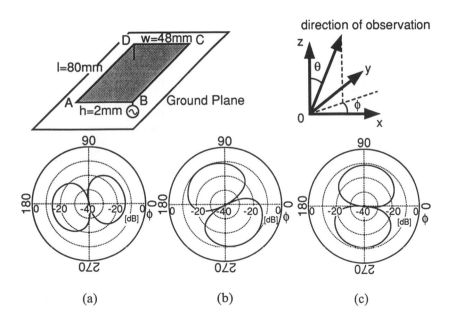

Figure 6.13 The mechanical method of polarization control [22]: (a) feed point is B, shorted point is A; (b) feed point is B, shorted point is C; (c) feed point is B, shorted point is D.

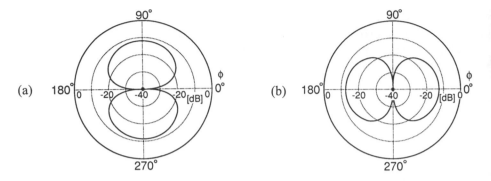

Figure 6.14 The electrical method of polarization control [22]: (a) polarization characterization with same amplitude and phase at A and B; (b) polarization characterization with same amplitude and opposite phase at A and B.

This section has described the small antennas of existing pagers. Public demand is for even smaller pagers that have higher performance. It is necessary, therefore, to develop very small but high-performance antennas.

6.3 DESIGN TECHNIQUES FOR PORTABLE PHONE ANTENNAS

The trend of the portable telephone technology in the last few years has been to dramatically decrease the size and the weight of the unit. In the initial offering (1984), the portable cellular phone was about 35 in^3 in volume (including batteries) and weighed about 30 oz (0.85 kg). Today, the most recent cellular portables have a volume of less than 100 cm^3 and weigh less than 80g. These dramatic weight and volume improvements have necessitated a rapid evolution of the antennas for the phones. The design efforts have been to maintain approximately the same antenna performance in terms of gain, coverage, and bandwidth in the face of rapidly decreasing size requirements. We will limit our discussion to antennas in the 800- to 1,000-MHz band, which covers most of the cellular systems in use at this time. These antennas can be used at higher frequencies by simply reducing their size by frequency scaling techniques.

6.3.1 Design Considerations

The main factors affecting the design of portable cellular phone antennas are the relatively large bandwidth (~10%), the desirability of a small size, and the need to provide a uniform coverage over the azimuthal angle, with a gain of about 0 dBi or higher. The coverage and size requirements are clearly contradictory in view of the fact that the antenna is held close to the head of the user, which is approximately an absorbing dielectric ellipsoid with the major axis one wavelength long. The head of the user absorbs and scatters the electromagnetic energy emanating from the nearby antenna, so that the azimuth coverage is never uniform if the head is close to the radiator. A simple way to achieve uniform azimuth coverage is to elevate the antenna above the head of the user. This solution requires a nonradiating supporting structure for the antenna about 6 in (about 15 cm) long. The antenna, in this case a half-wavelength dipole or similar radiator, also has a length of about 6 in (about 15 cm), giving a total length for the support and the radiator of about 1 ft (about 30 cm). This size is unacceptable, given that the largest dimension of cellular phones is no more than about 25 cm, which is approximately the distance between the mouth and the ears of an adult. A collapsible antenna of 30-cm length presents serious mechanical problems in the small and constantly decreasing volume of portable cellular phones. In addition, the result of dropping the radio with the antenna in the extended position may be an unacceptable degradation of performance.

Alternative structures would require separating the radio from the antenna, which would be located on a hat, a headband, or some type of head-mounted support. This solution requires a link between the radio and the antennas; in addition, the users may not accept any head-mounted solution of the coverage problem.

So far, only relatively small antennas have found acceptance in the rapidly growing cellular phone market. In the following sections, we will discuss the following antenna types: (1) the sleeve dipole, (2) the helical antenna, (3) the quarter-wavelength whip, and (4) the dipole helical antenna combination. These antennas represent most of the radiators used in cellular phone technology. The quarter-wavelength whip has not found favor with the cellular phone market, so it will be only briefly discussed in the section dedicated to the helical antenna.

6.3.2 Antenna Types

Sleeve Dipole

A sleeve dipole operating in the 800- to 900-MHz band is shown in Figure 6.15. It is essentially a half-wave dipole fed from one end by a coaxial line. The structure has cylindrical symmetry, so in free space it has excellent radiation pattern uniformity in azimuth.

The radiating structure is an asymmetric dipole made of conductors of different diameters and slightly different lengths. The thinner radiator is normally the inner conductor of the coaxial line feeding the antenna. This conductor must have an appropriate length to achieve good antenna match in the band of operation. The conductor of larger diameter has the critical function in proper antenna operation and must be designed with some care. The large-diameter conductor must provide effective choking of the RF currents at its own open end and also one-half of the radiating dipole. This conductor is shorted to the braid of the coaxial line feeding the sleeve.

To provide good RF current choking, it is essential that the sleeve part of the antenna have as large a diameter as possible compatible with the aesthetic acceptability of the

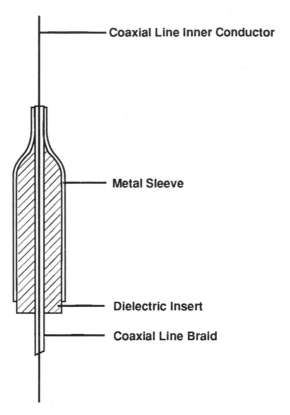

Coaxial Line Inner Conductor

Metal Sleeve

Dielectric Insert

Coaxial Line Braid

Figure 6.15 Sleeve dipole antenna cutaway.

antenna. The choke works most effectively if the sleeve transmission line formed by the braid of the coaxial line and the inner surface of the sleeve is resonant. In these conditions, the impedance presented by this line is

$$Z = jZ_0 \tan(kl) \sim \infty \quad kl = \pi/2 \tag{6.13}$$

$$Z_0 = \frac{60}{\sqrt{\epsilon_r}} \ln \frac{b}{a} \tag{6.14}$$

which also equals the characteristic impedance of sleeve line, where l is the length of the sleeve line, k is the propagation constant of the line, and ϵ_r is the relative dielectric constant of the insert. The sleeve line is normally dielectrically loaded. A low-loss, low-dielectric constant cylinder is needed to ensure the concentricity of line conductors, which is essential to the proper performance of the antenna. The dielectric material should have low ϵ_r to achieve maximum bandwidth of the antenna. In (6.14), a is the feeding coaxial line braid outer diameter and b is the metal sleeve inner diameter.

The outer surface of the metal sleeve is part of the radiating dipole, so its resonant length is slightly less than a quarter wavelength in free space, depending on its outer diameter. The differential length required for the outer and inner surfaces of the sleeve is made up by shorting the sleeve to the braid of the feeding coaxial line, as shown in Figure 6.15.

If the impedance of the sleeve line is reduced sufficiently, then RF currents will propagate down the feeding coaxial line and the radiator is no longer only the half-wavelength dipole. The radio case also becomes part of the radiating structure. Its pattern performance depends on the length of the exposed section of the feeding coaxial line and the size and shape of the metal surfaces of the radio. Rapid pattern deteriorations are seen within ±5% of the frequency given by (6.13) for relatively thin chokes (about 1-cm radius).

Radiation from Sleeve Dipoles

The radiation from half-wavelength dipoles has been the subject of a large body of literature [23–31], familiar to most readers. The sleeve dipole has a diameter discontinuity at the shoulder of the sleeve and at the feed point of the antenna. Sleeve dipoles with diameter discontinuity are dealt with in [32], which gives the radiation impedance of the antenna.

The bandwidth limitation of this antenna is dictated more by pattern performance than by impedance variation. If the choke is operated at a frequency of about ±5% away from the resonance of the sleeve line, RF currents will flow on the outer surface of the coaxial line feeding the antenna. These currents will excite the radio case, which radiates in phase opposition to the sleeve antenna, as shown in Figure 6.16. The pattern of the entire radiating structure has a minimum in the directions orthogonal to the axis of the antenna. This is normally the direction of the maximum desired coverage.

Figure 6.17 shows the radiation pattern of a sleeve antenna when the choke is near resonance, and so minimal RF currents excite the radio case. Note that the pattern of the antenna is practically that of a center-fed half-wavelength dipole. To check that the radio case has small or no RF currents flowing on it, the radio case was scanned with an isotropic RF-transparent, physically small (2.5-mm diameter) E-field probe. Figure 6.18 depicts the results of the scan, which show that the metal parts of the radio case have minimal RF currents or charges.

If the frequency of operation of the antenna is changed by +5% of the frequency of resonance of the choke, pattern lobing becomes clearly apparent, as shown in Figure 6.19, and similar pattern changes occur in Figure 6.20 for a 5% reduction in frequency. Clearly, these patterns are not as suitable for land mobile communication as that in Figure 6.17. A pattern notch in the direction of the horizon represents an effective coverage loss in most propagation environments. In addition, RF currents conducted on the radio case are partly absorbed by the hand of the user, causing an additional ohmic loss.

316

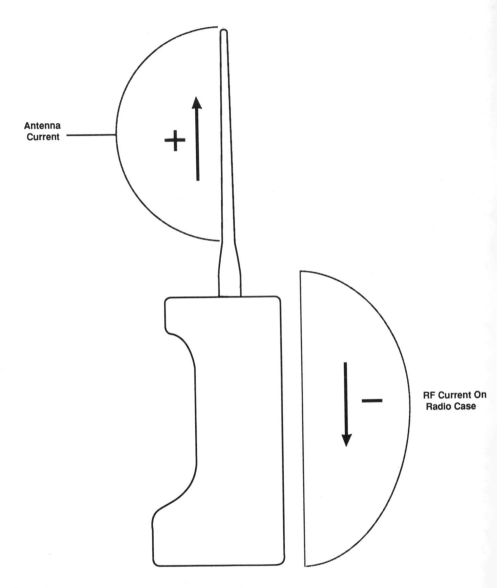

Figure 6.16 RF currents on a radio with sleeve dipole antenna.

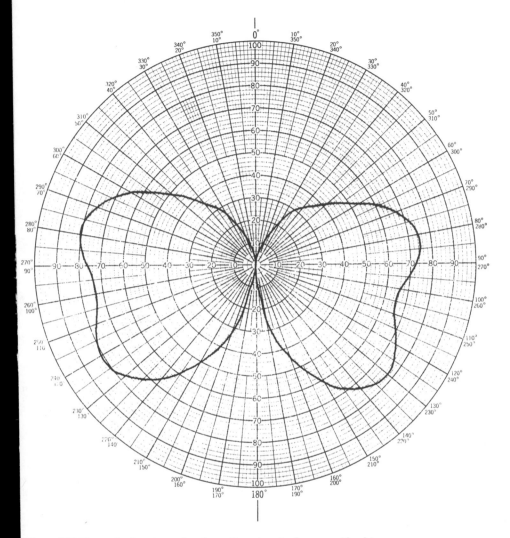

Figure 6.17 Sleeve dipole antenna elevation pattern at center frequency ($f = f_c$).

The gain of sleeve dipoles at the resonance frequency of the sleeve choke is almost the theoretical value for the half-wavelength dipole with absolute gain equal to 1.61 or 2.1 dBi. Ohmic losses in the thin braided coaxial line feeding the dipole (normally 0.05-in diameter or less) and in the thick dielectric molding used to protect the antenna reduce the absolute free-space gain of this dipole to 1 to 1.5 dBi.

End-fed half-wave dipoles are also used for portable phone systems. They normally consist of a half-wavelength wire antenna fed by a quarter-wavelength transformer, which

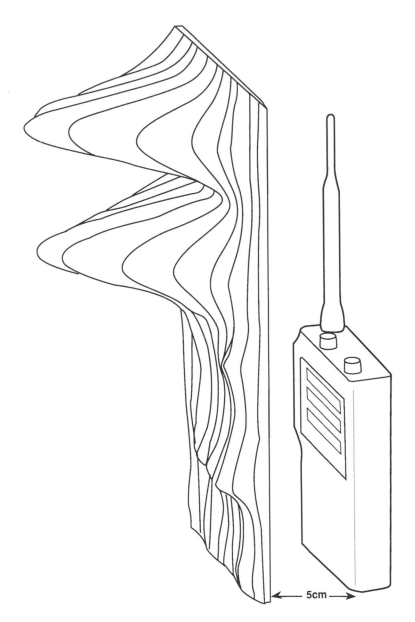

Figure 6.18 RF field near a radio with sleeve dipole at 840 MHz.

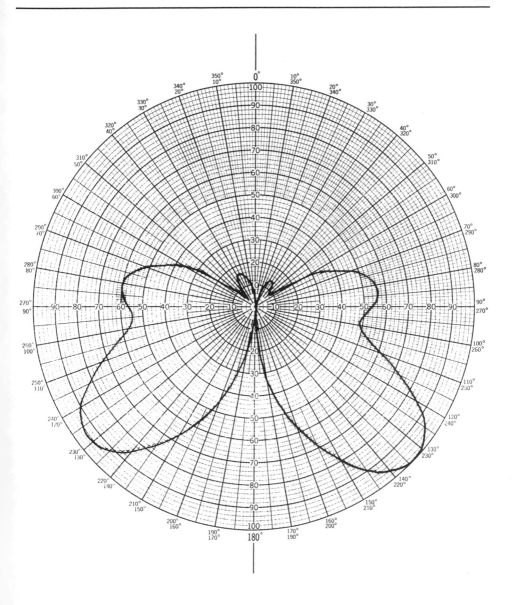

is necessary to match the high radiation resistance of the antenna to the 50Ω output impedance of the RF final amplifier in the radio. These antennas, depending on their construction and the bandwidth of the quarter-wavelength transformer, have performance characteristics similar to those just described for the sleeve dipole.

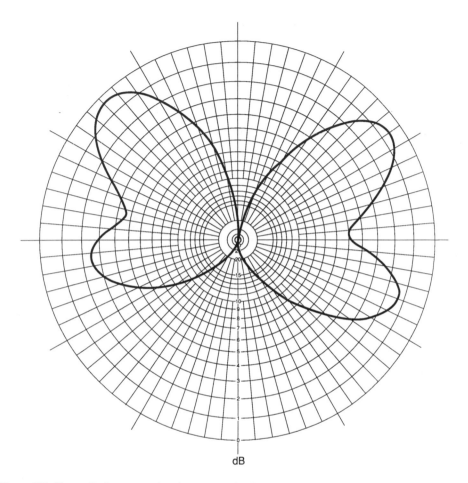

dB

Figure 6.20 Sleeve dipole antenna elevation pattern ($f = f_c - 5\%f_c$).

Helical Antennas

Helical antennas are used in portable communication radios at the low-frequency bands (30 to 150 MHz) to reduce the size of the radiator to comfortable lengths. With the proper selection of parameters, the normal-mode helical dipole is an efficient radiating structure, with pattern and gain performance similar to those of the half-wavelength dipole. The bandwidth of an efficient and short helical dipole is narrower than that of the half-wavelength antenna because of the higher Q factor, which depends on the number and the diameter of the turns of the antenna [33].

The radiation from the normal-mode helix, also called *omnidirectional,* has been the object of various works [33–36]. The reader can review the mathematical theory of the radiation of the helical antenna in the available literature. Here, we will summarize the limitations of the helical antenna in its applications to the two-way portable communication equipment, especially 800- to 900-MHz cellular phones.

The resonant quarter-wavelength helical antenna excites strong RF currents on the portable radio case, which is an integral part of the radiating system. The phenomenon can be easily detected by scanning the near field of the portable radio (about 5-cm distance or less). A picture of such a scan of a UHF portable radio is shown in Figure 6.21. Note the flow of the currents on the metal parts of the radio case. These RF currents are partly absorbed into the hand of the user, which must be considered as a lossy dielectric material wrapped around the radio case.

In most applications, the helical antenna is used to reduce by a factor of three or more the length of a resonant thin-wire antenna operating at the same frequency. The monopole helixes normally used in the portable two-way communication technology have a physical length of about 1/12 wavelength and an electrical length of about a quarter wavelength. The physical length can be reduced well below 1/12 wavelength, but the radiation performance loss is normally intolerable. The following considerations hold for approximately 1/12-wavelength-long resonant helical antennas; the losses are much higher for shorter antennas.

Gain measurements at the UHF band (450 MHz) have shown that a substantial gain loss with respect to the radio in free space is recorded when the radio is handheld. The RF currents on the radio case are conducted through the hand and dissipated through the arm of the user. In addition, some of the fields emanating from the helical antenna are directly coupled into the hand of the user, thus providing an additional loss mechanism. A gain loss of about 3 dB is commonly detected with the radio case held by a medium-sized person with dry hands.

The radiating structure helix plus radio case has enough bandwidth (about 50 MHz) to cover a substantial portion of the UHF two-way portable communication spectrum. With the gain losses due to RF absorption by the user, one finds that one helical antenna can cover the entire two-way UHF land mobile band (about 100 MHz), although the same helix exhibits a Q factor of about 60 if tested over a large conducting ground plane or in a helical dipole configuration.

The application of the helical antenna at the 800- to 900-MHz band has an additional disadvantage with respect to its use at UHF unless the radio case is about one-quarter wavelength long (about 9 cm). If the radio case is substantially longer than a quarter wavelength, the currents on part of the radio case are in phase opposition with that of the helix, producing pattern lobing with a gain loss at the horizon, if the radio case is substantially longer than a quarter wavelength. Early models of cellular portables had a radio case about 20 cm long, which was producing patterns of the type shown in Figure 6.22. Finally, the antenna is so small that it hardly radiates in the angular sector shadowed by the head of the user. A substantial pattern coverage gap is caused by the helix being held very close (2 to 4 cm) to the head of the user. From pattern measurements, the helical antenna on a portable phone near the head of a human shows an average gain loss of

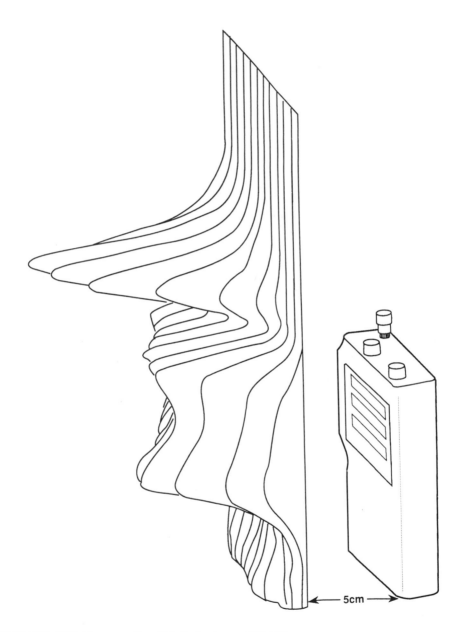

Figure 6.21 RF field near a radio with helical antenna at 814 MHz.

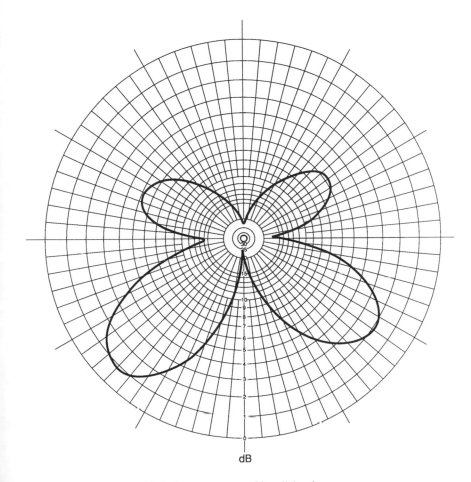

dB

Figure 6.22 Elevation pattern of helical antenna on portable cellular phone.

about 12 dB with respect to a half-wavelength free-space dipole. Pattern and RF losses cause the helical antenna to be hardly suitable for cellular radio application at 800 to 900 MHz, except for the small size of the antenna, which makes it very attractive for pocket-sized radios.

Most of the considerations regarding the helical antennas hold true for the quarter-wavelength whips. They radiate somewhat more efficiently than the foreshortened helical antennas, but do not overcome any of the pattern shape problems just discussed. Since there is no true advantage in using the quarter-wavelength whips over the helical antennas, the latter ones have found portable phone market acceptability. Practically no portable cellular phone model uses the quarter-wavelength monopole antennas, which have found acceptance in the mobile cellular technology.

6.3.3 Antenna Diversity

The poor performance of the helical antenna for pocket-sized portable phones at 800 to 900 MHz has forced designers to offer an alternative radiator to the short helix. The antennas currently used are two helixes: a primary fixed helix, approximately 2 cm long (quarter-wave electrical length) and a secondary helical resonator approximately 10 cm long (half-wave electrical length), which can be collapsed inside the radio case. When extended, the secondary helix becomes the dominant radiator because it is free from the losses caused by the hand of the user holding the radio case (Figure 6.23).

The secondary helix is end-fed by the primary antenna and exhibits excellent free-space performance, depending on its own length and the ohmic losses in the metal wire forming the helix. At the top of the secondary antenna there is a 2-cm-long dielectric rod to decouple the two helixes, when the secondary antenna is retracted, so minimal conducted RF is injected into the radio case through this path. This arrangement has provided some improvement to the communication range achievable by pocket phones with the helical antennas.

A summary of the performance of the antenna diversity arrangement just described is given in Table 6.2. The gain values were recorded in an anechoic chamber by taking elevation patterns of the radio case in free space and handheld at the head of a user like a normal household telephone. As is obvious from the results in the last column of Table 6.2, the performance of this radio telephone is far from optimal, although the pocket cellular phone has become extremely popular, which proves that engineering perfection is not always a condition for market success. Certainly, the next generation of pocket phones should provide some improvement on the results shown in Table 6.2.

6.3.4 Conclusions

The size reduction of cellular phones has driven designers to antenna choices that, using traditional portable radiators (dipoles and helices), have gain performance far from satisfactory. Considering the size of current pocket cellular phones (roughly $12 \times 6 \times 2$ cm), there is no fundamental physical limitation for why the performance summarized in Table 6.2 should be acceptable. A new set of basic radiator choices should be made and optimized for this application. The traditional antennas are all axially symmetric, so they radiate uniformly at the horizon. The necessity of this choice as the radiators become physically small with respect to the size of the head of the user should be questioned. Coverage in the directions shadowed by the head of the user is practically impossible and produces only RF ohmic losses in the head and neck of the subscriber to the phone service. The situation can be alleviated by using basic radiators with some form of directivity. In addition, the antenna should be decoupled from the RF absorbing hand of the user. With careful design and selection of the size and location of the radiator, it is possible to improve the current performance of pocket cellular phone antennas, even if their size is further reduced.

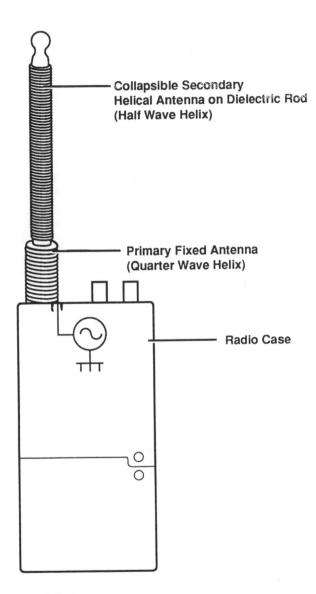

Collapsible Secondary
Helical Antenna on Dielectric Rod
(Half Wave Helix)

Primary Fixed Antenna
(Quarter Wave Helix)

Radio Case

Figure 6.23 Gain diversity helical antenna system.

6.4 PORTABLE PHONE ANTENNA SYSTEMS

6.4.1 System Design Aspects

The two chief design characteristics of any portable radio telephone antennas are that it must be mounted on the housing of a portable telephone and that during operation the

Table 6.2
Recorded Free Space and Handheld Gain
Loss of Diversity Helical Antenna*

	Free Space	*Handheld*
Secondary extended	−4	−7
Secondary retracted	−6	−13

*Gain referenced to a half-wavelength dipole.

set will be held by a human, who may randomly point the set in any direction. Because the antenna is forced into close proximity with the housing, the antenna current is induced not only into the antenna element, but also into the conductive housing. This current dispersion changes the shape of the original radiation pattern. Radiation efficiency is further degraded by the antenna's forced proximity to the human body, since the antenna is necessarily used near an operator. The polarization of radiation pattern is also changed by the changes in antenna direction caused by operator movements and habits. These difficult design constraints are complicated by the need to develop very small antenna elements to meet the demand for compact, portable equipment; it is common knowledge that radiation efficiency and bandwidth degrade as antenna element size is decreased. In addition to these design constraints, which are commonly found in portable radio telephone applications, system-based considerations such as the security of the specified frequency bandwidth and the need to support antenna diversity reception must also be taken into account. Figure 6.24 shows the interaction and impact of these technical subjects on the design of antennas mounted on portable radio telephone sets.

The principal subject that must be investigated is the effective gain of antennas in a multipath radio propagation environment. The key design task is to study these subjects synthetically and to maximize the effective gain in the desired frequency range as much as possible. High effective gain allows the size and weight to be reduced, while the usage time is increased. Not only does it permit the portable radio telephone to be used in areas of low electrical field strength, but it also allows transmitting power to be reduced. This reduction in transmitting power is extremely effective in reducing battery capacity (i.e., the weight of the portable radio telephone) or extending the usage time of a full battery. Therefore, the effective gain of antennas to be mounted on portable radio telephones must be maximized to develop small and high-performance portable telephone sets.

Simple Method for Evaluating Mean Effective Gain of Antennas in Multipath Propagation Environments

As described in Chapter 2, the mean effective gain of mobile antennas in a multipath propagation environment is determined by the joint contribution of the antenna pattern and the characteristics of the waves randomly arriving from the environment. The method

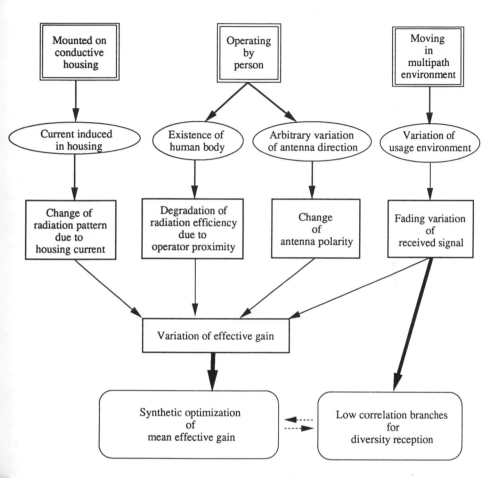

Figure 6.24 Technical subjects for designing antennas for portable telephone equipment.

of analyzing the mean effective gain also has been clarified. However, the statistical distribution of the arriving waves, which is needed in the analytical method, has not yet been sufficiently clarified. Furthermore, the three-dimensional radiation pattern is needed, but it is not easy in practice to measure such patterns. Hence, given the present situation, simplified method [37], derived from the analytical method described in Chapter 2, must be used to design antennas. In this simplified method, it is first assumed that the incident waves arriving at the mobile antenna are concentrated in the horizontal plane, and that the angular density function of incident waves follows the distribution function.

$$P_\theta = P_\phi = \frac{1}{2\pi} \delta\left(\theta - \frac{\pi}{2}\right) \tag{6.15}$$

where P_θ and P_ϕ are the θ and ϕ components of the angular density functions of arriving plane waves, respectively. This function indicates that the mean effective gain in represented by the following equation.

$$G_e = \frac{1}{2\pi} \int_0^{2\pi} \frac{\text{XPR}}{1 + \text{XPR}} \left[G_\theta\left(\frac{\pi}{2}, \phi\right) + \frac{1}{\text{XPR}} G_\phi\left(\frac{\pi}{2}, \phi\right) \right] d\phi \qquad (6.16)$$

where XPR is the average cross-polarization power ratio, and G_θ and G_ϕ are the θ and ϕ components of the antenna power gain pattern, respectively. Based on previous research and outdoor experiments, the XPR value used in the above equation (6.16) is assumed to be 9 dB [38,39]. This XPR value evaluates the lowest contribution of the ϕ component of the antenna power gain pattern to the effective gain. The mean effective gain of the target antenna with respect to a reference antenna is estimated by the following equation.

$$G_p = \frac{G_e(\text{testing mobile antennas})}{G_e(\text{reference antennas})} \qquad (6.17)$$

As described in Chapter 2, G_p presents the mean effective gain with respect to an isotropic antenna if the reference effective gain is the mean effective gain of a 55-deg inclined half-wavelength dipole antenna incremented by 3 dB. However, to evaluate the gain performance more easily, the mean effective gain of a vertically oriented half-wavelength dipole antenna (2.15 dBi) is normally used as the reference mean effective gain because it can be assumed that incoming plane waves are concentrated in the horizontal direction. It should be noted that, in general, this estimation value is not sufficiently accurate to assess gain performance in real-world propagation environments, because the mean effective gain of vertically oriented half-wavelength dipole antennas changes with propagation conditions and should be assumed to be lower than 2.15 dBi [40].

Since the mean effective gain evaluated by (6.17) is equal to the ratio of the average strengths of the radiation patterns of the target and reference antennas in the horizontal plane, it is also called the *pattern averaging gain* (PAG) [37]. PAG allows the practical evaluation of antennas to be simplified as follows.

- The performance of a target antenna can be easily evaluated by pattern measurement in a radio anechoic chamber.
- The gain degradation due to variations of radiation pattern can be easily evaluated irrespective of whether the variations are caused by the conductive housing or by human operation.
- The degradation of radiation efficiency due to the variation of input impedance caused by either proximity to a human body or the absorption of radiation power by the human body can be evaluated similarly.

• The frequency dependency of the effective gain can be evaluated by changing the measurement frequency.

Designing for Antenna Diversity

The correlation coefficient between antenna diversity branches can be investigated theoretically by calculating the complex radiation patterns of the antennas that form each branch in antenna diversity and applying them to the analytical method described in Chapter 2. Figure 6.25 shows a space diversity configuration using two parallel half-wavelength dipole antennas. The dipole antennas are inclined from the vertical (Z) direction around the Y-axis. It is also assumed that this antenna diversity branch is used in a postdetection diversity reception system (i.e., each branch is connected to a receiver whose input impedance is 50V). Because of the mutual coupling between the two antennas, the correlation coefficient depends on the antenna spacing, as shown in Figure 6.26. The XPR is the cross-polarization power ratio, m_V and m_H are, respectively, the mean elevation angle of the VP and the HP wave distributions observed from the horizontal direction, and σ_V and σ_H are, respectively, the standard deviation of the VP and HP wave distributions, as described in Chapter 2. When the inclination angle α is 0 deg, noncorrelation is achieved when the antenna spacing is larger than 0.3λ. When $\alpha = 90$ deg, the correlation can be strongly increased, as shown in Figure 6.27. This is because the overlap of the complex radiation patterns of the antenna branches increases with the inclination angle.

   The characteristics of the MEG of the antennas can be calculated similarly by using the analytical method described in Chapter 2. Figure 6.28 shows the MEG characteristics

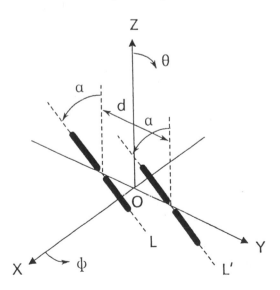

Figure 6.25 Space diversity configuration of parallel dipole antennas and the coordinate system.

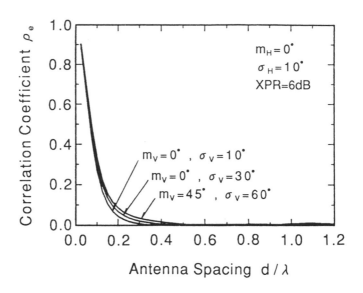

Figure 6.26 Correlation characteristics at antenna inclination angle α of 0 deg.

versus the antenna spacing at the antenna inclination angles α of 0 and 90 deg. MEG decreases rapidly when the antenna spacing is smaller than 0.2λ, because the radiation efficiency is decreased due to the mutual coupling loss. Figure 6.28 shows that the optimum antenna spacing for this type of space diversity with two parallel half-wavelength dipole antennas lies in the range 0.3λ to 0.4λ. As described above, the optimum configuration of antenna diversity can be designed by considering the characteristics of both the correlation coefficient and MEG.

In practice, the antenna diversity elements are mounted on the portable radio housing. In this case, if we calculate only the radiation pattern of the antennas by using the wire grid model and the moment method, and input the pattern so determined into the method described in Chapter 2, the MEG characteristics and the correlation characteristics are obtained for the portable diversity antennas. This approach can also be used to evaluate the effect of the operator. However, as described earlier, the statistical distribution of arriving waves has not yet been sufficiently studied, even though it is needed in the analytical method. Furthermore, the analytical method requires knowledge of the three dimensional radiation pattern, but measuring such patterns is not easy in practice. Hence in the design of antenna diversity systems, it is considered effective to apply a simplified method derived from the analytical method described in Chapter 2 to evaluate the correlation characteristics of antenna diversity. We must make an assumption similar to that used in the PAG estimation method in Section 6.4.1, whereby the angular density function of incident waves is assumed to be modeled by (6.15). This results in the following simplified formula for evaluating the correlation coefficient.

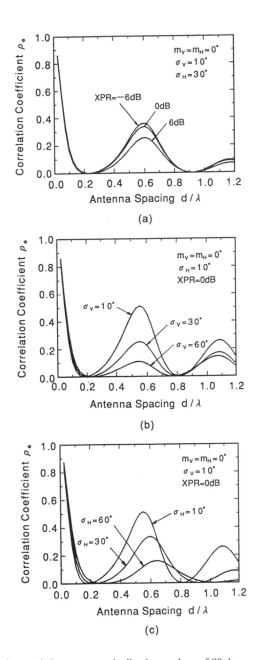

Figure 6.27 Correlation characteristics at antenna inclination angle α of 90 deg.

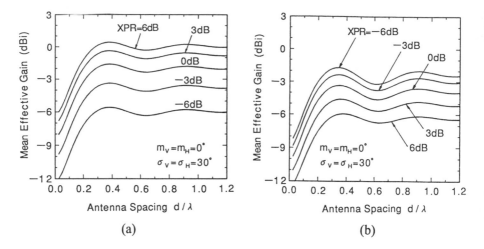

Figure 6.28 MEG characteristics of parallel dipole antennas: (a) $\alpha = 0$ deg; (b) $\alpha = 90$ deg.

$$\rho_e = \frac{\int_0^{2\pi} [\text{XPR} \cdot \overline{E}_{\theta 1}\overline{E}_{\theta 2}^* + \overline{E}_{\phi 1}\overline{E}_{\phi 2}^*]e^{-j\beta x}d\phi}{\int_0^{2\pi} [\text{XPR} \cdot \overline{E}_{\theta 1}\overline{E}_{\theta 1}^* + \overline{E}_{\phi 1}\overline{E}_{\phi 1}^*]d\phi \times \int_0^{2\pi} [\text{XPR} \cdot \overline{E}_{\theta 2}\overline{E}_{\theta 2}^* + \overline{E}_{\phi 2}\overline{E}_{\phi 2}^*]d\phi}$$

$$(6.18)$$

where $\overline{E}_{\theta n}$ and $\overline{E}_{\phi n}$ ($n = 1, 2$) are the complex envelopes of the θ and ϕ components of the electric field pattern in the horizontal plane, respectively. The contribution of the ϕ components of the electric field pattern should be rated lower, and thus it seems to be reasonable to assume that the XPR value used in (6.18) is 9 dB [38,39].

6.4.2 Handheld Systems

Antennas for Nondiversity Systems

Many kinds of handheld portable telephone units have been developed for the 900-MHz band land mobile communication systems. In the Japanese system MCS-L1 [41], which does not adopt diversity reception, the mobile stations use the 870 to 885-MHz band for receiving and the 925- to 940-MHz band for transmitting. Thus, to cover the whole frequency band, 870 to 940 MHz, the antennas for the mobile stations must have a relative bandwidth of about 7.8%. The most popular antenna for handheld units is the whip antenna, which not only has relatively broad bandwidth characteristics, but can also retract into the unit. Furthermore, the whip antenna has three other advantages: simple element structure, small element volume, and easy installation. However, antenna currents are impressed not only on the antenna element, but also on the conductive housing. This is why the radiation pattern depends on the size of the housing. Figure 6.29 shows the calculated radiation patterns of quarter-wavelength whip antennas mounted on a conductive housing 160 mm long, 30 mm wide, and 55 mm deep. In the case of the quarter-wavelength whip antenna, the currents in the antenna element and the housing are not in phase. This results in a radiation pattern that has four vertical radiation lobes, and the degradation of the directivity in the horizontal plane is considerable. Figure 6.30 shows the PAG

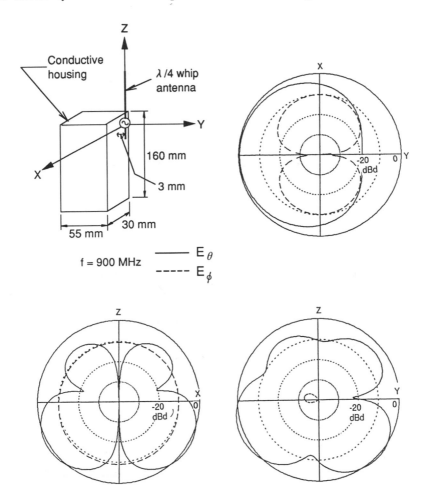

Figure 6.29 Calculated radiation patterns of quarter-wavelength whip antennas mounted on conductive housing 160 mm long, 30 mm wide, and 55 mm deep.

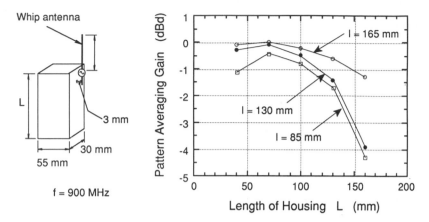

Figure 6.30 PAG characteristics of whip antenna mounted on conductive housing 30 mm wide and 55 mm deep.

characteristics, calculated at 900 MHz, for various-length whip antennas mounted on a conductive housing in the upright condition. It shows that the effective gain depends mainly on the length of the whip antenna element and the length of the conductive housing. Thus, in this type of handheld unit, we can optimize the antenna performance by selecting the optimum lengths. A recent commercial handheld unit uses a whip antenna, where its length, width, and depth are 178 mm, 40 mm, and 73 mm, respectively; that is, its volume is about 520 cc. The length of the antenna element is 160 mm in use, and this corresponds to about one-half wavelength. According to the PAG evaluation in Figure 6.30, this handheld unit was designed to yield the highest effective gain in the upright position. It can be expected that the operator would use this handheld unit in this position to improve the speech quality. The effective gain of this antenna, measured in a typical mobile radio environment, is about −1 dBd in the upright position, and this measured result corresponds to the result given in Figure 6.30, in which 0 dBd means the effective gain of a vertically oriented half-wavelength dipole antenna.

6.5 SAFETY ASPECTS OF PORTABLE AND MOBILE COMMUNICATION DEVICES

6.5.1 Exposure to RF Energy

Mobile and portable wireless communication devices emit RF energy, which is as low as a fraction of a watt in the case of two-way pagers, cellular phones, personal communication service (PCS) devices, and as high as 130W for certain mobile stations. In normal use, handheld radios and phones have the potential for causing higher exposure than mobiles because, although portable devices rarely emit more than 7W of RF power, a person's vital organs (e.g., the brain and eyes) are in the immediate vicinity of the RF source. Mobile radio antennas can be located a meter or more from car passengers, so the exposure is reduced due to the large distance. In addition, in metal-body cars, the vehicle's metal surfaces can form an effective RF shield for the car passengers if the antenna is mounted at the center of car roof, as discussed in Section 6.5.5.

In the current Institute of Electrical and Electronics Engineers RF Safety Standard C95.1-1999 [42] and the International Commission on Non-Ionizing Radiation Protection 1998 Guidelines [43], the exposure to RF electromagnetic fields (EMF) in the band of 3 kHz to 300 GHz is quantified in terms of electric field in volts/meter (V/m), magnetic field in amperes/meter (A/m), and incident power density in watts/square meter (W/m^2), or milliwatts/square centimeter (mW/cm^2). Two phenomena differentiate the absorption of RF versus infrared or visible light: coherent absorption and the structure of the EM fields in the vicinity of sources.

Coherent absorption is a typical resonance phenomenon. A human exposed in the far field to an EM wave with E-field parallel to the direction of body axis (E-polarization) supports a relatively strong coherent RF current if the wavelength of the incident field is about five times or two-and-a-half times the height of the human, depending on the grounding conditions. If the body is well grounded, the current in the human is similar to that of a quarter-wavelength dipole over a ground. If the body is poorly grounded, the coherent RF current resembles that of a resonant half-wave dipole. In both situations, a coherent current is excited and the peaks of the absorption of RF energy occur where the human anatomy has small cross sections (e.g., the ankles, the knees, and the neck area) [44,45], due to their increased current densities. At H- or k-polarization, the resonance effect is either less prominent or absent [46]. Therefore, field orientation and frequency, and body size are important parameters in the human absorption of RF energy.

Near RF sources, the fields are very complicated and the incident power density cannot be defined, because it requires a single direction of field propagation. The relevant exposure from low-power (7W or less) portable devices happens within a few centimeters' distance from the antenna, where some high EM energy density values may be found. At this short distance, the radiating source has a finite angular extent as seen from an observation point, so it is impossible to define a direction of power flow. In regions very close to a source, the direction of the RF energy propagation may not coincide with the direction of the Poynting vector ($1/2$ $\mathbf{E} \times \mathbf{H}^*$), which is normally associated with power flow in the far field of RF sources. In these conditions, we cannot simply define power flow, which makes any attempt to measure it practically impossible.

To quantify the human exposure to RF energy, it is necessary to invoke a basic physical law of Grotthuss-Draper, which states that a physical agent will have an effect only if it is inside a body. Given the large amount of reflection and scattering that a human body causes in the near field of a source of comparable size, it is necessary to measure the RF power deposited inside a human body rather than the external incident electromagnetic fields. Because the human body has practically irrelevant amounts of magnetic materials, it does not directly absorb magnetic energy. The human body is mainly composed of water, electrolytes, and complex molecules with a large net dipole moment [47]. With these constituents, the human body extracts energy from the RF E-field by ionic motion and oscillation of polar molecules. At present, other mechanisms of RF coupling with living tissue (e.g., Larmor precession of nuclei) are not considered relevant to the effects discussed in this section. The absorption of energy from the H-field is through Faraday's law ($\nabla \times \mathbf{E} = -\mu \mathbf{H}/\partial t$), whereby an electric field is generated by a temporal change of magnetic flux density.

6.5.2 Specific Absorption Rate of Electromagnetic Energy

From the considerations in the previous section, the evaluation of human exposure in the near field of RF sources, like portable and mobile phones and radios, can be accomplished by measuring only the E-field inside the body. We use the symbol σ to denote the conductivity of human tissue due to conduction and displacement currents (σ is measured in siemens per meter, S/m), then, by definition, the rate of RF energy deposited per unit volume is given by the product of the current density $\mathbf{J}$ and the conjugate of the electric field ($\mathbf{E}^*$). The power absorbed per unit volume of tissue is given by:

$$P_v = \frac{1}{2}\mathbf{J} \cdot \mathbf{E}^* = \frac{1}{2}\sigma|E|^2 \ [\text{W/m}^3] \tag{6.19}$$

If the absorption of RF energy per unit mass is desired rather than the absorption per unit volume, we need only to introduce the density ρ (in kilogram/cubic meter) of the tissue in (6.19), thus obtaining

$$P_g = \frac{1}{2}\frac{\sigma}{\rho}|E|^2 \ [\text{W/kg}] \tag{6.20}$$

where P_g is the *specific absorption rate* (SAR), which is defined as the time derivative of the incremental energy absorbed by (dissipated in) an incremental mass contained in a volume of a given density, $P_g = dW/dm$ [48]. To obtain the RF power P absorbed by a specific organ, one performs the integration: $P = \int_M P_g \, dm$, where M is the mass of the organ of interest, which can be the entire human body.

Due to the conduction and displacement current induced losses, the absorbed energy increases the tissue temperature. The SAR, P_g, is related to the temperature rise by the following equation, which is rigorously valid only for adiabatic heating, that is, at the beginning of the exposure (at $t = 0+$):

$$P_g \cong 4186\frac{c_H \Delta T}{\Delta t} \ (\text{W/kg}) \tag{6.21}$$

where

c_H is the specific heat capacity of the tissues (kcal/kg °C);
4186 is the conversion factor from kilocalories to joules;
ΔT is the temperature rise (°C);
Δt is the exposure duration in seconds.

RF dosimetry in the human body is very complex due to the many factors that affect energy absorption in tissues [49]. In (6.20), the conductivity Φ varies with tissue type and is a function of frequency [50], $\sigma = \sigma(x,y,z,f)$, where x,y,z are the coordinates of an arbitrary reference system connected with the human body, and f is the frequency. In addition, conductivity increases with temperature, about 2% per °C for muscle. The tissue density ρ also changes with tissue and is a function of water content, so $\rho = \rho(x,y,z)$. The internal electric field E is determined by the dielectric properties, shape, size, and orientation of tissue, polarization and frequency of fields, source configuration, and exposure environment. Equations (6.19) and (6.20) are valid for time-harmonic fields. The incident field may be time varying in a complex manner (e.g., a pulse train or several fields at different frequencies). In these conditions, it is possible to apply (6.20) to the spectral content of the internal field $E(t)$. If the signal has a broadband spectrum, care must be taken to use the appropriate value of σ at different frequencies. Dielectric properties of 45 tissues from 10 Hz to 20 GHz have been documented [50]. The results are widely used and are available on the Internet (www.brooks.af.mil/AFRL/HED/hedr/reports/dielectric/home.html). The U.S. Federal Communications Commission also has the same data between 10 MHz and 6 GHz posted on its web site (www.fcc.gov/fcc-bin/dielec.sh).

The time integral of the SAR is called *specific absorption* (SA) and represents the ratio of the amount of EM energy (dE) absorbed by an incremental mass of tissue (dm): $SA = dE/dm$. The concepts of SA and SAR have been borrowed from radiation medicine, where dose and dose rate have been used for many years in the treatment of tumors with ionizing radiation.

Using the SAR methodology in assessing the exposure of a human to incident EM energy totally bypasses the issue of near or far fields from sources. Although it is desirable to relate the SAR distribution of an exposure to the structure of the incident EM fields, this is difficult to accomplish, due to the complexity of dosimetry in the near field of RF sources. In the following section, the methods currently used for measuring SAR of wireless handsets and devices are described.

6.5.3 Measurement of SAR

According to (6.20) and (6.21), SAR can be determined either with E-field or temperature measurements. Using temperature measurements for SAR determination requires high-power exposures [51]. Handheld radios or cell phones are low-power devices, which cannot induce an adequate initial temperature rise for accurate measurements; therefore, only the E-field measurement method is described in this chapter. Techniques using nonperturbing temperature probes and infrared thermographic cameras for SAR measurements are described in [49,51,52].

The use of simulated tissue models probed with small electric field sensors is preferred for testing low-power wireless devices [53–57]. The E-field is measured with a computer-controlled robot operating a miniaturized E-field probe in a human head-shaped phantom or a full body phantom filled with tissue-simulating liquid when exposed to a mobile radio, telephone, or two-way pager. From the measured E-field values, the SAR distribution and the peak averaged SAR are obtained.

Currently, both the IEEE Standards Coordinating Committee 34 Subcommittee 2 [58] and the European Committee for Electrotechnical Standardization [59] are drafting standards for the measurement of peak SAR for mobile phone exposure. Parts of the two standards are included in the following section. For details of the SAR measurement, the readers should refer to the original standards.

SAR Measurement Equipment

In Figure 6.31, a typical SAR measurement system shows that a robot, an E-field probe, and a low dielectric material holder for phone positioning are used to perform SAR measurement in a head phantom. The E-field sensor consists of a thin dielectric substrate supporting an electrically short dipole, a diode to rectify the RF signal, and balanced high-resistance leads to extract the rectified voltage. The isotropic probe consists of three such devices arranged orthogonally in an I-beam shape or a triangular beam. The SAR measurement system consists of an E-field probe, DC voltage amplifiers, high-impedance cables connecting the amplifier outputs to a personal computer, a controlling robot, a simulated tissue phantom, and a holder assembly for placing the phone with respect to the phantom. An optical control system is also available for positioning the E-field probe.

The entire system, including the E-field probe, is calibrated in a controlled laboratory environment in each tissue-equivalent liquid at the appropriate operating frequency. The minimum detection limit of the system is lower than 0.01 W/kg and the maximum detection limit is slightly higher than 100 W/kg. The linearity is within 0.5 dB over the SAR range from 0.01 to 100 W/kg. The isotropy of the receiving pattern of the probe is within ±1 dB. Sensitivity, linearity, and isotropy are determined in the tissue-equivalent liquid. The length of the individual sensing elements in the E-field probe is less than 5 mm, and the outside diameter of the protective sleeve does not exceed 8 mm. When the measured signal is a pulsed signal (e.g., a TDMA signal), the integration and averaging time of the SAR measurement equipment (based on RMS detection) is able to yield results reproducible to within 5%.

The robot holding the probe scans the whole exposed volume of the phantom in order to evaluate the three-dimensional field distribution. The mechanical structure of the robot does not interfere with the SAR measurements, because the RF source is very close to the simulated tissue. The repeatability of the robot positioning of the E-field probe

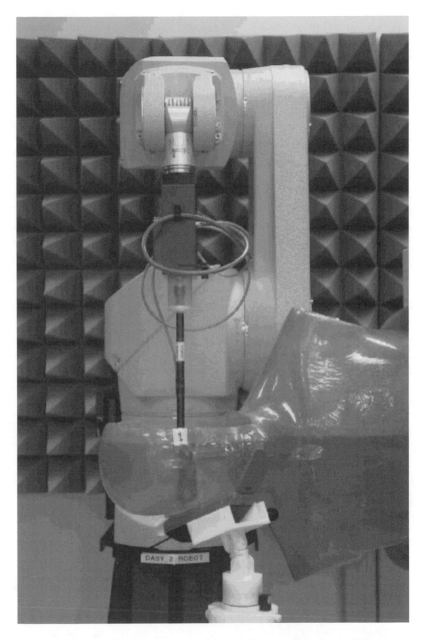

Figure 6.31 Cell phone SAR measurement system showing robot controlling electric field probe for SAR measurement in a head phantom exposed to a cellular phone at the right ear.

sensors over the measurement area is less than 0.1 mm. The sampling resolution is the step at which the measurement system is able to perform measurements of 1 mm or less. The mobile phone holder permits the phone to be positioned with a tolerance of $\pm 1°$ in the tilt angle. It is made of low-loss and low-permittivity material(s): $\tan(\delta) \leq 0.05$ and $\epsilon_r \leq 5$.

Phantom Model

To evaluate the near-field exposure produced by wireless handsets, phantom models have been constructed to simulate human exposures. The models vary from simple to anatomically correct shapes [53,54,60,61]. In addition, different dielectric properties of phantom tissue, size and shape of phantom models, and phone positions are used by test laboratories and manufacturers. For the same cellular telephone, the final results from different laboratories may vary over a range. To eliminate these discrepancies, the IEEE SCC34/SC2 and the CENELEC have defined a standard phantom model with specified size and shape, liquid compounds for tissue simulation, standard calibration techniques for E-field probes, and phone positioning and SAR measurement protocols so that laboratories can perform SAR tests in a consistent manner.

To simulate the human head, both committees agreed to use the dielectric property data of [50]. Based on a plane wave analysis, Table 6.3 lists the dielectric properties of the equivalent head tissue at various frequencies [62]. The dielectric constant is the average of all head tissues, and the conductivity is the larger of the 1-g or 10-g average calculated effective conductivity. Formulas for liquid head tissue phantoms for the various frequency bands can be found in the IEEE SCC34 Standard Annex 4C or the CENELEC Standard Annex 1. The dielectric constant and conductivity shall be within 5% of the target values at the specified frequencies, excluding instrument error. The dielectric properties should be periodically checked to ensure compliance. The measured dielectric properties of the

Table 6.3
Dielectric Properties of the Equivalent Head Tissue Between 300 and 3000 MHz

Frequency (MHz)	Relative Dielectric Constant (ϵ_r)	Conductivity (σ) (S/m)
300	45.3	0.87
450	43.5	0.87
835	41.5	0.90
900	41.5	0.97
1,450	40.5	1.20
1,800	40.0	1.40
1,900	40.0	1.40
2,000	40.0	1.40
2,450	39.2	1.80
3,000	38.5	2.40

liquid at the device operating frequencies should be used in SAR calculations instead of the target values at the listed frequencies shown in Table 6.3.

The 90% adult male head dimensions and shape, published by the U.S. army [63], were adopted by both committees. A hairless specific anthropomorphic mannequin (SAM) having a head and neck model with the army-specified dimensions and shape has been built. Figure 6.32 shows the front, back, and side views of SAM. The original ear protrusion was 28 mm, but it was reduced to 4 mm to simulate the compression of the ear during the use a cellular phone. This spacing brings the phone close to the head and gives results that are relevant to the exposure of people with smaller ears. The phantom shell is made of fiberglass from SAM with thickness no greater than 2 mm at the site of measurement. The dielectric constant of the shell is less than 5 and conductivity less than 0.01 S/m. A 4-mm lossless spacer plus a 2-mm shell thickness at the ear canal are used to simulate the ear. The spacer is tapered to the cheek as in a real human head. CAD files are contained on the compact disk included with the standard.

Right and left head models, which are obtained by bisecting the fiberglass SAM shell, are necessary because the asymmetric location of the antenna in many phones results in different exposure areas on the opposite sides of the head. The models are filled with the correct liquid mixture simulating head tissue at the desired frequency. The liquid is 15 cm ± 5 mm in depth measured at the ear canal, which is approximately equivalent to the distance between the ears of the phantom.

Measurement Procedures

Figure 6.33 is a flowchart of the SAR testing procedures. The chart labeled "TEST 5.6.2" is the basic scan procedure. E-field measurements are taken at a reference point (e.g., 15 mm above the ear reference point) to monitor power changes during testing. This measurement is conducted within 3 min after placing the device in operation with a fully charged battery. The surface scanned is that immediately adjacent to the phone. If the

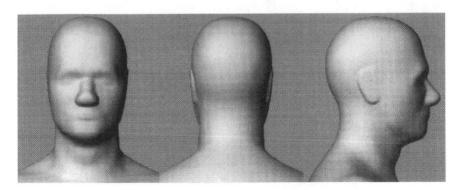

Figure 6.32 Front, back, and side view of the SAM phantom head [58].

342

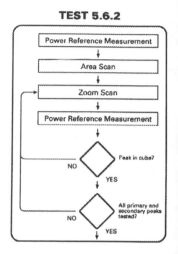

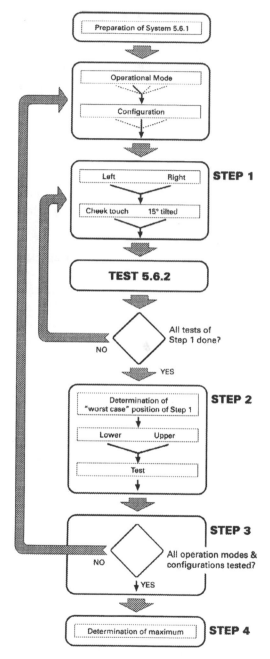

Figure 6.33 Flowchart of SAR testing procedures [58].

peak is at the border of the area, the area scan is repeated using an enlarged area when possible. These steps are repeated if the spatial peak touches any side of the zoom volume, which is the volume scanned for the purpose of SAR mass averaging. For any secondary peak found in the above "Area Scan" that is within 2 dB of the maximum peak and not within the "Zoom Scan," this step is repeated for each of these secondary peaks. At the end of the zoom scans, the field should be measured again at the same location of the initial reference point for power measurement. If the power has changed by more than 5%, the measurements should be repeated or the entire variation should be added to the assessed value.

In step 1, the above scan evaluation is conducted at two phone positions and at the center of the frequency band for left-hand usage and right-hand usage. The two positions are cheek touch and a 15-deg tilt angle. Figure 6.34 shows how to place the phone at the "cheek" position. The second position (15-deg tilt) is achieved by positioning the phone at the cheek touch position and then pivoting the device outward by 15 deg with the top of the phone against the pinna. In step 2, the same evaluation is done at low and high frequencies at the side and the position, which resulted in the largest spatial peak SAR. Step 3 calls for steps 1 and 2 to be repeated with the antenna fully extended and retracted, and in operational mode, that is, AMPS, TDMA, CDMA, and so on. Examine the data and determine the maximum spatial peak SAR, which is the largest value found in steps 1 to 3.

6.5.4 Safety Criteria

The methods described above help to characterize the RF exposure of a human to a reasonable degree of accuracy through SAR measurements. To decide whether the expo-

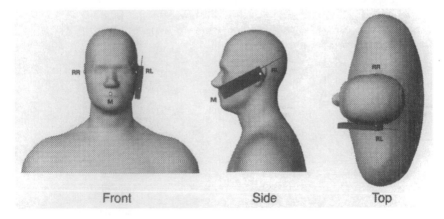

Figure 6.34 "Cheek" position of the mobile phone on the left side of the phantom. The 15-deg angle position (not shown here) is to rotate the phone off the cheek and pivot against the pinna by 15 deg [58,59].

sures to the mobile or portable phones or radios are safe or not, one is left with the fundamental question "What are the safe levels of SAR?" The answer goes beyond engineering and physics; it requires the confluence of knowledge from a variety of disciplines: pathology, biology, psychology, physiology, health physics, epidemiology, and veterinary medicine (experimental animals are exposed to establish threshold levels of adverse bioeffects). The process of establishing safety levels of human RF energy has involved hundreds of scientists around the world, and it is too long and would require extraneous engineering to be described in this chapter.

As of 2000, most countries have adopted or are in the process of adopting one of the two prominent safety standards for the exposure of humans to RF energy: (1) International Commission on Non-Ionizing Radiation and Protection (ICNIRP) 1998 Guidelines for limiting exposure in time-varying electric, magnetic, and electromagnetic fields (up to 300 GHz) or (2) the IEEE C95.1-1999 Standard for Safety Levels with Respect to Human Exposure to Radio Frequency Electromagnetic Fields, 3 kHz to 300 GHz.

Table 6.4 shows the basic SAR limits for both ICNIRP and IEEE standards. In the table, controlled environments represent locations where exposure occurs to persons who are aware of the potential for exposure as a concomitant of employment and by other cognizant persons. The uncontrolled environments are locations where individuals are exposed who have no knowledge of or control over their exposure. For the ICNIRP guidelines, the SAR values are averaged in any 10 g of contiguous tissue. In the IEEE standard, the SAR is averaged over a cube of 1-g tissue for the body except the extremities (i.e., hands, wrists, feet, and ankles), which are averaged over a cube of 10-g tissue. Whether the pinna should also be treated in the same category as the extremities is also being considered. For two-way pagers, handheld wireless phones, and portable radios, in the United States, the FCC adopted the IEEE SAR limits for compliance, that is, peak SAR due to exposure to cellular phones and two-way pagers follow the 1.6 W/kg limits in uncontrolled environments; and for licensed two-way radios used by professionals, the limit is 8 W/kg in controlled environments. Note that the IEEE SAR limits are more stringent than the ICNIRP guidelines. This is due to the fact that SAR averaged over

Table 6.4
Basic SAR Restrictions (W/kg)

Standard	Condition	Frequency	Whole Body	Local SAR (Head and Trunk)	Local SAR (Limbs)
ICNIRP	Occupation	100 kHz–10 GHz	0.4	10	20
	Gen. Pop.	100 kHz–10 GHz	0.08	2	4

Standard	Condition	Frequency	Whole Body	Local SAR (body Except Extremities)	Local SAR (Extremities)
IEEE	Controlled	100 kHz–6 GHz	0.4	8	20
	Uncontrolled	100 kHz–6 GHz	0.08	1.6	4

10 g is lower than a 1-g average and the ICNIRP limit of 2 W/kg is higher than the IEEE limit of 1.6 W/kg.

Table 6.5 lists the field strengths and power density limits of far-field exposures. For 100 kHz to 10 GHz in the ICNIRP guidelines, the occupational and general population limits are for a 6-min period. In the IEEE standard, the averaging time for controlled environments is 6 min for 3 kHz to 15 GHz, and for uncontrolled environments is 30 min for 3 MHz to 3 GHz. For other frequencies outside the mobile communication band, detailed averaging times can be found in the original standards. These reference limits are used for compliance of mobile products.

6.5.5 Exposure from Mobile Radios

The exposure of humans to mobile radios has been investigated by a variety of researchers [51,64–66]. Available data show that there are areas of relatively high E-field energy

Table 6.5
Reference Limits

Standard	Condition	Frequency	E-Field (V/m)	H-Field (A/m)	Power Density (mW/cm^2)
ICNIRP	Occupational	3–65 kHz	610	24.4	
		0.065–1 MHz	610	$1.6/f$	
		1–10 MHz	$610/f$	$1.6/f$	
		10–400 MHz	61	0.16	1
		400–2000 MHz	$3f^{1/2}$	$0.008f^{1/2}$	$f/400$
		2–300 GHz	137	0.36	5
	General Pop.	3–150 kHz	87	5	
		0.15–1 MHz	87	$0.73/f$	
		1–10 MHz	$87/f^{1/2}$	$0.73/f$	
		10–400 MHz	28	0.073	0.2
		400–2000 MHz	$1.375f^{1/2}$	$0.0037f^{1/2}$	$f/2000$
		2–300 GHz	61	0.16	1
IEEE	Controlled	3–100 kHz	614	163	
		0.1–3 MHz	614	$16.3/f$	
		3–30 MHz	$1842/f$	$16.3/f$	
		30–100 MHz	61.4	$16.3/f$	
		100–300 MHz	61.4	0.163	1
		300–3000 MHz			$f/300$
		3–15 GHz			10
		15–300 GHz			10
	Uncontrolled	3–100 kHz	614	163	
		0.1–1.34 MHz	614	$16.3/f$	
		1.34–3 MHz	$823.8/f$	$16.3/f$	
		3–30 MHz	$823.8/f$	$16.3/f$	
		30–100 MHz	27.5	$158.3/f^{1.668}$	
		100–300 MHz	27.5	0.729	0.2
		300–3000 MHz			$f/1500$
		3–15 GHz			$f/1500$
		15–300 GHz			10

located in and around the vehicles with RF transmitters operating at VHF and lower frequency bands. These locations of relatively high E-field energy are practically absent in vehicles with center-of-the-roof-mounted antennas operating at 450-MHz and higher bands. The same phenomena were detected with center-of-the-trunk-mounted antennas. Concerns about the possibility of sharp, localized high values of SAR were proven unfounded for the reasons given below.

Experimental results of [51] indicate that the vehicle passengers are significantly shielded by the metal surfaces of the car body. At high frequencies (450-MHz and higher bands), measured radiation patterns of antennas mounted on cars show little dependence on the number of passengers in the car or their location in the cabin. Stuffing the cabin of the car with RF-absorbing material or humans makes no measurable difference in the power levels radiated from mobile radios with center-of-the-roof-mounted antennas. This result points out that the total RF energy absorbed by the passengers is much less than 1% of the power radiated by the antenna. The main factors in determining RF emission from a mobile antenna are its location on the vehicle (often the center of the roof or the trunk lid) and the size and shape of the car. Measurements and computations show that the dimensions of the vehicle roof are an essential factor for the antenna efficiency and general pattern shape. Details of the pattern are determined by the height of the antenna above ground [67].

At low frequencies (150-MHz and lower bands), with wavelengths on the order of magnitude of a vehicle size, practically the entire metal body of the car participates in the radiation process by supporting the RF currents emanating from the antenna feed point. These currents terminate in electric charges where the conducting external surfaces of the car come to an abrupt end. These charges account for the strong E-fields of 100 to 200 V/m detected in the wheel wells and at the bottom of vehicles, when the transmitted power is 100W. Strong electric fields (about 100 V/m) have also been detected inside the vehicle near the top of the steering wheel, at the top of the foot pedals, and the transmission hump. These fields are probably due to RF currents that penetrate the vehicle and terminate at sharp metal edges. The E-fields do not induce SAR values larger than 0.2 W/kg in human phantoms, which is due to the fact that fields are connected with reactive phenomenon rather than with radiating currents, as pointed out in [68]. Using a quarter-wave antenna at 150 MHz mounted on the front fender at about a 50-cm distance from the driver, radiation pattern levels with or without the driver show that a person in the front seat on the side of the antenna absorbs less than 1% of the transmitted power. For a 100W mobile radio, such an exposure is about 1W of RF power absorbed.

High field intensities are found in the immediate vicinity of the antennas of mobile transmitters. At 150 MHz, measurements have shown that the E-fields of 380 V/m exist at a 20-cm distance from roof- or trunk-mounted antennas with 100W transmitted power. Exposure to these fields causes SAR values exceeding the IEEE C95.1-1999 limits. These strong E-fields decay almost linearly with distance from the antenna. Measurements show that for 100W radiated at 150 MHz, the E-field is 140 V/m at the edge of the roof or the trunk at about 60 cm from the antenna. A partial body exposure to this field is below the IEEE C95.1-1999 limits.

From the above considerations, based on measured data, it is clear that 100W mobile radio equipment emits strong EMFs. However, if the antennas are properly located at places that do not have easy access, these strong fields are attenuated by space propagation before incidence at places where people may be around the vehicle. Mobile radio operators and passengers are shielded from RF energy by the metal surfaces of the vehicle, if the antenna is located on the roof or on the trunk lid. Roof gutter and fender mounts are not recommended for high-power mobile radios (7W or more) because of the potential for overexposure of a bystander. Finally, for vehicles with plastic bodies, one should be very careful to locate the mobile radio antenna and its grounding surfaces, if any, at a sufficient distance from the driver and the passengers so as to avoid potential overexposure.

In a recent study, power densities and SAR in a sitting phantom exposed to 146- and 460-MHz fields at 100–120W were measured in the back seat of a compact car with trunk mount VHF and UHF antennas, at both center and edge of the trunk lid [66]. E- and H-fields were measured with survey probes inside of a half car and a full body car. SAR was measured with the robot controlled E-field probe as shown in Figure 6.35. The back half car was lifted at 45 deg to accommodate the liquid-filled phantom. Results show that E- and H-field intensities do not meet FCC compliance limits under the uncontrolled conditions (i.e., IEEE C95.1-1999 limits). The SAR data indicated that even though the external fields exceed the FCC limits, actual SAR tests conducted in simulated human tissue under the same conditions do meet the compliance limit of 1.6 W/kg.

6.5.6 Exposure of Portable Radio Operators

Because of the relatively low power of portable transmitters (usually less than 7W), the exposure of concern is only the area very close to the antenna, where the head of the user is located during the normal operation of the device. SAR methods described in Section 6.5.3 have been used to determine the peak SAR in operators of portable radios. A face-down full-body phantom, similar to the head and neck shown in Figure 6.31, was used for peak SAR measurements in head (during talking) and at abdomen (carrying position). For whole-body averaged SAR, the measurement is performed by contour plots of the radiation pattern of the radio in the presence and the absence of a standing human model filled with simulated muscle liquid (Figure 6.36). The difference between the two represents, as a first approximation, the total power absorbed. Some differences in the mismatch losses are neglected in this process, but these are not large, as detailed in [69]. The method yields a worst-case approximation.

The results of measurements of whole-body exposure conducted at various frequencies using 6.4W radios are summarized in Table 6.6 for various antenna types. This table also shows the power absorbed by the simulated head of a user. The total head absorption was measured as a difference of total power radiated with the radio held with a fully

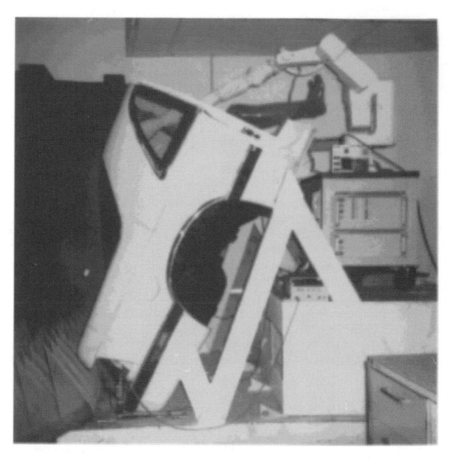

Figure 6.35 Car test setup with the robot probing the head of a phantom in a sitting position exposed to 100–120W VHF or UHF fields. (From [66], © 1999 IEEE).

extended arm, and at 5.0 cm from the face with the radio at mouth level. Cross-polarized components of the field are about 10 dB below the main radiated vertical polarization. In the case of the isolated radio, the cross-polarized radiation (horizontal polarization) is about 20 dB below the vertically polarized component.

Some comments are in order about the results at 30 and 800 MHz. The measurements at 30 MHz have been extremely difficult to perform. The isolated radio has poor radiation performance (especially with the helical antenna, which is only 40 cm long), making the measurements of total power emitted by the isolated radio imprecise. Only when the phantom operator holds the radio is there efficient RF emission, probably because of better antenna matching conditions. This phenomenon points out that the user is actually part of the antenna of the radio. The radio excites RF currents on the body of the user,

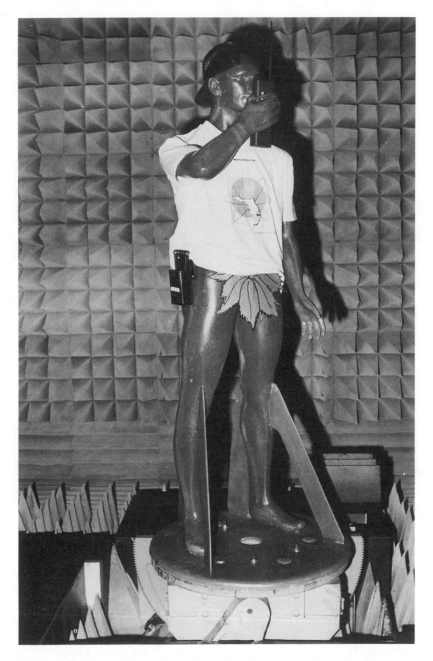

Figure 6.36 Measurement of radiation pattern of the radio in the presence of a standing human model filled with simulated muscle liquid.

Table 6.6
RF Absorption of Portable Radio Using 6.4W Transmitter Power

Frequency (MHz)	Antenna Type	Location	Power Absorbed (W)
30	Collapsible	Body and head	4
30	Helix	Body and head	5
150	Helix	Body	3.5
		Head	0.5
450	Dipole	Body	3.0
		Head	0.7
450	Helix	Body	3.5
		Head	0.5
800–900	Sleeve Dipole	Body	1.5
		Head	1.0

who has the physical dimension of a good antenna. In the 800-MHz band, a sleeve dipole was used in the measurements, with the specific aim to reduce the RF currents on the radio case. The limited role of the radio case as a source of EM energy explains the lower losses through the hand, arm, and body of the user than those at the lower frequencies. In all frequency bands, measurements of SAR in the hands and the arms of a simulated human have not detected a peak value greater than 0.5 W/kg.

6.5.7 SAR Values of Portable Radios

In the 30-MHz band, it has not been possible to perform SAR measurements either with a small implantable E-field probe or a temperature-sensing device with 0.005°C sensitivity. The radio used in the measurement was a 6.4W portable device. From the absence of a measurable temperature increase with a 0.005°C sensitive thermometer, it was possible to conclude that the peak SAR value in a user of a 6.4W 30-MHz radio is no more than 0.3 W/kg. From Table 6.6 the average body SAR for a 70-kg human is 57–71 mW/kg.

In the VHF frequency band (about 150 MHz), tests were conducted using 6.4W portable radios with helical antennas. Collapsible quarter-wavelength telescopic antennas have been used in the past at this frequency band, but they have been abandoned because of poor mechanical performance in drop tests. The exposure of users of VHF radios with extended collapsible antennas is expected to be lower than the values presented herein for the same power radiated and the same distance between the antenna and the user. The collapsible antennas are longer by a factor of 3 than the helical antennas commonly used at this frequency band. If all other conditions are the same, the energy density incident on the user is smaller in the case of the longer antenna.

Figure 6.37 shows the value and the location of the peak SAR value detected in a head phantom with a distance of 5 cm between the mouth and the radio case. The peak SAR is about 0.5 W/kg, located at 2.5 cm above the eyebrow on the side of the antenna. Anatomically, the peak SAR location is on the subcutaneous fat tissue or at the surface of the skull. The average brain exposure is 0.1 W/kg. Average head SAR value was obtained by dividing the power-absorbed value in Table 6.6 by 4.5 kg (approximate

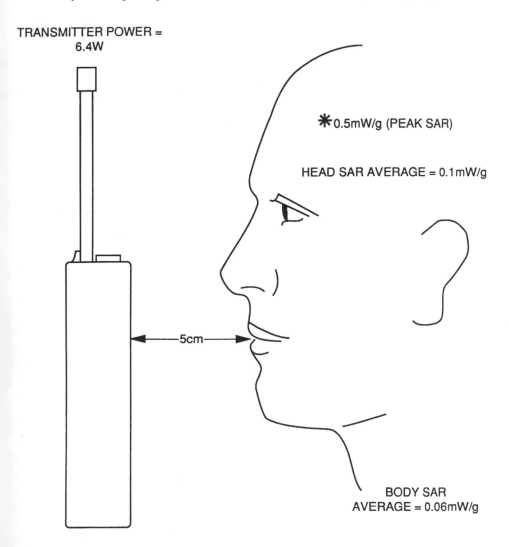

Figure 6.37 SAR in head and body of portable radio users at 150 MHz (helical antenna).

average weight of the head and neck of a 70-kg adult). The average body SAR is obtained by dividing the sum of the power absorbed by the body weight 70 kg. Radios operating in the 430- to 512-MHz band are normally equipped with a quarter-wavelength whip or a helical antenna. The results are shown in Table 6.7.

In both cases at 450 MHz, the peak exposure is at the surface of the eye that is closer to the antenna. It is interesting to note that the whip antenna, although physically longer, produces the larger peak SAR in the eye than the helical antenna. The cause of this phenomenon can be traced to the impedance (ratio $|E|/|H|$) of the fields incident on the surface of the eye. The fields generated by the whip have a relatively low impedance and so more readily penetrate the tissues rich in water like the eye, which are low-impedance media.

If the radio is brought closer to the face so that the radio case is practically touching the nose of the user (about a 2-cm distance from the mouth), the peak SAR at the surface of the eye is about 4 W/kg for both antenna types. The RF penetrating the eye decays following the approximately exponential falloff, as shown in Figure 6.38. Within 2 cm from the surface, the SAR is 50% lower than the peak value. The peak SAR averaged in 1-g tissue is about 3.5 W/kg, which is lower than the 8 W/kg for the controlled environments.

In the 800- to 900-MHz band, portable radios are equipped with sleeve dipoles, so the antenna is decoupled from the radio case. This arrangement substantially reduces the absorption of RF by the hand and arm, but increases the total amount of power absorbed by the head of the user, as can be seen from the values in Table 6.6. With this antenna, more than 50% of the available RF power is radiated for the purpose of communication. At the lower frequencies, more than 50% of the available RF power is absorbed by the body of the user. Results are shown in Table 6.7. The peak SAR is located at the head frontal lobe on the side of the antenna in correspondence with the feed points of sleeve dipole. The peak SAR was detected at the skull-brain interface. If the radio case is held very close to the face (at a 2-cm distance from the mouth), the peak SAR was found at the surface of the eye that is closer to the antenna; this peak value is about 3 W/kg. Note that the RF energy deposition in the eye in the 800- to 900-MHz band with a sleeve dipole is much lower than at 450 MHz with a whip antenna for the same exposure conditions.

Table 6.7
SAR (W/kg) in a Human Model Exposed to a 6.4W Portable Radio at a Distance of 5 cm from Mouth

Frequency (MHz)	Antenna Type	Head (Peak)	Head (Average)	Eye (Peak)	Body
150	Helical	0.5	0.1		0.06
450	Whip		0.15	1.2	0.05
450	Helical		0.12	0.9	0.06
840	Sleeve	0.7	0.2		0.04

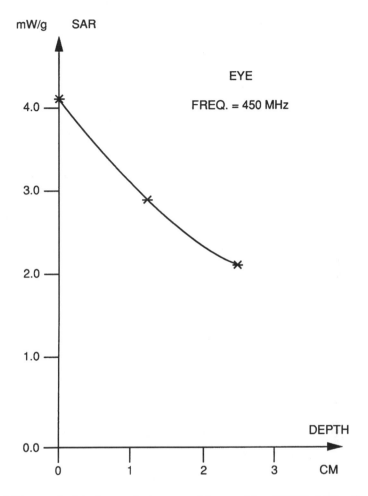

Figure 6.38 SAR versus depth in the eye of a human model exposed to a 450-MHz 6.4W radio.

The quarter-wave whip antenna is seldom used in the 800- to 900-MHz band because of its poor coverage properties for handheld radios. Measurements show that the 6.4W radio with a quarter-wavelength whip in proximity (radio held at 2 cm from the mouth) to the face produces a peak SAR of about 7 W/kg in the eye that is closer to the antenna [70,71].

6.5.8 SAR Values of Cellular Phones

Figure 6.39 shows the SAR patterns inside the human head phantom. For this particular phone, peak SAR occurs on the cheek near the ear lobe. Peak SAR increased from

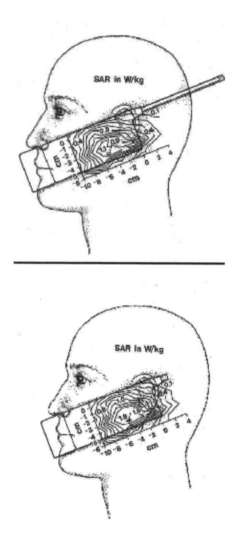

Figure 6.39 SAR patterns inside of a human head phantom for extended and retracted antenna configurations. (From [49], © 1996, Wiley-Liss, Inc.)

1.1 to 1.8 W/kg when the antenna is retracted. These peak SAR values are not averaged over 1-g tissue. SAR techniques described in Section 6.5.3 are used to obtain data for FCC-type acceptance application, that is, authorization to ship an RF-emitting product. Only those products below the FCC limits are authorized for shipment. Figure 6.40 shows a SAR pattern superimposed on the phone showing the loci of peak SAR in the human head; these are on the phone DTMF (dual-tone multiple frequency) keypad, at a distance away from the antenna feed point.

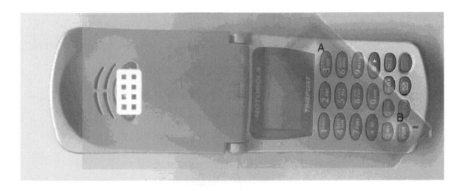

Figure 6.40 High SAR spots (A and B) are on the keypads for this model of cellular phone.

Eighteen GSM devices with helix, dipole, or patch antennas have been tested, as reported in [72]. Each device was tested three times at the same position. The average and standard deviations of SAR averaged over 10-g tissue are shown in Figure 6.41.

6.5.9 Discussion and Conclusions

Accurate RF dosimetry measurements in simulated humans are difficult to perform, yet are essential in establishing device compliance. It is essential to have both a universally accepted safety standard and test methodology. Currently, the IEEE and CENELEC are working on harmonized standards for handset SAR measurement. It will take several years to develop a worldwide harmonized safety standard, but the World Health Organization is pursuing this important objective. Controversy and lack of confidence in the safety standards will continue if the RF exposure limits vary from country to country.

Both IEEE and ICNIRP have evaluated the RF bioeffects database and found that RF heating is the only established cause of adverse health effects, and have set a standard based on 4 W/kg whole-body average SAR. Both IEEE and ICNIRP established a whole-body safety limit for the controlled environment and occupational group, respectively, to 1/10 of the 4 W/kg level, that is, 0.4 W/kg. The limit for uncontrolled environments and general population was obtained by incorporating a safety factor of 50, that is, 0.08 W/kg averaged over the whole body.

The peak SAR limits were derived differently by IEEE and ICNIRP. Based on an experimental study, the IEEE used the maximum ratio of peak to average SAR in human models exposed to a plane wave to derive the peak SAR limit. The ratio is 20, which results in 8 W/kg averaged over 1-g tissue for the controlled environments and 1.6 W/kg for the uncontrolled conditions. ICNIRP rationalized the limit by considering that the cataract formation threshold is 100 W/kg in the 10-g eye, and set 1/50 of the threshold value, that is, 2 W/kg averaged over 10-g tissue, as the limit.

While scientists continue to collect data on the various potential health effects of RF, it is important for the life scientists and physical scientists to collaborate. Due to the

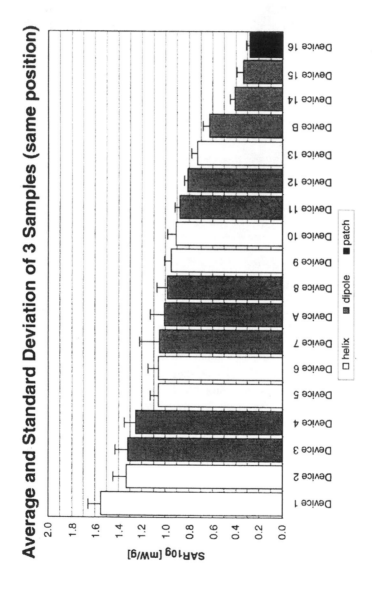

Figure 6.41 SAR of GSM phones tested [72].

complexity of RF dosimetry, life scientists without adequate understanding of physical aspects of the EM exposure can generate erroneous results and/or misinterpretations of their data. Only through interdisciplinary collaboration can the development of EM technology be fostered for the benefit of mankind while avoiding unreasonable restrictions.

In conclusion, the safety of mobile and portable communication devices has been regulated by government and health organizations. Based on current knowledge, safe field strengths and SAR limits have been set by international standards organizations. International committees are also formulating SAR testing methods. Before the end of 2000, manufacturers of cellular phone were to have universally accepted dosimetric test standards to prove that their products are within the compliance limits. Detailed knowledge and proper methods for making RF densitometry and dosimetry measurements are necessary to ensure that exposure to mobile communication devices is within the internationally established safety limits.

REFERENCES

[1] Gandhi, O. P., "Conditions of Strongest Electromagnetic Power Deposition in Man and Animals," *IEEE Trans. MTT*, Vol. MTT-23, No. 12, Dec. 1975, pp. 1021–1029.
[2] Fujimoto, K., A. Henderson, K. Hirasawa, and J. R. James, *Small Antennas*, Letchworth, England: Research Studies Press, 1987, pp. 75–116.
[3] Balzano, Q., and K. Siwiak, "The Near Field of Annular Antennas," *IEEE Trans. Vehicular Technology*, Vol. VT-36, No. 4, Nov. 1987, pp. 173–183.
[4] Smith, G. S., "Radiation Efficiency of Electrically Small Multiturn Loop Antennas," *IEEE Trans. Ant. Propag.*, Vol. AP-20, No. 5, Sept. 1972, pp. 656–657.
[5] Kraus, J. D., *Antennas*, 2nd ed., New York: McGraw-Hill, 1988, pp. 46–49.
[6] Johnson, C. C., and A. W. Guy, "Nonionizing Electromagnetic Wave Effects In Biological Materials and Systems," *Proc. IEEE*, Vol. 60, June 1972, pp. 642–718; IEC 489-6, 2nd ed., Appendix H, "Methods of Measurement for Radio Equipment Used in Mobile Services," 1987.
[7] Wheeler, H. A., "Small Antennas," *IEEE Trans. Ant. Propag.*, Vol. AP-23, No. 4, July 1975, pp. 462–469.
[8] Fujimoto, K., "Small Antennas," *J. IECE Japan*, Vol. 60, No.4, April 1977, pp. 391–397 (in Japanese).
[9] Fujimoto, K., "A Review of Research on Small Antennas," *J. IEICE Japan*, Vol. 70, No. 8, Aug. 1987, pp. 830–838 (in Japanese).
[10] Newman, E. H., "Small Antenna Location Synthesis Using Characteristic Modes," *IEEE Trans. Ant. Propag.*, Vol. AP-27, No. 4, July 1979, pp. 530–531.
[11] Ito, H., H. Haruki, and K. Fujimoto, "A Small-Loop Antenna for Pocket-Size VHF Radio Equipment," *National Technical Review*, Vol. 19, No. 2, April 1973, pp. 145–154 (in Japanese).
[12] King, H. D., et al., "Effects of Human Body on a Dipole Antenna at 450 and 900 MHz," *IEEE Trans. Ant. Propag.*, Vol. AP-25, No. 3, May 1977, pp. 376–379.
[13] Ebinuma, T., et al., "The Properties of Dipole Antenna Located Near the Human Body in the UHF Band," *Trans. IECE Japan*, Vol. J62-B, No. 9, Sept. 1979, pp. 1066–1067 (in Japanese).
[14] Krupka, Z., "The Effect of the Human Body on Radiation Properties of Small-Sized Communication Systems," *IEEE Trans. Ant. Propag.*, Vol. AP-16, No. 2, March 1968, pp. 154–163.
[15] King, H. E., "Characteristics of Body-Mounted Antennas for Personal Radiosets," *IEEE Trans. Ant. Propag.*, Vol. AP-23, No. 2, March 1975, pp. 242–244.
[16] Kraus, J. D., *Antennas*, 2nd ed., New York: McGraw-Hill, 1988, pp. 241–242.
[17] IECE Japan, *Antenna Engineering Handbook*, Ohm, 1980, pp. 474–477 (in Japanese).

[18] Devore, R., and P. Bohley, "The Electrically Small Magnetically Loaded Multiturn Loop Antenna," *IEEE Trans. Ant. Propag.*, Vol. AP-25, No. 4, July 1977, pp. 496–504.

[19] Sato, S., and Y. Naito, "Dipole Antenna Covered with Ferrite Sleeve," *Trans. IECE Japan*, Vol. J58-B, No. 6, June 1975, pp. 285–292 (in Japanese).

[20] Uda, S., *Radio Engineering, Transmission Volume*, Maruzen, 1964, pp. 79–82 (in Japanese).

[21] Haruki, H., H. Ito, and Y. Hiroi, "A Small-Loop Antenna for Pocket-Size VHF, UHF Radio Equipment," *Paper of Technical Group, IECE Japan*, Vol. AP74-29, Aug. 1974, pp. 7–12 (in Japanese).

[22] Ishii, N., and K. Itho, "A Consideration on the Numerical Method for a Card-Sized Thin Planar Antenna," *Paper of Technical Group, IEICE Japan*, Vol. AP91-36, June 1991, pp. 9–14 (in Japanese).

[23] Hallen, E., "Transmitting and Receiving Qualities of Antennas," *Nova Acta Reg. Soc. Sc. Ups.*, Vol. 11, No. 4, Series 4, Uppsala, Sweden, Nov. 1938, pp. 2–44.

[24] Schelkunoff, S. A., and H. T. Frus, *Antennas, Theory and Practices*, London: Chapman and Hall, 1952.

[25] King, R. W. P., *The Theory of Linear Antennas*, Cambridge, Mass.: Harvard University Press, 1956.

[26] Neff, N. P., et al., "A Trigonometric Approximation of the Current in the Solution to Hallen's Equation," *IEEE Trans. Ant. Propag.*, Vol. AP-17, No. 6, Nov. 1969, pp. 804–805.

[27] King, R. W. P., and C. W. Harrison, *Antennas and Waves: A Modern Approach*, Cambridge, Mass.: MIT Press, 1969.

[28] Jordan, E. C., and K. G. Balmain, *Electromagnetic Waves and Radiating Systems*, 2nd ed., Englewood Cliffs, N.J.: Prentice-Hall, 1968, pp. 333–338.

[29] Kraus, J. D., *Antennas*, New York: McGraw-Hill, 1950, pp. 232–238.

[30] King, R. W. P., and C. W. Harrison, "The Distribution of Current Along a Symmetrical Center-Drive Antenna," *Proc. IRE*, Vol. 31, Oct. 1943, pp. 548–567.

[31] Balzano, Q., et al., "The Near Field of Dipole Antennas," *IEEE Trans. Vehicular Technology*, Vol. VT-30, No. 4, Nov. 1981, pp. 161–181.

[32] Jasik, H., ed., *Antenna Engineering Handbook*, New York: McGraw-Hill, 1961, Sect. 3.5, pp. 3017–3018.

[33] Balzano, Q., et al., "The Near Field of Omnidirectional Helical Antennas," *IEEE Trans. Vehicular Technology*, Vol. VT-31, No. 4, Nov. 1982, pp. 173–185.

[34] Cha, A. G., "Wave Propagation on Helical Antennas," *IEEE Trans. Ant. Propag.*, Vol. AP-20, No. 5, Sept. 1972, pp. 556–560.

[35] Collin, R. E., Ch. 9 in *Field Theory of Guided Waves*, New York: McGraw-Hill, 1960, Sec. 9.8.

[36] Fujimoto, K., A. Henderson, K. Hirasawa, and J. R. James, *Small Antennas*, Letchworth, England: Research Study Press, 1987, pp. 59–71.

[37] Taga, T., and K. Tsunekawa, "Performance Analysis of a Built-In Planar Inverted F Antenna for 800 MHz Band Portable Radio Units," *IEEE Trans. Selected Areas in Communication*, Vol. SAC-5, No. 5, June 1987, pp. 921–929.

[38] Takegawa, I., and T. Taga, "Effective Gain Estimation of Portable Antenna," *Conv. Rec. IECE of Japan* (in Japanese), March 1984, p. 2454.

[39] Lee, W. C. Y., and Y. S. Yeh, "Polarization Diversity System for Mobile Radio," *IEEE Trans. Communications*, Vol. COM-20, No. 5, Oct. 1972, pp. 912–923.

[40] Taga, T., "Analysis for Mean Effective Gain of Mobile Antennas in Land Mobile Radio Environments," *IEEE Trans. Vehicular Technology*, Vol. VT-39, No. 2, May 1990, pp. 117–131.

[41] "Public Land Mobile Telephone Systems," CCIR SG-8, Report 742-3.

[42] IEEE C95.1-1999, Institute of Electrical and Electronics Engineers, Inc., "IEEE standard for safety levels with respect to human exposure to radio frequency electromagnetic fields, 3 kHz to 300 GHz," Piscataway, NJ, 1999.

[43] ICNIRP, "Guidelines for Limiting Exposure to Time-Varying Electric, Magnetic, and Electromagnetic Fields (Up to 300 GHz)," *Health Phys.* Vol. 74, No. 4, 1998, pp. 494–522.

[44] Guy, A. W., M. D. Webb, and C. C. Sorensen, "Determination of Power Absorption in Man Exposed to High Frequency Electromagnetic Fields by Thermographic Measurements On Scaled Models," *IEEE Trans. Biomed. Eng.*, Vol. 23, No. 5, 1976, pp. 361–371.

[45] Gandhi, O. P., "Conditions of Strongest Electromagnetic Power Deposition in Man and Animals," *IEEE Trans. Microwave Theory and Techniques*, Vol. MTT-23, No. 12, Dec. 1975, pp. 1021–1029.

[46] Durney, C. H., H. Massoudi, and M. F. Iskander, *Radiofrequency Radiation Dosimetry Handbook*, 4th ed., USAFSAM-TR-85-73, Brooks AFB, Texas, 1986.

[47] Pethig, R., *Dielectric and Electronic Properties of Biological Materials*, New York: John Wiley, 1979.

[48] NCRP, "Biological Effects and Exposure Criteria for Radiofrequency Electromagnetic Fields," Report 86, National Council on Radiation Protection and Measurements, Bethesda, Md., 1986.

[49] Chou, C. K., et al., "Radio Frequency Electromagnetic Exposure: A Tutorial Review on Experimental Dosimetry," *Bioelectromagnetics*, Vol. 17, 1996, pp. 195–208.

[50] Gabriel S., R. W. Lau, and C. Gabriel, "The Dielectric Properties of Biological Tissues: 2. Measurement in the frequency range 10 Hz to 20 GHz," *Phys. Med. Biol.*, Vol. 41, No. 11, 1996, pp. 2251–2269.

[51] Guy, A. W., and C. K. Chou, "Specific Absorption Rates of Energy in Man Models Exposed to Cellular UHF-Mobile-Antenna Fields," *IEEE Trans. Microwave Theory and Techniques*, Vol. 34, No. 6, 1986, pp. 671–680.

[52] Guy, A. W., "Analyses of Electromagnetic Fields Induced in Biological Tissues by Thermographic Studies on Equivalent Phantom Models," *IEEE Trans. Microwave Theory and Techniques*, Vol. 19, 1971, pp. 205–214.

[53] Kuster, N., and Q. Balzano, "Energy absorption Mechanism by Biological Bodies in the Near Field of Dipole Antennas above 300 MHz," *IEEE Trans. Vehicular Technology*, Vol. 41, No. 1, 1992, pp. 17–23.

[54] Balzano, Q., O. Garay, and T. Manning, "Electromagnetic Energy Exposure of the Users of Portable Cellular Telephones," *IEEE Trans. Vehicular Technology*, Vol. 44, No. 3, 1995, pp. 390–403.

[55] Hombach, V., et al., "The Dependence of EM Energy Absorption upon Human Head Modeling at 900 MHz," *IEEE Trans. Microwave Theory and Techniques*, Vol. 44, 1996, pp. 1865–1873.

[56] Watanabe, S., M. Taki, , and O. Fujiwara, "Characteristics of the SAR Distribution in a Head Exposed to Electromagnetic Fields Radiated by a Handheld Portable Radio," *IEEE Trans. Microwave Theory and Techniques*, Vol. 44, 1996, pp. 1874–1883.

[57] Meier, K., et al., "The Dependence of Electromagnetic Energy Absorption upon Human-Head Modeling at 1800 MHz," *IEEE Trans. Microwave Theory and Techniques*, Vol. 45, 1997, pp. 2058–2062.

[58] IEEE SCC34/SC2, "IEEE Recommended Practice for Determining the Spatial-Peak Specific Absorption Rate (SAR) in the Human Body Due to Wireless Communications Devices: Experimental Techniques," Institute of Electrical and Electronics Engineers, 2000.

[59] CENELEC, "Basic Standard for the Measurement of Specific Absorption Rate Related to Human Exposure to Electromagnetic Fields from Mobile Phones (300 MHz–3 GHz)," TC211 WGMBS, European Committee for Electrotechnical Standardisation, 2000.

[60] Kato, A., and S. Fujiwara, "Determining Localized SAR of Microwave Frequency Using a Real Head Model," *Shingaku Eng. J.*, EMCJ92-67, 1992.

[61] Yu, Q., et al., "An Automated SAR Measurement System for Compliance Testing of Personal Wireless Devices," *IEEE Trans. Electromagnetic Compatibility*, Vol. 41, No. 3, 1999, pp. 234–245.

[62] Drossos, A., V. Santomaa, and N. Kuster, "The Dependence of Electromagnetic Energy Absorption upon Human Head Tissue Composition in the Frequency Range of 300–3000 MHz," to be published.

[63] Gordon, et al., "1988 Anthropometric Survey of U. S. Army Personnel: Methods and summary Statistics," Technical Report NATICK/TR-89/044, 1989.

[64] Balzano, Q., "The Near Field of Portable and Mobile Transmitters and the Exposure of the Users," report submitted to the U.S. Federal Communication Commission in response to the Notice of Inquiry, General Docket No. 79-144, 1979.

[65] Lambdin, D. L., "An Investigation of Energy Densities in the Vicinity of Vehicles with Mobile Communication Equipment and Near a Handheld Walkie Talkie," ORP/ EAD 79-2, U.S. Environmental Protection Agency, March 1979.

[66] McCoy, D. O., D. M. Zakharia, and Q. Balzano, "Field Strengths and Specific Absorption Rates in Automotive Environments, *IEEE Trans. Vehicular Technology* Vol. 48, No. 4, 1999, pp. 1287–1303.

[67] Davidson, A. L., and W. J. Turney, "Mobile Antenna Gain in the Multi-Path Environment at 900 MHz," *IEEE Trans. Vehicular Technology*, Vol. VT-26, No. 4, Nov. 1977, pp. 345–348.

[68] Balzano, Q., et al., "Energy Deposition in Biological Tissue Near Portable Radio Transmitters at VHF and UHF," *Conf. Record of the 27th Conf. IEEE Vehicular Technology Group*, Orlando, Florida, Mar. 16–18, 1977, pp. 25–39.

[69] Balzano, Q., et al., "Investigation of the Impedance Variation and the Radiation Loss in Portable Radios," *IEEE Symp. Ant. Propag.*, Urbana, Illinois, June 2–4, 1975, pp. 89–92.

[70] Balzano, Q., et al., "Energy Deposition in Simulated Human Operators of 800 MHz Portable Transmitters," *IEEE Trans. Vehicular Technology*, Vol. VT-27, No. 4, Nov. 1978, pp. 174–181.

[71] Cleveland, R. F., and T. W. Athey, "Specific Absorption Rate in Models of the Human Head Exposed to Hand-held UHF Portable Radios," *Bioelectromagnetics*, Vol. 10, No. 2, 1989, pp. 173–186.

[72] Kuster, N. "Radiation Performance and Evaluation of Human Exposure from Mobile Handsets using Near-Field Measurements," *International Symposium on Electromagnetic Compatibility*, 19P404, 1999, pp. 480–483.

Chapter 7

Antennas and Humans in Personal Communications

Y. Rahmat-Samii, K.W. Kim, M. Jensen, K. Fujimoto,
and O. Edvardsson

7.1 APPLICATIONS OF MODERN EM COMPUTATIONAL TECHNIQUES

7.1.1 Introduction

Recent market evaluations and projections point to continuing growth in the popularity of personal terrestrial and satellite communications systems. Among the diversified components involved in the operation of these systems, handset units are perhaps the most visible part of the system that must be designed to satisfy user needs. Design issues range from the aesthetic look of the unit to health considerations and antenna performance. Depending on the user's sophistication and operational needs, it is projected that a wide variety of designs will be used.

Antennas play a paramount role in optimal design of wireless personal communication devices. Current evolution of wireless personal communications has necessitated a comprehensive understanding of electromagnetic (EM) interactions between handset antennas and the nearby human body. These human–antenna interactions influence the electromagnetic performance of the antenna by altering antenna input impedance, modifying the antenna radiation patterns and polarization state, absorbing antenna-delivered power, and so on. Also questions concerning health hazards have necessitated a more thorough evaluation and characterization of the specific absorption rate (SAR) in the human tissue. Recently, significant progress in understanding human–antenna interactions has been achieved due to advanced computational techniques as well as accurate near-field measurements. The reader is referred to [1–4] for a review of recent developments and a collection of many pertinent references that are not duplicated here.

Popular numerical techniques currently used for EM interaction computations are the finite-difference time-domain (FDTD) method [1], the method of moments (MoM) [5], the eigenfunction expansion method (EEM) [6], and their hybridization as shown in Figure 7.1. Among them, the most popular technique is the FDTD method because of its computational flexibility in modeling complex antenna geometries and the nearby biological tissues. For example, the radiation and absorption characteristics of various antenna configurations—such as the monopole, side-mounted dual PIFA (planar inverted F antenna), and back-mounted PIFA (see Figure 7.2)—in the presence of the human head have been presented [1]. In this reference an anatomical human head is modeled based on actual magnetic resonance image (MRI) scans. In addition, significant work has been performed to evaluate power absorption in biological tissues using various antenna configurations [1–4].

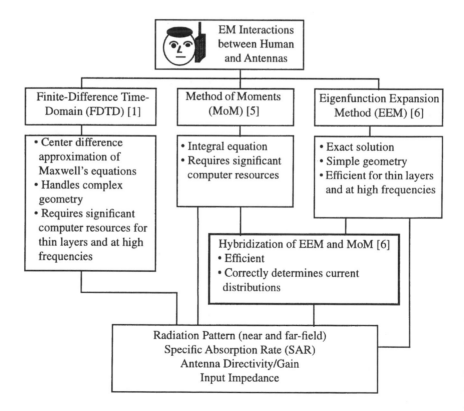

Figure 7.1 Various numerical approaches to study EM interactions between human and antennas.

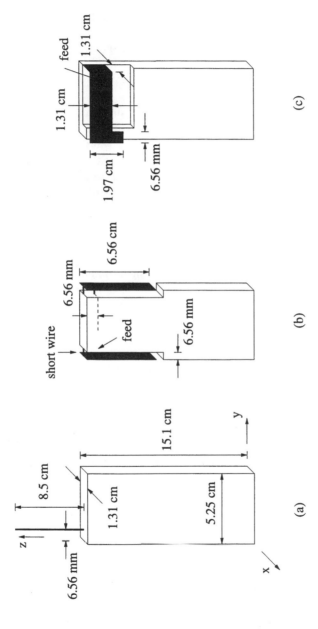

Figure 7.2 Various antenna configurations on typical handsets (~105 cc): (a) monopole; (b) side-mounted dual PIFA; (c) back-mounted PIFA. Frequency of operation is 915 MHz.

Nevertheless, in certain situations the use of FDTD may not be practical and may not allow generation of parametric engineering data. In these situations, the EEM has been used to efficiently evaluate the EM interaction between various antennas and the human head [6]. The EEM is based on the exact scattering solution of infinitesimal dipoles in the presence of a multilayered, lossy dielectric sphere. The EEM has facilitated systematic and parametric studies of EM interactions, resulting in useful engineering curves for antenna design. The results from the EEM can also be used to validate results computed by other numerical techniques.

In wireless communication systems, the ideal antenna for the handheld unit will be low profile and unobtrusive to the user. In principle, it must provide a good impedance match and offer radiation characteristics, such as pattern and polarization, that meet the requirements of the particular communications system. Difficulties in achieving these objectives arise from the presence of an operator whose biological tissues perturb the antenna's electromagnetic properties [7–10]. Investigations of these effects lend insight into new antenna designs that are less susceptible to human tissue influence. The practical considerations for the design of handset antennas are as follows: (1) impedance match and bandwidth, (2) biological tissue effect on the input impedance and radiation pattern, (3) antenna gain, (4) antenna polarization, (5) power absorption in human biological tissue, and (6) antenna diversity performance to minimize signal fading.

This chapter focuses on the biological tissue effects on the input impedance and radiation patterns, the power absorption in the human biological tissue, and the role of antenna diversity. The results presented have been obtained using both the FDTD and EEM. After a brief review of both methods, representative calculated results are discussed and many engineering-oriented design charts and tables are presented. Some of the major observations developed in this chapter can be summarized as follows:

1. The casual manner in which users hold their handset units impacts the antenna design.
2. The proximity of the head and hand can considerably influence the radiation patterns, polarization states, impedance match, and efficiency of handset antennas.
3. On average, between 30% and 60% of the total power radiated by the antenna is absorbed in the biological tissues.
4. Penetration depth within the tissue strongly depends on the frequency.
5. An important factor in determining power absorption is the proximity of the "hot" current spots near the head and hand, an observation which suggests that novel antenna designs may help reduce the interaction.
6. The health hazard aspects of handset units are not yet fully understood, and research and data collection efforts are ongoing.
7. As lower powered units are introduced, the peak average SAR levels are reduced.
8. Even a moderately larger separation between the antenna and tissues can reduce SAR considerably.

9. Internal integrated antennas can become useful if they are kept away from the hand and head.
10. Diversity and directive antennas can play an important role in certain applications.
11. When proper numerical models are used, most of the existing computational techniques predict similar results within variations on the order of 10%.
12. Antenna–human interactions must be evaluated in various configurations in order to obtain statistical averages of the overall performance for system link evaluations.
13. Elaborate measurement techniques have been developed for antenna–tissue interaction characterization.

7.1.2 Definition of Design Parameters for Handset Antennas

In this section, definitions of the important design parameters—specific absorption rate (SAR), antenna gain, input impedance, and so on—for handset antennas are briefly reviewed.

7.1.2.1 Absorbed Power and Specific Absorption Rate

If the handset antenna is located in the vicinity of human biological tissue, some portion of the antenna delivered power is absorbed in the tissue. The total power absorbed in the lossy tissues can be defined as

$$P_{abs} = \frac{1}{2} \int_V \sigma |\mathbf{E}|^2 dV \tag{7.1}$$

where σ is the conductivity of the tissue, $\mathbf{E}$ is the electric field intensity, and V is the volume of the biological tissue. The radiated power to the far-field region is obtained as

$$P_{rad} = \frac{1}{2} \text{Re} \left\{ \int_S \mathbf{E} \times \mathbf{H}^* \cdot \hat{\mathbf{n}} dS \right\} \tag{7.2}$$

where $\mathbf{E}$ and $\mathbf{H}$ are the electric and magnetic field intensities on a surface S completely enclosing the antenna and tissue, and $\hat{\mathbf{n}}$ is the outward unit vector normal to the surface. The efficiency of the antenna/tissue system can be defined as

$$\eta_a = \frac{P_{rad}}{P_{rad} + P_{abs}} = \frac{P_{rad}}{P_{del}} \tag{7.3}$$

where P_{del} is the antenna delivered power. The total antenna efficiency η_0 is the product of the reflection efficiency $(1 - |\Gamma|^2)$, conduction efficiency, dielectric efficiency, and antenna–tissue efficiency.

Specific absorption rate (SAR) is one of the most widely used parameters when discussing the health risk associated with electromagnetic power absorption. It is defined as

$$\text{SAR} = \frac{\sigma}{2\rho}|\mathbf{E}|^2 \tag{7.4}$$

where ρ is the material density. The ANSI/IEEE standard C95.1-1992 RF Safety Guideline [11] suggests that the 1-g averaged peak SAR should not exceed 1.6 W/kg and the whole-body averaged peak SAR should be less than 0.08 W/kg. These guidelines are applicable to uncontrolled situations and therefore must be satisfied for personal handsets.

7.1.2.2 Directivity and Gain

The directivity of an antenna, D, is the ratio of the radiated power density $p(\theta, \phi)$ to the radiated power density averaged in all directions, or

$$D = \frac{p(\theta, \phi)}{(1/4\pi r^2)\iint p(\theta', \phi')r^2 \sin\theta' d\theta' d\phi'} \tag{7.5}$$

The gain of an antenna, G, takes into account the total efficiency η_0 of the antenna as well as the directivity. It is defined as

$$G = \frac{4\pi r^2 p(\theta, \phi)}{P_{\text{in}}} = \eta_0 D \tag{7.6}$$

7.1.2.3 Antenna Impedance and S_{11}

The antenna input impedance is defined as the ratio of the voltage (V) to the current (I) at the antenna feed point; that is,

$$Z_{\text{in}} = \frac{V}{I}\bigg|_{\text{at antenna feed point}} \tag{7.7}$$

S_{11} of the antenna is the reflection coefficient of the antenna at the feed point and is defined as

$$S_{11} = \frac{Z_{\text{in}} - Z_0}{Z_{\text{in}} + Z_0} \tag{7.8}$$

where Z_0 is the characteristic impedance of the feeding transmission line.

7.1.3 Finite-Difference Time-Domain Formulation

The FDTD algorithm can be derived from Maxwell's time-domain integral equations given as

$$\frac{d}{dt} \int_S \epsilon(\mathbf{r}) \mathbf{E}(\mathbf{r}, t) \cdot d\mathbf{S} = \oint_C \mathbf{H}(\mathbf{r}, t) \cdot dl - \int_S \sigma(\mathbf{r}) \mathbf{E}(\mathbf{r}, t) \cdot d\mathbf{S} \qquad (7.9)$$

$$\frac{d}{dt} \int_S \mu(\mathbf{r}) \mathbf{H}(\mathbf{r}, t) \times d\mathbf{S} = -\oint_C \mathbf{E}(\mathbf{r}, t) \times dl \qquad (7.10)$$

where $\mathbf{E}$ and $\mathbf{H}$ are the electric and magnetic field intensities and ϵ, μ, and σ are the permittivity, permeability, and conductivity, respectively. By placing the field components on the "Yee" unit cell as shown in Figure 7.3 [12], these equations can be discretized to provide second-order accuracy [13]. For example, z components of the $\mathbf{E}$ and $\mathbf{H}$ fields are obtained from (7.9) and (7.10) as

$$E_{z,i,j,k}^{n+1} = \alpha_{z,i,j,k} E_{z,i,j,k}^{n} + \beta_{z,i,j,k} \left[\frac{H_{y,i+1,j,k}^{n+1/2} - H_{y,i,j,k}^{n+1/2}}{\Delta x} - \frac{H_{x,i,j+1,k}^{n+1/2} - H_{x,i,j,k}^{n+1/2}}{\Delta y} \right]$$

$$(7.11)$$

$$H_{z,i,j,k}^{n+1/2} = H_{z,i,j,k}^{n-1/2} + \gamma_{z,i,j,k} \left[\frac{E_{x,i,j,k}^{n} - E_{x,i,j-1,k}^{n}}{\Delta y} - \frac{E_{y,i,j,k}^{n} - E_{y,i-1,j,k}^{n}}{\Delta x} \right]$$

$$(7.12)$$

where

$$\alpha_{p,i,j,k} = \frac{\dfrac{\overline{\epsilon}_{p,i,j,k}}{\Delta t} - \dfrac{\overline{\sigma}_{p,i,j,k}}{2}}{\dfrac{\overline{\epsilon}_{p,i,j,k}}{\Delta t} + \dfrac{\overline{\sigma}_{p,i,j,k}}{2}}, \quad \beta_{p,i,j,k} = \frac{1}{\dfrac{\overline{\epsilon}_{p,i,j,k}}{\Delta t} + \dfrac{\overline{\sigma}_{p,i,j,k}}{2}}, \quad \gamma_{i,j,k} = \frac{\Delta t}{\overline{\mu}_{p,i,j,k}}$$

$$(7.13)$$

and

$$\overline{\epsilon}_{p,i,j,k} = \frac{1}{\Delta S} \int_{\Delta S} \epsilon(\mathbf{r}) \hat{n} \cdot d\mathbf{S}, \quad \overline{\sigma}_{p,i,j,k} = \frac{1}{\Delta S} \int_{\Delta S} \sigma(\mathbf{r}) \hat{n} \cdot d\mathbf{S}, \quad \overline{\mu}_{p,i,j,k} = \frac{1}{\Delta S} \int_{\Delta S} \mu(\mathbf{r}) \hat{n} \cdot d\mathbf{S}$$

$$(7.14)$$

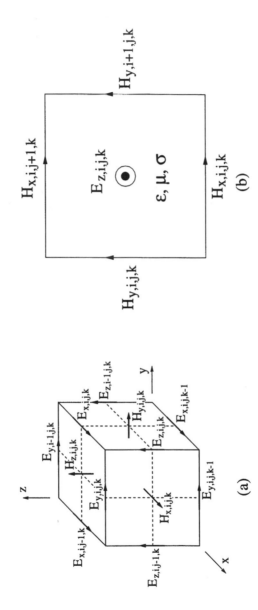

Figure 7.3 Location of the electric and magnetic field vectors for the FDTD methodology: (a) on a cubic unit cell (Yee cell); (b) on one face of unit cell.

where ΔS represents the area of the cell face normal to the unit vector $\hat{n}$. Other field components can be obtained in a similar fashion [13].

In the FDTD calculation, all field values are initially set to zero. A source is introduced by setting a voltage at the antenna feed point and computing from it the source electric field [14]. The electric and magnetic field values are then calculated using a *leap-frog scheme:* The magnetic field is first calculated at a time $t = n + 1/2$, and the electric field is subsequently calculated at $t = n + 1$. Continuity of fields is naturally enforced at a dielectric boundary in this algorithm. However, at the conductor interface, the tangential E-field must be explicitly set to zero. At the outer boundary of the computation, Mur's absorbing boundary conditions (second order) [15] or perfectly matched layer (PML) boundary conditions [16] are typically applied.

Once the time-domain calculation is complete, frequency-domain quantities such as input impedance, patterns, and gain can be obtained using the fast Fourier transform (FFT). By carefully choosing the frequency content of the time-domain excitation, this approach can be used to obtain the antenna response over a broad frequency band.

7.1.4 Eigenfunction Expansion Method

7.1.4.1 EEM Implementation

The eigenfunction expansion method (EEM), which uses the dyadic Green's function, is based on the exact scattering solution of infinitesimal dipoles. The scattered field is expressed as a series expansion of the spherical vector wave functions. Figure 7.4 illustrates the geometry of a multilayered sphere consisting of N regions with the corresponding

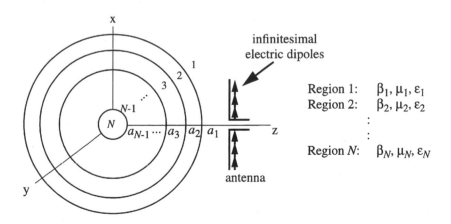

Figure 7.4 Geometry for EEM: A multilayered, lossy dielectric sphere with $(N - 1)$ layers; N discrete regions can be defined to describe the scattered field in each region; a_p is the radius of each layer in the concentric sphere.

constitutive parameters. For example, infinitesimal dipoles are assumed to be oriented in the x direction and located outside of the multilayered sphere.

The total electric field can be expressed using the dyadic Green's function [17] as

$$\mathbf{E}(\mathbf{r}) = -j\omega\mu_1 \iiint \overline{\mathbf{G}}_e^{(p1)}(\mathbf{r}, \mathbf{r}') \cdot \mathbf{J}(\mathbf{r}')dv' \tag{7.15}$$

where $\overline{\mathbf{G}}_e^{(p1)}(\mathbf{r}, \mathbf{r}')$ is the dyadic Green's function for each region and μ_1 is the permeability for region 1. In the superscript of $\overline{\mathbf{G}}_e^{(p1)}(\mathbf{r}, \mathbf{r}')$, p denotes for the pth region and 1 is used for the source region.

For an infinitesimal electric dipole located at $\mathbf{r}' = (r_0, \theta_0, \phi_0)$ with a current moment $\mathbf{c} = Il\hat{\mathbf{x}}$, the electric field can be obtained from (7.15) as

$$\mathbf{E}(\mathbf{r}) = -j\omega\mu_1 Il \cdot \overline{\mathbf{G}}_e^{(p1)}(\mathbf{r}, \mathbf{r}') \cdot \hat{\mathbf{x}} \tag{7.16}$$

Using the dyadic Green's function for the scattered field in region $p (a_p \le r \le a_{p-1})$ where $p = 1, 2, \ldots, N$, the scattered electric field can be obtained as

$$\mathbf{E}_{(p)}^s(\mathbf{r}) = -\left(\frac{\omega\mu_1 Il}{4\pi} \cdot \beta_1\right) \sum_{n=1}^{\infty} \sum_{m=0}^{n} (2 - \delta_{m0}) \frac{(2n+1)}{n(n+1)} \frac{(n-m)!}{(n+m)!} \tag{7.17}$$

$$\cdot \left\{ \begin{array}{l} \left(A_{n,p} \mathbf{M}_{e_{mn}}^{(4)}(\beta_p) + C_{n,p} \mathbf{M}_{o_{mn}}^{(1)}(\beta_p) \right) \left[\mathbf{M}_{o_{mn}}^{(4)'}(\beta_1) \cdot \hat{\mathbf{x}} \right] \\ + \left(B_{n,p} \mathbf{N}_{e_{mn}}^{(4)}(\beta_p) + D_{n,p} \mathbf{N}_{o_{mn}}^{(1)}(\beta_p) \right) \left[\mathbf{N}_{e_{mn}}^{(4)'}(\beta_1) \cdot \hat{\mathbf{x}} \right] \end{array} \right\}$$

where $\beta_p = \omega\sqrt{\mu_p \epsilon_p}$ and δ_{m0} is the Kronecker delta function. $\mathbf{M}_{e_{mn}}^{(i)}$ and $\mathbf{N}_{e_{mn}}^{(i)}$ are the even or odd spherical vector wave functions, which are solutions of the source-free vector wave equation $\nabla \times \nabla \times \mathbf{E} - \beta^2\mathbf{E} = 0$ [17]. $A_{n,N} = B_{n,N} = 0$ in the above equation to avoid infinite field at the origin. The expansion coefficients $A_{n,p}$, $B_{n,p}$, $C_{n,p}$, and $D_{n,p}$ in (7.17) are obtained by applying boundary conditions to the tangential electric and magnetic fields at the dielectric interfaces ($r = a_p$).

7.1.4.2 Hybridization of the EEM and MoM

To accurately account for EM interactions between antennas and the multilayered sphere, the hybridization of the EEM and the method of moments (MoM) has been performed. Using this hybridization, the current distribution on the antenna can be determined efficiently. The unknowns are limited only to the surface of the antenna since the scattered dyadic Green's function is provided by EEM.

The electric field integral equation (EFIE) for any antenna in the presence of a multilayered dielectric sphere can be written as

$$\mathbf{E}(\mathbf{r}) = \mathbf{E}^i(\mathbf{r}) + \mathbf{E}^s(\mathbf{r}) = \mathbf{E}^i(\mathbf{r}) - j\omega\mu \iiint \overline{\mathbf{G}}_e^{(11)}(\mathbf{r}, \mathbf{r}') \cdot \mathbf{J}(\mathbf{r}')dv' \qquad (7.18a)$$

$$\hat{\mathbf{n}} \times \mathbf{E}(\mathbf{r}) = 0 \text{ on the surface of the antenna} \qquad (7.18b)$$

where $\mathbf{E}^i(\mathbf{r})$ is the incident field due to the localized source (modeled as a delta gap or a magnetic frill), $\mathbf{E}^s(\mathbf{r})$ is the scattered field due to the induced current on the antenna, and $\overline{\mathbf{G}}_e^{(11)}(\mathbf{r}, \mathbf{r}')$ is the dyadic Green's function for the region outside the sphere (region 1).

In this treatment, a thin wire dipole antenna oriented in the x-direction is considered. On the surface of the thin wire, tangential components of the electric field vanish such that

$$E_x = E_x^i + E_x^s = 0 \qquad (7.19)$$

The tangential field can approximately be evaluated along the center of the thin wire, while the induced current is confined to the surface of the wire (thin wire approximation). Then, one obtains

$$
\begin{aligned}
-E_x^i = E_x^s &= E_x^{s,fs} + E_x^{s,sob} \\
&= -j\omega\mu \iiint \hat{\mathbf{x}} \cdot \overline{\mathbf{G}}_{e0}(\mathbf{r}, \mathbf{r}') \cdot \hat{\mathbf{x}}J_x(\mathbf{r}')dv' \qquad (7.20) \\
&\quad - j\omega\mu \iiint \hat{\mathbf{x}} \cdot \overline{\mathbf{G}}_{es}^{(11)}(\mathbf{r}, \mathbf{r}') \cdot \hat{\mathbf{x}}J_x(\mathbf{r}')dv'
\end{aligned}
$$

where $E_x^{s,fs}$ is the field due to induced current through free-space propagation, $E_x^{s,sob}$ is the field due to scattering from the multilayered sphere, $\overline{\mathbf{G}}_{e0}(\mathbf{r}, \mathbf{r}')$ is the free-space dyadic Green's function, and $\overline{\mathbf{G}}_{es}^{(11)}(\mathbf{r}, \mathbf{r}')$ is the dyadic Green's function for the scattered field that is provided by EEM. The above equation can be solved for the unknown current coefficients on the surface of the antenna by applying the MoM with pulse basis functions and point matching weighting functions.

7.1.5 Results Using EEM

7.1.5.1 Human Head Model

A six-layered, lossy dielectric sphere is used to simulate the biological tissues of a human head. Identified biological tissues include skin, fat, bone, dura, cerebrospinal fluid (CSF), and brain. The electrical parameters (permittivity and conductivity) of the biological tissues at 900 MHz and 1.9 GHz are shown in Table 7.1. The electrical parameters of the tissues at 900 MHz are taken from [10], and at 1.9 GHz from [8]. These parameters can also be obtained using the multiple Cole-Cole dispersion equation and the corresponding parameters in [18].

7.1.5.2 EM Interaction Characterizations

First, we assume that a $\lambda/2$ antenna is located 2 cm away from the surface of the spherical head at 900 MHz. The gain patterns are shown in Figure 7.5(a). As can be seen in the figure, the far-field patterns are significantly modified by the presence of the spherical head. In the x-z plane, the peak gain with the spherical head is slightly enhanced ($\sim$0.28 dB) in the $\theta = 0$ direction, but is reduced by 2.28 dB toward the spherical head direction as compared with the free-space antenna gain.

The unaveraged and 1-g averaged SAR distributions inside the spherical head along the z-axis are shown in Figure 7.5(b). With the unaveraged SAR distribution, the main peak SAR occurs at the cerebrospinal fluid (CSF) layer just outside of the brain region; the second peak occurs at the skin layer. With the averaged SAR distribution (averaging over a $1.1 \times 1.1 \times 1.1$ cm^3 cubical volume with eleven 1-mm cells each side of the cube), the peak values are much lower than those of the unaveraged SAR. For example, with 1W of delivered power, the peak values are 4.5 W/kg with the 1-g averaged SAR near the CSF layer and 11.2 W/kg with the unaveraged SAR. Also, we observe that the peak SAR locations are shifted, due to the averaging process, from those of the unaveraged SAR. The SAR peaks at the biological layers, which possess high conductivity such as the skin and CSF layers.

Three-dimensional surface plots of the SAR distribution are shown in Figures 7.6(a) and (b) in the x-z plane and in the y-z plane, respectively. As can be seen in the figures, peak SAR lies along the z-axis. In the figures, it is interesting to observe small ripples

Table 7.1
Electrical Parameters of a Six-Layered Head Model at 900 MHz and 1.9 GHz*

Biological Tissues	Radius a_p (cm)	900 MHz ϵ_r	900 MHz σ	1.9 GHz ϵ_r	1.9 GHz σ	Mass Density (10^3 kg/m^3)
Skin	9.00	40.7	0.65	37.21	1.25	1.01
Fat	8.90	10.0	0.17	9.38	0.26	0.92
Bone	8.76	20.9	0.33	16.40	0.45	1.81
Dura	8.35	40.7	0.65	37.21	1.25	1.01
CSF	8.30	79.1	2.14	77.30	2.55	1.01
Brain	8.10	41.1	0.86	43.22	1.29	1.04

*a_p is the radius, ϵ_r the permittivity, and σ the conductivity of each spherical layer.

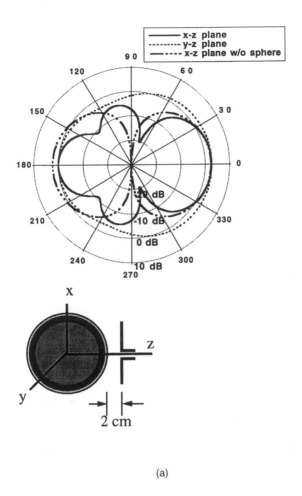

(a)

Figure 7.5 (a) Far-field gain patterns of a $\lambda/2$ antenna with and without a six-layered spherical head. (b) SAR distribution—1-g averaged and unaveraged—along the z-axis. Operating frequency of antenna is 900 MHz. Abbreviations in part (b): BR (brain), C (CSF), D (dura), BN (bone), F (fat), and S (skin).

near the center of the brain region due to the focusing effect of the sphere. These ripples may be more conspicuous when the head–antenna separation distance is increased. Contour plots of the electric field intensity are shown in Figures 7.6(c) and (d). The numbers in the contour plots designate contour lines for $20 \log_{10} |\mathbf{E}|$.

As another example, EM interaction results are obtained at an operating frequency of 1.9 GHz. Figure 7.7(a) shows the gain patterns. As compared with the free-space antenna gain in the x-z plane, the peak field intensity for the spherical head is enhanced by 2.21 dB in the $\theta = 0$ direction, but is reduced by 8.27 dB toward the spherical head

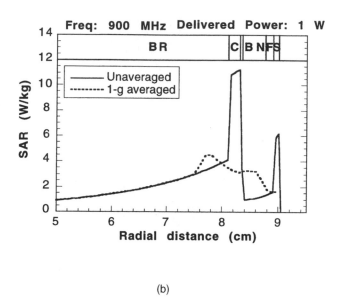

(b)

Figure 7.5 (continued).

direction. The SAR distribution along the z-axis is shown in Figure 7.7(b). The magnitudes of both 1-g averaged and unaveraged SAR peaks near or at the skin layer and CSF region are greater than those of the 900-MHz case [see Figure 7.5(b) and Table 7.2]. The reason may be the shorter $\lambda/2$ antenna length (7.9 cm) at 1.9 GHz as compared with that (16.7 cm) at 900 MHz. Also, the SAR value at the skin layer is greater than that at the CSF layer, which implies the importance of the skin layer at higher frequencies. With 1-g averaged SAR, the main SAR peak is near the CSF layer rather than near the skin layer.

The total power absorption in the spherical head at 900 MHz is plotted as function of the head–antenna separation distance and the head size in Figure 7.8(a). As shown in the figure, the power absorption is strongly dependent on the head–antenna separation distance (d). A significant portion of the antenna delivered power is absorbed in the spherical head (40% to 50% at $d = 2$ cm) when the head is located near the antenna, but becomes smaller at large distances ($\leq$10% at $d = 10$ cm). In this case, it is observed that the absorbed power decays roughly as $d^{-0.75}$ at close distances from the sphere. Also, the total power absorption variation in the spherical head at 1.9 GHz is shown in Figure 7.8(b). As compared with the 900-MHz case, the power absorption drops more rapidly when the antenna separation distance is small; that is, $P_a \propto d^{-0.85}$. It is observed that the size of the head does not strongly influence the total power absorption at this frequency.

The results of the EM interactions between half-wave antennas and a six-layered spherical head at 900 MHz and 1.9 GHz are summarized in Table 7.2.

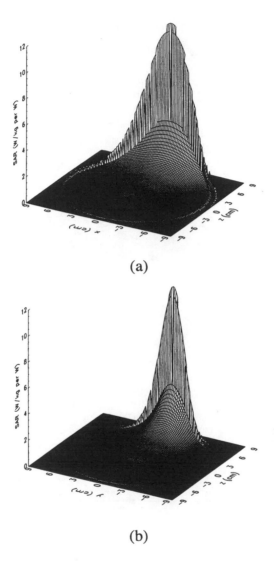

(a)

(b)

Figure 7.6 3-D surface plots of SAR distributions inside the spherical head (a) in the *x-z* plane and (b) in the *y-z* plane; contour plots of the electric field distributions (log scale) (c) in the *x-z* plane and (d) in the *y-z* plane. Operating frequency of the $\lambda/2$ antenna is 900 MHz. The numbers in the contour plots designates contour lines for $20 \log_{10}(|\mathbf{E}|)$.

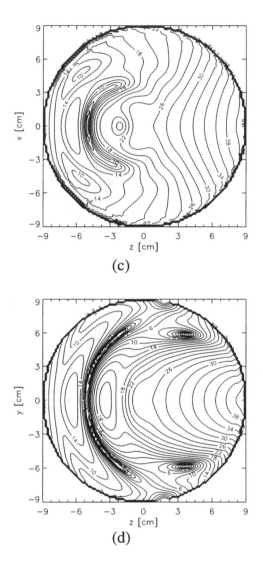

Figure 7.6 (continued).

7.1.5.3 Effects of Size of the Head Model: Adult and Child

Head sizes vary widely among individuals. Recently, for a smaller size head (such as a child's head), higher and larger in-depth penetration of the absorbed energy has been reported [8]. To investigate this phenomenon, a $\lambda/2$ dipole antenna at 900 MHz is assumed to be located 2 cm away from the spherical head. Also, it is assumed that the radius of

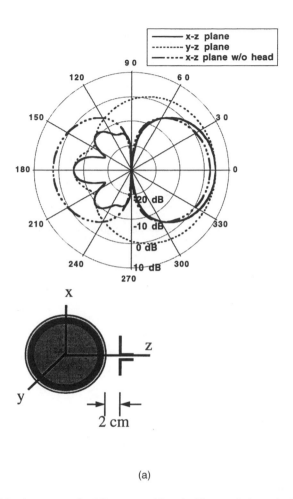

(a)

Figure 7.7 (a) Far-field gain patterns of a $\lambda/2$ antenna with and without a six-layered spherical head. (b) SAR distribution—1-g averaged and unaveraged—along the z-axis. Operating frequency of antenna is 1.9 GHz. Abbreviations in part (b): BR (brain), C (CSF), D (dura), BN (bone), F (fat), S (Skin).

an adult head is 10 cm, and that of a child is 7 cm with a six-layered head model as described in Table 7.1. Figure 7.9(a) shows the comparison of SAR distributions along the z-axis. Clearly, the SAR in the interior region of the child head is greater even though the SAR at the outer surface of the brain region is roughly the same as that of the adult head. This is because the size of the child's brain region is smaller than that of the adult head. Contour plots of the electric field distribution in the x-z plane are shown in Figures 7.9(b) and (c). As can be seen in the figures, the electric field penetrates more deeply into the child head than the adult head. However, according to Figure 7.8(a), the total

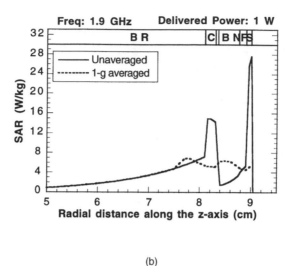

(b)

Figure 7.7 (continued).

Table 7.2
Summary of EM Interactions Between a Half-Wave Dipole Antenna and a Six-Layered Spherical Head
at 900 MHz and 1.9 GHz*

$\lambda/2$ Antenna	900 MHz	1.9 GHz
Power absorption (%)	43.9%	35.3%
Peak SAR (W/kg)		
Unaveraged	11.2 (at CSF layer)	27.6 (at skin layer)
1-g Averaged	4.5 (near CSF layer)	6.9 (near CSF layer)
Gain (dBi)	2.54	4.46
Directivity (dBi)	4.88	6.32
Input impedance (Ω)		
without head	$99.9 + j51.9$	$99.6 + j51.9$
with head	$128.0 + j51.9$	$104.0 + j30.0$

*The antenna-delivered power is 1W. Antenna is located 2 cm away from the surface of the spherical head.

absorbed power with the child head (38.4%) is smaller than that of the adult head (47.6%) at $d = 2$ cm.

7.1.5.4 Comparison Between Homogeneous and Multilayered Spheres

The differences in SAR distributions between the homogeneous and multilayered spherical heads are discussed in this section. For a homogeneous head, the electrical parameters of brain in Table 7.1 are used. Figure 7.10 compares 1-g averaged SAR distributions between

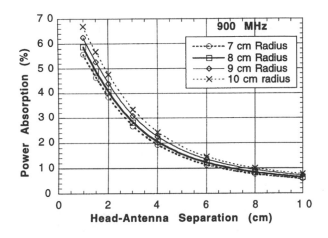

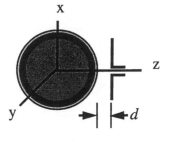

(a)

Figure 7.8 Power absorption in the six-layered spherical head as function of the head–antenna separation distance and the size of the head at (a) 900 MHz and (b) 1.9 GHz. The power absorption drops roughly as $d^{-0.75}$ at 900 MHz and $d^{-0.85}$ at 1.9 GHz for small d.

the six-layered and homogeneous spheres together with unaveraged SAR distributions at 900 MHz. As can be seen, near the surface of the spherical head, the 1-g averaged SAR with the homogeneous sphere is much larger than that with the six-layered spherical head. Since SAR at the fat or bone layer with the six-layered spherical head is much smaller than that of the homogeneous sphere, the 1-g averaged SAR is lower near the skin layer. Therefore, the homogeneous sphere overestimates the SAR values near the surface of the spherical head.

7.1.5.5 Vertical Location of Antennas

Because the vertical handset position will differ for different users, it is interesting to assess the influence of antenna vertical location on the observed interaction. Gain patterns

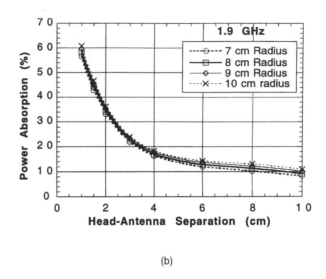

(b)

Figure 7.8 (continued).

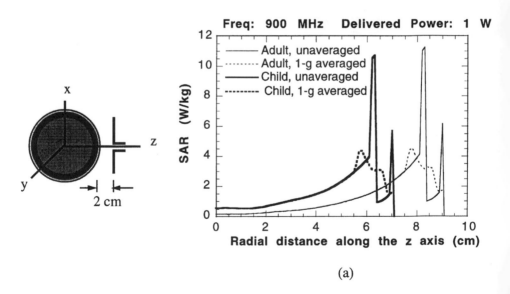

(a)

Figure 7.9 (a) Comparison of SAR distributions in an adult head and a child head. Contour plots of the electric field (b) in an adult head and (c) in a child head. Operating frequency is 900 MHz. The numbers in the contour plots designate contour lines for $20 \log_{10}(|\mathbf{E}|)$.

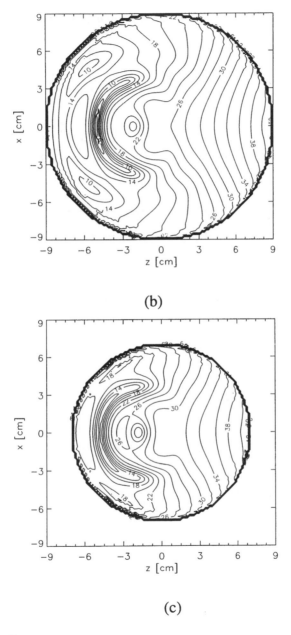

(b)

(c)

Figure 7.9 (continued).

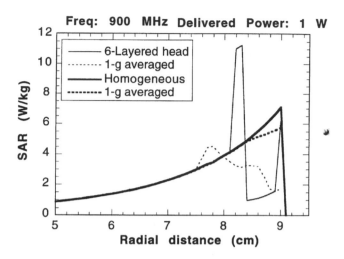

Figure 7.10 Comparison of 1-g averaged SAR distributions between multilayered and homogeneous spherical heads at 900 MHz. The antenna delivered power is 1W.

at 900 MHz without and with the head in the horizontal plane (y-z plane) are shown in Figure 7.11(a). In the figure, the gain patterns are obtained as the distance (x_1) between the center of the dipole antenna and the head center is varied. One can observe that, in the presence of the head, the antenna gain can be lowered as much as ~5.7 dB (with $x_1 = 0$) as compared with the free-space dipole antenna gain. This reduction of antenna gain is seen along $\theta = 120$ deg in Figure 7.11(a). This may be an important factor in evaluating the communication link budget. As x_1 increases, the gain converges to the free-space antenna gain. Similar changes are observed in the antenna gain in the x-z direction in Figure 7.11(b). The total absorbed power in the head is 43.9% with $x_1 = 0$ cm, 30.1% with $x_1 = 5$ cm, and 14.8% with $x_1 = 10$ cm.

Antenna gain patterns in the horizontal plane (y-z plane) at 1.9 GHz without and with the head for several dipole positions are shown in Figure 7.12(a). In this figure, we can observe that the antenna gain is significantly reduced (~13.7 dB with $x_1 = 0$ at $\theta = 150°$) when the spherical head is present as compared to the free-space antenna gain. The antenna gain again approaches to the free-space gain as x_1 increases. The gain pattern changes in the x-z plane are shown in Figure 7.12(b). Also, the total absorbed power decreases as x_1 increases: that is, 35.3% with $x_1 = 0$, 24.6% with $x_1 = 5$ cm, 13.3% with $x_1 = 10$ cm.

7.1.5.6 Comparison with EEM and FDTD

In Figures 7.13(a) and (b), the gain patterns of a half-wave dipole antenna (in the x-z and y-z planes) using the FDTD method in the presence of a homogeneous spherical head are

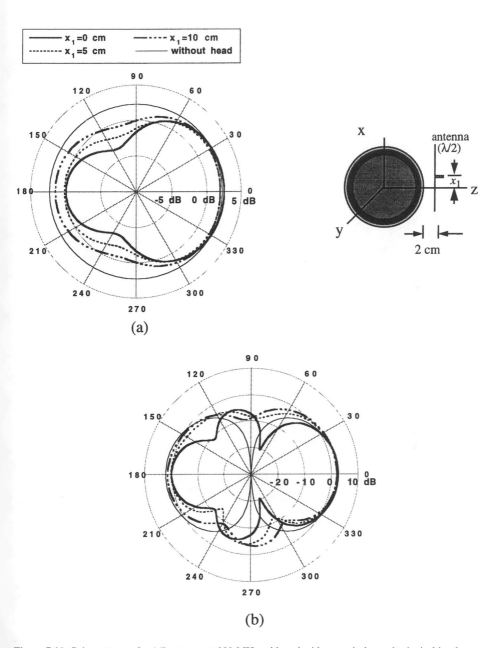

Figure 7.11 Gain patterns of a $\lambda/2$ antenna at 900 MHz with and without a six-layered spherical head as changing the antenna vertical location (x_1) (a) in the y-z plane and (b) in the x-z plane. The antenna delivered power is 1W.

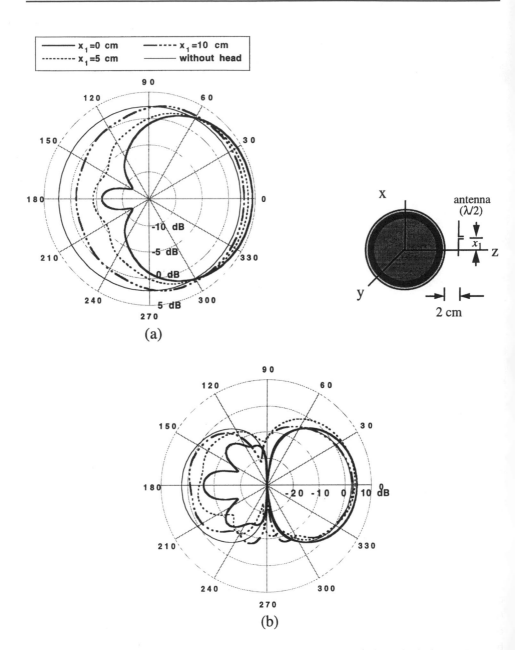

Figure 7.12 Gain patterns of a $\lambda/2$ antenna at 1.9 GHz with and without a six-layered spherical head as changing the antenna vertical location (x_1) (a) in the y-z plane and (b) in the x-z plane. The antenna delivered power is 1W.

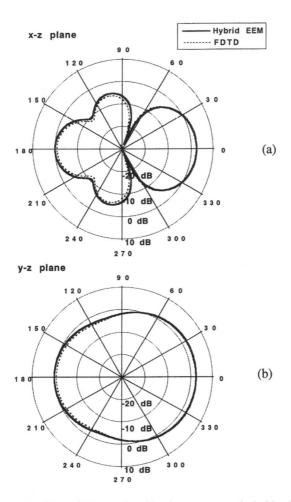

Figure 7.13 Comparison of FDTD and EEM results with a homogeneous spherical head ($\epsilon_r = 41.0$, $\sigma = 0.86$) and a half-wave dipole antenna at 900 MHz: gain patterns (a) in the x-z plane and (b) in the y-z plane; (c) total electric field magnitude ($20 \log(|\mathbf{E}|)$ inside the spherical head along the z-axis. The antenna delivered power is set at 1W.

compared with those using the EEM. The relative permittivity (41.0) and conductivity (0.86 S/m) of the spherical head are taken as those of brain at 900 MHz. The diameter of the head is 17.3 cm and the antenna is located 2 cm away from the surface of the head. For the FDTD calculation, the size of the cubical grid cell is $0.02\lambda_0$ (free-space wavelength) or $0.128\lambda_d$ (dielectric wavelength). The gain patterns from both techniques agree very well, with only minor differences observed in the backward direction. Also, the electric field distributions inside the spherical head using both techniques are compared

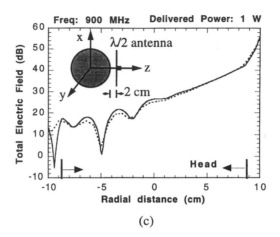

(c)

Figure 7.13 (continued).

in Figure 7.13(c). As can be seen in the figure, field distributions obtained using both techniques agree well, even for the resonant field structures. Better agreement could be obtained with finer FDTD grid cells.

7.1.5.7 Anatomical Head Versus Spherical Head

A simple spherical head can be used to predict the peak SAR in the anatomical head. The electromagnetic interaction computations with anatomical head models [see Figures 7.21(a) and (b)] using FDTD are described in detail in Section 7.1.6. Electrical parameters of the anatomical head model are given later in Table 7.5. In the following computations, a dipole antenna is assumed to be located 2 cm away from the outer edge of the ear. To compare with the anatomical head results, antenna interactions with a homogeneous spherical head with a diameter of 17.3 cm are analyzed using the hybrid EEM. Also, EM interactions with a homogeneous head with an anatomical head shape are analyzed using FDTD. The gain patterns in the horizontal plane (y-z plane) and the vertical plane (x-z plane) are shown in Figures 7.14(a) and (b). The gain patterns are similar even with different head configurations except for minor differences in the direction of the head.

In evaluating the amount of power absorption when using the handset antenna, SAR is a useful and pertinent parameter. However, SAR values change with electrical conductivity of biological tissues, and SAR distribution may not be intuitively understood. In Figure 7.15, the electric field distributions with the different configurations of the head models—a homogeneous spherical head, an anatomical head with different tissues, and a homogeneous head with an anatomical head shape—are shown. An interesting observation is that the magnitudes of the electric field near the surface of the outer edge of the different head models are very similar. Total electric field inside the head is smoothly

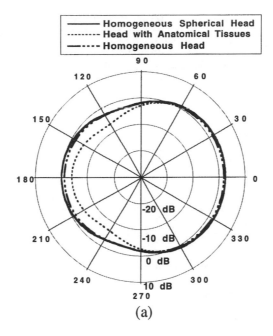

(a)

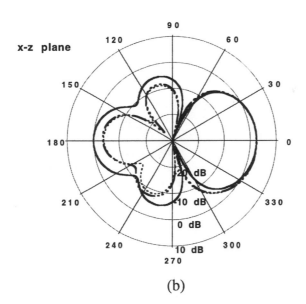

(b)

Figure 7.14 Gain patterns of a $\lambda/2$ dipole antenna at 900 MHz in the presence of a homogeneous spherical head, anatomical head with different tissues, and a homogeneous head with an anatomical head shape: (a) in the y-z plane and (b) in the x-z plane.

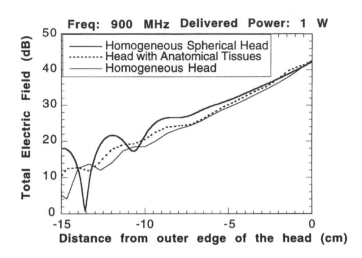

Figure 7.15 Electric field magnitude inside three different head configurations: a homogeneous spherical head, an anatomical head with various biological tissues, and a homogeneous head with an anatomical head shape. The antenna delivered power is 1W.

decreasing even at the interface between different biological tissues due to the tangential continuity of the electric field. The peak SAR value in the anatomical head can be roughly estimated by calculating the electric field using a simple homogeneous spherical head and multiplying by the electric conductivity of the biological tissue [see (7.4)].

7.1.5.8 Directional Antennas

An array of antennas with appropriate amplitude and phase will produce radiation patterns that reduce the EM radiation toward the human head, thus resulting in less power absorption in the head. Previously, omnidirectional antennas were conceived as better antennas for better signal reception for mobile telecommunication equipments (MTE). However, as cell areas decrease, directional antennas may have advantages over the omnidirectional ones since they reduce the absorption in the human head and therefore prolong battery life.

As an example, an end-fire directional antenna array consisting of two $\lambda/2$ dipole antennas at 1.9 GHz is considered in Figure 7.16(a). The gain patterns with and without the six-layered spherical head in the y-z plane (horizontal plane) are compared in Figure 7.16(b). As compared with the free-space dipole antenna gain (2.23 dB), the antenna gain of the end-fire antenna is significantly reduced in the direction of the head (−7.53 dB), but enhanced in the opposite direction (~5.61 dB). Note that, in the presence of the spherical head, the gain of a nondirectional dipole antenna at 1.9 GHz in the direction of the head is about −6.11 dB, and the gain in the opposite direction is 4.46 dB. With this

389

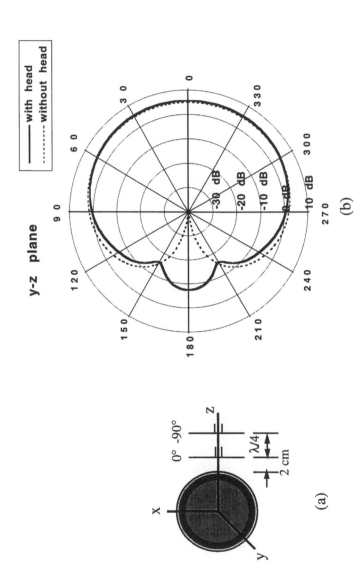

Figure 7.16 (a) Schematic of arrangement; (b) far-field gain patterns of an end-fire array (with two half-wave dipoles) in the presence of the spherical head in the y-z plane. Operating frequency is 1.9 GHz.

end-fire antenna array, however, the total power absorption is only 5.5% of the antenna delivered power, while the total power absorption with a single $\lambda/2$ dipole antenna at $d = 2$ cm is ~34.8%. Various directional antenna configurations with additional antenna elements or with the different current amplitude/phase arrangements have also been investigated.

7.1.5.9 High-Frequency Effect

To accommodate transmission of the vast amount of multimedia data in wireless communications, wideband operations at Ka-band (e.g., 30 GHz) are drawing considerable attention in both wireless terrestrial and satellite communications. At these high frequencies, the popular numerical techniques such as FDTD have severe limitations in EM interaction calculations due to huge computer storage requirements and excessive computation time (see Table 7.3). This is because the typical size of the adult head is 18 cm in diameter, which corresponds to ~75 dielectric wavelengths in the head at 30 GHz. On the other hand, the hybrid EEM/moment method technique is ideally suited to critically assessing human–antenna interactions at these high frequencies [19].

Head Model

As will be demonstrated, most of the EM interaction at high frequencies occurs near the exterior region of the head. Therefore, application of the exact anatomical head model will not appreciably change the outcome of the computations. The electrical parameters for the six biological tissues at 30 GHz (see Table 7.4) were calculated using the multiple Cole-Cole dispersion equation [18]. It is observed that the conductivities of the tissues are high (e.g., 27.1 S/m for skin), whereas the permittivities are relatively low as compared with those at lower frequencies (see Table 7.1). Dimensions of the layered head model are also listed in Table 7.4. The outer radius of the spherical head is 9 cm, and the thickness of the skin layer is 1 mm.

Table 7.3

Required Number of Cells for the Computation of EM Interactions Using FDTD Calculation at Different Frequencies*

Frequency	Head Diameter	Skin	Number of Cells	ϵ_r	σ (S/m)
900 MHz	$3.5\lambda_d$	$0.02\lambda_d$	42,875	41.1	0.65
1.9 GHz	$7.5\lambda_d$	$0.04\lambda_d$	421,875	43.2	1.29
30 GHz	$75.6\lambda_d$	$0.39\lambda_d$	432×10^6	17.62	27.78

*λ_d is the dielectric wavelength. Number of cells for FDTD calculation is based on a cell size of $0.1\lambda_d$. Skin layer ~ 1 mm; outer radius of the spherical head ~ 18 cm.

Table 7.4

Electrical Parameters of a Six-Layered Head Model at 30 GHz*

Biological Tissues	Radius a_p (cm)	30 GHz ϵ_r	σ	Mass Density $(10^3 \ kg/m^3)$
Skin	9.00	15.52	27.10	1.01
Fat	8.90	5.91	5.33	0.92
Bone	8.76	6.12	7.21	1.81
Dura	8.35	15.52	27.10	1.01
CSF	8.30	30.72	57.81	1.01
Brain	8.10	17.62	27.18	1.04

*a_p is the radius, ϵ_r the permittivity, and σ the conductivity of each biological tissue.

Nondirectional Antennas

First, we assume that the multilayered spherical head is irradiated by a half-wave dipole antenna located 2 cm away from the sphere surface at 30 GHz as shown in Figure 7.17(a). The antenna is oriented in the x-direction. Figure 7.17(a) shows the unaveraged SAR and 1-g averaged SAR distributions in the spherical head along the z-axis (where the maximum SAR occurs). In the figure, we observe that extremely high unaveraged peak SAR (457 W/kg per 1W of antenna delivered power) occurs at the thin skin layer (~1 mm), and most power absorption is localized near the skin ($\leq$1 mm). This high peak SAR results from the high conductivity of the skin layer at 30 GHz. Because the head is located in the far-field region of the half-wave antenna at this frequency, the peak electric field at the outer skin layer can be roughly estimated using a plane wave approximation as shown in Figure 7.18. The peak SAR value obtained by this method for 1W of antenna delivered power is 417 W/kg, which is very close to the accurate result using EEM (457 W/kg). With 1-g averaging over a specified volume ($1.1 \times 1.1 \times 1.1$ cm^3 cubical volume with eleven 1-mm cells each side of the cube), the averaged peak SAR becomes much lower (10.6 W/kg), although significant averaged SAR values are spread to a broader region in the head. The total power absorption in this case is 14.7% of the antenna delivered power, which is much smaller than those at 900 MHz (43.9%) and 1.9 GHz (35.3%). The gain patterns of the half-wave dipole antenna in the y-z plane with and without the head are shown in Figure 7.17(b). It is observed that, in the presence of the head, the gain pattern of the half-wave antenna is significantly distorted and clearly loses its omnidirectionality in this plane.

Directional Antennas

The peak SAR may be significantly reduced by utilizing directional antennas. For example, we consider an end-fire configuration with two half-wave dipole antennas; one antenna

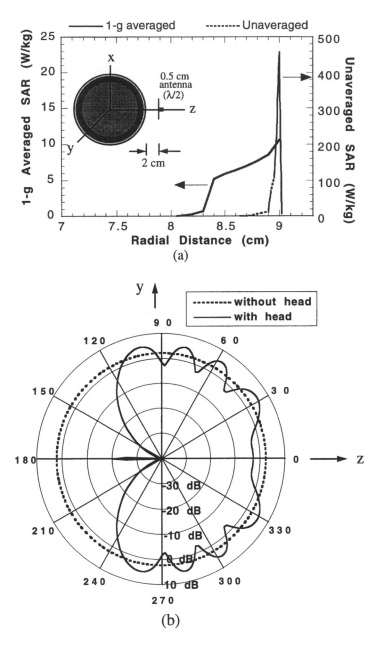

Figure 7.17 Nondirectional antenna (a half-wave dipole antenna): (a) unaveraged and 1-g averaged SAR distributions along the z-axis in the six-layered spherical head; (b) gain patterns in the y-z plane with and without the spherical head. The antenna delivered power is 1W. Operating frequency is 30 GHz.

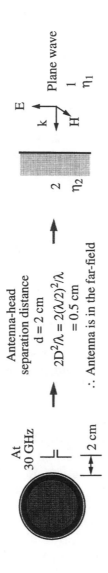

Figure 7.18 At high frequencies (e.g., 30 GHz), peak electric field magnitude at the outer edge of the skin layer can be roughly estimated using plane wave approximation.

At
30 GHz

2 cm

Antenna-head
separation distance
d = 2 cm
$2D^2/\lambda = 2(\lambda/2)^2/\lambda$
$= 0.5$ cm
∴ Antenna is in the far-field

2
η_2

E
k
H

Plane wave
1
η_1

is located 2 cm away from the head surface and the other separated by $\lambda/4$ (2.5 mm) with a −90-deg phase shift. The unaveraged and 1-g averaged SAR distributions are shown in Figure 7.19(a). [Note that the scales for this figure are different from those for Figure 7.17(a).] As shown in the figure, the unaveraged peak SAR reduces to 2.8 W/kg, and the 1-g averaged SAR reduces to 0.06 W/kg with 1W of the delivered power. Also, the total power absorption is dramatically reduced to ~1% of the delivered power. The reason for these striking reductions of SAR and power absorption is that, at 30 GHz, the head is located in the far-field region of the antenna and the incident field due to the end-fire antenna array is very small in the direction of the head. At lower frequencies, the handset antennas are usually located in the reactive near-field region, and the directional antennas may not be as effective (see Section 7.1.5.8). The gain patterns of the directional antenna in the y-z plane with and without the head are shown in Figure 7.19(b). In this case, the gain pattern of the directional antenna does not change appreciably in the presence of the head except for small changes in the direction of the head.

7.1.6 Results Using the FDTD Method

The results from the EEM outlined in the previous section provide a wide variety of engineering data that can be used to assess the behavior of many different antenna–tissue configurations. In all of these results, however, the antenna was limited to simple geometries such as a dipole radiator. While theoretically any antenna structure could be modeled using the EEM in conjunction with the MOM as discussed, modeling requirements of complicated handsets often become very difficult using this approach. Therefore, this section focuses on using the FDTD method to demonstrate the electromagnetic behavior of realistic handset configurations operating in the presence of human tissues. The computations will concentrate on the three handset configurations depicted in Figure 7.2.

7.1.6.1 Tissue Models

The derivations provided in Section 7.1.3 illustrate that modeling of anatomical features within the FDTD framework can be accomplished by assigning appropriate values for the permittivity and conductivity at each spatial grid cell. The human hand is simply modeled as a layer of bone surrounded by a layer of muscle that covers three sides of the handset as depicted in Figure 7.20. An anatomical head model based on MRI scans of a human head has also been constructed. Figure 7.21(a) shows the full head model with the hand and the handset, and Figure 7.21(b) illustrates a midsagittal cross section of the head. The cell size for the computations is 3.28 mm. Clearly, more refined cell sizes can be used if necessary. The tissues utilized in the models are listed in Table 7.5. Figures 7.22(a) and (b) show two different views of the computational model along with relevant dimensions. It is noteworthy that in order to allow accurate modeling of realistic operator/handset configurations, it is important to allow rotation of the handset. This is

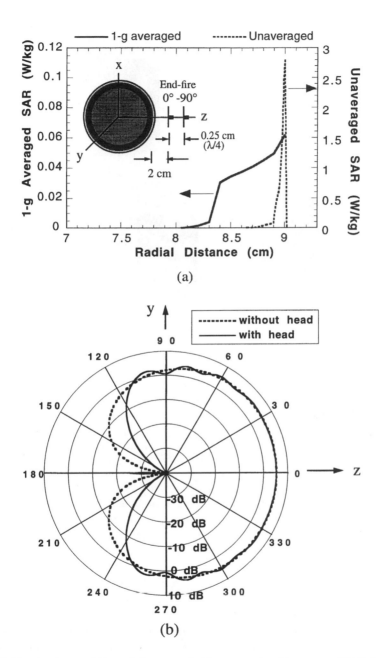

Figure 7.19 Directional antenna (the end-fire antenna): (a) unaveraged and 1-g averaged SAR distributions along the z-axis in the six-layered spherical head; (b) gain patterns in the y-z plane with and without the spherical head; z direction points away from the head. The antenna delivered power is 1W. Operating frequency is 30 GHz.

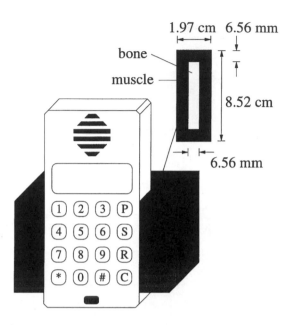

Figure 7.20 Model of the human hand used in the FDTD simulation of antennas on a handset.

accomplished by rotating the position of each cell in the head model in the y-z plane about the head center and reconstructing the grid based on these rotated tissue locations as implied in Figure 7.22(a).

7.1.6.2 Input Impedance and the Importance of the Hand Position

A first investigation into the influence of biological tissue on handset-mounted antennas involves a study of the effect of the operator on the antenna input impedance behavior. This study was conducted for each handset in Figure 7.2 by first computing the antenna reflection coefficient $|S_{11}|$ for each configuration (assuming a 50Ω transmission line) with no tissue and comparing the result to that for the handset with the tissue. For each case, the hand was introduced first and moved to different vertical positions along the handset (denoted by the parameter d as shown in Figure 7.22).

The impedance results for the monopole antenna as a function of frequency are shown in Figure 7.23. As can be seen, the introduction of the hand slightly shifts the antenna resonant frequency. However, the results indicate that the impedance behavior is relatively insensitive to the absolute hand position. The addition of the head also produces a shift in the antenna resonant behavior due to the loading by the high-permittivity tissue. It is interesting to contrast these results with those from the side-mounted PIFA antennas. Reflection coefficient curves for this configuration are provided in Figure 7.24 for several

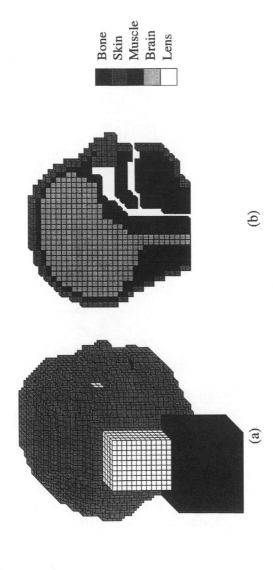

Figure 7.21 (a) Head model with hand and the plastic covered handset; (b) a sagittal cut of the discrete head model at the head center.

Table 7.5

Relative Permittivity, Conductivity, and Density of the Tissues in the Hand and Head Near 900 MHz

Tissue	Permittivity	Conductivity (S/m)	Density ($\times 10^3$ kg/m^3)
Bone	8.0	0.105	1.85
Skin/fat	34.5	0.60	1.10
Muscle	58.5	1.21	1.04
Brain	55.0	1.23	1.03
Humour	73.0	1.97	1.01
Lens	44.5	0.80	1.05
Cornea	52.0	1.85	1.02

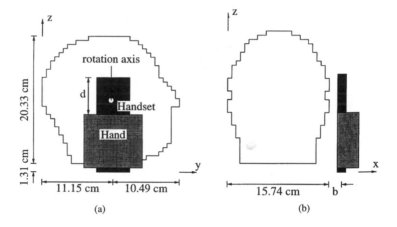

Figure 7.22 (a) Side and (b) rear views of the FDTD head/hand/handset model showing dimensions.

different hand positions. In this case, because of the element locations, the operator may very easily mask part or all of the antenna with his hand while using the device. Most significant is the high impedance mismatch that occurs when the hand begins to mask the antenna. This mismatch will result in decreased communication range due to the decreased amount of signal power transmitted or received through the antenna. These results illustrate the importance of minimizing antenna masking through proper antenna placement. The dots in the figure indicate experimentally obtained results for the antenna when the hand is absent and when the hand is at $d = 6.56$ cm. This comparison shows good correlation between the experimentally and computationally obtained results.

The curves in Figure 7.25 represent $|S_{11}|$ with the hand at $d = 7.21$ cm and the head at $b = 1.97$ cm. Two different models of the head have been used in this example. The curve labeled "sphere" corresponds to the case of a homogeneous spherical ball of muscle ($\epsilon_r = 58.5$, $\sigma = 1.21$ S/m) with a radius of 9 cm. The second head is the anatomical inhomogeneous head model. As can be seen, for this particular configuration the choice

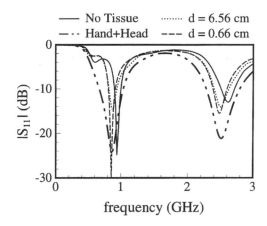

Figure 7.23 $|S_{11}|$ versus frequency for the monopole of Figure 7.2(a) with no tissue, with the hand at two locations, and with the head and hand ($b = 1.97$ cm, $d = 5.90$ cm).

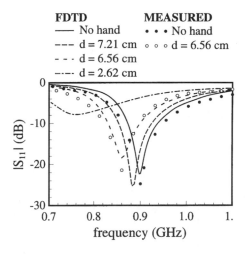

Figure 7.24 Computed value of $|S_{11}|$ for the side-mounted PIFA on the handset without the hand and with the hand for three different values of d. Measured values appear for the configurations with no hand and with the hand at $d = 6.56$ cm.

of models exercises little influence on the antenna input impedance. This insensitivity occurs because the input impedance is a reasonably local phenomenon that is influenced most significantly by structures in the near vicinity of the antenna rather than objects such as the head, which are displaced somewhat from the antenna feed point. Once again, the measured data given in Figure 7.25 show that the FDTD accurately predicts the effects of the human operator on the antenna performance.

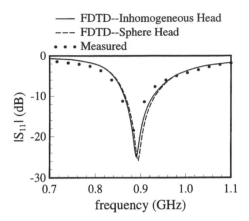

Figure 7.25 $|S_{11}|$ versus frequency for the side-mounted PIFA on the handset with the head and the hand ($b = 1.97$ cm, $d = 7.21$ cm). Computed results obtained using both head models are compared with measured data.

Figure 7.26 illustrates the $|S_{11}|$ behavior of the back-mounted PIFA, where once again the different curves illustrate the performance when no tissue is present, when the hand is at four positions, and when the hand ($d = 4.26$ cm) and the head ($b = 1.97$ cm) are included. These results again demonstrate the change in resonance frequency and resonant resistance with changing tissue locations. For this configuration, once the hand comes above the step in the conducting chassis, significant degradation to the antenna performance begins to occur.

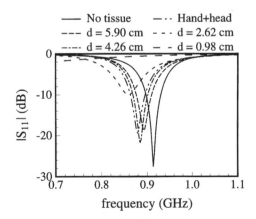

Figure 7.26 $|S_{11}|$ versus frequency for the back-mounted PIFA on the handset for different handset/ tissue configurations. When the head is included, $b = 1.97$ cm and $d = 4.26$ cm.

7.1.6.3 Gain Patterns

As expected, the presence of the operator tissue can have a pronounced effect on the radiation properties of the antenna because it impacts the total radiated power, the shape of the pattern, and the polarization characteristics of the radiated field. As an example, consider the gain patterns shown in Figure 7.27 for the monopole antenna on the handset at 915 MHz. For these computations, the handset is rotated 60 deg from upright in the y-z plane. Results for both the handset alone and the handset in the presence of the tissue are provided (b = 1.97 cm, d = 5.90 cm). As can be seen, the peak gain, pattern shape, and radiated field polarization are all impacted by the presence of the tissue. Figure 7.28 shows similar patterns for the back-mounted PIFA for the same tissue configuration and handset rotation. It is interesting that in this case, while the tissue does influence the radiation pattern, it appears that less gain loss occurs for this geometry as compared to that for the monopole antenna.

Figure 7.29 provides a thorough study of the pattern characteristics for the side-mounted PIFA configuration. In this set of plots, the patterns in the principal vertical planes are shown for no tissue, for the homogeneous spherical head model, and for the inhomogeneous head model (b = 1.97 cm, d = 7.21 cm). In this case, it is apparent that while there is some difference in the results obtained from the two different head models, these differences are not drastic. This provides confidence in the ability to approximately assess the effects of tissue on antenna radiation characteristics using simplified tissue models.

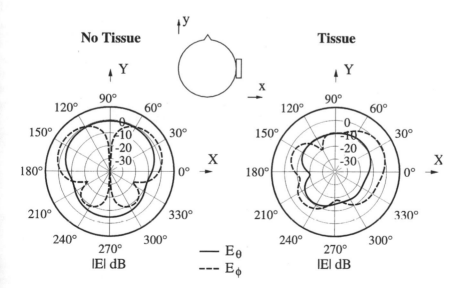

Figure 7.27 Gain patterns for the monopole antenna on the handset at 915 MHz without and with the head and hand (b = 1.97 cm, d = 5.90 cm). The antennas are rotated 60 deg from upright.

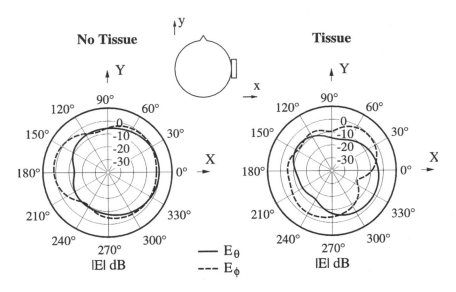

Figure 7.28 Gain patterns for the back-mounted PIFA on the handset at 915 MHz without and with the head and hand ($b = 1.97$ cm, $d = 5.90$ cm). The antennas are rotated 60 deg from upright.

7.1.6.4 Near Fields and SAR

Figure 7.30 compares the field variation around the handset, head, and hand at 915 MHz for the side-mounted PIFA (both inhomogeneous and spherical head models), the monopole, and the back-mounted PIFA. Figure 7.31 presents the SAR distribution for the same configurations. For each computation, $d = 7.21$ cm, $b = 1.97$ cm, and the handset is upright for simplicity in data presentation. The data plane is located at the center of the head/handset combination, and the plots are viewed from the $-y$-direction (see Figure 7.22). The values in Figure 7.30 represent the total squared magnitude of the peak electric field per watt of total power delivered to the antenna. Similarly, the values in Figure 7.31 represent the SAR averaged over 1 g of tissue and normalized to the power delivered to the antenna.

Table 7.6 provides a numerical comparison of the peak SAR values occurring in the head and hand for each of the configurations in Figures 7.30 and 7.31. The two sets of data correspond to the configurations where the handset is rotated 60 deg and the handset is upright with respect to the head. As can be seen, each of the antenna structures result in very similar values of the peak SAR for both handset orientations with the exception of the side-mounted PIFA geometry. In this case, the rotated handset places the antenna nearly in contact with the ear tissue, resulting in a somewhat higher SAR value as compared to the upright handset. For all of the configurations, the peak SAR in the head occurs either in the ear tissue or in the skin/fat layer in the antenna vicinity. Table 7.6 also

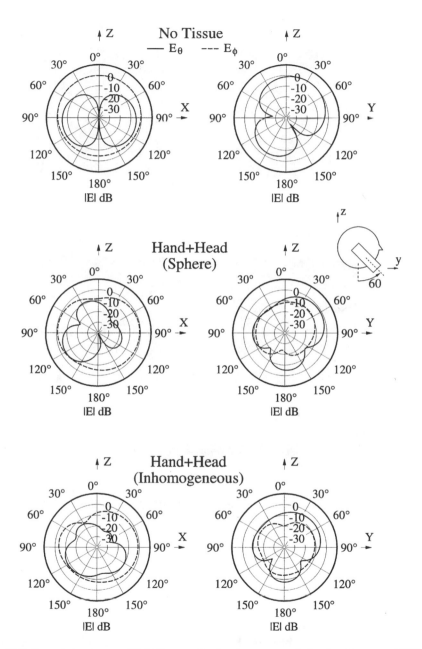

Figure 7.29 Computed patterns at 915 MHz normalized to the antenna gain for the side-mounted PIFA on the handset rotated 60 deg from upright. Results are shown for the handset alone, with the spherical head model, and with the inhomogeneous head model ($b = 1.97$ cm, $d = 7.21$ cm).

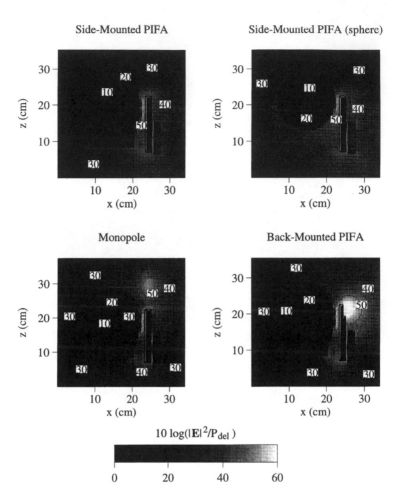

Figure 7.30 Computed normalized near-field distribution at 915 MHz in a plane through the center of the head with $d = 7.21$ cm, $b = 1.97$ cm. The configurations are the side-mounted PIFA, the side-mounted PIFA with the spherical head, the monopole, and the back-mounted PIFA.

provides the SAR averaged over the entire head, which as expected is considerably lower than the peak SAR levels.

A very interesting phenomenon is observed for the back-mounted PIFA, which has a peak SAR in the head that is considerably reduced in comparison to the values for the other antennas. This occurs because the conducting handset chassis lies between the antenna and the head, providing some degree of shielding from exposure. These numbers and plots also show that the spherical and inhomogeneous head models predict slightly different absorption and SAR characteristics. It is important to note that the numbers

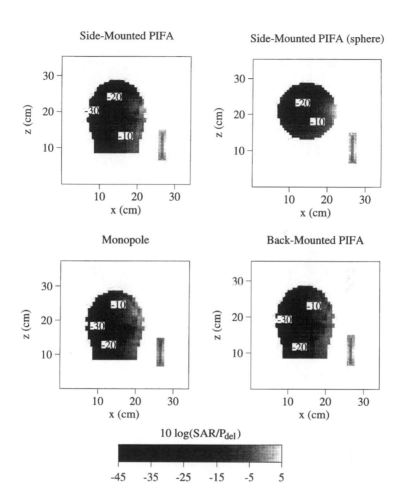

Figure 7.31 Computed normalized SAR distribution at 915 MHz in a plane through the center of the head with $d = 7.21$ cm, $b = 1.97$ cm. The configurations are the side-mounted PIFA, the side-mounted PIFA with the spherical head, the monopole, and the back-mounted PIFA.

provided here are consistent with previously reported measured and computed results [20–23]. Comparisons such as these are very useful for determining the suitability of different radiators for personal communications applications.

Table 7.6 also presents the fraction of power absorbed in the head and hand for the different antenna topologies, along with the radiation efficiency η_a of the configuration. As can be seen from these results, the large amount of power dissipation considerably reduces the antenna gain, with the efficiencies dropping below 50% for most cases. This gain loss is an extremely important issue that should be considered when planning link budgets for wireless communications systems involving handheld subscriber units.

Table 7.6

Computed Normalized Power Absorption and Peak SAR (mW/g/W) in the Head and Hand, Average SAR (mW/g/W) in the Head, and Radiation Efficiency for the Different Handset/Body Configurations Shown in Figure 7.31*

Configuration	$\dfrac{P_{abs}^{head}}{P_{del}}$	$\dfrac{P_{abs}^{hand}}{P_{del}}$	η_a	$\dfrac{SAR_{max}^{head}}{P_{del}}$	$\dfrac{SAR_{max}^{hand}}{P_{del}}$	$\dfrac{SAR_{ave}^{head}}{P_{del}}$
Handset rotated 60 deg with respect to head						
Monopole	0.359	0.169	0.472	1.97	2.29	0.0875
Side-mounted PIFA	0.382	0.301	0.317	3.81	4.54	0.0931
Side-mounted PIFA[†]	0.212	0.318	0.470	3.14	5.42	0.1003
Back-mounted PIFA	0.260	0.222	0.518	1.32	3.58	0.0634
Handset upright with respect to head						
Monopole	0.351	0.184	0.465	2.06	2.43	0.0856
Side-mounted PIFA	0.332	0.324	0.344	2.07	4.91	0.0809
Back-mounted PIFA	0.258	0.225	0.517	0.90	3.53	0.0629

*All data are computed at 915 MHz and 1W delivered power.
[†]With homogeneous spherical head ($\epsilon_r = 58.5$, $\sigma = 1.21$).

The preceding results for radiation efficiency and peak SAR in the head have all been provided for a given separation between the head and the handset (b = 1.97 cm). However, it is interesting to examine the effect of this distance on these parameters. Figure 7.32 presents the variation of the antenna efficiency and peak SAR (1W delivered power) in the head versus the distance b for the monopole and the back-mounted PIFA configurations with the handset upright. As might be expected, the radiation efficiency increases with distance, while the peak SAR decreases in a nearly exponential fashion.

7.1.7 Assessment of Dual-Antenna Handset Diversity Performance

The majority of handsets currently in use for mobile communications systems utilize only a single radiating element. However, there is increasing interest in the placement of two antennas on the handset to allow signal diversity combining for multipath fading mitigation [24,25]. In this case, it is interesting to assess the effect of the tissue on the diversity performance of the dual-antenna configuration [26–28].

One difficulty in predicting this diversity performance is its dependence on the propagation environment in which the handset operates. While formulations exist for assessing diversity gain given envelope correlation coefficients in a Rayleigh distributed fading scenario, they are only approximate in nature and rely on the different antennas having matched radiation characteristics. Therefore, a complete characterization of antenna diversity performance must rely on computational as well as experimental studies. The goal of this section is to outline computational and experimental procedures and results for assessing the diversity performance of dual-antenna handsets operating in typical indoor propagation environments.

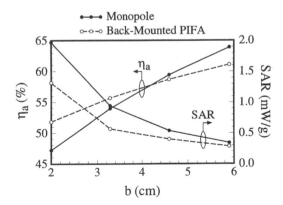

Figure 7.32 Antenna efficiency η_a (%) and peak SAR (P_{del} = 1W) in the head for the monopole and back-mounted PIFA versus the distance between the head and handset. The handset is in an upright position.

7.1.7.1 Dual-Antenna Handset Geometries

The two different handset geometries shown in Figure 7.33 have been used in the computational and experimental observations in this study. The first configuration, shown in Figure 7.33(a), consists of two quarter-wavelength monopole antennas, with one mounted on the handset top and another lying on a hinged mouthpiece at the bottom of the handset. The second geometry, shown in Figures 7.33(b) and (c), utilizes the same mouthpiece-mounted monopole structure, but replaces the top-mounted antenna with the back-mounted PIFA. For simplicity of simulation and consistency in the data collection, the mouthpiece-mounted antennas are positioned to be perpendicular to the handset chassis, as shown in Figure 7.33. The handset bodies have been constructed as aluminum and copper casings and are modeled as perfect conductors in the numerical simulations.

7.1.7.2 Simulated Assessment of Diversity Performance

The FDTD simulation technique allows computation of antenna parameters that can serve as key indicators of diversity performance in a multipath fading environment. The two parameters chosen for this study are the antenna mean effective gain (MEG) and antenna correlation coefficient (ρ_e). To compute these parameters, the FDTD methodology is

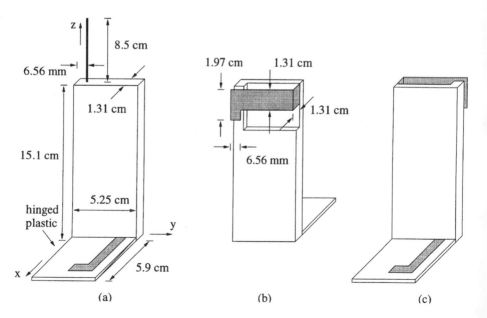

Figure 7.33 Two dual-antenna diversity handset configurations: (a) monopole/mouthpiece monopole: (b) PIFA/mouthpiece monopole; (c) front side of handset in part (b).

utilized to compute the antenna vector power gain pattern $\overline{G}(\theta, \phi)$, which has $\hat{\theta}$ and $\hat{\phi}$ vector components G_θ and G_ϕ, respectively. The MEG G_e is then computed using the expression [29]

$$G_e = \frac{P_{\text{rec}}}{S_\theta + S_\phi} = \frac{1}{1 + X} \int_0^{2\pi} \int_0^{\pi} [XG_\theta(\theta, \phi)P_\theta(\theta, \phi) + G_\phi(\theta, \phi)P_\phi(\theta, \phi)] \sin\theta \, d\theta \, d\phi$$

(7.21)

where S_θ and S_ϕ represent the average power contained in the $\hat{\theta}$ and $\hat{\phi}$ components of the incident field (with respect to the antenna coordinate system), respectively, and P_{rec} refers to the average power received by the antenna along a random route of the handset. The term $X = S_\theta/S_\phi$ is referred to as the cross-polarization discrimination ratio (XPR) and is dependent on the particular propagation environment of interest. Values P_θ and P_ϕ represent the angular density of plane waves polarized in the $\hat{\theta}$ and $\hat{\phi}$ directions in the environment. In this study, they are approximated using the expression [29]

$$P_\xi(\theta, \phi) = \frac{A_\xi}{2\pi} \exp\left[-\frac{(\theta - m_\xi)^2}{2\sigma_\xi^2}\right]$$

(7.22)

where $\xi = \theta$ or ϕ and m_ξ and σ_ξ denote the mean and standard deviation of the elevation angle for the ξ polarization. In this work, a value of $m_\xi = \pi/2$ (horizontal plane) will be assumed for all computations. Also, based on the findings in [24,25] that incident multipath signals arrive at elevation angles less than 40°, we will use $\sigma_\theta = \sigma_\phi = 40$ deg. The coefficient A_ξ is determined from the property

$$\int_0^{2\pi} \int_0^{\pi} P_\xi(\theta, \phi) \sin\theta \, d\theta \, d\phi = 1$$

(7.23)

The envelope correlation coefficient is a metric for assessing the temporal correlation of the signal envelopes at the two antenna terminals. Unfortunately, no simple formulation exists for obtaining this parameter directly based on the antenna gain characteristics and the average parameters of the multipath field. Therefore, a commonly used alternative approach that evaluates the *signal* correlation coefficient ρ_s will be employed in this study. The envelope correlation coefficient can then be approximated using the relation [30]

$$\rho_e \approx |\rho_s|^2$$

(7.24)

The details of this approach are provided in [14]. It has been shown that this relation is relatively accurate for urban environments where the multipath received signal is Rayleigh distributed. Therefore, it is interesting to compare the results of this simple approach to values obtained from the experimental measurements outlined below for the indoor environments assessed in this work.

7.1.7.3 Experimental Assessment of Diversity Performance

The platform used to acquire measurements of different propagation and antenna diversity characteristics is shown schematically in Figure 7.34. The transmitter consists of a Hewlett-Packard HP 8657A sweep oscillator at 915 MHz, which is connected to a vertically oriented balun-fed dipole antenna. The receiving subsystem consists of two independent branches, allowing simultaneous measurement of signals received on the two antennas. Each antenna signal is routed through a low-noise amplifier providing 23 dB of gain into a Hewlett-Packard HP 8590B spectrum analyzer operating in zero-span mode at the transmission frequency. The output voltage from the analyzer is sampled using a two-channel analog-to-digital converter (ADC), which offers a sample rate of 400 Hz per channel. During the measurement campaign, the system was calibrated by putting a known power level into each channel and properly weighting the ADC output to accurately represent the power level. A test performed by sweeping the input power from −40 to −70 dBm showed the system power reading to be linear to within 0.25 dB.

The data for this study were collected in two different buildings on a university campus. The first building has steel-reinforced concrete structural walls and cinder-block/brick partition walls. Two different measurement sites were used in this building. For site 1, the transmitter and receiver were placed about 50 m apart on the building fourth floor. For site 2, the transmitter was placed on the third floor with the receiver (approximately 40 m away) on the fourth floor. The second building contains steel-reinforced concrete structural walls and drywall/steel-stud partition walls. Only one measurement site (site 3) was used in this building, with the transmitter and receiver on the same floor and separated by a distance of approximately 30m.

The measurements at each site were performed by moving the transmitter over a circular path with a 3-m diameter. This pattern was chosen because (1) it localizes the

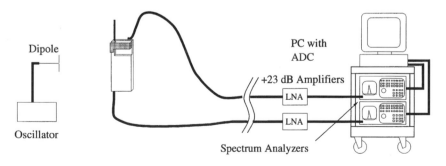

Figure 7.34 Data acquisition platform consisting of two-branch receiver and remote transmitter.

antenna position such that the mean signal envelope remains relatively constant over the measurement, and (2) it lends repeatability to the measurements to allow comparison between operation with and without human tissue. In each case, the handset was moved at a rate of approximately 0.5 m/s. In measurements involving human operators, data were collected for two different subjects (adult males). Simple measurements of XPR using two orthogonally oriented dipole antennas yielded values of $X_1 = X_2 = 5$ dB at sites 1 and 2 and $X_3 = 1.5$ dB at site 3.

The first evaluation of the antennas shown in Figure 7.33 consists of measuring the MEG for each element with and without the presence of the operator. In the measurement setup, the handset under test was mounted to a Styrofoam cross-member attached to a Styrofoam mast, approximately 2 m above the ground, affixed to the cart carrying the receiver subsystem. The handset was oriented at 60 deg from vertical in order to allow comparisons between performance with and without the tissue. One of the handset antennas was connected to the receiver, while the second was terminated in a 50Ω load. The second receiver branch was connected to a vertically polarized half-wave dipole antenna positioned 50 cm from the handset on the Styrofoam cross-member. Following completion of measurements for this configuration, the handset was removed from the mast and held by an operator to the right ear again at an angle of 60 deg. The operator walked in front of the cart (approximately 50 cm from the cart-mounted dipole antenna) as the measurement was performed.

From the acquired data, the MEG was computed using the formula [31]

$$\text{MEG} = \frac{\overline{P}_{\text{AUT}}}{\overline{P}_d} \tag{7.25}$$

where $\overline{P}_{\text{AUT}}$ and $\overline{P}_d$ represent the time-average power received by the antenna under test and the dipole, respectively. It should be noted that in the FDTD simulations, the computed MEG is normalized by the computed MEG of a dipole antenna to allow comparisons between measured and simulated data.

To measure the envelope correlation coefficient, the two handset antennas were simultaneously connected to the receiver system. The handset (rotated 60 deg) was either mounted on the mast or held by the user. If V_1 and V_2 represent voltage *envelopes* collected at antennas 1 and 2, respectively, the envelope correlation coefficient is given as

$$\rho_e = \frac{E[(V_1 - \overline{V}_1)(V_2 - \overline{V}_2)]}{\sqrt{E[(V_1 - \overline{V}_1)^2]E[(V_2 - \overline{V}_2)^2]}} \tag{7.26}$$

where $\overline{V}_i$ represents a time average of voltage envelope V_i.

7.1.7.4 Results

Table 7.7 provides a summary of the computed and measured MEG values for each of the different handset antennas in the three environments under consideration. The results show reasonable agreement between the two sets of data. The discrepancies most likely arise from incorrect modeling assumptions in the MEG simulation model. Additionally, the values provided in Table 7.7 are similar to values reported in the literature for antenna/handset configurations of similar construction [26,31,32]. The important finding of Table 7.7 is the uniformity of MEG for antennas on a single handset. With few exceptions, the values for different antennas on a handset remain within 1.5 dB and are generally within 0.5 dB. The similarity of these values is essential for achievement of high diversity gain. Additionally, the data show that the tissue reduces the MEG by between 3 and 6 dB. Finally, it is noteworthy that the widest variability in MEG occurs at site 2, most likely due to the multipath structure observed there.

Table 7.8 shows measured and computed values of ρ_e for the two different handsets at the three different measurement sites. Three different observations can be made concerning these data. First, both measurements and simulations show that correlation coefficients for each handset are very low (generally less than 0.2). This implies that both handsets will offer reasonable diversity performance. Second, while simulation shows that the operator tissue always reduces ρ_e, measurements indicate that the operator can actually increase the envelope correlation, perhaps because of strong mutual antenna interactions induced by the high-permittivity tissues in the head. Finally, while the simulation approach precludes the possibility of negative correlation coefficients, Table 7.8 indicates that measured values of ρ_e can be less than zero. This occurrence of negative correlation

Table 7.7

Measured and Simulated Values of MEG for Individual Handset Antennas With and Without the User Present

			Handset 1		Handset 2	
			Monopole	Mouthpiece	PIFA	Mouthpiece
Site 1	No user	Measured	−2.32	−2.90	−2.66	−2.01
		Simulated	−3.20	−3.26	−3.47	−4.24
	User	Measured	−5.44	−5.67	−6.55	−5.83
		Simulated	−7.63	−7.60	−6.71	−7.52
Site 2	No user	Measured	−1.27	−2.35	−2.00	−0.89
		Simulated	−3.20	−3.26	−3.47	−4.24
	User	Measured	−7.17	−5.46	−7.06	−7.33
		Simulated	−7.63	−7.60	−6.71	−7.52
Site 3	No user	Measured	−0.82	−0.99	−1.81	−1.61
		Simulated	−1.70	−1.81	−2.12	−2.41
	User	Measured	−4.89	−4.10	−5.61	−5.95
		Simulated	−5.92	−5.60	−5.36	−5.64

Table 7.8

Measured and Simulated Values of Envelope Correlation Coefficients for the Handset Branches With and Without the User Present

			Handset 1	*Handset 2*
Site 1	No user	Measured	0.1262	−0.0361
		Simulated	0.0377	0.1329
	User	Measured	0.0758	0.1246
		Simulated	0.0006	0.0008
Site 2	No user	Measured	0.2866	0.0138
		Simulated	0.0377	0.1329
	User	Measured	0.0908	0.0521
		Simulated	0.0006	0.0008
Site 3	No user	Measured	0.0965	0.1147
		Simulated	0.0614	0.1641
	User	Measured	0.2719	0.2979
		Simulated	0.0015	0.0043

coefficients implies the possibility of even higher diversity gain than what is predicted by standard theoretical approaches which assume that $\rho_e > 0$. However, it is notable that only one negative value appears, and its magnitude is relatively small. It is conceivable that this negative value would become positive if more data were used in the computation.

The preceding results indicate that the antennas of Figure 7.33 possess similar MEG values and low values of ρ_e, suggesting that they should offer good diversity performance. It is interesting to assess this performance by directly measuring the diversity gain for each handset in the different environments. This is accomplished by taking the experimental data collected for determination of ρ_e and forming signal-to-noise ratio (SNR) histograms for each antenna individually and then for a diversity combined system. In this case, only selection combining is used such that the sequence used for the diversity histogram is obtained by comparing the SNR reading at the two antennas for each sample and choosing the largest value. Diversity gain is then obtained by computing the cumulative density function (CDF) from each histogram. The gain is defined as the difference in SNR between the single-branch (highest of the two branch signals is chosen) and combined signals at a given probability level.

Table 7.9 illustrates the diversity performance of the two handsets at three different reliability levels. The values represent averages of three different measurements for each site. The theoretical results are taken from curves in [25] for zero branch correlation. The "average" values are obtained by averaging over all measurement sites and both users. As can be seen, selection combining offers a gain of 7 10 dB at the 99% level. It is noteworthy that while the tissue tends to lower the antenna MEG, it has little impact on the diversity gain of the system. Finally, the results indicate that the theoretical gains obtained from ideal branches are reasonably close to what has been observed in these measurements. Discrepancies in these values may arise from the nonequal effective antenna

Table 7.9
Averaged Diversity Gain Values for Both Handsets at Three Reliability Levels

		Handset 1			Handset 2		
		90%	95%	99%	90%	95%	99%
Site 1	No user	3.8	5.0	7.9	5.6	7.6	9.8
	User 1	4.7	5.6	6.6	4.1	4.7	5.6
	User 2	4.3	5.3	5.9	4.9	6.8	10.0
Site 2	No user	4.4	5.6	8.4	5.1	6.0	8.9
	User 1	4.4	5.4	5.4	4.4	4.7	4.5
	User 2	4.0	5.2	6.4	4.3	4.7	5.3
Site 3	No user	5.0	6.4	9.8	5.0	6.4	9.8
	User 1	4.6	5.5	9.1	5.7	6.9	11.5
	User 2	4.3	5.9	11.2	3.8	5.2	9.4
Average	No user	4.4	5.6	8.7	5.2	6.7	9.5
	User	4.4	5.4	7.1	4.9	5.8	8.1
Theory ($\rho_e = 0$)		5.6	6.7	10.0	5.6	6.7	10.0

gains as well as the nonzero envelope correlation coefficients characterizing the handset antennas.

7.2 DESIGN AND PRACTICE OF ANTENNAS FOR HANDSETS

7.2.1 Design Concept

General design concepts for mobile communications antennas are described in Chapter 1. The technical subjects pertaining to designing the handset antennas are discussed in Chapter 6 and summarized in Figure 6.24. The essentials of designing handset antennas have not changed. However, as mobile communication systems have advanced, so has the design of antennas. Accordingly new antenna systems have been developed for various mobile systems, and the design has evolved, also.

It can be said that typical trends in mobile communications are the personalization, the globalization, and the utilization of multimedia services. Development of advanced systems, and thus the progress of antenna systems, is mainly attributed to these trends.

Use of mobile systems on an individual scale has promoted downsizing of hardware, particularly of mobile terminals. Weight and dimensions of handheld terminals have been greatly reduced. The trend toward downsizing of handsets is shown in Figure 1.5 in Chapter 1, where the smallest mobile phone in 1999 is shown to be less than 60 cc in volume and less than 60 grams in weight.

As the size of handsets tends to get smaller, the size of their antennas need to get smaller as well. This has presented a big challenge to antenna engineers. Ideally, system performance should not be degraded for the smaller handsets, nor should their antenna performance suffer by being smaller. Note that the smaller the handset unit gets, the greater the contribution of the antenna in determining what the system's performance might be. In other words, system performance would almost depend wholely on the antenna performance. Regardless of the quality in other parts of the system design, it can hardly expect them to compensate for degraded antenna performance. Antenna technologies such as modification of the antenna structure, use of traveling-wave structures, and integration of reactances with the antenna structure are applied to overcome the problem.

The small handsets may have made a great contibution to the rapid increase in the number of mobile phone subscribers. This increase is remarkable, particularly in Asia. Figure 1.1 in Chapter 1 shows the statistics in Japan as an example. As a consequence, efficient use of the frequency spectrum and reduction of interference have become serious requirements. Increasing the number of channels has become a worldwide problem and hence the frequency bands such as 1.5 GHz, 1.9 GHz, and even higher frequency regions, have been allocated for mobile phones, cordless phones, and so forth. Use of higher frequencies is rather favorable for antenna design, because the size of antennas can be reduced and they are easier to mount on the smaller handset unit.

With the increasing number of mobile systems users, hybrid systems of PDC and PHS, GSM and DECT, and GSM and GPS, and so forth, have been required. Also handsets that incorporate two system protocols, called dual-mode handsets, have been developed. In a similar manner, handsets that can operate in two frequency bands, such as the 800-MHz and 1.5-GHz bands, called dual-band handsets, have been developed. In the early stages of these systems, two antennas were used for two systems each or two frequencies each. However, recent handsets employ only a single antenna, which is designed to operate in either wide bandwidth or two separate frequency regions. Various types of antennas have been developed for that purpose.

Other problems that require serious consideration are the influence of the operator's hand and head on antenna performance and also the effect of handset radiation on the human brain. Both of these effects become more severe when smaller handsets are used. They may be operated closer to the human head compared to larger handsets. The amount of power absorbed by the human head may increase, and at the same time the influence of the operator's hand and head on antenna performance may be increased.

Many studies and investigations have been conducted into the problem of possible EM hazard on the human head due to the handset fields. In Sections 6.5, 7.1, and 7.2, that problem is treated with both theoretical and experimental aspects. In practice, the fields from the handset against the human head should be reduced. So far, some trials to decrease the fields toward the human head have been made. One antenna that accomplishes this is an array antenna, which radiates mostly in the opposite direction of the human head, and another is a shielding plate, which is placed between the head and an antenna element, mounted behind the shielding plate. However, the problem is still controversial and no definitive solution has been found yet.

It has been an accepted concept when designing an antenna for a handset that the handset unit will be included as a part of the radiator, so that antenna performance can be enhanced even though a very small antenna element is used. This is because of the currents on the handset unit, which are excited by the antenna element mounted on the handset unit. In practice, the performance of the antenna system presently used in handsets should be considered to be that of an antenna element combined with that of the handset unit.

In turn, as the currents on the handset unit are varied by the operator's hand and head, the antenna performance may degrade due to their influence. Thus, to reduce the influence of the operator's hand on the antenna performance the current distributions on the handset unit should be reduced. This concept contradicts the conventional design concept commonly used during the 1990s.

Use of an antenna that has balanced terminals and is fed with a balanced line has been studied as one of the effective ways to avoid the degradation of antenna performance due to the influence of the operator's hand. In this case, care should be taken that the decrease in current distributions on the handset unit will not result in lowering the gain and narrowing the bandwidth.

The globalization in mobile communications also has promoted advances in personalization and development of new types of handsets as well. The mobile satellite systems aim at the global communication link combined with terrestrial systems such as cellular systems, by using GEO, MEO, LEO, and HEO systems, and hence handset systems are required. The design concept of such handset systems is similar to that of cellular systems mentioned earlier. However, there are some differences for mobile satellite systems. Among these are requirements for the radiation pattern to have wide coverage, circular polarization, and be wideband (for some applications), in addition to the requirements for the ordinary handset such as downsizing. These problems are discussed and practical antennas are described in Section 9.6.2.

The multimedia systems in mobile communications developed so far are voice, data, and video transmission systems. Typical examples of these are PDA (personal data assistant), mobile computing systems, PHS installed video system, and so forth. In the PDA system, a monopole antenna has generally been used as the simplest and lowest cost antenna element. But the noise interference caused by the clock oscillator, which generates high harmonic frequencies and disturbs the system performance, should be eliminated. A similar problem observed in cellular handsets, in which the reduction of currents flowing on the handset unit should be reduced, must be seriously taken into consideration. Then it is possible that the concept developed to solve the cellular handset design problem could be applied to avoid the noise problem in the PDA systems. A half-wave sleeve dipole, which was used in early versions of mobile phone handsets, is very useful, because the current flow on the handset unit can be kept very small. The application of antenna systems using balanced terminals and fed with a balanced line is also useful for this purpose.

Video transmission requires wideband performance for the system, so corresponding wideband performance may be forced on the antenna systems. It would be essential, then, to select an antenna structure that has wideband characteristics, and to apply wideband technology throughout the antenna system. However, the technology of developing such wideband antennas is still considered to be immature, so further study on the problem is needed.

In designing antennas for handsets, the first considerations are generally that the antennas should have:

- small size;
- light weight;
- compact structure;
- low profile;
- robustness;
- flexibility;
- low cost;

plus additional requirements for specific applications. Parameters pertaining to the communication system, propagation problems, and environmental conditions are summarized in Figure 7.35. In fact, all of these parameters, more or less, should be taken into consideration in handset antenna design. Systems require particular specifications, so the antenna should be designed specifically to satisfy the requirements, depending on the systems. Propagation problems have presently been diversified to include complicated environments such as indoors, subway stations, underpasses, and so forth. Also, the problems now include frequency regions that reach into microwaves and millimeter waves.

To overcome various problems, functional antenna systems such as adaptive array antennas and signal processing antenna systems, in addition to diversity systems, have been studied and developed. Diversity systems applied to handset systems have been investigated intensively by many people, and parameters for obtaining effective performance have been discussed. Application of the adaptive array technology to handsets is the next problem.

Typical antenna elements used to date for handsets are monopole (MP), dipole, normal mode helix (NMHA), planar inverted-F (PIFA), microstrip patch (MSA), and meander line. The majority of handsets use an MP as the main element and a PIFA as the subelement, which is the pair element of the diversity antenna. Variations of these elements modified from the original structure have been developed and used in practice. Combined use of two elements, such as an MP and an NMHA or a meander line, is rather common in most of the handsets.

In the following sections, practical antennas used in handsets for mobile systems such as GSM, PDC, and PHS are described.

7.2.2 Antennas for GSM

7.2.2.1 General

Basically the antenna requirements are independent of the phone systems but user demands seem to differ anyway. GSM in Europe seems to use less extendible antennas than PDC

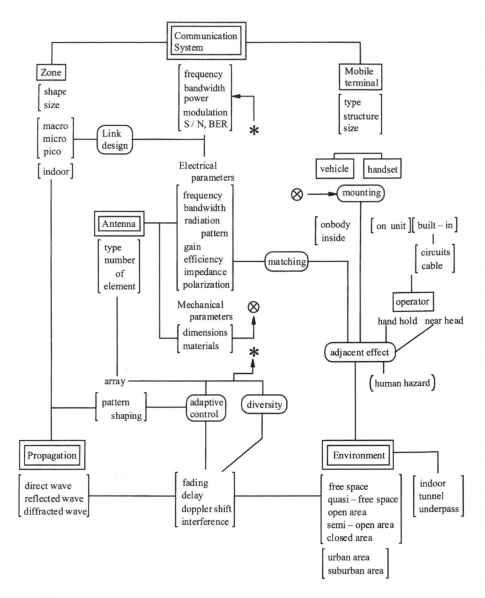

Figure 7.35 Parameters considered in designing handset antennas.

in Japan and AMPS in the United States. Cellular handset antennas from the beginning were developed from communication radio antennas at lower frequencies. A quarter-wave whip was the original type, but it was realized early on that the quarter-wave whip could be shortened considerably by a distributed inductive loading. A typical antenna of that type is made of a spiral enclosed in plastic or rubber with a total length of 4–15% of a wavelength, and on communication radios it is known under the nickname "rubber duck." Electrically these are still $\lambda/4$ whips, which are tuned to a shorter length by distributed inductive loading (=helically wound wire). Normal mode helical antenna (NMHA) is the conventional name for this common antenna. For all such basically quarter-wave antennas, the body of the phone (and partly also the user) is a very important part of the antenna function. Electrically this antenna is an asymmetrically fed half-wave antenna, where the feeding point is located where the antenna element enters the phone body. At the widely used phone bands within 800–1,000 MHz, the phone itself has a length fairly similar to $\lambda/2$ and a short stubby antenna element is commonly used. For a low frequency (starting with communication radios below 100 MHz), this is a necessary opportunity to extend the electrical size of the antenna, but the obvious disadvantage is that the main antenna current is flowing through the phone body and thus also partly through the user. This causes losses and it has also been questioned whether at high powers this could cause undesirable medical influences. Because of these reasons, some manufacturers have instead used half-wave whips having a high feeding impedance and corresponding low currents along the telephone body. The half-wave whip can also be made shorter than $\lambda/2$ by a distributed inductive loading at the "cost" of a smaller but still sufficient bandwidth. The still rather long whip can usually be retracted into the phone body to simplify transportation. A number of methods have been used to maintain a reasonable antenna function at least for paging purposes with the antenna retracted, and typically some coupling device is used to switch to a short NMHA antenna element mechanically integrated with the extendible whip. A very basic division of cellular handset antennas after their function is shown in Table 7.10.

Table 7.10
Basic Classification of Cellular Antennas

	Movable Geometry	Fixed Geometry
Passive phone body (i.e., small or no currents on the phone body)	Extendible $\lambda/2$ whip passing through a stub; the $\lambda/2$ whip can be shortened by helical shape	Fixed $\lambda/2$ whip or fixed sleeve antenna
Active phone body (=carrying antenna current)	Extendible $\lambda/4$ whip combined with a stub or a built-in antenna; short helix (stub) atop an extendible whip	Fixed $\lambda/4$ monopole; short helix (NMHA) or some other inductive loading; built-in antenna element

The switching function mentioned in Table 7.10 is in some constructions made galvanic but in others capacitively or by changing a resonance. With the more widespread use of cellular phones and with the steadily improved cellular networks, more emphasis is presently being placed on a handy design rather than on maximum performance, and thus most new developments within GSM are put in the lowest square to the right in Table 7.10, that is, fixed antennas where the phone body itself is an active part of the antenna. Still a few decibels can be gained by principles indicated in the upper left corner, that is, extendible antennas, especially of the $\lambda/2$ type. This is of course especially useful in countries or areas with poor cellular coverage. When a NMHA and a whip are combined, the helical part can either be located at the top or the bottom of the whip. With an extended whip the NMHA will either be detuned or decoupled.

NMHA antennas are classically done by a wire helically wound on a dielectric core. Any whip or NMHA can be seen as a kind of transmission line with the second conductor being the corresponding ground plane or more figuratively the phone body. With a straight wire without dielectric material the wave velocity along this transmission line will be like that of light in a vacuum. With the helix shape the combined inductive and dielectric loading decreases the wave velocity considerably making the tuning like that of a $\lambda/4$ whip in spite of the physical length being reduced to 6–13% of a wavelength. The radiation resistance, on the other hand, depends mainly on the exterior physical length and only slightly on details of the helical winding. The axial current distribution will be a bit different giving up to 2.5 times higher radiation resistance for the NMHA as compared to a straight monopole of the same physical length. The NMHA is thus more wideband than a straight monopole of the same length which is tuned by an internal inductance. To illustrate the influence of the length a comparison can be made with the radiation resistance of a monopole on a big ground plane having an electrical length $x = kL$ expressed in radians, where L is the length of the monopole and k ($= 2\pi/\lambda$) the wave number. The radiation resistance R_s is approximately $10x^2(1 + 0.19x^2)$ ohms for a monopole having a length from very short up to a quarter of a wavelength [33]. With $x = 1.57$ this corresponds to a $\lambda/4$ whip having 36Ω radiation resistance. Figure 7.36 shows the variation given as a function of the length as compared to a wavelength. For quarter-wave dipole (25%) the radiation resistance is 36Ω and the figure clearly indicates the impractical small resistances for a very short monopole. On a phone the expression for the radiation resistance will be more complicated but still rapidly decreases for a short antenna element. The radiation resistance as measured on a phone will typically be several times bigger than the values given in Figure 7.36, which is very important for the bandwidth. With the shorter length the matching to 50Ω may be more demanding but the most important is that the bandwidth will decrease in proportion to the radiation resistance. Practically, the length of a NMHA is 20–40 mm for AMPS or GSM 900 and the minimum length is determined by the need for bandwidth (8–10%). Because the phone body is a part of the radiating structure, the tuning of the $\lambda/4$ stub will depend on the size and shape of the phone. Designwise the length is sometimes disguised in the shape of the phone.

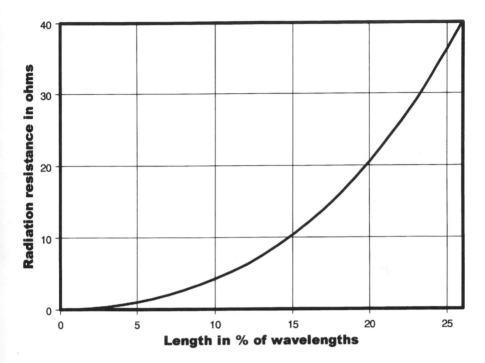

Figure 7.36 Radiation resistance for a monopole on a big ground plane.

7.2.2.2 Helical Antennas with and Without a Whip

Figures 7.37 and 7.38 shows a typical helical antenna with an extendable whip having the bottom helix configuration. This is a classical and still common antenna type as it combines performance in extended position regardless to how the phone is gripped with a handy retracted position optimized for paging. On this particular telephone, the antenna is developed for dual-band service. The top of the whip is nonmetallic so when the whip is retracted the function is identical to that of a fixed NMHA. The whip has a metallic connection in its bottom so when the whip is extended the whip will be connected in parallel to the helix. A part of the whip passes through the helix and will detune it so only the whip itself will be fed in that case. By this arrangement the antenna will be tuned to approximately a real input impedance in both retracted and extended positions, but the radiation resistance is lower in the retracted position as can be expected from the approximate formula given above.

By a suitable matching network located on the printed circuit board of the phone typically made as shown in Figure 7.39, the matching can be made to an average value of retracted and extended to give an acceptable VSWR in both cases. Specified VSWR may be 1:2 or 1:2.5 over the bands and modes of use. In practical use, there are two

(a)

Figure 7.37 Bottom helix: (a) exploded (ext) view and (b) retracted (ret). (Courtesy of Allgon.)

(b)

Figure 7.37 (continued).

Figure 7.38 Bottom helix antenna (extended position). (Courtesy of Allgon.)

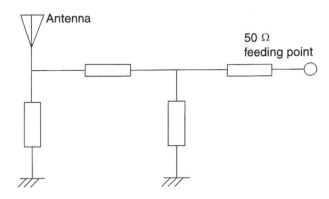

Figure 7.39 Typical antenna matching network.

further application cases as the phone has to work in both free space (FS) and in real use close to the head of the user (TP = "talk position"). The matching network is determined to give an acceptable worst case of the four combinations of TP/FS and extended/retracted. Most new phones are made for two or more bands, further increasing the need for versatility. The antenna design of the antenna in Figures 7.37 and 7.38 are shown in a number of diagrams condensed in Figure 7.40 (for 915 MHz) and Figure 7.41 (for 1,785 MHz). Each figure is arranged into four rows (a, b, c, and d) showing FS extended, TP extended, FS retracted, and TP retracted, respectively. For each case "elevation" and "azimuthal" (= seen from the top of the phone) diagrams are shown to the left and right, respectively.

At the low frequency (915 MHz) the antenna pattern in free space is very similar to that of a dipole antenna and the peak antenna gain is similar in the two positions. The talk position is of course more important for practical use and it can be noted that most practical phones suffer from a several decibel loss as compared to free space. Another change is that the head in talk position gives a considerable "shadow" as can be seen from the azimuthal diagrams Figures 7.40(b) and (d). At the higher frequency (1,785 MHz) the similarity with a simple dipole is less obvious simply because the phone now is much bigger than $\lambda/2$. Figure 7.42 is a 3D plot at the higher frequency further illustrating this.

Many measurements of average gains on phones on the market have been done using various methods. As compared to an ideal case (like a $\lambda/2$-dipole in free space) a very big spread in the average gain has been verified from different sources. Variation between different phones and different users in the range of −4 to −16 dB as compared to the ideal case has been verified. To improve performance, a small attenuation is desirable but only a few conclusions can be drawn from practical phones so far. The higher frequency bands are slightly less sensitive to the influence of the user and this is due to the fact that the distance to the skin of the user expressed in wavelengths is an important parameter. Bigger phones and extended antennas also have a few decibels lower attenuation but

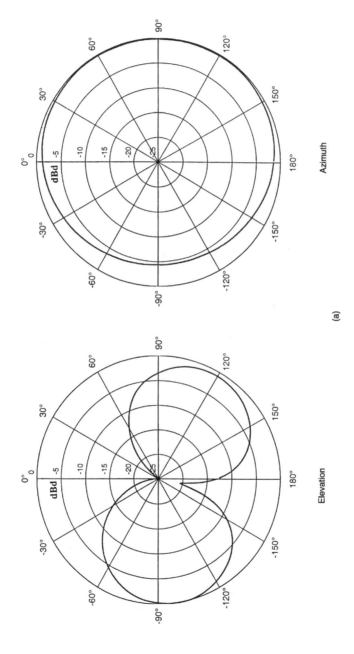

Figure 7.40 Antenna patterns at 915 MHz for (a) FS ext, (b) TP ext, (c) FS ret, and (d) TP ret. (Courtesy of Allgon.)

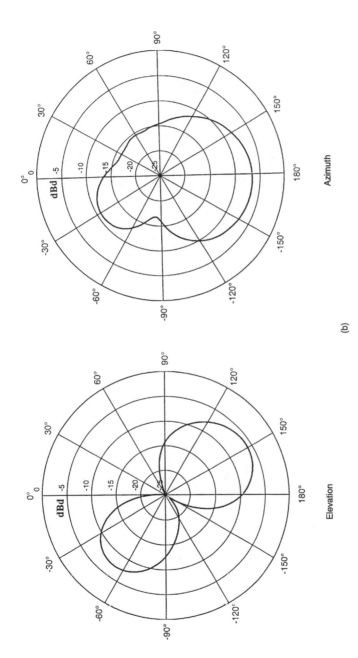

Azimuth

(b)

Elevation

Figure 7.40 (continued).

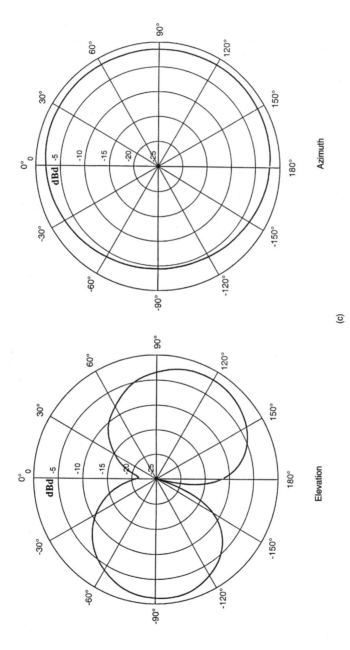

(c)

Figure 7.40 (continued).

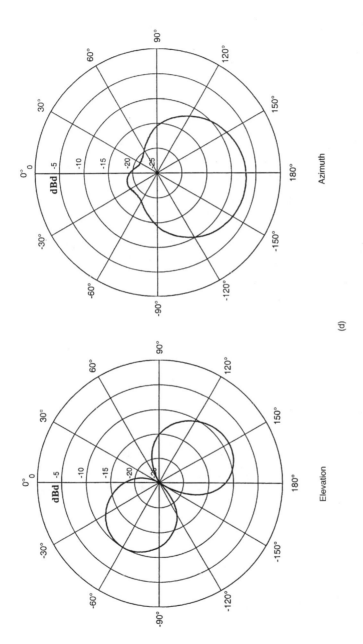

(d)

Azimuth

Elevation

Figure 7.40 (continued).

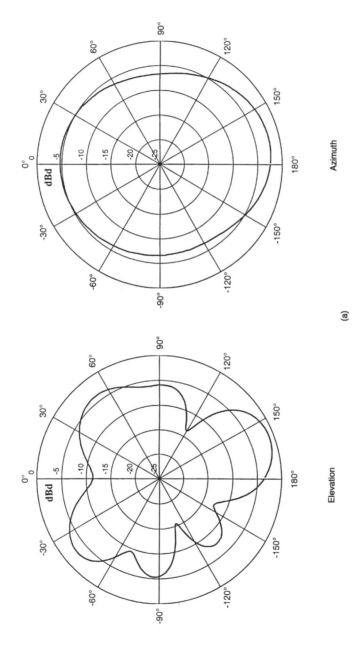

Elevation

Azimuth

(a)

Figure 7.41 Antenna patterns at 1,785 MHz for (a) FS ext, (b) TP ext, (c) FS ret, and (d) TP ret. (Courtesy of Allgon.)

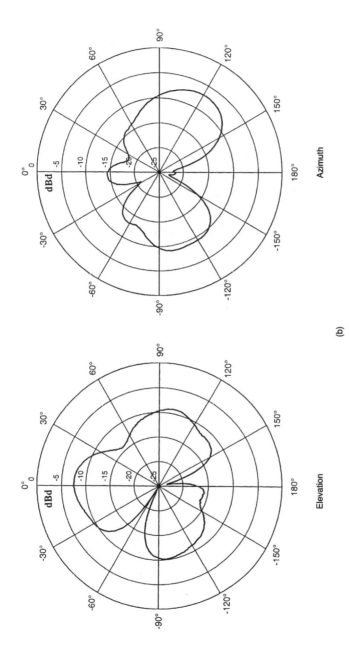

(b)

Azimuth

Elevation

Figure 7.41 (continued).

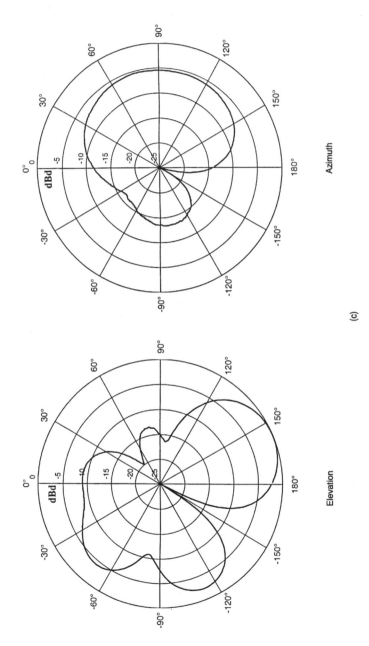

(c)

Figure 7.41 (continued).

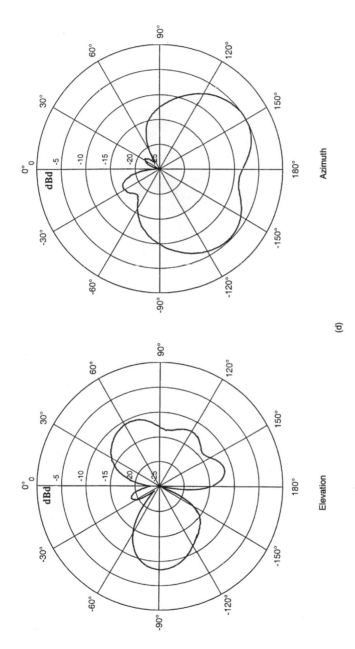

(d)

Figure 7.41 (continued).

3D antenna gain pattern in talk position, θ pol.

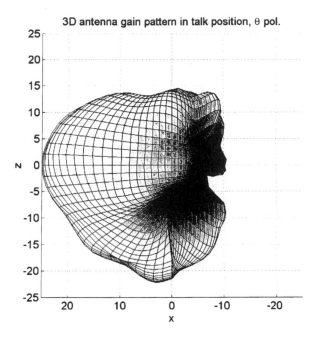

Figure 7.42 3D antenna pattern for extended position.

obviously this is a compromise to other user requirements. A slight azimuthal asymmetry away from the head is further useful because it tends to decrease the losses in the head.

7.2.2.3 Meanders and Dual-Band Antennas with and Without a Whip

The classical helical shape of the conductor inside a NMHA is an inductive loading to simplify the tuning in spite of the short length, but the same slow-wave effect can be obtained in many other ways. Instead of a helical wire, a printed meander pattern has proven to be very useful. Due to the fabrication technique, virtually any meander shape can be obtained at the same cost and this can be used to create multiband operation, matching networks, and so on. One of the ways to accomplish multiband operation is to connect two or more $\lambda/4$ meanders in parallel, each tuned to its own frequency. In recent years two or three frequency bands are very common, with GSM 900, 1,800, and 1,900 being a typical case, allowing, for instance, both dual-band operation in GSM countries as well as in PCS 1,900 areas in the United States. Due to the quadratic growth of the radiation resistance with the frequency not only the absolute but also the relative bandwidth at higher frequencies will generally be bigger. Thus both GSM 1,800 and GSM 1,900 may very well be contained within the higher band while GSM 900 may be the biggest difficulty to contain if the stubby antenna is short.

Figure 7.43 shows a dual-band meander antenna corresponding to a fixed NMHA. The meander pattern is printed on a small piece of flexible board (shown flat in Figure 7.44), which is rolled on a core as shown in Figure 7.43. Due to the many possibilities for shaping the pattern, better optimization can be obtained as compared to an antenna

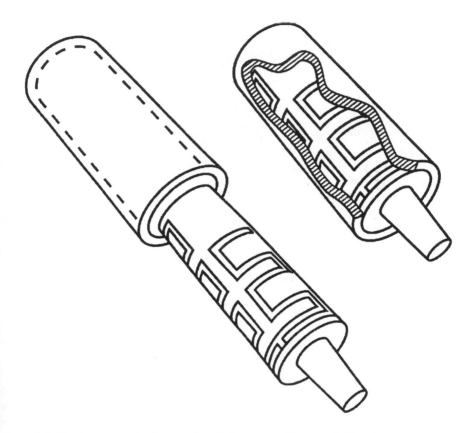

Figure 7.43 Fixed antenna made of a meander rolled on a core. (Courtesy of Allgon.)

Figure 7.44 The meander patch of Figure 7.43 on a ribbon before assembly. (Courtesy of Allgon.)

made of helical wire, especially for multiband operation. Different patterns can be used such as meander patterns, fractal patterns, and patterns generated by genetic algorithms. In all of these cases the added inductance is the important factor for making the small antenna usable at the lowest frequency band, while details of the pattern are more important for getting good multiband performance. There are helical wire antennas too for a few dual-band phones but as the tuning is more dependent on the shape of the phone than on the antenna element the possibilities of getting a good match for a pure helical configuration are not always present. The number of possible parameters to adjust for adapting the antenna element to the shape of the phone is simply too small for a common helical antenna.

7.2.2.4 Built-In Antennas

Many new GSM phones have built-in antennas in the sense that the antennas are not visible from the exterior shape of the phone. Starting from the radiating function of the full telephone there is (at least on the 800- to 1,000-MHz bands) very little difference in whether the antenna elements are visible of not because in both cases it is the phone body that is the main radiation source. And, at any rate, the RF radiating is not an optical phenomenon depending on the visibility. In both cases the antenna element is a kind of feeding structure to induce currents on the phone body in order to get the RF radiation. At the higher frequency bands (GSM 1,800/1,900) the antenna element might work with less help from the phone body, but generally the same antenna element is used both for the lower and higher frequency band.

The helical or meandered stubby antenna can be seen as the top of the phone and a similar function can be obtained by some kind of patch on the phone. The length of the stubby antenna can be said to correspond to the surface of the patch conformal to the phone. The very top of the phone is not the best place but generally a surface on the upper back of the phone is preferred. The main reason is that the near field very close to the antenna element should be kept away from the user to avoid unnecessary losses there. The antenna element should not be located too low on the back of the phone either as such a location can increase the losses in the hand of the user. The losses in the user are generally measured as loss density (loss per gram), which makes a bigger antenna element more favorable in this way simply by spreading the same power out but also by increasing the radiation resistance (or conductance). A similar improvement applies to the radiation efficiency.

Various design requirements go in the opposite direction so the size of the antenna element will always be a compromise between function and design. The radiation properties (radiation resistance or conductance) as well as the near field around the antenna elements are mainly dependent on the surface of the antenna element but for the bandwidth and losses the most important characteristic is the volume of the antenna element as formulated by Wheeler in [33]. The bandwidth for any radiating structure can be deduced from the radiated power compared to the stored energy (Bode-Fanos law as in [34] and the volume as in [33]). Thus for the same fields on the outside (i.e., the same radiated power and also the same near fields) of the radiating structure the stored energy behind the radiating structure will be inversely proportional to the volume occupied by the internal field. In Chapter 6 of [33] the term power factor p is used for the product of efficiency and relative bandwidth. If a is the radius of the smallest sphere enclosing the antenna structure the maximum power factor p is deduced to be $p = (ka)^3/4.5$ where k is the wave number $2\pi/\lambda$. As a consequence of the Bode-Fanos law [34] matching circuits can give limited improvement only. For a VSWR limit of 1:2, the bandwidth may be improved by a factor of 2 at most compared to single tuned resonant circuit. A trivial observation is that the volume occupied by an internal antenna element will always be bigger than the internal volume occupied by an external antenna.

Figure 7.45 shows a built-in antenna that is basically a meandered PIFA element. The meander pattern adds inductance along the PIFA and can be seen as a better way to shrink the physical length than adding capacitance by using dielectrics. The typical PIFA can be seen as a quarter-wave stub, which when seen from the open end gives a real admittance at quarter-wave resonance. The physical length (electrical $\lambda/4$) can be decreased either by a capacitive loading (high ϵ) or by adding inductance (meandering, etc.). The characteristic impedance of the quarter-wave stub will be increased by the inductive loading but decreased by the capacitive loading while the radiating conductance will be the same. Thus the bandwidth will be much wider for the inductively loaded case (meander,

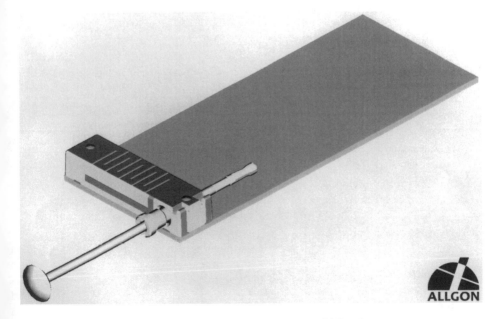

Figure 7.45 Built-in meander antenna with optional whip. (Courtesy of Allgon.)

etc.) compared to the case with high ϵ. Again the volume will be the critical parameter and in both cases the bandwidth will be reduced compared to the full-size PIFA. The meandering technique has a limitation in that losses will increase if lines that are too narrow are used. The built-in antenna shown in Figure 7.45 also has an optional whip, which can be extended to decrease the influence of the hand of the user. In this case the whip is fully retracted into the phone body combining the advantages of the built-in antenna with the performance of the extended whip.

Figure 7.46 shows an exploded view of a generic phone with a built-in antenna. The antenna element covers a part of the upper back of the phone. The volume between the antenna element and the screenings on the PCB is one important quantity used to characterize the antenna. As a consequence of Wheeler's antenna bandwidth limitation (discussed above and in Chapter 6 of [33], including many references) the product of bandwidth and efficiency is related to said volume.

7.2.3 Antennas for PDC

PDC stands for the personal digital cellular system, which is named for the Japanese mobile phone systems. Two bands are used; the lower bands are 810–828 MHz and 940–958 MHz for receiving and transmitting, respectively, and the higher bands 1,477–1,501 MHz and 1,429–1,453 MHz for receiving and transmitting, respectively. An example of the handset is shown in Figure 7.47. Antennas used for PDC handsets are mainly composed of a whip, which is a monopole, and a built-in PIFA, as Figure 7.48 shows.

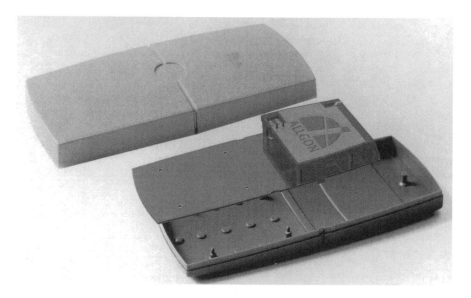

Figure 7.46 Generic telephone with built-in antenna. (Courtesy of Allgon.)

Figure 7.47 An example of PDC handset. (Courtesy of DoCoMo, Japan.)

Unlike other contemporary mobile phone systems, the PDC system improves reception quality by using a second antenna for diversity reception. As can be seen in Figures 7.47 and 7.48, the MP element (whip) has an NMHA on its top, which is encapsulated in a plastic cover. This NMHA is sometimes called a coil antenna. No electrical connection is made between the MP element and the NMHA element. In the transmitting/receiving mode, the MP element is extracted outside the handset case and fed at the bottom. The NMHA is not active in this situation, but only the MP element works. In the receiving mode, as the MP element is retracted inside the handset case and its bottom is disconnected from the feed terminals, only the NMHA acts as an antenna. These are shown in Figure 7.49, where (a) shows the MP element (whip) extracted from the unit and (b) retracted inside the unit.

The length of the MP element is preferably near $(1/2)\lambda$ (λ = wavelength), because the current flow on the handset unit may be small. This is discussed in [35] and [36], and also shown in Figure 7.50. However, the typical length used in practice is either $3/8\lambda$ or $5/8\lambda$, because with these lengths, relatively small current flows on the handset unit and better input impedance to match the load impedance are achieved than in the cases of $(1/2)\lambda$ and other lengths [36].

The typical length (extended) of the NMHA is about $(1/4)\lambda$. A length other than $(1/4)\lambda$ can be used if, for instance, a particular input impedance is specified. The gain of

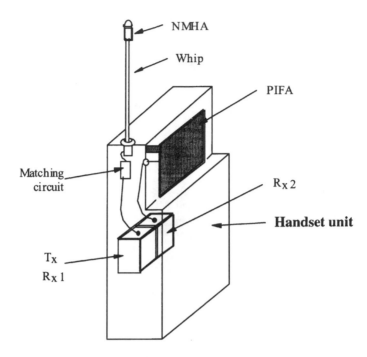

NMHA

Whip

PIFA

Matching circuit

$R_X 2$

Handset unit

T_X

$R_X 1$

Figure 7.48 Antenna elements used in a PDC handset.

this type of antenna is generally about −0.5 dBd at its peak and about −6 to −8 dBd average in the horizontal plane, when the length is about $(1/4)\lambda$. However, it varies depending on the length of the handset unit.

A PIFA element, a built-in antenna, is placed on the ground plane, which is actually a shielding plate for the RF circuits in the handset unit. The PIFA is employed as a pair of diversity elements with the MP element. The PIFA is an antenna of rather narrow bandwidth; that is, 1% to 2% in the relative bandwidth in free space. However, when it is mounted on a finite ground plane, the bandwidth usually increases. In practice, the bandwidth of an antenna system, which uses a built-in PIFA element in a handset unit, can be designed to have wider bandwidth than the whole spectrum bandwidth of the PDC system. This is attributed to the handset unit that acts as a part of the radiator due to the current flow excited by the PIFA element.

Consequently, the impedance of the antenna system, which consists of an antenna element and the handset unit, is varied by the effect of the handset unit. Figure 7.50 illustrates an example of the current distributions on an antenna model [37] that is composed of an MP element, a PIFA element, and the handset unit. Figure 7.51 shows the antenna model used for the analysis and the dimensions. In the analysis the handset unit is simulated by a rectangular conducting box and modeled with wiregrids; the frequency is 900 MHz

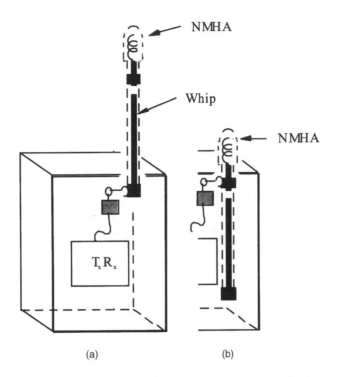

Figure 7.49 MP element: (a) extracted from the handset unit; (b) retracted in the handset unit.

[38]. Figure 7.50 shows the current distributions for cases where (a) the length Lw of the MP = 83 mm ($\lambda/4$) and the length Lz of the handset unit = 125 mm, (b) Lw = 180 mm (0.54λ) and Lz = 125 mm, and (c) Lw = 83 mm ($\lambda/4$) and Lz = 83 mm. The PIFA, 60×35 (mm) and placed 12 mm apart from the handset unit surface, is assumed to be the same for all cases. Amplitudes of the current distributions are illustrated on each wire of the wire grid. The method of moments is applied to obtain these distributions. It has been shown that with the MP element of $\lambda/4$, much current flow on the handset unit is observed, while with that of 0.54λ (nearly a half-wavelength) almost no current flow is observed. A similar result is mentioned in Section 7.2.1 [35]. This is a very important matter to consider when designing handset antennas.

An example of the input impedance of a similar antenna model is shown in Figure 7.52(a) [39]. The equivalent circuit of this antenna model is shown in Figure 7.52(b). It is interesting to note that in the equivalent circuit there is a reactance jX. The bandwidth enhacement of an antenna system, being composed of a PIFA element placed on the handset unit, may attribute to this reactance. This reactance has constant impedance over much wider bandwidth than that of the antenna system.

The frequency range of the PDC systems in the 800-MHz band is very wide, beginning from 810 MHz, the lowest frequency used in the base stations and ending at

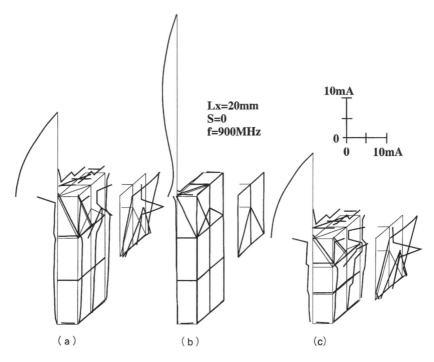

Figure 7.50 Current distributions on the antenna structure (after [37]): (a) Lw = 83 mm ($1/4\lambda$); Lz = 125 mm; (b) Lw = 180 mm (0.54λ); Lz = 125 mm; (c) Lw = 83 mm and Lz = 83 mm. The PIFA has the size of 60 × 35 (mm) and is placed 12 mm apart from the handset unit (parameters are shown in Figure 7.51).

958 MHz, the highest frequency used in the mobile stations. Thus the antenna system should cover the total bandwidth of 148 MHz, which corresponds to the relative bandwidth of about 17%. Matching circuits and various modified structures of PIFA element have been contrived in order to achieve this wide bandwidth.

Some examples are shown in Figure 7.53 [40]. Figure 7.53 illustrates (a) a matching circuit placed at the input terminals of the antenna structure, (b) a perturbation element added to the end of the planar element, (c) short pin(s) used at the edge(s) of the planar element, (d) a coaxial line attached to the edge of the planar element, instead of a lumped impedance, (e) two planar elements for the separate resonance of two frequencies, and (f) an additional parasitic planar element placed over the planar element. Variations of these are used in practice.

One of the recent trends in mobile communication is to have dual-mode handsets that can be used for both 800-MHz and 1.5-GHz bands and thus dual-frequency antennas are needed. To achieve dual-frequency operation, technologies similar to those for achieving wide bandwidth are used. Some of them are shown in Figure 7.54 and 7.55. Figure

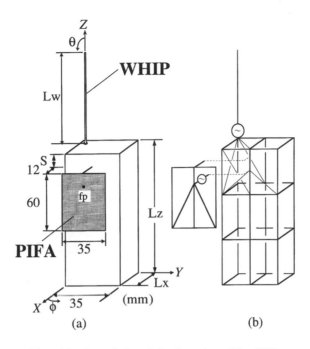

Figure 7.51 Antenna model used for the analysis and the dimensions. (After [38].)

7.54 shows (a) is a matching circuit and (b) is the typical VSWR characteristic. The two components L2 and C2 are specifically selected for dual-band operation. When the bandwidth is too wide to cover by a matching circuit, a switching circuit is used. An example of such a switching circuit is shown in Figure 7.55, where (a) shows a varactor circuit and (b) VSWR's for two frequencies.

As the size of handsets becomes smaller, antenna elements can also be made smaller. There are several ways to decrease antenna size. One way is to use traveling-wave structures such as a helix, meader line, as an antenna element so that the length of the antenna element is shorter. Another way is to design the matching circuit to match the impedance of a small sized antenna. Figure 7.56 [41] shows other examples that are variations of the antenna structure: (a) slit(s) on the planar element, which make(s) the resonance frequency lower, and (b) a metal plate attached to the end of the planar element, by which the length of the planar element can be shortened. Figure 7.56(c) is a case where a matching circuit is used. There are many variations of these in practical applications.

Antenna performance of a handset usually degrades at the talk position mainly due to the effects of the operator's hand and head. These problems were discussed earlier. The major reason for the degradation is the variation in the current flow on the handset unit. A typical example of distorted radiation patterns caused by the body effect is shown in Figure 7.57, where (a) shows a handset placed in free space and (b) at the talk position.

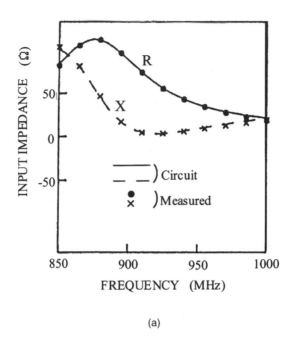

(a)

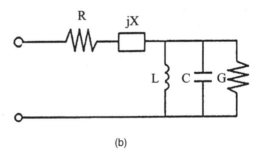

(b)

Figure 7.52 Input impedance of a PIFA on the ground plane (After [39].): (a) impedance characteristics; (b) equivalent circuit.

Because the current on the handset unit is produced by excitation of the built-in PIFA, and variations are caused mainly by the body effects, the degradation of antenna performance can be avoided by decreasing such current flow. The reduction of such current flow can be realized by means of an antenna system that has balanced terminals and is fed with a balanced line [42]. An example, which uses a loop antenna placed on

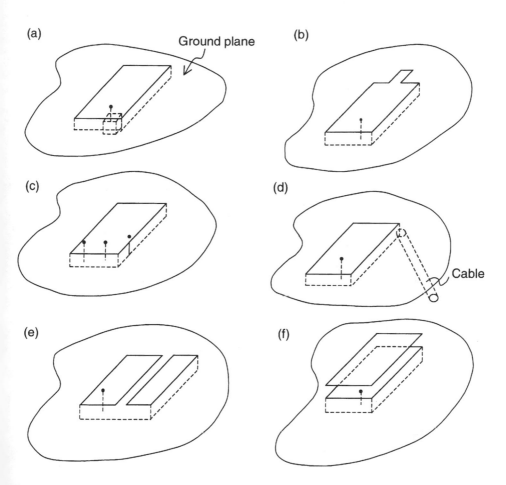

(a)

Ground plane

(b)

(c)

(d)

Cable

(e)

(f)

Figure 7.53 Configurations of planar element for achieving wide bandwidth. (After [40].)

a ground plane and fed with a two-wire line, is shown in Figure 7.58, in which the feed with an unbalanced line is expressed in (a), but a balanced line is used in (b). Figure 7.59 shows current distributions in an unbalanced structure (a) on the antenna element, (b) on the ground plane (the Y-axis), and (c) on the ground plane (the X-axis). Figure 7.60 shows similar current distributions in a balanced structure for the sake of comparison with the unbalanced structure. Figure 7.61 illustrates magnetic field distributions on the surface of the antenna system that correspond to current distributions. Figure 7.61(a) shows the distributions in the unbalanced structure and (b) that in the balanced structure [42].

We can see in Figure 7.61 that the significant reduction of current flow can be realized by means of the balanced antenna structure. The antenna element is not necessarily a loop for this purpose, but any other element such as NMHA, meander line, and so forth,

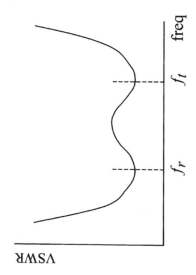

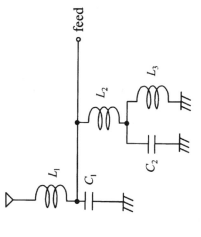

Figure 7.54 Example of a matching circuit for dual bands.

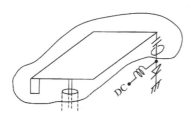 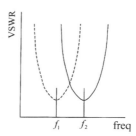

Figure 7.55 Switching circuit for dual-frequency resonance.

can be used, whenever they have balanced input terminals and are fed with a balanced line. One should be careful if an antenna system has very asymmetric structure, even when the terminals have balanced structure and are fed with a balanced line. With the asymmetric structure, there may be some difficulty in achieving really electrically balanced structure, and thus some unbalanced currents flow at the terminals.

Current antenna elements used for mobile terminals are MP, NMHA, PIFA, ceramic chip, meander line, MSA (microstrip antenna), and so forth. The majority of PDC mobile terminals employs an MP element, and a built-in PIFA as a pair element for a diversity antenna. An NMHA element placed on the top of an MP element was formerly used as an extension of the MP element in order to reduce the length of the MP; however, the design concept has been changed to use an NMHA element as the replacement of the MP, when the MP element is retracted into the handset unit. An NMHA element has been used as the main antenna since the time when high frequencies such as 1.5 GHz and higher were allocated.

PIFA elements vary greatly in practice. In fact, some PIFA elements can hardly be identified as such because of extensive modifications from the prototype.

A ceramic chip antenna, which is an NMHA element molded in the ceramic chip, has been used in some types of handsets. This is a type of small, compact, lightweight antenna. The radiation efficiency of this type of antenna is not very high and the bandwidth is inherently narrow. However, when it is placed on a conducting body of finite size, the gain and the bandwidth will be increased as a result of the radiation currents excited on the conducting body. This type of antenna has been practically applied to PHS handsets.

7.2.4 Antennas for PHS

PHS stands for the personal handy phone systems, which have been used in Japan as an extension of the indoor use of the cordless phone system to outdoor applications. The communication areas are limited to small zones; for example, zones several hundred meters in diameter. The transmitter power of the mobile terminals is kept as low as 10 mW. The frequency band is 1,895–1,918 MHz. Because the PHS mobile terminals are inexpensive, and are handled very simply and easily, the number of subscribers has

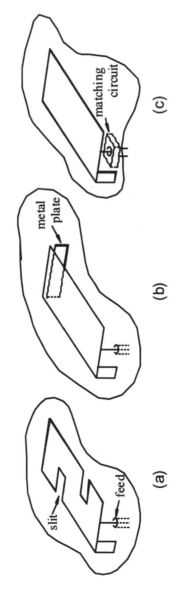

Figure 7.56 Small sizing of planar element. (After [41].)

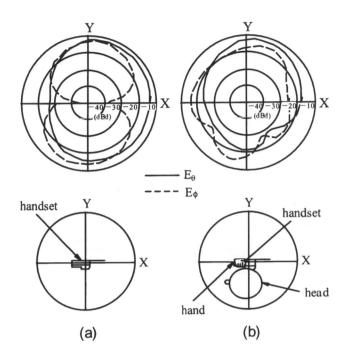

Figure 7.57 Radiation patterns: (a) in free space; (b) at talk position.

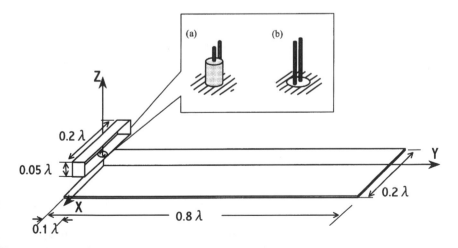

Figure 7.58 Loop antenna system: (a) unbalanced feed; (b) balanced feed. (After [42].)

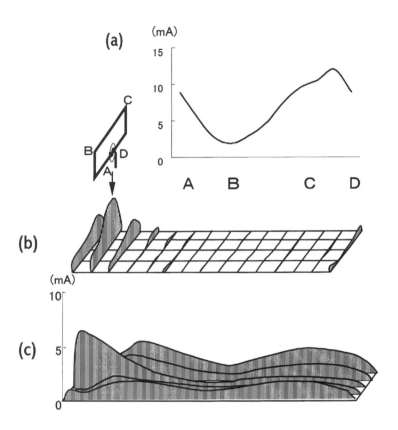

Figure 7.59 Current distributions in an unbalanced antenna system: (a) on the loop element; (b) on the ground plane (the *Y*-axis); (c) on the ground plane (the *X*-axis). (After [42].)

rapidly increased. However, the increasing trend is currently at a standstill, because the size and cost of PDC terminals have become comparable with those of PHS terminals. Note that faster data transmission is possible with a PHS compared with the PDC.

The PHS system employs the TDD system, and hence the same frequency can be used for both transmitting and receiving. The mobile terminals do not employ the diversity system. Therefore the mobile terminals use a single antenna element, which is typically an MP, an NMHA, or a small chip element.

The handset units are in general smaller than those of the PDC. Because the operating frequency is higher and the bandwidth is narrower than those in the PDC system, a smaller antenna element can be used. The antenna design is easier than that in the PDC systems. Presently, almost all of the PHS handsets use either an MP or an NMHA element. The MP element has a structure similar to that of the PDC; that is, the element is retracted down in the handset unit during the receiving mode, but extracted for the receiving/

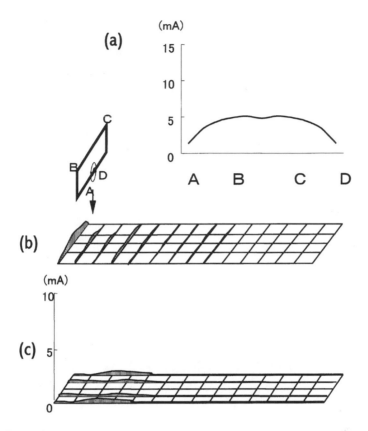

Figure 7.60 Current distributions in a balanced antenna system. (Similar to expressions in Figure 7.59.)

transmitting mode. The length of it is shorter than that of the PDC. The MP elements may be longer than $(3/8)\lambda$ or $(5/8)\lambda$, depending on the size of the handset unit.

When an NMHA is used, it can be fixed to the handset unit, as the length is not very long. A cross-sectional view of an NMHA is shown in Figure 7.62. A helical element is wound along the groove on the plastic top and covered by the plastic cap. The latest models do not use wire, but print the element on the groove. There is a case where a plastic sheet, on which the element is printed, is attached on the plastic top. A meander line can be used instead of an NMHA, an example of which is described in the previous section.

Another interesting element used in PHS is a chip antenna, which is a small normal mode helical element molded in a ceramic chip. Ceramic materials, having a high relative permeability of 20 or higher, are used. There are many variations in this type of antenna. Some examples are shown in Figure 7.63. An example of the inside view of a chip antenna is shown in Figure 7.64, where the chip dimensions are given. A model of a chip antenna

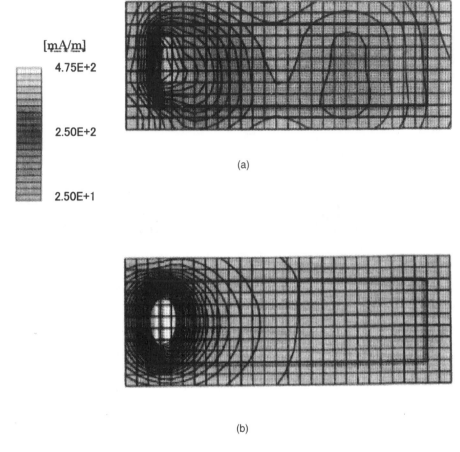

[mA/m]

4.75E+2

2.50E+2

2.50E+1

(a)

(b)

Figure 7.61 Measured current distributions of loop antenna system: (a) an unbalanced system; (b) a balanced system. (After [42].)

placed on a corner of the ground plane is shown in Figure 7.65. Figure 7.66 shows the return loss characteristic of this model, which indicates sharp resonance performance. Wider bandwidth may be possible by using either a matching circuit or a PIN diode circuit. A PIN diode circuit switches frequency so as to cover the necessary bandwidth. Figure 7.67 shows the radiation patterns. The gain is evaluated to be about −2 to −4 dBd at maximum and −4 to −7 dBd in average, depending on the plane measured. An example of a chip element mounted on a handset unit is illustrated in Figure 7.68. An NMHA is a pair element of the diversity antenna.

7.2.5 Diversity Performance in PDC Handsets

Diversity reception is accomplished by using an MP element and a PIFA element and has been applied to PDC handsets; it has also long been used in the analog mobile phone

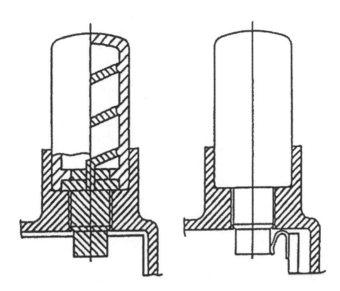

Figure 7.62 Cross-sectional view of an NMHA element. (Courtesy of Nippon Antenna Co., Ltd.)

systems. However, analysis of the correlation coefficient taking both MP and PIFA elements into consideration has been done rather recently [43,44]. In the analysis, the correlation coefficient ρ_e has been analyzed in relation to antenna structure, the placement of antenna element on the handset unit, incident wave, and handset operation status, in the frequency range of 900 MHz. The model used in the analysis is the same as that shown previously in Figure 7.51.

It has been discussed [44] that the correlation coefficient ρ_e as well as the radiation efficiency varies with the length of the MP, the size and position of the PIFA, the dimensions of the handset unit, the operation status of the handset, and so forth. The ρ_e is also concerned with the parameters of incident waves such as amplitude and phase, incident angle, cross-polarization ratio, and the slanted angle α of the handset unit at the talk position. The incident waves of both vertical and horizontal polarizations are assumed to have phase of uniform distribution and amplitudes of Rayleigh distribution and arrive from various directions with uniform distribution in azimuth and with Gaussian distribution in elevation, which has mean values of mV and mH, respectively, and standard deviations of σV and σH, respectively, where V and H denote vertical and horizontal, respectively. The cross-polarization ratio of the incident waves XPR is considered to be 6 dB here, which is taken from the statistics evaluated in the city area propagation in Japan.

Because two antenna elements, an MP and a PIFA, are placed very closely, with a separation of only about 0.1 to 0.2 wavelength, on the ground plane of the handset, the mutual coupling between them may degrade the radiation efficiency and also diversity function. The analyses have shown that there are optimum values for the length Lw of an MP and the length Lz of the handset unit in order to achieve low correlation coefficient ρ_e without causing much degradation in the radiation efficiency. When the arrival angle mV of the incident waves is equal to mH and lower than 20 deg, ρ_e is lower than 0.4 without

Figure 7.63 Chip antennas. (Courtesy of Murata Mfg. Co., Japan.)

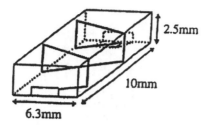

Figure 7.64 An example of the inside view of a ceramic chip antenna and the dimensions. (Courtesy of Murata Mfg. Co., Japan.)

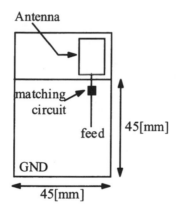

Figure 7.65 An antenna model placed on the ground plane. (Courtesy of Murata Mfg. Co., Japan.)

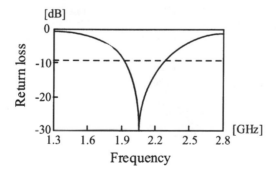

Figure 7.66 Return loss characteristics. (Courtesy of Murata Mfg. Co., Japan.)

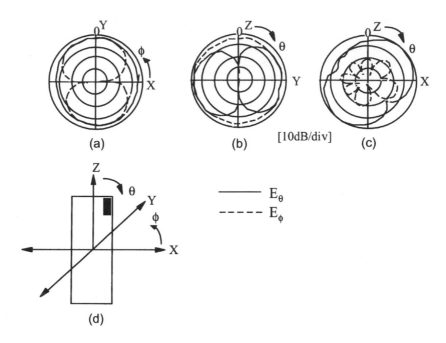

Figure 7.67 Radiation patterns. (Courtesy of Murata Mfg. Co., Japan.)

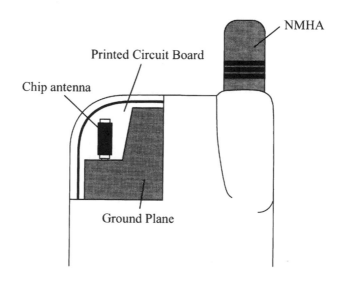

Figure 7.68 A chip antenna mounted at a corner of the handset unit. (Courtesy of Murata Mfg. Co., Japan.)

regard to the length Lw of the MP; however, when $mV = mH$ is greater than 40 deg, ρ_e becomes higher than 0.5 when Lw is longer than $(3/8)\lambda$, and is highest when Lw is 0.5λ.

Figure 7.69 shows the dependence of the correlation coefficient ρ_e on the slanted angle α of the handset at the talk position. As α increases, ρ_e decreases independently on mV and mH. In practice, $mV = mH$ is said to be less than 60 deg, so ρ_e is estimated to be less than 0.2 for α of nearly 60 deg, the average angle at the talk position. The length Lz of the handset unit does not affect ρ_e very much for any length Lw of the MP.

The diversity antenna performance depends not only on the correlation coefficient ρ_e, but also on the mean effective gain (MEG) of the antenna system. It has been shown in [43] that MEG of the antenna system with an MP depends much on the length Lw, slanted angle α of the handset, and the standard deviations σV and σH of the incident wave arrival angle distributions. The analysis has shown that the maximum MEG with the MP is -3 to -1.3 dBd when Lw of the MP is 0.5λ to 0.57λ, while the MEG with the PIFA is greater than -6 dBd and the maximum is -3.5 to -1.3 dBd when Lw is shorter than 0.5λ. The MEG does not depend on the slanted angle α when the cross-polarization ratio XPR is -2 to -1 dB. The MEG varies with the slanted angle α. This is shown in Figure 7.70, where (a) is for $Lw = (1/4)\lambda$ and (b) is for $(1/2)\lambda$, for cases with either an MP or a PIFA.

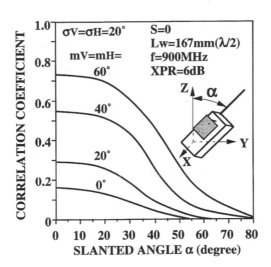

Figure 7.69 Correlation coefficient versus slanted angle of handset. (After [44].)

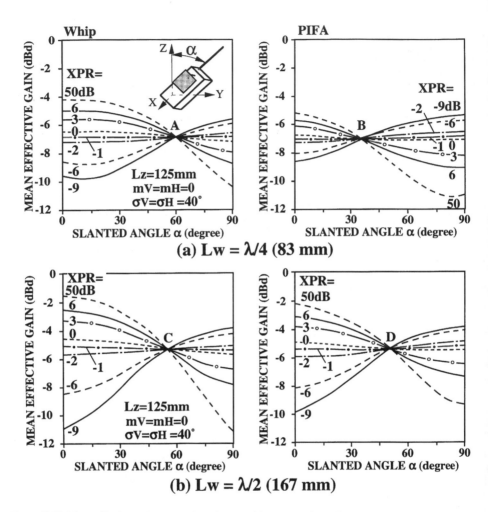

Figure 7.70 Mean effective gain versus slanted angle of handset. (After [43].)

REFERENCES

[1] Jensen, M. A., and Y. Rahmat-Samii, "EM Interaction of Handset Antennas and a Human in Personal Communications," *Proc. IEEE*, Vol. 83, No. 1, pp. 7–17, Jan. 1995.

[2] Rahmat-Samii, Y., and W. L. Stutzman, eds., "Special Issue on Wireless Communications," *IEEE Trans. Ant. Propagat.*, Vol. 46, June 1998.

[3] Rosen, A., and A. Vander Vorst, eds., "Special Issue on Medical Application and Biological Effects of RF/Microwaves," *IEEE Trans. on Microwave Theory and Tech.*, Vol. 44, Oct. 1996.

[4] Gandhi, O. P., "FDTD in Bioelectromagnetics: Safety Assessment and Medical Applications," in *Computational Electromagnetics: The Finite difference Time Domain Method*, A. Taflove, ed., Norwood, Mass.: Artech House, 1995.

[5] Colburn, J. S., and Y. Rahmat-Samii, "Electromagnetic Scattering and Radiation Involving Dielectric Objects," *J. Electromagnetic Waves and Applications*, Vol. 9, No. 10, pp. 1249–1277, 1995.

[6] Kim, K. W., and Y. Rahmat-Samii, "Antennas and Humans in Personal Communications: An Engineering Approach to the Interaction Evaluation," *Proc. IEEE Engineering in Medicine and Biology Society*, Chicago, Oct. 1997, pp. 2488–2491.

[7] Kim, K. W., and Y. Rahmat-Samii, "EM Interactions between Handheld Antennas and Human: Anatomical Head vs. Multi-layered Spherical Head," *IEEE-APS Conference on Antennas and Propagation for Wireless Communications*, Waltham, Mass., Nov. 2–4, 1998, pp. 69–72.

[8] Gandhi, O. P., G. Lazzi, and C. M. Furse, "Electromagnetic Absorption in the Human Head and Neck for Mobile Telephones at 835 and 1900 MHz," *IEEE Trans. Microwave Theory Tech.*, Vol. 44, pp. 1884–1897, Oct. 1996.

[9] Okoniewski, M., and M. A. Stuchly, "A Study of the Handset and Human Body Interaction," *IEEE Trans. Microwave Theory Tech.*, Vol. 44, pp. 1855–1864, Oct. 1996.

[10] Hombach, V., et al., "The Dependence of EM Energy Absorption upon Human Head Modeling at 900 MHz," *IEEE Trans. Microwave Theory Tech.*, Vol. 44, pp. 1865–1873, Oct. 1996.

[11] ANSI/IEEE C95.1-1992, *American National Standard—Safety Levels with Respect to Exposure to Radio Frequency Electromagnetic Fields, 3 kHz to 300 MHz.* New York: IEEE.

[12] Yee, K. S., "Numerical Solution of Initial Boundary Value Problems Involving Maxwell's Equations in Isotropic Media," *IEEE Trans. Antennas Propagat.*, Vol. AP-14, pp. 302–307, May 1966.

[13] Taflove, A., *Computational Electromagnetics: The Finite-Difference Time-Domain Method*, Norwood, Mass.: Artech House, 1995.

[14] Jensen M. A., and Y. Rahmat-Samii, "Performance Analysis of Antennas for Hand-Held Transceivers Using FDTD," *IEEE Trans. Antennas Propagat.*, Vol. 42, pp. 1106–1113, Aug. 1994.

[15] Mur, G., "Absorbing Boundary Conditions for the Finite-Difference Approximation of the Time-Domain Electromagnetic Field Equations," *IEEE Trans. Electromagnetic Compatibility*, Vol. 23, pp. 377–382, 1981.

[16] Berenger, J.-P., "A Perfectly Matched Layer for the Absorption of Electromagnetic Waves," *J. Computational Physics*, Vol. 114, pp. 185–200, 1994.

[17] Tai, Chen-To, *Dyadic Green Functions in Electromagnetic Theory*, 2nd ed., IEEE Press Series on Electromagnetic Waves, New York: IEEE Press, 1994.

[18] Gabriel, S., R. W. Lau, and C. Gabriel, "The Dielectric Properties of Biological Tissues: III. Parametric Models for the Dielectric Spectrum of Tissues," *Phys. Med. Biol.*, Vol. 41, pp. 2271–2293, Nov. 1996.

[19] Kim K. W., and Y. Rahmat-Samii, "Handset Antennas and Humans at Ka-band: The Importance of Directional Antennas," *IEEE Trans. Ant. Propagat.*, pp. 949–50, June, 1998.

[20] Toftgård, J., S. N. Hornsleth, and J. B. Andersen, "Effects on Portable Antennas of the Presence of a Person," *IEEE Trans. Antennas Propagat.*, Vol. 41, pp. 739–746, June 1993.

[21] Dimbylow, P. J., "FDTD Calculations of the SAR for a Dipole Closely Coupled to the Head at 900 MHz and 1.9 GHz," *Phys. Med. Biol.*, Vol. 38, pp. 361–368, Feb. 1993.

[22] Chuang, H. R., "Human Operator Coupling Effects on Radiation Characteristics of a Portable Communication Dipole Antenna," *IEEE Trans. Antennas Propagat.*, Vol. 42, pp. 556–560, Apr. 1994.

[23] Mumford, R., Q. Balzano, and T. Taga, "Land Mobile Antenna Systems II: Pagers, Portable Phones, and Safety," Ch. 4 of *Mobile Antenna Systems Handbook*, K. Fujimoto and J. R. James, eds., Norwood, Mass.: Artech House, 1994.

[24] Lee, W. C. Y., *Mobile Communications Engineering*, New York: John Wiley & Sons, 1982.

[25] Jakes, W. C., Jr., *Microwave Mobile Communications*, New York: John Wiley & Sons, 1974.

[26] Pedersen G. F., and S. Skjaerris, "Influence on Antenna Diversity for a Handheld Phone by the Presence of a Person," *Proc. 1997 47th IEEE Vehicular Technology Conference*, Phoenix, Ariz., May 4–7, 1997, Vol. 3, pp. 1768–1772.

[27] Green, B. M., and M. A. Jensen, "Diversity Performance of Personal Communications Handset Antennas Near Operator Tissue," *1997 IEEE AP-S Intl. Symp. Digest*, Montreal, Canada, July 13–18, 1997, Vol. 2, pp. 1182–1185.

[28] Colburn, J. S., et al., "Evaluation of Personal Communications Dual-Antenna Handset Diversity Performance," *IEEE Trans. Vehicular Technology*, Vol. 47, pp. 737–746, Aug. 1998.

[29] Taga, T., "Analysis for Mean Effective Gain of Mobile Antennas in Land Mobile Radio Environments," *IEEE Trans. Vehicular Technology*, Vol. 39, pp. 117–131, May 1990.

[30] Pierce, J. N., and S. Stein, "Multiple Diversity with Nonindependent Fading," *IRE Proc.*, pp. 89–104, Jan. 1960.

[31] Murase, M., Y. Tanaka, and H. Arai, "Propagation and Antenna Measurements Using Antenna Switching and Random Field Measurements," *IEEE Trans. Vehicular Technology*, Vol. 43, pp. 537–541, Aug. 1994.

[32] Arai, H., N. Igi, and H. Hanaoka, "Antenna-Gain Measurement of Handheld Terminals at 900 MHz," *IEEE Trans. Vehicular Technology*, Vol. 46, pp. 537–543, Aug. 1997.

[33] Johnson, R. C., *Antenna Engineering Handbook*, 3rd ed., McGraw-Hill, 1980.

[34] Matthaei, G., L. Young, and E. M. T. Jones, *Microwave Filters, Impedance-Matching Networks and Coupling Structures*, Norwood, Mass.: Artech House, 1980.

[35] Hirasawa, K., and K. Fujimoto, "Characteristics of Wire Antennas on a Rectangular Conducting Body," *Trans. IEICE*, Vol. J65-B, No. 9, 1982, pp. 1133–1139 (in Japanese).

[36] Fujimoto, K., et al., *Small Antennas*, Chap. 4, Research Studies Press, 1987.

[37] Ogawa, K., and T. Uwano, "Analysis of a Diversity Antenna Comprising a Whip Antenna and a Planar Inverted-F Antenna for Portable Telephones," *Trans. IEICE*, Vol. J79-B, No. 12, 1996, p. 1008 (in Japanese).

[38] Page 1004 in [37].

[39] Satoh, K., et al., "Characteristics of a Planar Inverted-F Antenna on a Rectangular Conducting Body," *Trans. IEICE*, Vol. J71-B, No. 11, 1988, pp. 1237–1243 (in Japanese).

[40] Fujimoto, K., Y. Yamada, and K. Tsunekawa, *Antenna Systems for Mobile Communications*, 2nd ed., Denshi Sohgou Publishing Co., Japan, 1999, p. 126 (in Japanese).

[41] Page 125 in [40].

[42] Morishita, H., and K. Fujimoto, "A Balance-Fed Loop Antenna System for Handset," *Trans. IEICE*, Vol. E82-A, No. 7, 1999, pp. 1138–1143.

[43] Ogawa, K., and T. Uwano, "Mean Effective Gain Analysis of a Diversity Antenna for Portable Telephones in Mobile Communication Environments," *Trans. IEICE*, Vol. J81-B-11, No. 10, 1998, pp. 897–905 (in Japanese).

[44] Pages 1003–1012 in [37].

Chapter 8

Land Mobile Antenna Systems III: Cars, Trains, and Intelligent Transportation Systems

H. K. Lindenmeier, L. Reiter, J. Hopf, K. Fujimoto, and K. Hirasawa

8.1 ANTENNA SYSTEMS FOR BROADCAST RECEPTION IN CARS

It has been common practice for several decades to fit commercial vehicles and private cars with radios to receive national broadcast programs. The performance of these radios has now become more critical in view of the demand for high-fidelity audio, the increasing number of local radio stations, and, not least, the recent escalation in the electronic complexity of automobile information systems, which include the reception of Television, navigation, and traffic control data.

The typical problems in the design of modern car radio antennas are (1) matching the antenna to the amplitude modulation (AM) receiver in the frequency range of 526.5 to 1,606.5 kHz and to the TV receiver for wide-frequency bands covering both VHF and UHF bands; (2) reduction of multipath fading in FM and TV broadcast reception; and (3) avoiding the reception of noise generated by the engine and electronic circuits installed in the car. In the AM frequency bands, exact matching is impossible, since the antennas commonly used are electrically very small and thus have extremely small radiation resistance and very large reactance. This problem has been solved by placing an amplifier with low gain (about 2 dB) close to the antenna terminals. The amplifier has a high input impedance and an output impedance matched to the receiver in order to reduce transmission losses due to mismatch. The same concept is applied to the design of antennas for AM broadcast reception that are printed on window glass.

Ignition noise from the engine enters the antenna by both direct radiation and coupling to the vehicle body. Digital electronic circuits in a car may also act as a source of interference. The current in digital circuits contains many harmonic frequency components in both the FM and TV broadcast bands. When either electromagnetic or electrostatic coupling exists between digital circuits and antenna or receiver circuits, digital currents may be induced on the antenna element, thus appearing in the FM or TV receiver circuits and interfering with broadcast reception. Thus, the place to mount an antenna on the body of a car must be selected carefully in order to avoid noise interference. When a monopole element is used, the best place to mount it is often considered to be the rear fender. For an antenna printed on the window glass, the chosen location depends on the need to maintain unobstructed vision and the nature of the ambient noise interference.

Section 8.1 places emphasis on the FM reception and the improvements obtained by diversity techniques in the presence of multipath fading. Section 8.2 illustrates the additional complexity required to include the reception of TV and further complements the technical details given in Section 8.1. The practical examples in both sections are taken from systems being manufactured in Germany and Japan.

8.1.1 Introduction

The well-known standard whip antenna used for car radios since the early 1950s is inadequate nowadays. The electrical performance of new types of car antennas, such as active window antennas, has been proved to be at least equivalent to the whip. Those modern antennas make use of improved antenna structures; they profit from the progress in semiconductor technology and the dramatically reduced price of active elements; and they take advantage of much research done with respect to low noise and highly linear amplifier circuitry.

A multitude of car antennas for AM and FM reception with different principles of operation have been developed and implemented mainly by German and Japanese car producers. Section 8.1.2 refers to antennas that have been developed at the Institute of High-Frequency Techniques, University of the Bundeswehr Munich, Germany, in cooperation with car radio and antenna producers and that are in mass production with several German car producers.

The AM reception quality mainly depends on the propagation path between the transmitting broadcast station and the mobile car radio. In the morning hours and the early afternoon, only a few stations with comparatively small service area can be received due to the surface wave propagation along the ground. Receiving distortions are caused mainly by motor ignition pulses and manmade noise.

In the evening hours and the night, especially, the AM radio waves are reflected and diffracted at the ionosphere. The service area of the radio stations is therefore large, and a multitude of waves superimpose at the receiving antenna with different amplitudes and phases, leading to signal fadings. With a signal level exceeding the receiver noise level, this effect does not reduce the reception quality because of the automatic gain control (AGC) control circuits in the radio. In this case, the main distortions are caused by cochannel interference due to the long-range propagation conditions in the night.

Newly developed active AM antennas are discussed in Section 8.1.2. Due to the long AM wavelength antenna (space), diversity is not an appropriate means to improve the AM reception quality in cars. With FM reception, however, the FM distortions well known from the reception with the whip antenna still occur with the new active window antennas; these distortions result from the field situation in the FM range and not from an inferiority of the antennas. The advantage of the new antennas is of a different type, because unlike the whip, the window antennas completely meet the requirements as far as mechanical and other automotive aspects are concerned. There is no corrosion, no need for cleaning, no danger of breaking, no handicap with respect to styling, and no noise due to air turbulence.

For a considerable improvement of the quality of reception in the FM range, however, more complex techniques are required, such as multiantenna arrangements applied to scanning diversity systems. The overall costs are moderate, since only one receiver is required, and very compact antenna designs with up to four antennas in a single window have been developed. Such narrowly spaced window antennas with high efficiency within the diversity system can only be realized if the basic principle of active antennas is applied. Multiantenna diversity systems have been in mass production since 1989 with a German car producer.

Neither modern active window antennas, nor compact multiantenna arrangements, nor the extremely fast distortion detector implemented in the diversity processor could have been developed without a deep knowledge of the basic field situation in the FM range and the distortions associated with it [1].

8.1.2 Reception for AM and FM

Antennas for car radios are required for reception of the AM and FM bands. AM (long-, medium-, and short-wave frequency band) covers the frequency range from 150 kHz to approximately 6 MHz. In Europe and North America, the FM band [2] is located within 87.5 and 108 MHz. In the AM band, the polarization is vertical. In the FM range in Europe, most stations are horizontally polarized, whereas the United States, for example, vertically or circularly polarized stations broadcast, too.

One of the main problems with mobile reception results from the fact that the maximum antenna output voltage delivered to the receiver input is many orders of magnitude higher than the noise level. Taking the FM range and a receiver with 150Ω input impedance, for example, the maximum output voltage was found to be up to 125 dB above the noise level of the system, which is typically at -5 dBμV, if a receiver bandwidth of 120 kHz is taken as a basis. In fact, in the very vicinity of strong transmitting stations, for example, in downtown New York, output signals of 1V (effective value) at the receiver input have been registered. On the other hand, a signal with a mean level of only 25 dBμV has been found to be still satisfactory [4]. In the AM range, the basic situation is similar.

Consequently, all components of a mobile receiving system containing active and therefore nonlinear elements are endangered by intermodulation and cross-modulation distortion. Receiver and active antennas therefore have to meet very hard requirements, not only with respect to sensitivity, but also with respect to large signal behavior.

Passive Whip Antennas

Since the early 1950s, a passive whip about 1m long has been in use for radio reception in cars. The antenna is mostly mounted on the fenders or, for example, in France, on the roof of the car. The received signal is delivered to the receiver by means of a coaxial cable with a low value of capacitance per unit length (about 30 pF/m) and therefore with a high value of the nominal impedance of about 150Ω. The standard length of the cable is about 1.5m, which fits an antenna mounted in the front of a car. If the antenna is mounted in the rear, an additional cable of up to 6m length is required.

In contrast to the FM range, no matching of the antenna impedance with the nominal impedance of the cable is attempted in the AM range, since there is no chance of arriving at a broadband solution. This results from an antenna impedance with an extremely low radiator bandwidth of only 2.8×10^{-7}, due to an antenna impedance of approximately $(0.0044 - j15600)\Omega$ (for example [5], for a 1m whip antenna at 1 MHz).

In the FM range, the antenna impedance matches the nominal impedance of the cable with a VSWR value of about 2. The mainly vertically oriented whip, however, does not fit the horizontally polarized FM waves almost exclusively broadcast in Europe, since a rod antenna vertically mounted above ground delivers no output signal with horizontally polarized waves. As a result, the excitation of any antenna mounted on a point of symmetry on a car (e.g., in the middle of the roof or in the middle of the trunk) is small. Therefore, the combination of the car and the whip form the antenna arrangement with its special characteristic. It is obvious from this that the horizontal antenna pattern cannot be expected to be omnidirectional, as can be seen from Figure 8.1.

Figure 8.1(a,b) refer to a whip antenna and Figure 8.1(c,d) refer to a screen anenna. All the diagrams have been measured on a turntable. With the pattern displayed in Figure 8.1(a,c), the excitation was an ideal plane wave radiated from a transmitting antenna positioned nearby. A comparison between the diagrams of the whip and the screen antenna shows that there is no fundamental difference. Both diagrams are not at all omnidirectional, but have minima and sidelobes.

The pattern of Figure 8.1(b,d) has been recorded by receiving an FM station broadcasting from about 100 km from a high mountain. In spite of the fact that the location of the turntable was in a rural area and in spite of the optical view to the transmitter, the pattern shapes of Figure 8.1(b,d) are not similar to those measured with the ideal plane

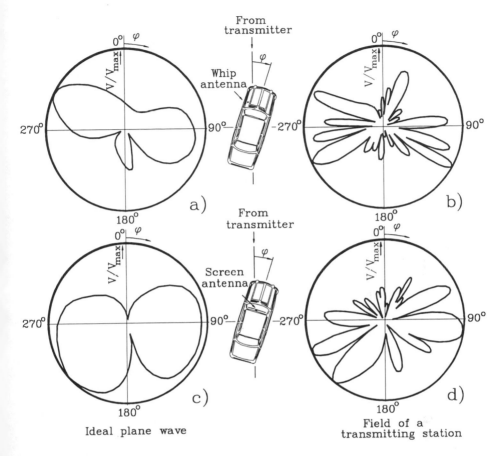

Figure 8.1 Antenna pattern obtained with a whip (a,b) and with a screen antenna (c,d), measured with an ideal plane wave (a,c) and with a high-power broadcasting station (b,d).

wave. This results from the inevitable multipath propagation. In consequence, no conclusions should be derived from the pattern measured with an ideal plane wave. Only statistical evaluations of the time-dependent signal level and of the time-dependent distortions are adequate for comparing the performance of different car antennas. It is well known that with vertical and circular polarization, multipath propagation reception occurs in a similar way. Therefore, the shape of the antenna diagram is of minor importance, even with those polarizations.

From the beginning of mobile reception there was a strong demand for antennas smaller than the 1m-long whip. In Figure 8.2, the measured reduction of sensitivity is displayed versus antenna length for a passive whip having a 1.5m-long coaxial cable for both the AM and FM bands. We see that other solutions are required to arrive at smaller antennas, since no decrease in sensitivity can be tolerated.

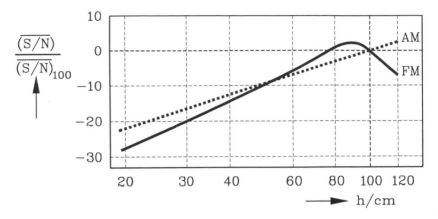

Figure 8.2 Related field strength sensitivity in the AM and FM bands versus the height of a passive whip antenna.

Passive Screen Antennas

Passive screen antennas meet the mechanical automotive requirements excellently. Two antenna types have been introduced in the market, the antenna structures of which are displayed in Figure 8.3. Neither is still in production, due to an intolerably reduced sensitivity. The antenna structure of Figure 8.3(b), for example, comes out to between −5 and −6 dB compared to a standard whip antenna in the AM and FM range. The unsatisfactory performance of those passive screen antennas, however, has been misinterpreted in the past: the shape of the diagram is not more directional than that of the whip antenna. The antenna directional diagram is not responsible for the degradation; rather, the reduced average field strength sensitivity is resulting in deeper absolute signal values during the minima of reception.

Active Antenna in the Rear Window

With modern cars, electromagnetic interference due to microprocessor-operated components applied, for example, in motor control units, in the instrument panel, or in the

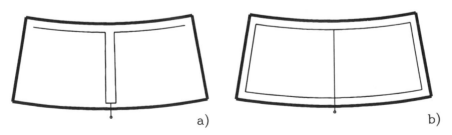

Figure 8.3 Antenna structures applied for passive windscreen antennas.

onboard computer, is a great problem for all kinds of antennas. Antennas mounted in the front of the car are mainly subject to this interference. In the rear, the interference is present, too, but is typically about 6 to 12 dB smaller. Consequently, an antenna with the same excellent overall performance as the above front screen antenna was required for the rear window. The new antenna design, which is displayed in Figure 8.4, has been applied to BMW cars since 1989 and meeets the requirements. Extensive tests and measurements of this antenna were also performed in Europe and in the United States before its introduction to series production.

Different antenna structures are used for AM and FM. Both structures are printed onto the glass with the same technology as the heaters. Thus, both the heater and the antenna conductors can be simply printed onto the glass in one stage of manufacture. The FM antenna structure containing vertical antenna conductors can be in galvanic contact with the horizontal heater conductors. Since the connecting points between the antenna structure and heaters are arranged on equipotential lines, the defrosting of the rear window is not influenced. Only in the upper part of the antenna structure, above the uppermost heater, does a shunt with a comparatively high resistance reduce the heater current in the middle, but only to an insignificant extent.

The AM structure is located in the free strip above the heaters without any galvanic contact with them. Therefore, the problems involved with antennnas making use of the heaters for AM reception are avoided, since there is no need for broadband RF insulation by large coils (for up to 30A of heating current). However, with the expenditure of greater technical effort, an excellent antenna can be realized by using the heaters for AM, too. This has been explained in [6] in detail.

From research done with respect to the AM structure of Figure 8.4, the conclusion was drawn that the optimum signal-to-noise ratio is obtained if the AM structure is centered with respect to the height of the free strip, and that the optimum width of the

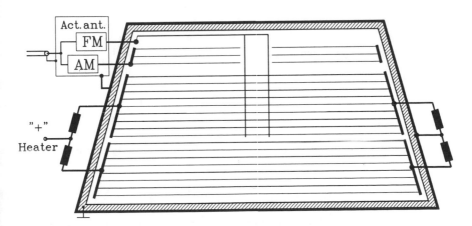

Figure 8.4 Active rear screen antenna for all FM polarizations and for AM.

structure depends on the capacitive load of the antenna amplifier [6]. With a load of 10 pF on the antenna amplifier, the optimum width is 40% of the height of the free strip. Taking a free strip of about 14 cm high, for example, the ultimate sensitivity of the antenna is about 3 dB better than with a standard whip in the rear of the car.

8.1.3 FM Antenna Diversity Systems

The minima of RF field amplitude with different antenna locations on a car do not generally occur at the same time. In the past, many attempts have been made to make use of this effect by selecting the receiving signal from that antenna which momentarily provides the best signal. In the absence of a sufficiently fast operating distortion detector, no antenna diversity was presented in the past that could make use of the advantage of two or more antennas on a fast car.

An antenna diversity system implementing a distortion detector of extremely short distortion-detection time is described here. In the following, the efficiency of such a system using such a distortion detector is investigated as a function of the number of antennas applied [3].

Available Improvement with a Multitude of Antennas

Multipath propagation amplitude fadings create distinct amplitude minima in the received signal during mobile reception. As displayed in Figure 8.5, different kinds of distortions, such as noise and cochannel, adjacent channel, and intermodulation inteference may occur during fading periods. The superposition of waves with a time-delay difference greater than 3μs, as previously discussed, also leads to audible distortions if an amplitude condition is satisfied, which is marked by the shadowed range. Therefore, it is obvious that the distortions are strongly correlated with amplitude minima.

To obtain a realistic judgment of the distortion, the definition of a subjectively acceptable upper limit of audible distortion is necessary for a reference. The presence of a distortion is detected by comparing the instantaneous value with this reference, as shown in Figure 8.6 for two antennas. With p_1 and p_2 representing the fraction of time for a so-defined distortion in signal 1 and signal 2, respectively, the likelihood of the simultaneous occurrence of a distortion in both signals is the joint probability $p_d = p_1 \cdot p_2$. Assuming an extremely fast operating distortion indicator, a scanning diversity system may be applied, now making use of a multitude of receiving antennas on a car with only one receiver. The signal quality Q_s available with reception from only one antenna can be defined as

$$Q_s = 20 \cdot \log(1/p) \tag{8.1}$$

with p representing the related fraction of time with distorted reception. Under the assumption of equal signal quality Q_s of each antenna and with negligible correlation between

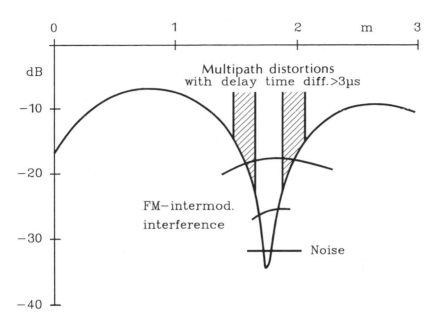

Figure 8.5 Different kinds of distortions in a level minimum.

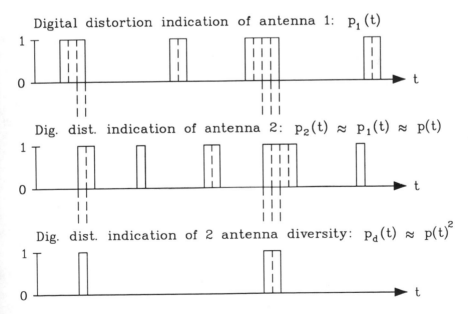

Figure 8.6 Digital distortion indication and distortion reduction with antenna diversity.

the distortions on the N antennas, the related fraction p_d of time under distorted reception is found by the joint probability

$$p_d = \prod_{i=1}^{N} p_i \tag{8.2}$$

If the improvement factor q_{dB} is defined as p/p_d, its logarithmic value

$$q_{dB} = 20 \cdot \log(p/p_d) \tag{8.3}$$

becomes

$$q_{dB} = (N - 1) \cdot Q_s \tag{8.4}$$

This factor shows that the improvement factor is considerably increased by each additional antenna and that the fraction of distortion time is reduced exponentially by the number of antennas applied.

For this reason, up to 12 antennas installed on a car have been measured. The installation of these antennas is shown in Figure 8.7, where four antennas (1, 2, 5, 6) are mounted on the fenders, six antennas (7 to 12) are mounted on the shown locations on the roof of the car, and an additional pair of windshield antennas (3, 4) was fixed onto the front screen. With space diversity, it is well known that the signal correlation factor between similar antennas increases with decreasing distance between adjacent antennas. In order to find the improvement factor for distortion suppression as a function of different antenna locations on a car, special measurement equipment had to be developed.

In order to obtain objective measuring results, a radiated test signal was required, because multipath distortions are a function of the modulation. For this reason, in cooperation with the Bavarian Broadcasting Corporation, a 1-kHz modulation signal with a maximum frequency deviation of 35 kHz has been radiated during the night with a power of 100-kW ERP from several stations along the north rim of the Bavarian Alps. With the flat area north of the Alps and with Munich being a densely built up area and with a

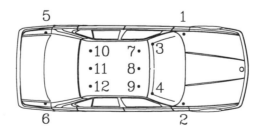

Figure 8.7 Test car with 12 antennas.

mountainous topography near the Alps, all important different receiving conditions relevant in practice were available. Therefore, studies could be made in the flat areas with low multipath distortions and in high multipath city areas with small delay-time differences between the superimposed waves, and in high multipath areas in the mountains with great delay-time differences between superimposed waves.

The measuring device in the car consists of a scanner, which sequentially switches the antennas to the receiver for one period of the audio frequency. The receiver contains a broadband, highly linear FM demodulator and an AM demodulator of a wide dynamic range for amplitude detection. The mixed-down FM signal is demodulated and filtered by a low-pass of 500 kHz and a signal trap of 1 kHz, and this AF signal is then integrated. The instantaneous amplitude is also integrated and is available by means of a sample and hold circuit at the same time at the input of the selector. This selector alternately reads the FM and AM values into the system voltmeter, which digitizes the values for the system computer. After each scanning cycle, the instantaneous error of the received signal of each antenna is stored in the computer for statistical evaluation.

Figure 8.8 shows the improvement factor q_{dB} versus the total number N of antennas applied. Antenna 1 of Figure 8.7 serves as the main antenna, while the auxiliary antennas were antennas 2, 5, and 6 for $N = 2$, 3, and 4 respectively. Curve 1 describes the q_{dB} value for different values of N and is found to be almost a straight line, as calculated by (8.4). This curve has been measured in the Munich area with a radio signal being radiated from a weak transmitting station 80 km away from Munich. The most interesting aspect is the fact that the improvement factor increases with a median value of roughly 17 dB per additional antenna; hence, more than two antennas are beneficial. Curve 2 in Figure 8.8 represents a measured result obtained in the city area at a different location and on a different day with the additional antenna 7 on the roof of the car. The results show a curve which shows improvements with increasing N, but the increase of the q_{dB} factor is less pronounced with N greater than 3. In this context, it should be noted that the available improvement factor is a function of the signal quality being received with the signal antenna system.

In city areas, the time-delay difference between superimposed waves is comparatively small, and the number of waves of different time delays contributing to multipath distortions is usually comparatively low. This is in contrast to the situation in mountain areas, where the median value of the difference in delay time between the superimposed waves is on the order of 40 μs corresponding to a detour of 12 km in propagation. Due to this fact, the average multipath distortions are much more severe in mountain areas. The improvement factor of curve 3 has been measured in a heavily multipath-distorted region around the lake of Tegernsee in Bavaria. This curve is almost a straight line, indicating the law of equal increase of the logarithmic improvement factor per additional antenna (8.4). Even though there is only a little space between adjacent antennas, the cross-correlation factor between the output voltages is still low enough to get the measured increase of q_{dB} of about 8 dB per additional antenna. In this case, the signal supply is very poor, and therefore the signal quality, which is the percentage of time in which the

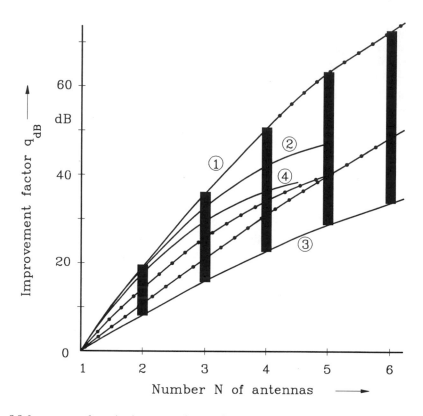

Figure 8.8 Improvement factor in city areas and mountainous areas versus the number of antennas.

reception is undistorted, is low. As a result of the poor signal supply, the increase of the improvement factor per additional antenna is less than that in city areas.

In a mountain valley where good signal supply was found, curve 4 has been measured. With low numbers of N, the increase of the improvement factor is roughly 20 dB per additional antenna and is found to be as large as that in city areas. This is true in spite of the large differences in delay time between superimposed waves. As a result of the good signal quality, however, the further increase of the improvement factor per additional antenna is less if N is a large number.

Among the measured results, the curves marked with dots in Figure 8.8 have been found in other receiving situations. Therefore, the improvement factor q_{dB} for a given number N of antennas is presented by the dark columns, which can be expected in a large variety of receiving areas of different characteristics.

8.1.4 FM Multiantenna Systems

The results presented in Section 8.1.3 recommend the application of more than two antennas with antenna diversity systems. It is obvious that antenna diversity on the basis of two or more standard whip antennas would not be accepted, due to esthetic, optical, and aerodynamic constraints. Therefore, investigations were made with multiantenna arrangements in windows in spite of the fact that most theories on diversity systems claim a certain distance between antennas in order to obtain the necessary decorrelation between the antenna signals [7].

The correlation factor between the signal levels of two antennas, however, is not the only relevant criterion to estimate the obtainable diversity improvement. With the quality of reception being at least satisfactory, audible distortions occur only within 1% of the time. Consequently, the contribution of those small time periods to the correlation factor is only 1%, too. Therefore, the probability of a simultaneous occurrence of interference with both antennas has to be considered.

With weak signals and in regions with small values of delay time between superimposed waves, the AF distortions mainly result from noise if the signal level fades (Figure 8.5). For these, an analysis of the signal level versus time obtained with different antennas under test is suitable for judgment of diversity performance.

With multipath distortion due to time-delay differences, with τ_i greater than 3 μs, and with cochannel or adjacent-channel interference, a high antenna output signal may actually be more disturbed than a weaker signal, so the AF distortions versus time and not the signal level have to be measured and analyzed.

The antenna systems presented below consist of two, three, or four antennas per window. In most cases, the mutual coupling between the antenna structures mounted on a single screen cannot be neglected; therefore, active antennas are advantageously used to provide antennas with the required decoupling.

Four-Antenna System Arranged in Front and Rear Windows

Figure 8.9 shows the thoroughly optimized antenna configuration that has been installed in several types of cars for further investigations. In the rear window, two FM antennas are obtained by subdividing the heater structure horizontally into two partial heating areas, and four FM resonant decoupling networks are required. The connection points for the FM amplifiers on the bus bars are applied at the opposite limits of the bus bars on the same side. AM reception is performed by a separate structure corresponding to that previously described.

In the windshield, a second structure with an approximately 40-cm-long wire at a distance of 5 cm from the frame is used. Both windshield antenna amplifiers are located near the dashboard: one on the left side, the other on the right side.

First, an analysis of the signal level of the respective antennas will be discussed, from which the obtainable signal-to-noise ratio improvement in weak signal areas can be predicted. The results displayed in Figure 8.10 are obtained from longer test drives on different frequencies. The quality of reception, which is defined as the inverse of the percentage of time with disturbed reception, is plotted versus the actual required signal level. If, for example, a signal level below the mean signal level of each of the antennas

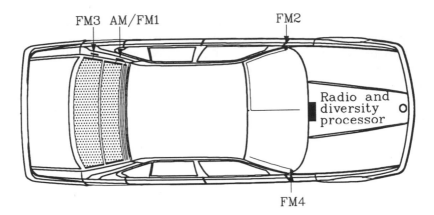

Figure 8.9 Scanning diversity system with a four-antenna arrangement.

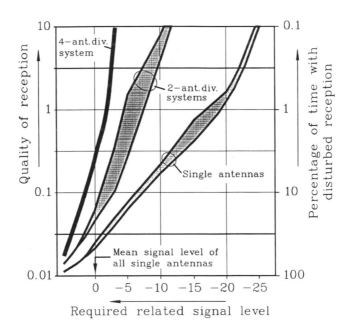

Figure 8.10 Reception quality versus required related signal level.

(0 dB) is assumed to be the threshold between undisturbed and disturbed reception (for example, by noise), a single-antenna system would be disturbed typically 41% to 46% of the time; the actual value within this range depends on the actual antenna. Considering all two-antenna diversity systems, which can be realized by any two of the four available

antennas, it is found that the system would be disturbed 16% to 22% of the time, whereas the four-antenna diversity system would be disturbed only 3% of the time.

With a required signal level of −3 dB, the percentage of time there is disturbed reception is as follows: no distortion at all with the four-antenna system, 2.8% to 9% with any of the two-antenna systems, and 20% to 24% with any of the single antennas.

In the following, the improvement achieved by the antenna system of Figure 8.9 is discussed. At the received station, the mean output signal level was good and the recorded distortions displayed in Figures 8.11(a–d) result from multipath propagation with large τ_i and from adjacent-channel interference in a car driven in a defined narrow circle.

In Figure 8.11(a), only one antenna was applied, and the curve displayed represents the distortions versus time. The scale on the vertical axis is related to the threshold of audibility of any distortion. Another region in this figure marks medium, and a further region indicates severe distortions. In Figure 8.11(a), several spikes exceed the threshold of audibility leading to the well-known short and hard distortions. If a second antenna is applied to the system [Figure 8.11(b)], the number of spikes is already significantly reduced. With the three-antenna diversity system [Figure 8.11(c)], only three short spikes exceed the threshold. In the recorded situation with four antennas applied to the system [Figure 8.11(d)], the distortions have been reduced to the extent that none of the distortion spikes exceeds the threshold of audibility. These measured curves are a good illustration of the contribution of any additional antenna to the reduction of distortion.

Four-Antenna System in the Rear Window

The four-antenna diversity system in Figure 8.9 consists of a large number of components distributed over the car and connected via complex wiring. For car production, the antenna assemblies would be subcontracted out to a component supplier, and more appropriate solutions better adapted to the manufacturing process of cars are required.

To that end, the antenna arrangement of Figure 8.12 was designed and optimized, because for the scanning diversity system, the technical effort on the diversity processor side is independent of the number of antennas applied. Consequently, the additional costs per antenna have to be kept small in order to allow the application of a large number of antennas. A cost-effective arrangement of antennas is to place them close together on a window, which results in minimal costs for housing and wiring. Under this assumption, Figures 8.13(a,b) give an impression of how costs and the reception quality will increase with the increasing number of antennas applied. As previously discussed in Figure 8.13(a), it is assumed that the distortion percentage for each of the antennas is 30%, corresponding to a signal quality of 10 dB. The respective values for the related time with audible distortions can be seen from Figure 8.13(a), too.

In contrast to this, the cost of additional technical effort increases considerably if any diversity system is to be used, since a distortion detector, additional wiring, and a second antenna are required. The additional technical effort for additional antennas, however, is small if an arrangement as displayed in Figure 8.12 (or similar) is used.

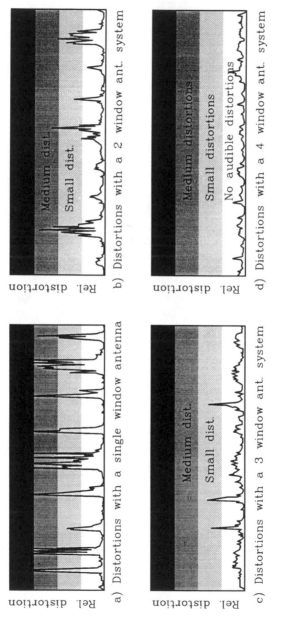

Figure 8.11 AF distortions versus time with (a) single-window antenna; (b) two-window antenna; (c) three-window antenna; (d) four-window antenna diversity system. (Black band represents severe distortions.)

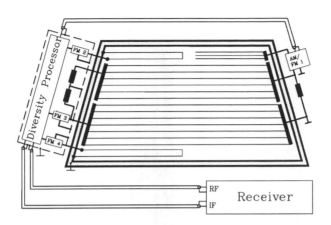

Figure 8.12 Four-antenna arrangement in the rear window.

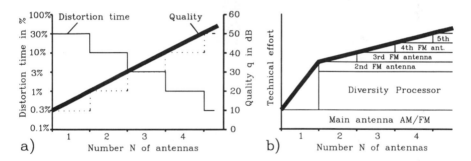

Figure 8.13 Improvement in signal quality and corresponding technical effort as function of antennas applied.

The antennas have different receiving signals due to different types of antenna structures and different RF loads at the heater bus bar opposite the respective amplifier. In the following, the receiving performance of a car equipped with four rod antennas is investigated in comparison to the antenna arrangement in Figure 8.12 in the rear window. In Figure 8.14(a), the related distortion level of a single standard whip antenna is plotted versus time, and Figure 8.14(b) shows the distortion level of one of the active window antenna systems of Figure 8.12. Figure 8.14(c) presents the audio distortions with the diversity system using four whips, and Figure 8.14(d) displays the curves with the four-window antenna system. The probability of a simultaneous occurrence of interference on all four closely packed screen antennas is not greater than that with the four whip antennas considerably spaced apart.

As a criterion for judging the capability of an antenna diversity system in mobile communication, the correlation factor of the time-dependent output signal levels of the

478

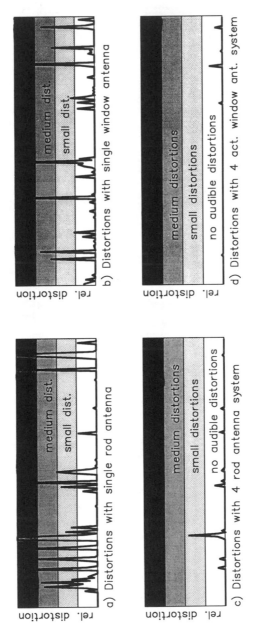

Figure 8.14 AF distortions with different kinds of single antennas (a,b) and with four-antenna diversity systems (c,d). (Black band represents severe distortions.)

various antennas is considered significant. This has caused the working group of the International Committee on Radio (CCIR) to recommend a distance of 2.8m between the antennas for FM antenna diversity [7]. The above documented results contrast with this recommendation. In [7], a time-invariant noise level is assumed and a distortion occurs if the noise threshold exceeds the instantaneous signal level. With mobile broadcasting reception, however, the interference due to time-variant intermodulation, adjacent-channel and cochannel interference, and multipath distortion are important. For the reduction of audio distortions by means of antenna diversity, the correlation factor between the actual distortions in the received signals is most significant.

In Figure 8.15, the correlation factor c_{ij} has been evaluated with respect to the signal levels and to the distortions. With the distortions, the correlation factor c_{ij} is

$$c_{ij} = \frac{\int_0^{t_0}[(V_i(t) - \overline{V_i(t)})] \cdot [(V_j(t) - \overline{V_j(t)})]dt}{\sqrt{\int_0^{t_0}[(V_i(t) - \overline{V_i(t)})]^2dt \cdot \int_0^{t_0}[(V_j(t) - \overline{V_j(t)})]^2dt}} \tag{8.5}$$

where V_i is the distortion level of the ith and V_j of the jth antenna. $\overline{V_i}$ and $\overline{V_j}$ represent the mean values of the appertaining AF distortion levels during the drive time t_0. In Figure 8.15, c_{ij} is plotted for measured values of V_i and V_j. This correlation factor of the distortion level is represented by the dark columns. The other columns show the correlation factor of the appertaining signal level for comparison, which, in FM broadcasting, is regularly much greater than the correlation factor of the distortion levels. Therefore, in

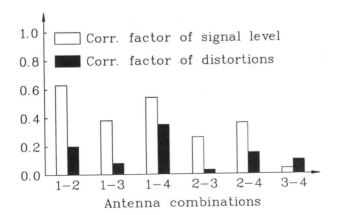

Figure 8.15 Correlation factor of signal level and corresponding AF distortions with various antenna combinations.

broadcast reception, the correlation factor of signal levels is not a valid criterion for judging the capability of an antenna arrangement.

Three-Antenna System in the Rear Window

Having successfully passed extensive tests comparable to those performed with the single-window antennas, antenna diversity systems were introduced in series production in the new BMW car types 5 and 7 in Europe in autumn of 1989 and in autumn 1990 in the United States. The designs of the antenna structures for Europe and the United States are not identical. Here, the U.S. version will be described.

The structures printed on the glass are the same as those of the single antenna of Figure 8.4, so an easy upgrading from a single-window antenna to the three-antenna diversity system is possible by an exchange of the electronic components right and left of the window and by adding two further coaxial lines as displayed in Figure 8.16. The two additional FM antennas are provided by the AM antenna structure and the lower part of the heater element. The FM1 amplifier, the AM/FM2 amplifier, the feeding and decoupling elements for the dc current-FM signal separation with respect to the heaters, the diversity processor, and the pin-diode switches are arranged in one housing (size $15 \times 4.5 \times 1.2$ cm) mounted on the left C-post of the car. The FM3 amplifier and the respective feeding elements are arranged in a second housing mounted on the right C-post of the car. Additional coaxial cables are required as the connection from the FM3 amplifier to the diversity processor and from the radio (IF signal) to the diversity processor. Thus, with a small technical effort, maximum performance is obtained.

Naturally, the performance of an optimized three-antenna diversity system is inferior to a four-antenna diversity system. However, in comparison to the performance of a

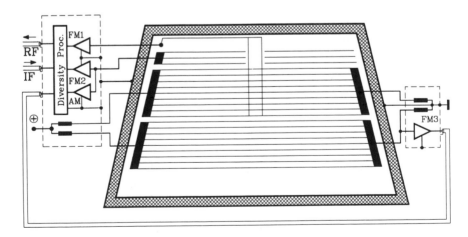

Figure 8.16 Three-antenna diversity system now in series production.

standard system with only one antenna [Figure 8.17(a)], the reduction of distortions on the same drive is impressive [Figure 8.17(b)]. The mean signal level value for Figures 8.17(a,b) is about 45 dBμV. In situations with strong multipath reception and with adjacent-channel or intermodulation interference, the improvement in signal quality is very high.

8.2 ANTENNA SYSTEMS FOR TV RECEPTION IN CARS

A printed antenna system in conjunction with a monopole mounted on the rear fender has been used for AM, FM, and TV reception where the diversity reception for FM and TV broadcasts is performed by combining the monopole and the printed elements. An example of an antenna system composed of four printed elements and a monopole is shown in Figure 8.18, where the table shows the function of each element. Space diversity is achieved by combining both a monopole element and a printed element for FM broadcast reception and by combining a monopole element and printed elements for TV broadcast reception in the VHF and UHF bands. Glass is not perfectly transparent to high-frequency

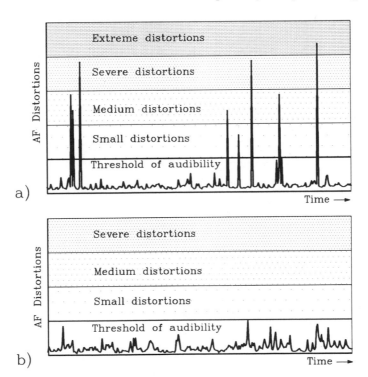

Figure 8.17 (a) AF distortions with a single antenna and (b) with three-antenna diversity system of Figure 8.16.

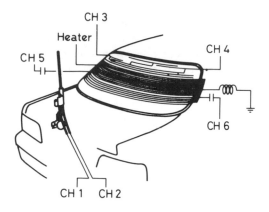

Antenna function

Ant CH	Radio		TV		
	A M	F M	VHF 1 ~ 3 ch	VHF 4 ~12ch	UHF 13~62ch
1	◎	◎			
2			◎	○	○
3			○	○	◎
4		○			
5			○	◎	○
6			○	○	○

◎ Main
○ Sub

Figure 8.18 A monopole (motor driven) and four printed antennas on the rear window.

radio waves. An analysis has shown [8] that the transmissivity through a glass sheet in the VHF and UHF frequency region is 95% or greater when the thickness d of the glass is less than 6 mm. Figure 8.19 shows the transmissivity versus frequency, where the thickness d of the glass is taken as the parameter and the dielectric constant ϵ_r of the glass is assumed to be 6.5. Since the thickness of glass currently used in ordinary automobiles is 3.5 mm, glass transmissivity is not a problem.

8.2.1 Antennas Printed on the Rear Quarter Windows

The antenna structure is shown in Figure 8.20, along with the amplifier and the TV tuner [8]. The antenna system consists of four elements. Output of these antenna elements are

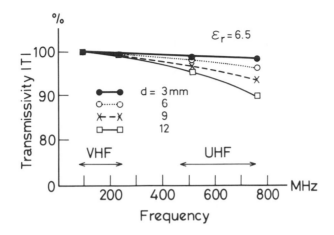

Figure 8.19 Glass transmissivity against frequency. (From [8], © IEEE.)

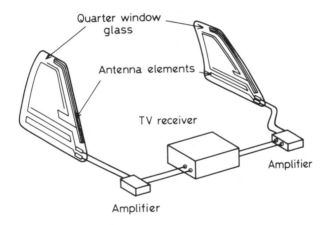

Figure 8.20 Antenna structure printed on the quarter-window glass, antenna amplifier, and TV receiver. (From [8], © IEEE.)

compared, and the one that provides the highest quality is selected by the switching circuit. Figure 8.21 is a simple block diagram of the receiver. The switching is performed during each blanking period of the TV video signals to prevent the picture from fluttering.

The antenna patterns printed on the rear quarter window are illustrated in Figure 8.22. Antenna 1 is composed of two elements: a slanted element consisting of three thin lines, to which a trap is attached to maximize sensitivity, and a horizontal element, to which a slanted element is connected. These elements, connected in parallel at the antenna output terminals, are used for the VHF band. Another, shorter horizontal element is attached to antenna 1 for the UHF band. Antenna 2 consists of a horizontal element with a trap and a slanted element. Antennas 1 and 2 are capacitively coupled to make the

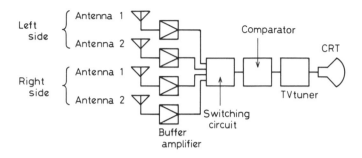

Figure 8.21 Antenna circuit and TV receiver to perform diversity reception. (From [8], © IEEE.)

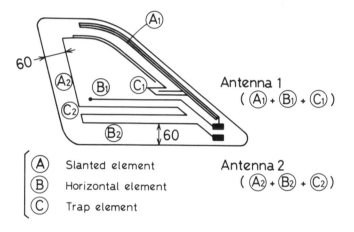

Figure 8.22 Antenna patterns printed on the quarter-window glass. (From [8], © IEEE.)

sensitivity difference in the two antennas as small as possible. The antenna elements are printed symmetrically with the same patterns on the right and left quarter windows. These antenna patterns are based on extensive analyses performed to determine the specified sensitivity. The measured antenna sensitivity is shown in Figure 8.23, where a comparison is made with the sensitivity of a conventional monopole antenna. The antenna sensitivity is defined as the induced voltage at the antenna's open terminals under a field strength of 60 dBμ (0 dBμ = 1 μV/m).

The directional characteristics of the antenna are shown in Figure 8.24. To determine the antenna sensitivity, the averaged value of the receiving pattern over the entire horizontal plane is taken.

For TV reception in a car, antenna sensitivity is specified to be greater than 45 dBμ in the VHF ranges (90 to 108 MHz and 170 to 220 MHz) and greater than 30 dBμ in the UHF bands (470 to 766 MHz). A conventional monopole antenna has an antenna

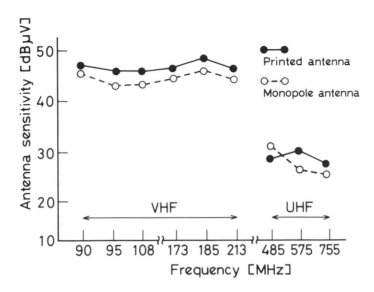

Figure 8.23 Antenna sensitivity for horizontal polarization. (From [8], © IEEE.)

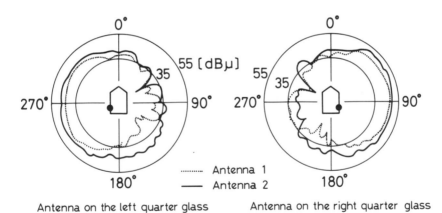

Antenna on the left quarter glass   Antenna on the right quarter glass

Figure 8.24 Antenna sensitivity patterns. (From [8], © IEEE.)

sensitivity of 43 to 45 dBμ in VHF bands and 25 to 30 dBμ in UHF bands. The VSWR characteristics are shown in Figure 8.25. The quality of TV reception in the field is evaluated by using a five-rank rating, the results of which are shown in Table 8.1 [8], where the evaluations of both sound and video quality are provided. The rank represents the quality with interference: rank 5 corresponds to "no interference at all," rank 4 to "slight interference but not distractive," rank 3 to "interference being distractive but not

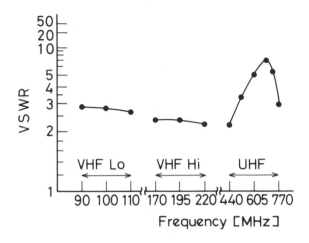

Figure 8.25 VSWR characteristics. (From [8], © IEEE.)

Table 8.1
Five-Rank Evaluation (Average) of TV Broadcast Reception by the Five-Rank Method

	Audio		Video	
	Printed Wire	Monopole	Printed Wire	Monopole
VHF	4.4	3.5	3.8	3.5
UHF	4.4	3.7	3.5	3.5

Note: Evaluation for the audio quality was performed during driving, while that for the video quality was performed when the car made a stop.
Source: After [8].

obstructive," rank 2 to "interference being heavily distractive and obstructive," and rank 1 to "unable to receive."

8.2.2 A Three-Element Antenna System for AM, FM, and TV Reception

An example of an antenna system for AM, FM, and TV reception will be described [9]. All the technical data are provided courtesy of the Nissan Motor Co., Japan. The antenna system is composed of three elements: a monopole element and a printed element on the rear window are used for AM and FM reception, and printed elements on both rear and quarter windows are used for TV reception. Diversity reception for FM broadcasts is performed by the monopole and printed elements on the rear window, and TV diversity reception is achieved by the two printed elements. The antenna pattern printed on the rear window glass is shown in Figure 8.26, where a scale shows the actual length.

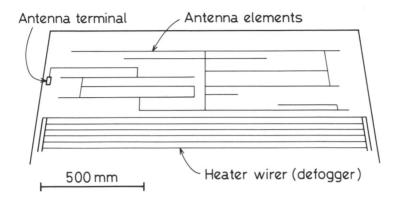

Figure 8.26 Antenna pattern printed on the rear-window glass for FM/AM and TV reception.

AM Reception

The sensitivity of the receiver is defined to be the minimum input field strength at which the specified output S/N is obtained. The S/N performance at 525 kHz is shown in Figure 8.27, where variations of the output signal power S and noise power N are given against

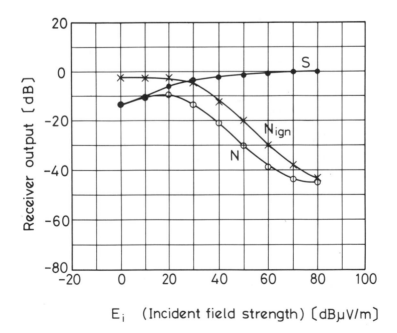

Figure 8.27 S/N characteristics in the AM reception.

the input signal field strength. The signal level gradually varies under the effect of automatic gain control (AGC). When the engine is operating, the output noise power increases due to the addition of ignition noise as shown by the variation of N_{ign}, the dotted line in the figure. The input to the receiver is an AM wave with a 30% tone modulation of 1 kHz, and the received power is measured at the speaker terminals of 4Ω (0 dB in this figure means the output power is 0.5W at 1 kHz). The sensitivity varies depending on the type of receiver, but typical values specified for ordinary receivers are 43 to 48 dBμ to achieve the S/N_{ign} of 12 dB. The S/N performance is measured in a shielded room with a nearly homogeneous electromagnetic field, and with a car placed in the middle of the room.

FM Reception

The gain versus frequency is shown in Figure 8.28, where dark lines express mean values and dotted lines the standard deviation with respect to the mean values, which were taken from more than 70 measured data. The gain is defined as the ratio of the received power of an antenna mounted on a car to that of a standard dipole antenna in free space. Since the antenna system has directional characteristics, the receiver output varies as the direction of the car changes. Hence, in order to determine the gain, the averaged receiver output over the entire horizontal plane, taken from the receiving pattern, is compared to the received level of the standard dipole antenna (0 dB in the receiving pattern). The measured

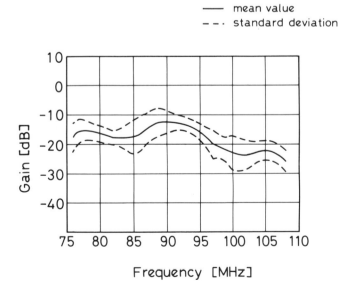

Figure 8.28 Gain characteristics in the frequency range of 76 to 108 MHz (horizontal polarization).

receiving patterns at four frequencies are shown in Figure 8.29. The receiving pattern is measured in an anechoic chamber containing a car, and the output (received) power of the antenna mounted on the car is measured. The impedance characteristics of the antenna are shown in Figure 8.30. The gain and impedance characteristics of the higher VHF bands (90 to 110 MHz and 170 to 220 MHz) are shown in Figures 8.31 and 8.32, respectively.

TV Reception

The gain, receiving patterns, and impedance characteristics in the UHF bands (470 to 770 MHz) are shown in Figures 8.33, 8.34, and 8.35, respectively.

Diversity Reception

FM receiver performances of a monopole element (antenna 1) and a printed element (antenna 2) on the rear window are shown in Figure 8.36. They illustrate receiver outputs

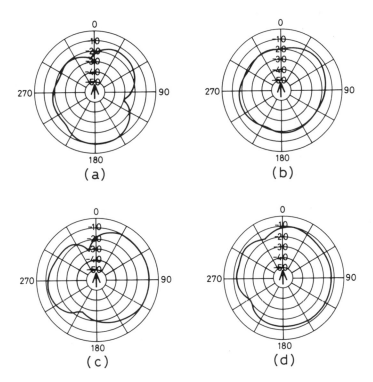

Figure 8.29 Receiving patterns in VHF (FM) bands: (a) 76 MHz (−17.7 dB); (b) 82 MHz (−18, 1 dB); (c) 87 MHz (−15 dB); (d) 93 MHz (−13.6 dB). Number in parentheses is the gain.

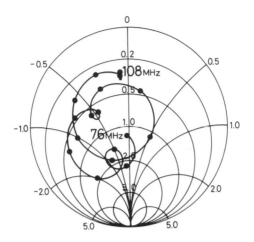

Figure 8.30 Impedance characteristics in the frequency range of 76 to 108 MHz (2-MHz intervals).

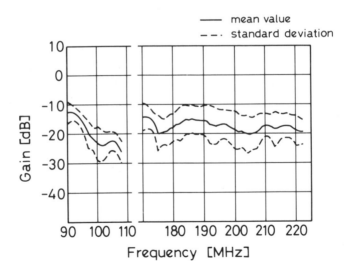

Figure 8.31 Gain characteristics in VHF (TV) bands (90 to 110 MHz and 170 to 222 MHz).

which fluctuate heavily due to the effect of multipath propagation. The dark line shows the received level of antenna 1 and the dotted line that of antenna 2. The selective diversity scheme has been used in which the antenna branch with the higher output signal is selected by means of switching. Figure 8.37 shows the cumulative probability distribution of the received power against the input signal strength for three cases in which a monopole (dark line), printed elements (dotted line), and diversity antenna (chained line) are employed. The effect of diversity can be clearly observed in the figure.

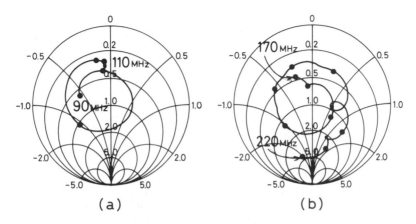

Figure 8.32 Impedance characteristics of VHF (FM) bands: (a) low channel, 90 to 110 MHz; high channel, 170 to 220 MHz.

8.3 ANTENNA SYSTEMS FOR SHINKANSEN (NEW BULLET TRAIN)

8.3.1 Introduction

Leaky coaxial cables are used for the train communication systems of Tohoku [Tokyo-Morioka (535 km)] and Joetsu [Tokyo-Niigata (334 km)] Shinkansen [Japan Railways (JR) new bullet trains] [10–14]. They are coaxial cables whose outer conductors have slots (an example is shown in Figure 8.38) radiating a part of the transmitted energy [15–18]. LCX cables were first developed for train communication systems in a tunnel for Tokaido [Tokyo-Osaka (553 km)] and Sanyo [Osaka-Fukuoka (624 km)] Shinkansen. The advantages of LCX cable systems compared to radio communication systems were then realized. They offer more stable and better train communications compared to radio communication systems because LCX cables radiate weak electromagnetic energy and the environmental effects on their radiation characteristics are small [19,20]. Once LCX cable is installed their radiation characteristics are small [19,20]. One LCX cable is installed along each railway; thus, there are two LCX cables along the railway for inbound and outbound train communications, as shown in Figure 8.39. Trains have slot antennas that transmit and receive electromagnetic energy to and from the LCX cables.

8.3.2 Train Radio Communication Systems

In the LCX system there are 24 radio communication channels. They are divided into four systems: train operation commands (4 channels), commands related to passengers (2 channels), business and public telephones (6 channels), and data communications (3 channels). The rest (9 channels) are for spare channels. Train operation commands are

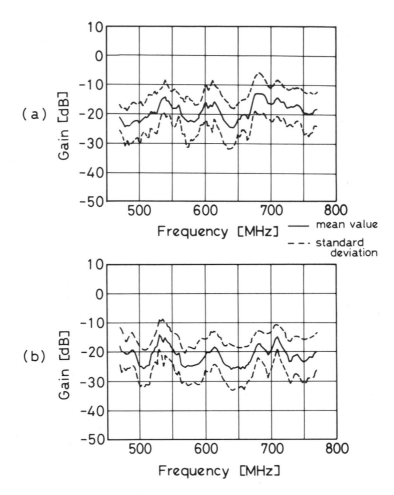

Figure 8.33 Gain characteristics in UHF (TV) bands (470 to 770 MHz): (a) horizontal polarization; (b) vertical polarization.

sent through direct channels between the central command station in Tokyo and motor men. Business commands are also sent through direct channels between the central command station and train conductors. Two public telephones in a train are connected to NTT public telephones all over Japan through JR's telephone exchange networks. Data channels are frequency-shift keying (FSK) 1,200 bps and are used to monitor train running conditions in order to deal with an emergency. Each train can always use six channels to cover the above four communications systems. The bandwidths are 900 and 700 kHz from base station to trains and from trains to base station, respectively. Each channel occupies 25 kHz. The advantages of the LCX system are stable channel quality, efficient channel

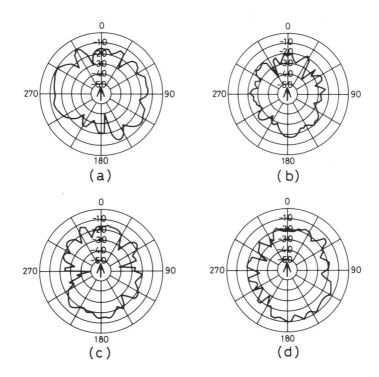

Figure 8.34 Receiving patterns in UHF (TV) bands: (a) 470 MHz (−18.2 dB); (b) 569 MHz (−25.5 dB); (c) 671 MHz (−21.7 dB); and (d) 770 MHz (−20.1 dB). Number in parentheses is the gain.

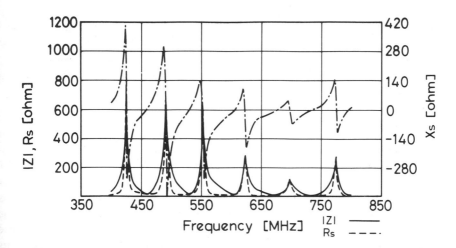

Figure 8.35 Impedance characteristics in UHF (TV) bands.

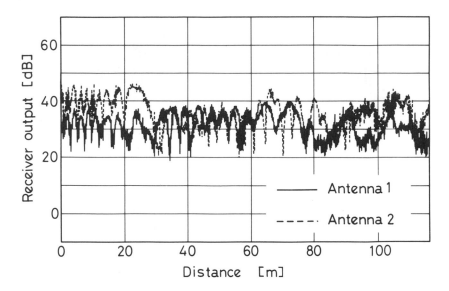

Figure 8.36 Receiver outputs at each of two elements.

utilization, and small environmental effects. The last two advantages are, as mentioned
due to weak radiation from the LCX slots.

Each base station is located at each train station and one base station covers on
service area (average 20 km). The LCX relay system is shown in Figure 8.40. When ther
is a communication problem in one LCX route, a route is changed to the other LCX a
the relay station to prevent the trouble from spreading to other places. There is a maximum
of 20 relay stations between the two train stations and four kinds of LCX cables wit
different coupling loss; #488 (loss = 75 dB), #487 (loss = 65 dB), #486 (loss = 55 dB)
and #485 (loss = 50 dB), between a train antenna and an LCX cable. They are connecte
properly to reduce the received signal level change (less than 10 dB with respect to th
center level) between the two relay stations, as shown in Figure 8.41. The maximum
distance between the two relay stations is 1.5 km. Each relay station has a 400-MH
bilateral amplifier with a 42-dB gain, and the output is about 1W.

8.3.3 Antenna System

The slot array shown in Figure 8.42 consists of an array of three slots with differen
inclination angles. The separation of two sets of three slots is P. They are designed t
work in a 400-MHz band, and in future operation they will work in 800-MHz bands [21
LCX cables and train antennas are located as shown in Figure 8.43. There are two LCX
cables along the railway and an array of four folded slot antennas on each side of th
train. The LCX cables at the left side of the train toward the direction of travel (train

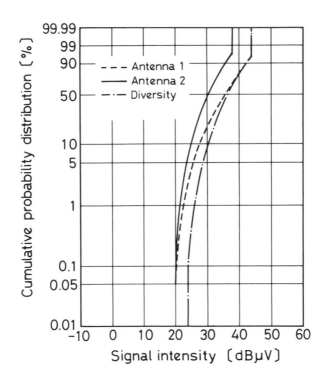

Figure 8.37 Diversity performance.

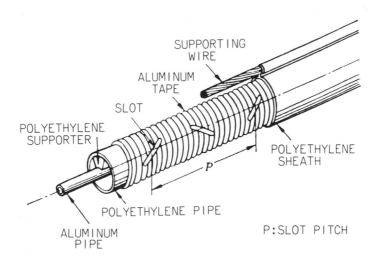

Figure 8.38 LCX cable structure. (From [11], © *IECEJ*.)

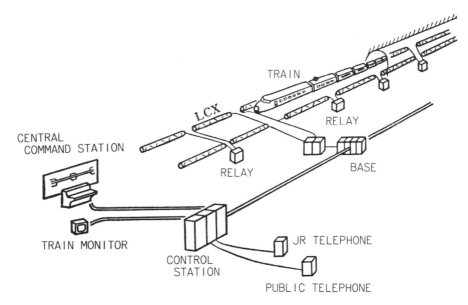

Figure 8.39 Outline of an LCX cable communication system. (From [11], © *IECEJ.*)

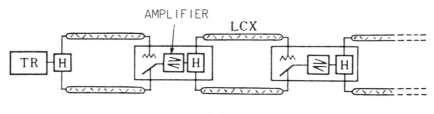

TR: TRANSMITTER, H: HYBRID CIRCUIT

Figure 8.40 LCX relay system. (From [11], © *IECEJ.*)

keep to the left in Japan) is usually used for communication. When there is a fault, the communication channels are changed to the right-side cable.

The vertical radiation pattern of LCX cables has a broad half-power beamwidth, and a measured vertical radiation pattern is shown in Figure 8.44. The maximum radiation of LCX cables in the horizontal direction occurs slightly backward (i.e., toward a transmitter direction), and the peak radiations of the LCX cables for inbound and outbound trains are slightly different. Thus, the radiation pattern of an antenna has two peaks corresponding to those of the LCX cables, as shown in Figure 8.45(a). On the other hand, vertical radiation patterns of a train antenna must have a broad half-power beamwidth [Figure 8.45(b)] coping with LCX cable locations having different heights and the inclination of

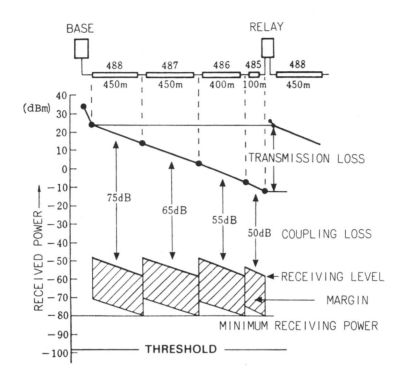

Figure 8.41 LCX cable combination for signal attenuation grading. (From [12], © *IECEJ.*)

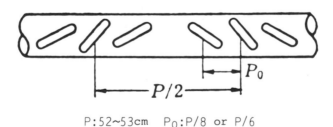

P:52~53cm P_0:P/8 or P/6

Figure 8.42 LCX slot configuration. (From [12], © *IECEJ.*)

the train on a curve. The half-power beamwidth of the vertical plane of the train antenna is about 110 deg. The gain is about 5 dB with respect to a half-wave dipole antenna.

Antennas on the train are as follows. One folded slot antenna is shown in Figure 8.46 and the four equiphase folded slot array assembly shown in Figure 8.47 is placed on each side of the train, as in Figure 8.43. The four elements are equally fed by coaxial cables of the same length. The spacing between the adjacent elements is 50 cm. A matching

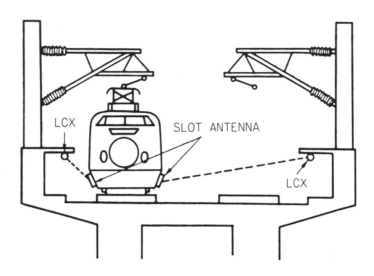

Figure 8.43 LCX cable and train antenna location. (From [12], © *IECEJ*.)

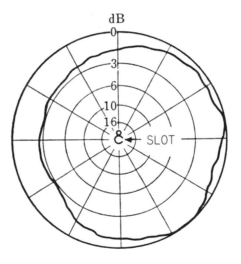

Figure 8.44 Vertical radiation pattern of an LCX cable. (From [11], © *IECEJ*.)

circuit is needed for different frequencies: 412 MHz (transmitting) and 452 MHz (receiving).

There are two four-element folded slot arrays on each side of a front car. A train transmitter will be connected to arrays on the side where the received power level is higher. Thus, communication can be maintained when one of the LCX cables has a communication problem. The power level of an auxiliary array on each side is 20 dB

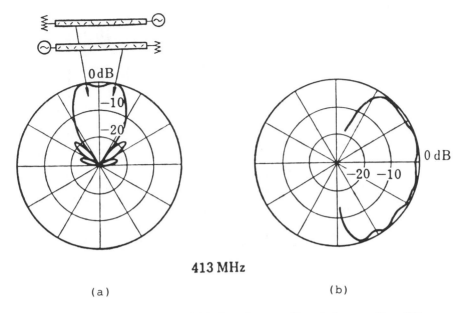

413 MHz

(a)   (b)

Figure 8.45 Train antenna radiation pattern: (a) horizontal pattern; (b) vertical pattern. (From [11], © *IECEJ*.)

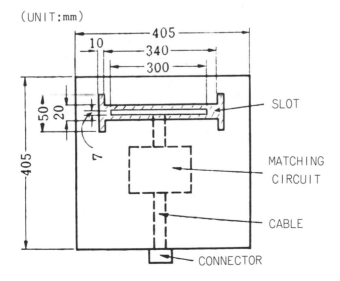

Figure 8.46 Radiating element structure of a train antenna. (From [12], © *IECEJ*.)

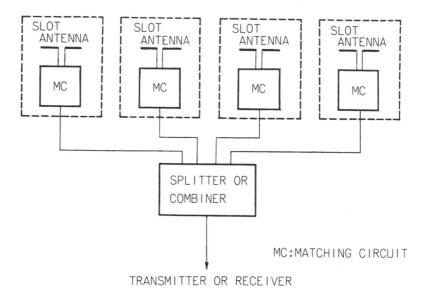

MC:MATCHING CIRCUIT

Figure 8.47 Train antenna configuration. (From [12], © *IECEJ*.)

lower than that of the main array. The auxiliary array is combined with the output of the main array to prevent the power level from abruptly dropping due to environmental effects, such as steel poles along the railways.

8.4 INTELLIGENT TRANSPORTATION SYSTEMS

8.4.1 General

Intelligent transportation systems (ITS) are defined under a broad concept of traffic systems and are aimed at improvement of road traffic, safety, efficiency, and comfort in transportation by introducing electronics and communication systems, control and information systems, signal processing, and computer hardware and software into transportation systems. Expected benefits are the reduction of energy consumption and exhaust emissions, with a consequent improvement of environmental conditions, thus aiding the health and welfare of general society.

The significance of this concept has been recognized worldwide, and research and development of ITS in various aspects have been promoted in many countries of the world. ITS history can be traced from at least the early 1960s when various trials of electronic route guidance systems and traffic information systems were made. Since then, some transportation systems have been developed based on the results obtained via trials and experiences.

Wireless communication systems play one of the most essential roles in ITS. By means of communications between vehicles and roadside facilities and also vehicle-to-vehicle communications, various new services such as information systems, traffic control and management, and vehicle operation and management systems, for example, can be realized. Communication systems employed in ITS are essentially the same as what wer call "mobile communications." However they differ from the conventional mobile communications, because they deal mainly with data, not voice. In most cases the operators are not aware of the communication. In the conventional sense, operators operate systems. For instance, in order to transmit and receive voice, an operator would expect to interact with a mobile communication system, but in ITS, by contrast, data are transmitted and received automatically, perhaps with only derived results being displayed or communicated to the operator.

Antennas used for communication systems in ITS require special considerations for the design in most cases. Antennas should be designed individually to satisfy the specific requirements for the systems. For instance, a type of phased array has been designed for electronic toll collection (ETC) systems, where the radiation pattern should illuminate a specified area on the lane at the toll gate. In an information system such as a vehicle information and communication systems (VICS), a two-beam antenna is specifically designed so that the direction of vehicle motion is identified and vehicles can get information depending on which lane the vehicle is using.

Inductive communication systems have been used for traffic information and control systems since at least the late 1960s. The operating frequency is near 100 kHz. Antennas used are a loop element embedded in or near the surface of the road and a coil element installed on the vehicles. In early 1970, ferrite antennas were applied to both the roadside and vehicles. An example is the antennas used in the bus management systems in Japan. Because inductive communication is a system that operates at the low-frequency bands, and handles only low speeds and low data rates, it can hardly be applied to modern systems. The bus management system, to which the digital multichannel access (MCA) system is applied, has been developed and practical operation began in early 1998. The operating frequency is 1.9 GHz. Yagi-type antennas used for the roadside systems and a monopole for the buses. Presently, use of higher frequency bands such as 2.4 and 5.9 GHz, in addition to the cellular frequency bands, has been allocated for ITS systems.

Antennas for ETC and VICS are described in the following sections.

8.4.2 Antennas for ETC

Design of Antennas for ETC

ETC is a system that enables vehicles to pass through a toll gate without stopping for the toll payment. The toll collection is processed automatically while vehicles pass through the toll gate by means of wireless communications between vehicles and the toll gate facilities. The non-stop driving and cashless payment at the toll gate should reduce traffic congestion, fuel consumption, and exhaust-gas concentration. This also improves services for drivers and reduces the costs of toll management. The ETC system is composed of the wireless communication system and the toll processing system.

The communication system consists of the wireless system between vehicle and roadside, its control systems, and the data processing system pertaining to the toll processing. This communication system is an application of the dedicated short-range communication (DSRC) systems, which are spot-area or small-area communications performed primarily between the roadside facilities and vehicles.

The ETC system is specified to use the operating frequency of the 5.8-GHz bands [22]. The modulation of the transmitted signal is ASK, and the data rate is 1024 kbps. The power input to the antenna is 10 mW for a range of less than 10m, and 300 mW for between 10 and 30m. The antenna gain is less than 20 dBi for the roadside and less than 10 dBi for the vehicle. Right-hand circular polarization is used. The number of channels available is eight and the TDMA-FDD is employed for channel access. The system should work well for a maximum vehicle speed of 180 km/h, and the error in the data processing should be less than 10^{-6} per vehicle per gate.

A brief sketch of the ETC system is shown in Figure 8.48. There are two areas where data communication between vehicles and the toll office at the entrance gate is performed for the toll processing, so that two antennas are used as Figure 8.49 shows. The data necessary for the toll settlement at the entrance gate are processed automatically while the vehicles pass through the two communication areas. At the exit gate there are also two areas where communication between vehicles and the toll office is performed in order to complete the toll process.

The antenna used at the toll gate is specifically designed to have radiation patterns that illuminate the communication area uniformly and have low sidelobes in order to avoid interference caused by the reflected, diffracted, and/or scattered waves that are generated by the fixed objects around the gate, and also to eliminate noise and signals from other vehicles on the same and/or adjacent lanes. This pattern must be precisely designed because errors could not be allowed in the toll process. The antenna is usually mounted on the gantry as shown in Figure 8.48. The height of the antenna over the lane is about 5m and the communication area is specified for an area 4m long and 3m wide at an elevation of 1m over the lane as illustrated in Figure 8.50 [22]. Circular polarization (right-hand) is used, because the waves reflected back may have inverse polarization and not interfere with the signals.

An example of the radiation pattern is depicted in Figure 8.51, which shows the 3-dB beamwidth of over 32 deg and the sidelobe level of less than −30 dB in the horizontal plane, while in the vertical plane they are over 20 deg and less than −20 dB, respectively. The axial ratio is less than 3 dB. To achieve this kind of pattern, a sector beam antenna is preferred and a design example is described next.

Method of Design

A sector beam antenna, which ideally has uniform radiation over the main beam (a sector of space) and zero sidelobes, is considered to be the most appropriate antenna for the

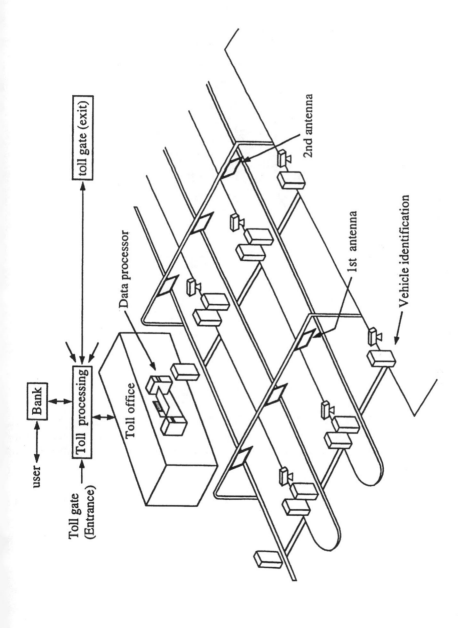

Figure 8.48 Brief sketch of toll gate with ETC installed.

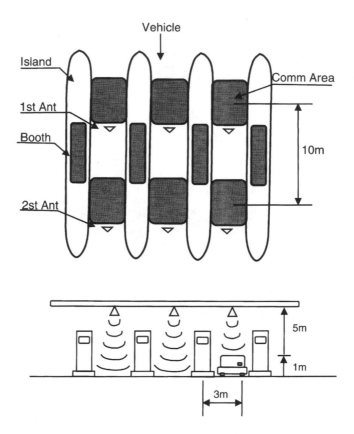

Figure 8.49 Antenna arrangement and communication areas at the ETC gate.

ETC gate. Several methods are used to synthesize the sector beam, for example, the Fourier transform method and the Woodward-Lawson sampling method. Here an application of Remez's algorithm [23] is introduced. This algorithm is usually used for designing a digital filter, which has very sharp skirt attenuation characteristics.

The concept of Remez's algorithm is to obtain the function that can approximate the desired function by minimizing the maximum value of the difference between the synthesized function and the desired function. Let $f(\theta)$ approximate the desired radiation pattern $f_0(\theta)$. Here $f(\theta)$ is considered to be a function with N degrees of freedom, which is expressed by

$$f(\theta) = \sum_{n=1}^{N} a_n \psi_n(\theta) \tag{8.6}$$

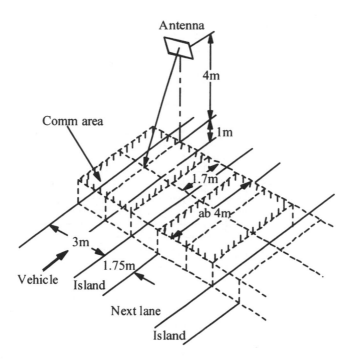

Figure 8.50 Communication areas at the ETC toll gate.

where a_n are the unknown amplitude constants to be determined and $\psi_n(\theta)$ denotes the nth basis function. Now the deviation $\xi(\theta)$ of the synthesized pattern $f(\theta)$ from the desired pattern $f_0(\theta)$ is expressed by

$$\xi(\theta) = \sum_{n=1}^{N} a_n \psi_n(\theta) - f_0(\theta) \quad (a \le \theta \le b) \tag{8.7}$$

where the function here is defined in the region $[a,b]$.

Then the synthesizing process is to find a_n when the maximum value of $|\xi(\theta)|$ will become minimum with respect to θ in the range $[a,b]$.

1. Set θ_m ($m = 1, 2, \ldots, N + 1$) as the initial values for the points, where $|\xi(\theta)|$ takes the extreme value, as follows:

$$a \le \theta_1 < \theta_2 < \ldots < \theta_{N+1} \le b \tag{8.8}$$

2. By using θ_m, solve the following equation of a_n ($n = 1, 2, \ldots, N$) and α

$$\sum_{n=1}^{N} a_n \psi_n(\theta_m) + (-1)^m \alpha = f_0(\theta_m) \ (m = 1, 2, \ldots, N + 1) \tag{8.9}$$

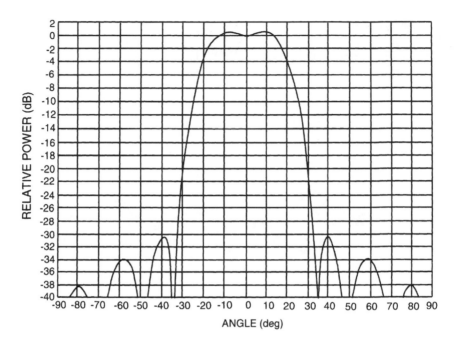

Figure 8.51 An example of the radiation pattern for ETC gate antenna. (Courtesy of Yagi Antenna Corp., Japan.)

where α is a value that corresponds to $|\xi(\theta_m)|$. In the antenna synthesis, there are often cases where a_n is determined so as to achieve the desired sidelobe level α, with the main radiation level being unity.

3. By using a_n $(n = 1, 2, \ldots , N)$ obtained in step 2, set new θ_m for the points $|\xi(\theta)|$ takes the extreme value as in step 1 as follows.

$$a \le \theta_1 < \theta_2 < \ldots < \theta_{N+1} \le b \tag{8.10}$$

In the process, first obtain θ'_m by differentiating $\xi(\theta'_m)$ with respect to θ and equating it to zero; that is,

$$d(\theta'_m)/d\theta = 0, \ (m = 1, 2, \ldots N + 1) \tag{8.11}$$

Then take up $(N + 1)$ of $\xi(\theta)$ among $\xi(\theta'_m)$, $\xi(a)$, and $\xi(b)$ in order of increasing value. These are the θ values expressed in (8.10).

4. If $|\xi(\theta_m)|$ with respect to m is less than the desired value, the process ends. If not, return to step 2 and repeat the process.

After completion of the iteration, $f(\theta)$ is determined.

Example of Antenna

A 4 × 8 microstrip array is introduced as an example of an antenna for the ETC gate. The MSA array pattern is shown in Figure 8.52, where the amplitude distribution of the driving voltages is given, as obtained from Remez's algorithm. The size of the ground plane is 295 × 162.5 mm. The separation of the elements is $0.67\lambda_0$ (λ_0 = the wavelength in free space) for the horizontal array and $0.73\lambda_0$ for the vertical array. In Figure 8.53, the synthesized radiation pattern in the horizontal plane is shown by dots and the experimental result by a line [24]. The 3-dB beamwidth is 36 deg, the sidelobe level is less than −29 dB over ±30 deg, and the axial ratio is 1 dB at the beam maximum. These approximate the desired values closely.

8.4.3 Vehicle Information and Communication System

The Vehicle Information and Communication System (VCIS) provides real-time road and traffic information to drivers. It is expected that VICS contributes to reducing traffic congestion, decreasing fuel consumption, and improving driving conditions. Four media are used for providing information to drivers. They are the radio beacon, the infrared beacon, the FM multiple broadcast, and the GPS as shown in Figure 8.54.

The system guides drivers on a route, by which drivers can reach the destination through the shortest route or with the shortest time. Drivers are informed about traffic congestion, the road regulation, road work, accidents, and so forth, on their way, so that a better route can be selected. This will contribute to reducing fuel consumption, save driving time, and aiding driving safety.

The information is given by voice and images on the display installed on the vehicle. Drivers know precisely where they are by means of GPS, if GPS information is combined with the VICS system.

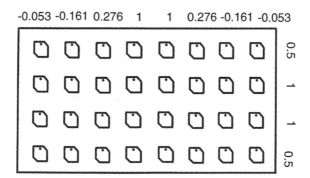

Figure 8.52 MSA array pattern and feed amplitude distributions of the array elements. (Courtesy of Yagi Antenna Corp., Japan.)

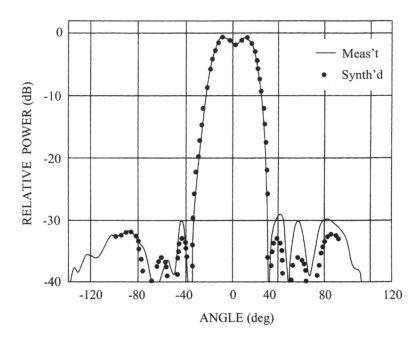

Figure 8.53 Radiation patterns: (a) synthesized pattern; (b) measured pattern. (Courtesy of Yagi Antenna Corp., Japan.)

The radio beacon employs a spot zone for communication on the lane, and links the information center with vehicles by a wireless system. The spot zones are areas about 70m long and 10–15m wide, located intermittently along the road within a distance of 2–5 km, as Figure 8.55 shows.

The system operates in the 2.4-GHz band. The roadside system transmits two signals to vehicles, a 64-kbps GMSK data signal and a 1-kHz pulsed AM signal. At first, these two signals enable a vehicle to determine its position (the location of the roadside) and the direction of the travel. Then the vehicle extracts information from the data signals pertaining to the direction of travel. The information includes route guidance, road traffic, regulations, accident, road work, and so forth, along the path the vehicle is going to take.

Two antennas are used as shown in Figure 8.56. Each antenna produces a beam directing the driver to the right-hand side or the left-hand side of the lane, respectively. These two beams are combined in space to cover both sides of a lane. The radiation pattern in the horizontal plane is shown in Figure 8.57(a) and that in the vertical plane in Figure 8.57(b). The data signals are transmitted by two antennas at the same time, while the AM signal is transmitted by two antennas, but separately with inverse phases relative to each other. As a result, regions where in-phase and anti-phase signals are received are produced on the lane as shown in Figure 8.58(a). Figure 8.58(b) shows the

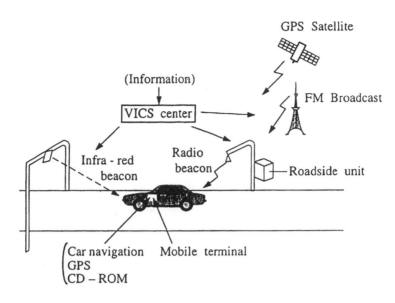

Figure 8.54 Four media for providing information to vehicles in VICS.

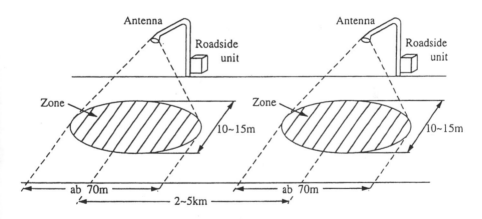

Figure 8.55 Communication zones in VICS system.

pattern produced by the two antennas transmitting the data signals, and also the pattern produced by the two antennas transmitting the AM signals with inverse phase relationship. The "in-phase" here is defined for a case where the beginning of the data frame is synchronized with the rising edge of the AM pulse, while the "anti-phase" region is that where the beginning of the data frame is synchronized with the declining edge of the AM pulse as Figure 8.58(c) illustrates.

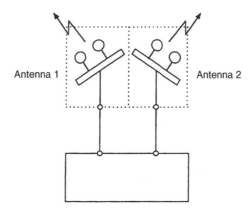

Figure 8.56 Two-antenna arrangement.

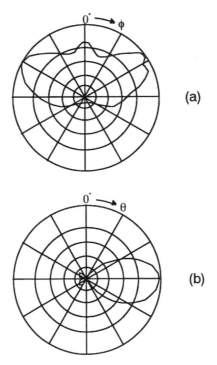

Figure 8.57 Radiation patterns: (a) in the horizontal plane achieved by spatially combining two beams; (b) in the vertical plane. (Courtesy of Matsushita Communication Ind. Co., Japan.)

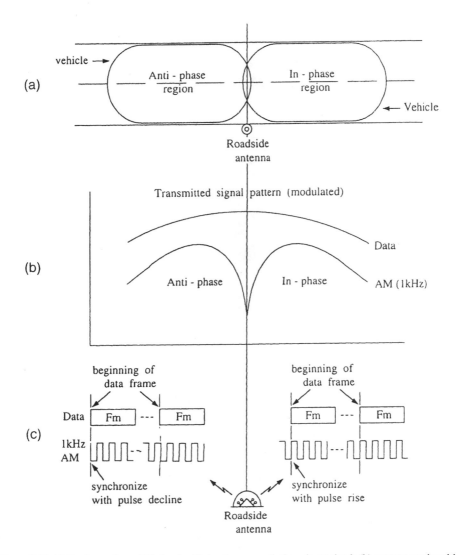

Figure 8.58 (a) Regions where AM signal with in-phase or anti-phase is received; (b) patterns produced by data transmission and by AM signal transmission; (c) phase relationship between a data frame and an AM pulse for identifying direction of vehicle travel.

When a vehicle passes through the anti-phase region, the in-phase region identifies the signal phase and the vehicle can then know the direction of motion, which is from the left-hand side to the right-hand side in Figure 8.58(a). It is necessary for the vehicle to identify the direction of movement, because the information differs depending on the direction of travel. An example of the antenna is shown in Figure 8.59, which shows (a) an antenna mounted on a pole, (b) an inside view, and (c) a printed element pattern.

(a)

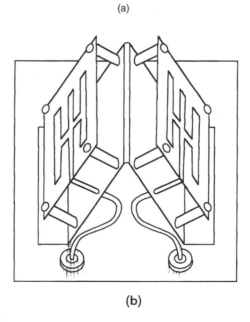

(b)

Figure 8.59 Roadside antenna: (a) an antenna mounted on a pole; (b) an inside view; (c) a printed pattern of the antenna element. (Courtesy of Matsushita Communication Ind. Co., Japan.)

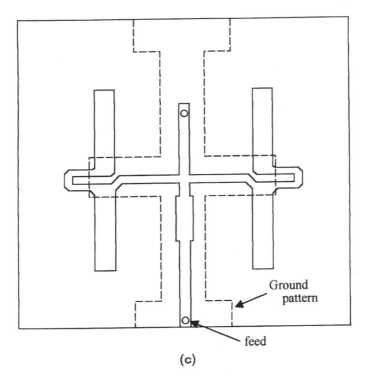

(c)

Figure 8.59 (continued).

An antenna used on a car must be small and compact, because it is usually mounted on the narrow dashboard in the car. Consequently, planar antennas are generally preferred. Typical ones include a PIFA or a microstrip antenna (MSA). Figure 8.60 illustrates an example of a car-mounted PIFA: (a) a sketch, (b) a front view, and (c) a side view. The antenna is directly connected at the feed point to a LNA (low-noise amplifier), which is located beneath the ground plane as shown in Figure 8.60(c). The gain of the antenna is specified to be over −1 dB with VSWR less than 2. The LNA has a gain of typically 15 dB with a noise figure of 2 dB. Figure 8.61 shows an example of a car-mounted MSA, which uses high permeability ceramic substrate: (a) a sketch, (b) the radiation pattern, (c) the axial ratio, which is given by the difference between the relative power of the right-hand circular polarization (RHCP) and that of the left-hand circular polarization (LHCP), and (d) impedance characteristic and VSWR. The permeability of the substrate is 20.

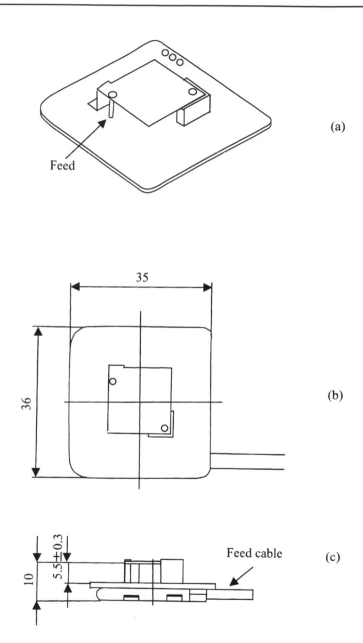

Figure 8.60 An example of a car-mounted PIFA: (a) a sketch; (b) a front view; (c) a side view. (Courtesy of Nippon Antenna Co., Japan.)

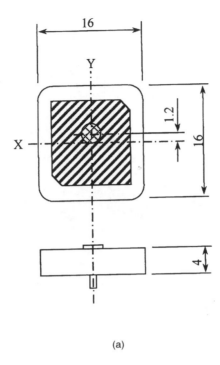

(a)

Figure 8.61 An example of a car-mounted MSA: (a) a sketch; (b) the radiation pattern; (c) the axial ratio (in terms of relative power ratio of RHCP and LHCP); (d) the input impedance and VSWR. (Courtesy of Miyazaki Matsushita Electric Co., Japan.)

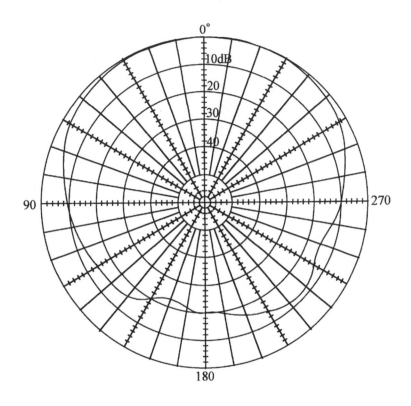

(b)

Figure 8.61 (continued).

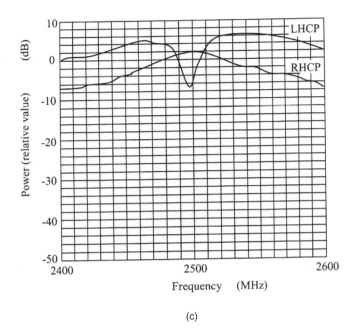

(c)

Figure 8.61 (continued).

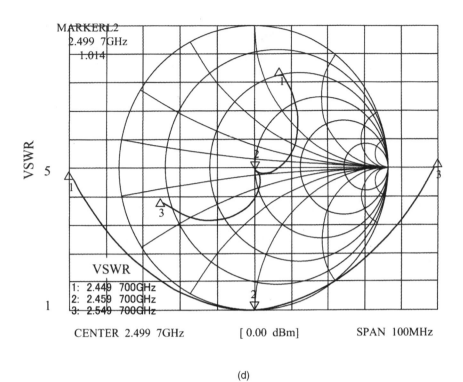

(d)

Figure 8.61 (continued).

REFERENCES

[1] Fujimoto, K., and J. R. James, eds., *Mobile Antenna Systems Handbook*, Norwood, Mass.: Artech House, 1994, pp. 277–293.

[2] CCIR, "Planning Standards for FM-Sound Broadcasting at VHF," *Recommendations and Reports of the CCIR*, Vol. X-1, Rec. 412-3, ITU, Geneva, 1982.

[3] Manner, E. J., "Distortions and Their Reduction by Antenna Diversity with Mobile FM Reception," doctoral thesis, University of the Bundeswehr Munich, 1985 (in German).

[4] Lindenmeier, H. K., J. F. Hopf, "Investigations for the Evaluation of the Minimum Required Mean Signal Level with Mobile FM Radio Reception," *Rundfunktechn. Mitteilungen*, Jahrg. 28, 1984 (in German).

[5] Lindenmeier, H. K., "Optimum Bandwidth of Signal-to-Noise Ratio of Receiving Systems with Small Antennas," *AEÜ*, Band 30, Heft 9, 1976.

[6] Lindenmeier, H. K., J. F. Hopf, and L. M. Reiter, "Active Window Antennas for Radio Reception in Cars for Single Antenna and for Antenna Diversity Application," *Journées Internationales de Nice sur les Antennes* (JINA), 1986, Nice, France.

[7] CCIR Working Group 10-B, "Diversity Rec. in Automobiles for Frequency Modulated Sound Broadcasts in Band 8 (VHF)," Doc. 10/160-E.

[8] Toriyama, H., et al., "Development of Printed-on-Glass TV Antenna Systems for Car," *IEEE VT-S Conf. Digest*, 1981, pp. 334–342.

[9] Technical data described in this part are provided by courtesy of Nissan Motor Co., Japan.

[10] Kishimoto, T., and S. Sasaki, "Train Telephone System," *Proc. IECEJ*, Vol. 63, No. 2, Feb. 1980, pp. 128–133.

[11] Kishimoto, T., and S. Sasaki, "LCX Communication Systems," *IECEJ*, Tokyo, 1982.

[12] Watanabe, H., "Electronic Control and Communication System of Shinkansen," *IECEJ*, Tokyo, 1982.

[13] Hayashi, Y., "Train Radio System of Shinkansen," *Proc. IECEJ*, Vol. 65, No. 5, May 1982, pp. 541–543.

[14] Taguchi, K., et al., "Recent Train Radio Communication Systems," Mitsubishi Electric Co. Technical Report, Vol. 61, No. 2, Feb. 1987, pp. 33–37.

[15] Amemiya, Y., "Surface Wave Coaxial Cable," *Proc. IECEJ*, Vol. 48, No. 12, April 1965, pp. 131–142.

[16] Mikoshiba, K., Y. Nurita, and S. Okada, "Radiation From a Coaxial Cable and Its Application to a Leaky Coaxial Cable," *Trans. IECEJ*, Vol. 51-B, No. 10, Oct. 1968, pp. 499–505.

[17] Yoshida, K., "New Communication Systems by Leaky Coaxial Cables," *Proc. IECEJ*, Vol. 55, No. 5, May 1972, pp. 655–663.

[18] Cree, D. J., and L. J. Giles, "Practical Performance of Radiating Cables," *Radio & Electronic Engineering*, Vol. 45, No. 5, May 1975, pp. 215–223.

[19] Mikoshiba, Y., S. Okada, and S. Aoki, "Near Electromagnetic Fields Around a Leaky Coaxial Cable," *Trans. IECEJ*, Vol. 54-B, No. 12, Dec. 1971, pp. 789–796.

[20] Delogne, P., *Leaky Feeders and Subsurface Radio Communication*, London: Peter Peregrinus, Ltd., 1982.

[21] Kurauchi, N., K. Yoshida, and Y. Miyamoto, "Wideband Leaky Coaxial Cable," *Trans. IECEJ*, Vol. 54-B, No. 10, Oct. 1971, pp. 682–686.

[22] ARIB STD-T55

[23] Temes, D. C., et al., "The Optimization of Bandlimited Systems," *Proc. IEEE*, Vol. 61, No. 2, 1973, p. 196.

[24] Arai, Y., T. Iino, and K. Fukai, "Sector Beam Array Antenna for Electronic Toll Collection System," *Proc. '97 IEICE General Conf.*, B-1-61, 1997 (in Japanese).

Chapter 9

Antennas for Mobile Satellite Systems

T. Shiokawa, S. Ohmori, T. Teshirogi, M. Williamson,
and O. Edvardsson

9.1 INTRODUCTION

The decade of the 1990s has been an era of mobile satellite communications serviced on a commercial basis. Since 1982, the INMARSAT system has provided international maritime satellite communication services and is expanding the services to aircraft and land mobiles. On the other hand, systems such as American Mobile Satellite Corporation (AMSC) in the United States, MSAT in Canada, and AUSSAT in Australia [1] provide domestic satellite communication services mainly for land mobiles using dedicated satellites. The research and development activities on mobile satellite communications have continued since the mid-1970s in many countries and organizations. Typical research programs are MSAT in Canada [1], the ETS-V in Japan [2], MSAT-X in the United States [3], and PROSAT in Europe [4].

The mobile communication systems mentioned above are using geostationary earth orbiting (GEO) satellites, but recently new concepts have been proposed by several private sectors in the United States. The new systems are called low earth orbiting (LEO) satellite communication systems, which use a group of low-altitude orbiting satellites. The typical examples of LEO satellite systems are Iridium, Odyssey, and Globalstar, which use 66, 12, and 48 satellites in low earth orbits, respectively. The main advantage of LEO satellite systems from the point of view of antenna design is that mobile and handy terminals can use low-gain omnidirectional antennas, because of smaller values of free-space propagation loss compared with those of GEO satellite systems.

In implementing mobile satellite communications, a vehicle antenna is one of the most important key technologies. In this chapter, a vehicle antenna means an antenna system that is mounted on a mobile for satellite communications. This chapter describes system requirements and antenna design considerations for PROSAT, ETS-V, MSAT-X, MSAT, INMARSAT (International Maritime Satellite Organization), GPS, and broadcasting systems.

To assist the reader, an outline of mobile satellite communication systems is briefly mentioned. The allocation of frequency, which is an essential factor in designing and analyzing antennas, is defined by the Radio Regulation on an international basis, depending on service types such as communication, navigation, and broadcasting.

Communication

Typical mobile satellite communication systems are INMARSAT, AMSC, MSAT, and AUSSAT, in which the L-band (1.6/1.5 GHz) is used. The only exception is Omni-TRACS in the United States, which is serviced in the Ku-band (14/12 GHz). In Japan, use of the S-band (2.5/2.4 GHz) has been studied for a domestic system. In recent years, research on the feasibility of advanced mobile satellite communications using the Ka-band (30/20 GHz) has started, and typical research and development programs are the ACTS in the United States [5] and the COMETS in Japan [6].

Navigation

A typical system is the GPS, in which L-band (1.6 and 1.3 GHz) frequencies are used from a satellite to the earth.

Broadcasting

This service is to broadcast TV and radio programs from a satellite to the earth. Although the present systems are designed for fixed terminals, not for mobiles, an antenna system for trains and ships has been developed to receive TV programs from the broadcasting satellite (BS) satellite in Japan.

9.2 SYSTEM REQUIREMENTS FOR VEHICLE ANTENNAS

9.2.1 Mechanical Characteristics

Compact and lightweight equipment is a self-evident requirement for vehicle antennas in addition to ease of installation and mechanical strength. In the case of shipborne antennas, the installation requirement is not as severe compared to that of aircraft and cars, because even in small ships there is a little space to install an antenna system. However, in the case of cars, especially for small, private cars, low-profile and lightweight equipment is an essential requirement. Although the demands are the same for aircraft, more severe conditions are required to satisfy the standards for avionics. The low air drag must be one of the most important requirements for aircraft antenna [7].

9.2.2 Electrical Characteristics

Frequency and Bandwidth

In almost all present and forthcoming systems using GEO satellites, the L-band (1.6/1/5 GHz) is used in communication links between the satellite and mobiles. The required frequency bandwidth in L-band communication systems is about 8% to cover transmitting and receiving channels. So in using a narrow-band antenna element such as a patch antenna, some efforts have to be made to widen the bandwidth.

Gain, Beamwidth, and Beam Coverage

The required gain is decided by a link budget, which is calculated by taking into account the required channel quality (expressed as C/No, which is the carrier-to-noise power density ratio) and the satellite capability.

Figure 9.1 shows the relationship between gain and aperture size, and Figure 9.2 shows the relationship between gain and half-power beamwidth (HPBW), respectively. From Figure 9.2, the HPBW of a 15-dBi antenna is about 30 deg, and that of a 4-dBi antenna is about 100 deg. In general, since an antenna beam is required to cover 0 to 90 deg in elevation and 0 to 360 deg in azimuth directions (upper hemisphere), the former antenna needs a tracking capability. Table 9.1 shows a classification of directional and omnidirectional antennas. In the INMARSAT system, as shown in Table 9.1, terminals have been classified by their receiving capabilities and services. The outline specification of INMARSAT terminals is given in Table 9.2.

Polarization, Axial Radio, and Sidelobes

Circular polarized waves are used in order to eliminate the need for polarization tracking. In the INMARSAT system, right-hand circular polarization has been used. In the case of aperture-type antennas such as the parabolic antenna, which is commonly used as a shipborne antenna in the current INMARSAT-A terminal, an axial ratio of below 1.5 dB in a boresight direction is so easy to achieve that polarization mismatch loss is almost negligible. However, in the case of phased-array antennas, it must be taken into account because of degradation of the axial ratio caused by beam scanning.

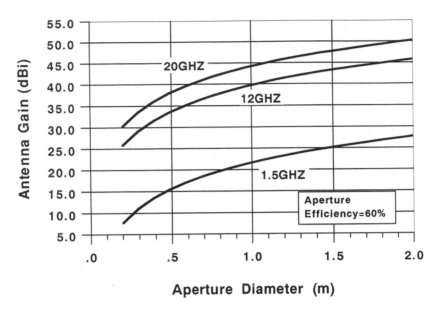

Figure 9.1 Relationship between gain and aperture size.

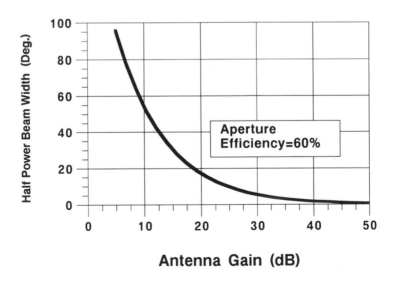

Figure 9.2 Relationship between gain and half-power beamwidth.

Table 9.1
Classification of L-Band Antennas in Mobile Satellite Communications

Antenna	Gain Class	Typical Gain (dBi)	Typical G/T (dBk)	Typical Antenna (Dimensions)	Typical Service
Directional	High	20–24	−4	Dish (1 mϕ)	Voice/high-speed data Ship (INMARSAT—A,B)
Directional	Medium	12–15	−10 to −13	SBF (0.4 mϕ) Array Phased array	Voice/high-speed data Ship (INMARSAT—M) LM (INMARSAT—M) Aircraft (INMARSAT—Aero)
Omnidirectional	Low	0–4	−23	Quadrifilar Drooping-dipole Patch	Low-speed data Ship (INMARSAT—C) Aircraft Land mobile

Note: LM = land-mobile; SBF = short backfire antenna.

Table 9.2
Specification of INMARSAT Terminals

Terminal	G/T (dBk)	EIRP (dBw)	Gain (dBi)	Main Communication Service	Main Service
A	−4	36	20–23	Voice, telex	Ship
B	−4	33	20–23	Voice,* telex, data	
C	−23	16	0–3	Message	Very small ship, land mobile
M	−10	25	15[†]	Voice	Small ship
	−12	25	12[‡]	Data	Land mobile
Aero	−13	25.5	12	Voice	Aircraft
	−26	13.5	0	Data	Aircraft

*Digital.   [†]Maritime.   [‡]Land.

Figure 9.3 shows the relationship between polarization mismatch loss and axial ratios of a receiving antenna, in which ARa and ARb denote axial ratios of transmitting and receiving antennas, respectively. From Figure 9.3, it is found that in order to get the polarization mismatch loss below 0.5 dB, an axial ratio of a vehicle antenna is required under 5 dB in all directions, because an axial ratio of a satellite antenna is, in general, below 2 dB. This requirement must be considered for a phased-array antenna in scanning beam over wide angular areas.

For large antennas, with diameters over 100λ (wavelengths), a reference radiation pattern is recommended by the CCIR [8] for interference to and from other satellite and terrestrial communication systems. However, the diameters of the vehicle antennas under

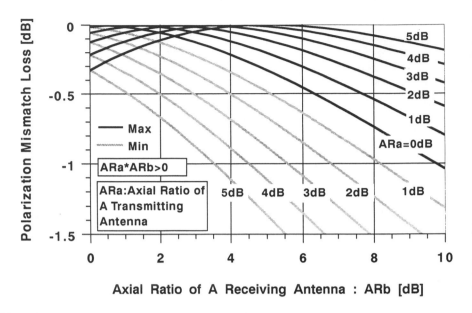

Figure 9.3 Relationship between polarization mismatch loss and axial ratio of receiving antenna.

discussion are, in many cases, below five wavelengths in the L-band. Further CCIR action is expected to define a reference radiation pattern for vehicle antennas in mobile satellite communications.

G/T and EIRP

Although gain is an essential factor in considering antennas, the figure of merit G/T (which is the ratio of gain to system noise temperature) is more commonly specified from the standpoint of satellite communications. System noise temperature T at an input port of a low-noise amplifier (LNA) is defined by (9.1) [9], and the system gain G is defined at the same port, taking account of losses caused by tracking, feed lines, and a radome.

$$T = T_a/L_f + T_0(1.0 - 1.0/L_f) + T_r \qquad (9.1)$$

T_a is an antenna noise temperature which comes from such effects as ionosphere and the earth. Although T_a depends on factors such as frequency and beamwidth, a typical value in the L-band is about 200K [10]. T_0 is the temperature of the environment, usually 300K. T_r is the noise temperature of a receiver (LNA), and its typical value is about 80K to 100K in the L-band. L_f is a total loss of feed lines and components such as diplexer, cables, and phase shifters if a phased-array antenna is used. Figure 9.4 shows a relationship between G/T and feeder loss in the case of a 15-dBi antenna, which is a typical candidate

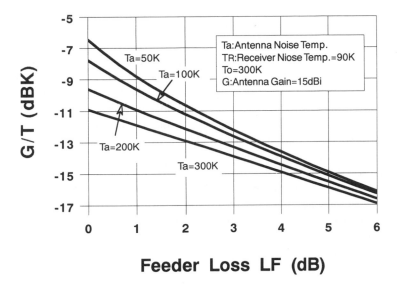

Figure 9.4 Relationship between G/T and feeder loss.

in the forthcoming systems. It is found that if G/T is required over −13 dBK, feeder loss must be under about 3 dB.

Equivalent isotropically radiated power (EIRP) is an important value in evaluating the transmitting performance of a terminal including an antenna. EIRP (dBW) is defined by a sum of the antenna gain (dB) and output power of the high-power amplifier (dBW), taking account of feed losses such as feed lines, cables, and a diplexer.

9.2.3 Propagation Problems

Telecommunications over earth-space links for maritime mobile satellite systems lead to propagation problems substantially different from those arising in the fixed satellite service [11]. For instance, the effects of reflections and scattering by the sea surface become quite severe, particularly where antennnas with wide beamwidths are used. Furthermore, maritime mobile satellite systems may operate on a worldwide basis, including paths with low elevation angles. Moreover, due to the use of L-band frequencies for INMARSAT systems, the effect of ionospheric scintillation is not negligible, particularly in equatorial regions during the years of high solar activity of an 11-year cycle. On the other hand, tropospheric effects such as rain attenuation and tropospheric scintillation will be negligible for the said frequency band. Signal level attenuation due to blocking by ship superstructure is considerable problem. In the following sections, brief explanations are presented.

Multipath Fading Due to Sea Surface Reflection

Multipath fading due to sea reflection is caused by the interference between direct and reflected radio waves. The reflected waves are composed of a coherent (specular reflection) component and an incoherent (diffuse) component that fluctuate due to the motion of the sea waves. The coherent component is predominant under calm sea conditions, and the incoherent component plays an important role in rough sea conditions. If the intensity of the coherent component and the variance value of the incoherent component are both known, the cumulative time distribution of the signal intensity can be determined by statistical considerations [12,13].

The amplitude of the coherent component decreases rapidly with increasing wave height, elevation angle, and frequency. Figure 9.5 shows estimates of amplitude of the coherent component for an omnidirectional antenna as a function of significant wave height for low elevation angles; the frequency is 1.5 GHz and polarization is circular. The incoherent component is random in both amplitude and phase, since it originates from a large number of reflecting facets on the sea waves. The amplitude of this component follows a Rayleigh distribution and the phase has a uniform distribution.

Since the theoretical model concerning the incoherent component is not suitable for engineering computations using a small calculator, simpler prediction models are useful for approximate calculation of fading. Such simple methods for predicting multipath power or fading depth have been developed [14,15]. Figure 9.6 shows the relationship between multipath power and elevation angle for different antenna gains based on the method in [15]. Although fading depth depends slightly on sea conditions, even if the incoherent

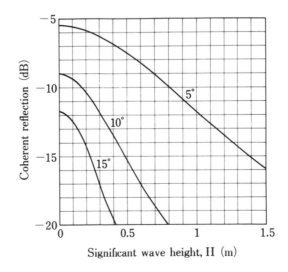

Figure 9.5 Relationship between coherent reflection and significant wave height.

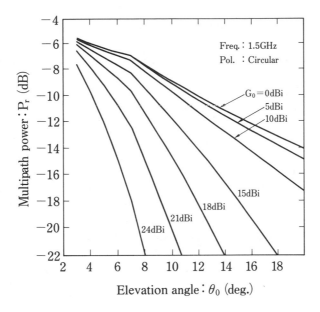

Figure 9.6 Relationship between multipath power and elevation angle.

component is dominant, as shown in [13], the simple model is useful for the rough estimate of fading depth.

Blocking by Ship Superstructure

Blocking is caused by ship superstructures such as the mast and various types of other antennas. The geometry for blocking by a mast is shown in Figure 9.7. Attenuation due to blocking depends on various parameters such as diameters of the column, distance

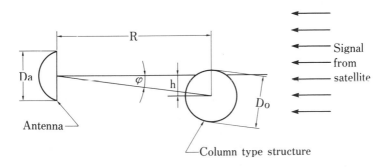

Figure 9.7 Geometry of blocking.

between antenna and column, and size of antenna. Based on experimental data reported so far, attenuation due to blocking caused by a column-type structure is given in Figure 9.8 for antenna gains of 20 and 14 dBi [11].

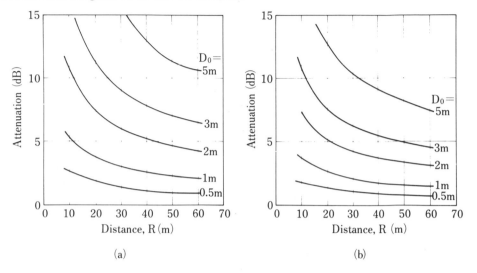

(a)

(b)

Figure 9.8 Attenuation due to blocking: (a) antenna gain: 20 dB; (b) antenna gain: 14 dB.

9.2.4 Fading Reduction Techniques

For low- and medium-gain systems, the effect of multipath fading due to sea surface reflection is a severe problem, especially at low elevation angles, as pointed out previously. In this section, possible fading reduction techniques for low- and medium-gain ship earth station (SES) antennas are surveyed, and field experimental results on the reduction effects are presented [16]. Fading reduction techniques applicable to these SES antennas are discussed in the following sections.

Diversity Method

Diversity techniques such as space, polarization, and frequency diversities have already been used practically in radio communication systems subject to severe fading. The space diversity technique needs two or more antennas, while other diversity techniques can be effected using a single antenna. In any case, the fading reduction effect largely depends on the correlation of signals with different branches concerning frequency, polarization, time difference, and so on. Figure 9.9 shows the principle of space diversity with a switch-and-stay algorithm [17]. As can be seen in this figure, the diversity output $[R(t)]$ is

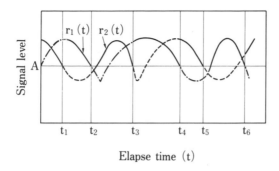

Figure 9.9 Principle of space diversity.

selected by switching two signals through antenna 1 [$r_1(t)$] and antenna 2 [$r_2(t)$]. With this technique, the greatest reduction effect is expected when the correlation of the signals between the two antennas is near zero or lower.

A space diversity experiment applying the switch-and-stay algorithm was carried out by setting up two short backfire antennas 20 cm in diameter on a small vessel under conditions with wave heights of about 1.0m to 1.5m [17]. These antennas were located with a three-wavelength separation in the vertical direction. Figure 9.10 shows the space

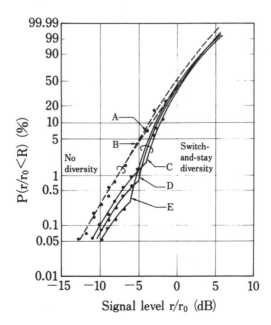

Figure 9.10 Space diversity effect.

diversity effect obtained by a computer simulation using actual data obtained by two antennas. From this figure it can be seen that the cumulative time distributions through each branch are nearly equal to each other, and these distributions almost correspond to a Rician distribution with carrier-to-interference ratio (C/I) of 6.25 dB. It can also be seen that the fading depth (=1% value in the figure) of about 8 dB at an elevation angle of 9 deg can be reduced to about 5 dB when the threshold level is set to −5 dB.

Frequency diversity can produce a good reduction effect when the frequency difference between two signals is longer than the correlation bandwidth of the multipath channel. Assuming that the height of the ship antenna is 15m and the elevation angle is 5 deg, the correlation bandwidth is about 20 MHz [18], the value of which is larger than the allocated bandwidth of the current maritime mobile satellite services (i.e., 15 MHz or so). Therefore, frequency diversity does not seem promising at this stage. However, if the frequency bands become widened in the future, this method could be applicable.

Polarization Shaping Method [19]

For reflections from the sea surface at 1.6/1.5 GHz, the horizontally polarized wave is almost perfectly reflected, while the vertically polarized wave is reflected with a large attenuation at grazing angles below 20 deg. Thus, the polarization of the reflected wave becomes elliptical with the opposite sense of rotation with respect to the incident circular polarization when the elevation is above 6 deg, and its major axis is nearly horizontal. (An elevation angle of around 6 deg is the Brewster angle at 1.5/1.6 GHz over sea paths.) Accordingly, if the polarization ellipse of a shipborne antenna in the direction of the reflected wave could be adjusted so that it always stays orthogonal to that of the reflected wave, the reflected wave can be suppressed. This principle can be easily applied to cross-dipole-fed antennas, such as a short backfire antenna [19]. For instance, if the cross-dipole elements are inclined by 45 deg with respect to the horizontal line and the phase is adjusted by a phase shifter inserted in one of the ports of the short backfire antenna feed, as shown in Figure 9.11(a), the axial ratio of this antenna can be arbitrarily controlled. In this case, the major axis of the polarization ellipse is always vertical, as shown in Figure 9.11(b).

Another reduction technique applying a polarization method has been presented in [20,21]. This technique suppresses the coherent component of sea-reflected waves, which is the dominant component needed for the calm sea condition [see Figure 9.12(a,b)].

Pattern Shaping Method

It is possible to suppress the reception of a reflected signal by using a shaped pattern antenna that has low-gain radiation characteristics in the direction of sea-reflected waves. This method may be realized by an array antenna or a shaped-reflector antenna. The radiation pattern is usually shaped so as to be flat in the main beam and to be suppressed in other directions [19]. This antenna is expected to have a good reduction effect when

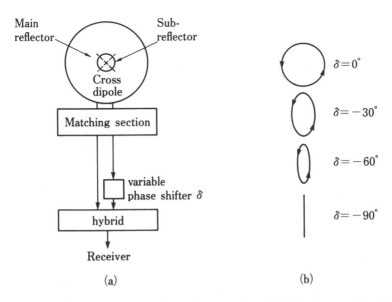

Figure 9.11 Short backfire antenna with polarization shaping method: (a) block diagram; (b) polarization ellipse of ship antenna.

the reflected waves come from directions away from the main beam. The shaped pattern antenna, however, has the disadvantage that the aperture efficiency of the antenna generally becomes comparatively low.

Maximum-Level Tracking Method

If the radiation pattern of a receiving antenna is controlled so that the received signal intensity is always maintained at a higher level, the fading could be substantially suppressed. This principle can be easily applied to array antennas. The phases of variable-phase shifters inserted in the feed circuit for each antenna element are varied by a small amount in order to see whether the resultant signal level increases. The phase has to be adjusted at a rate sufficiently fast to track the speed of the fading. If the signal level increases, the control voltage is allowed to change continuously in the same direction. If not, the polarity of the control voltage has to be reversed to control the variable-phase shifter in the opposite direction. Fading can be reversed by repeating this operation.

A field experiment has been carried out around a bay to receive a signal transmitted from the shore [19]. In this experiment, a quad-helical antenna was used for the signal reception and had a gain of about 13 dBi and an axial ratio of about 1.0 dB. The elevation angle was 7.5 deg and the wave height was 15 to 20 cm, corresponding to fairly calm sea conditions. The receiving antenna was fixed at the minimum-level point of the height

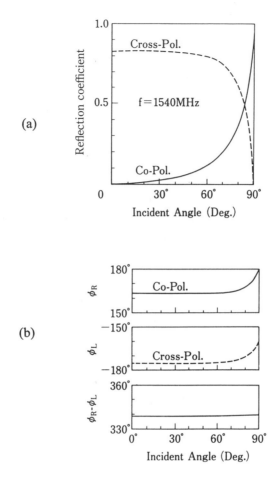

Figure 9.12 Reflection coefficient of sea surface: (a) amplitude; (b) phase.

pattern to evaluate the worst-case fading, where the received signal level is minimum. Figure 9.13 shows the cumulative time distribution of the received signal intensity. The fading depth of about 11 dB, corresponding to 99% of the time, has been reduced to about 7.5 dB.

9.2.5 Mount Systems

In the case of mobile communications, the antenna is always required to be pointed toward the satellite in spite of any vehicle's motions. Accordingly, the mount system is one of the key problems in designing the mobile antenna system from the technical and economic

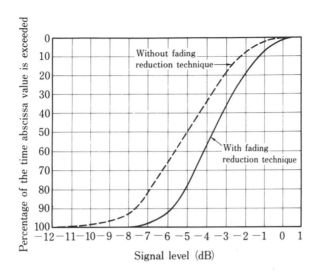

Figure 9.13 Cumulative time distribution of fading.

viewpoints. In this section, several types of mount system are introduced, especially those concerned with shipborne antenna systems. On the other hand, in the case of aeronautical and land mobile communications, electrical pointing systems are preferable.

Two-Axis Mount (El/Az, Y/X)

An antenna mount is a mechanically moving system that can maintain the antenna beam in a fixed direction. In maritime satellite communication systems, the mount must have a function to point in any direction on the celestial semisphere, because the ship will sail across the sea. It is well known that the mount of the two-axis configuration is the simplest mount providing such functions. There are two typical mounts of the two-axis configuration: one is the El/Az mount and the other is the Y/X mount. Simplified stick diagrams of both mounts are shown in Figure 9.14. In the El/Az mount, a full steerable function can be obtained by choosing the rotation range of the azimuth axis (Az-axis) from 0 to 90 deg. In the Y/X mount, a full steerable function is achieved by permitting the rotation angle from −90 to +90 deg to both the X-axis and Y-axis.

In general, the ship motion consists of seven elements: turn, roll, pitch, yaw, heave, surge, and sway. Turn is change of headway. The components are illustrated in Figure 9.15. In these components, heave, surge, and sway are motions with the acceleration. The pointing errors due to acceleration can be made small enough for practical use by a careful adjustment of weight balance around each mount axis. The characteristics of yaw are the same as turn.

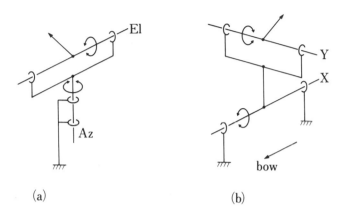

(a)   (b)

Figure 9.14 Two-axis mount systems: (a) El/Az; (b) Y/X.

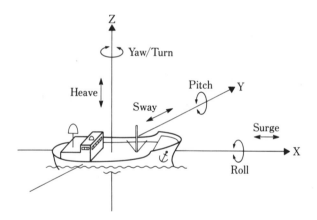

Figure 9.15 Ship's motion components.

Three-Axis Mount (El/Az/X, El'/El/Az, X'/Y/X)

The three-axis mount system is considered to be a modified two-axis mount which has one additional axis. The typical configurations of the three-axis mount are shown in Figure 9.16. The three-axis mount of an El/Az/X type shown in Figure 9.16(a) is the El/Az mount with one additional X-axis [22]. The function of the X-axis is to eliminate the rapid motion of the two-axis mount due to roll. In this system, however, the possibility of the gimbal lock for pitch is still left near the zenith, when the El-axis is parallel to the X-axis. The three-axis mount of an El'/El/Az type shown in Figure 9.16(b) is the El/Az mount with an additional cross-elevation axis El [17]. In the mount system, the change of the azimuth angle is tracked by rotating the Az-axis, and the change of the elevation

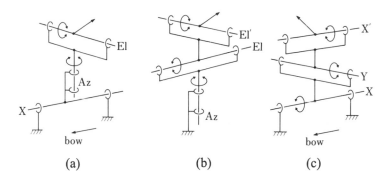

Figure 9.16 Three-axis mount systems: (a) El/Az/X; (b) El'/El/Az; (c) X'/Y/X.

angle is tracked by a combined action of the El and El' axes. As can be seen clearly in the figure, the El' and El axes allow movements in two directions at a right angle. With an approximate axial control, this mount is free from the gimbal lock problem near both the zenith and the horizon.

The three-axis mount of an X'/Y/X type is the two-axis Y/X mount system with the X'-axis on it to remove the gimbal lock at the horizon. The configuration is shown in Figure 9.16(c). When the satellite is near the horizon, the X'-axis takes out the rapid motion due to yaw and turn. Note that the X'-axis can rotate within ±120 deg, so the X'-axis can only eliminate the rapid motion within the angular range. In general, the axis control of the three-axis mount is rather more complex than that of the four-axis mount, because steering and stabilization interact with each other.

Four-Axis Mount (El/Az/Y/X)

In the mount, the stabilized platform is made by the X-axis and Y-axis to take out roll and pitch, and the two-axis mount of the El/Az type is settled on the stabilized platform. The axis configuration is shown in Figure 9.17. The tracking accuracy of the four-axis mount is, of course, the best of all mount systems because the stabilization function is separated from the steering function; that is, four major components such as roll, pitch, azimuth, and elevation are controlled by its own axis individually. Therefore, the four-axis mount has been adopted in many SESs of the current INMARSAT-A system. However, the cost would increase by introducing servo mechanisms and attitude sensors for stabilization.

9.2.6 Tracking/Pointing Systems

Pointing/tracking is another important function required of the mount system. It should be noted that the primary requirements for SES tracking systems are that they be

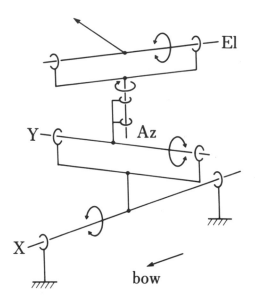

Figure 9.17 Four-axis mount system.

economical, simple, and reliable. Tracking performance is a secondary requirement when an antenna beamwidth is broad.

9.3 OMNIDIRECTIONAL ANTENNAS FOR MOBILE SATELLITE COMMUNICATIONS

9.3.1 Overview

In this section, three types of basic omnidirectional antennas are introduced. As described in the previous section, antennas for mobile satellite communications are classified into omnidirectional and directional antennas. The gain of omnidirectional antennas is generally from 0 to 4 dBi in the L-band, which does not require the capability of satellite tracking. In a family of omnidirectional antennas for mobile satellite communications, there are three basic antenna elements. The first is a quadrifilar helical, the second is a crossed-drooping dipole, and the third is a microstrip patch antenna. In this section, these omnidirectional antennas, which were mainly studied and breadboarded by the Jet Propulsion Laboratory (JPL) are introduced by quoting from [23]. These omnidirectional antennas are very attractive, owing to the possible small size, light weight, and circular polarization properties. These basic antennas are also used as elements of directional antennas, which will be mentioned in later sections.

9.3.2 Quadrifilar Helical Antenna

The quadrifilar helical antenna [24] is composed of four identical helixes wound, equally spaced, on a cylindrical surface. The helixes are fed with signals equal in amplitude and 0, 90, 180, and 270 deg in relative phase. Figure 9.18 shows the elevation patterns that can be achieved from three different quadrifilar antennas for various angular coverage requirements. General characteristics are: gain, −4 dBi minimum; axial ratio, 3 dB maximum; and height, 40 cm.

9.3.3 Crossed-Drooping Dipole Antenna

The crossed-drooping dipole antenna is the most interesting for land mobile satellite communications, where required angular coverage is narrow in elevation and is almost constant in azimuth. By varying the separation between he dipole elements and the ground plane, the elevation pattern can be adjusted for optimum coverage for the coverage region of interest. Figure 9.19 [23] shows patterns of a crossed-drooping dipole antenna. By

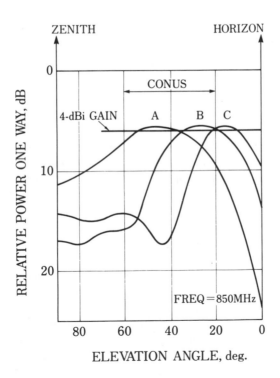

Figure 9.18 Quadrifilar helical antenna patterns. (Courtesy of JPL.)

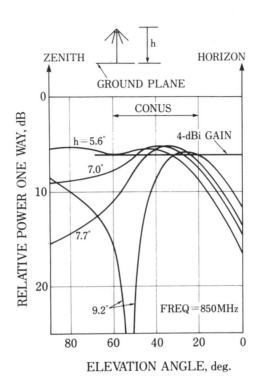

Figure 9.19 Crossed-drooping dipole antenna patterns. (Courtesy of JPL.)

adjusting the height of the antenna from the ground plane, a 4-dBi gain can be achieved over the entire continental United States (CONUS). General characteristics are: gain, 4 dBi minimum; axial ratio, 6 dB maximum; and height, 15 cm.

9.3.4 Patch Antenna

The circular microstrip disk antenna [25] has a circular metallic disk supported by a dielectric substrate material over a ground plane. In order to produce circularly polarized waves, a patch antenna is, in general, excited at two points orthogonal to each other and fed with signals equal in amplitude and 0 and 90 deg in relative phase. A higher mode patch antenna [26] can also be designed to have a similar radiation pattern to the drooping dipole. To produce conical radiation patterns (null on axis) suitable for land mobile satellite applications, the antenna is excited at higher order modes. Figure 9.20 shows the radiation patterns for different modes. General characteristics are: gain, 3.5 dBi minimum; axial ratio, 4 dB maximum; and height, 1 cm.

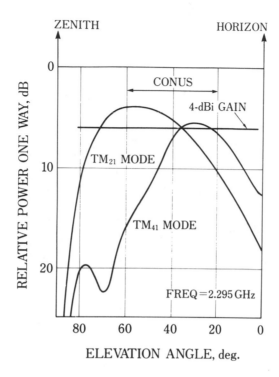

Figure 9.20 S-band circular microstrip disk antenna patterns. (Courtesy of JPL.)

RT/Duroid with dielectric constant of 2.3 (thickness = 3.2 mm) is used as a dielectric substrate because of its good temperature characteristics. Since the available frequency bandwidth of a patch antenna is very narrrow, the two-layer patch antenna [27] is adopted in which the upper and lower parts play a role for transmission and reception, respectively. Each layer antenna is individually fed at two points with a phase difference of 90 deg in order to obtain the circular polarization. The upper layer antenna is the well-known conventional circular microstrip antenna (MSA) [28], while the lower one is a circular MSA with an electric shielding ring which supplies enough space for the upper antenna to be easily fed [29]. Another useful printed antenna element is the cavity-backed cross-slot antenna (XSA). Each slot antenna is fed with an equal amplitude and in-phase condition at two points located equally from the center [30]. As a result, the input impedance can be matched for a wider frequency band than in the case of the slot antenna. In the case of the MSA, the antenna gain at boresight is about 15.2 dBi and a gain of about 13.5 dBi is obtained at a scanning angle θ of 45 deg, while for the XSA, the boresight gain is about 15.7 dBi and the gain at $\theta = 45$ deg is about 14 dBi. It can be shown from these figures that the degradation of antenna gain for both antennas is not so large, even if the scanning angle is 45 deg.

9.4 DIRECTIONAL ANTENNAS FOR MOBILE SATELLITE COMMUNICATIONS

9.4.1 Antennas for INMARSAT

9.4.1.1 Antennas for INMARSAT-A and -B

For INMARSAT-A and -B systems with antenna gains of about 20 dBi, an aperture antenna such as a paraboloidal antenna is suitable because of the high aperture efficiency. It is simple in its structure and can be designed flexibly with respect to its antenna gain when the gain is more than around 20 dBi.

9.4.1.2 Antennas for INMARSAT-M

INMARSAT-M has been operated with a small antenna to meet the requirements of a small-size and low-cost SES, because INMARSAT-M is mainly for small ships, such as fishing boats, and land vehicles. Assuming the same satellite EIRP as that of the current CFM channel, the G/T is −10 dBK for the maritime mobile version and −12 dBK for the land mobile version in order to get a sufficient quality of planned service. Under these conditions, the antenna gains ranging from 13 to 16 dBi are necessary for getting the required G/T. This is compatible with one of the baseline assumptions on INMARSAT-M concerning the efficient utilization of satellite power.

Short Backfire Antenna

The SBF antenna (see Figure 9.21) was developed experimentally by H. W. Ehrenspeck in the 1960s [31,32]. This antenna is well known as a highly efficient antenna of simple and compact construction. Its high-directivity and low-sidelobe characteristics make this antenna a single antenna with high-gain characteristics, which is applicable to satellite communications, tracking, and telemetry. Moreover, due to its excellent radiation characteristics, this antenna and a modified SBF antenna (mentioned below) have been recently proposed for shipborne antennas of INMARSAT-M.

In Figure 9.21, D_l, D_s, H_r, H_s, H_d, and D_d are the diameter of large reflector, the diameter of small reflector, the rim height, the distance between small and large reflectors, the distance between large reflector and feed dipole, and the antenna length of feed dipole, respectively.

Half-power Beamwidth

Figure 9.22 shows the HPBW of a SBF antenna with D_s as parameters. Calculated results of the E-plane are presented with a solid line and of the H-plane with a broken line.

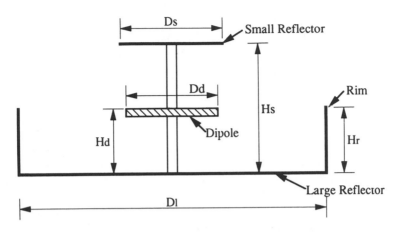

Figure 9.21 Short backfire antenna.

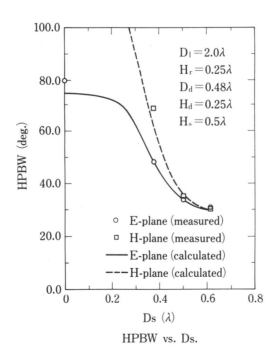

HPBW vs. Ds.

Figure 9.22 HPBW of SBF antenna.

Measured results are also shown. The other antenna parameters are $D_l = 2.0\lambda$, $H_r = 0.25\lambda$, $H_s = 0.5\lambda$, $H_d = 0.25\lambda$, and $D_d = 0.48\lambda$. From this figure, we can see that the HPBW is remarkably affected by the diameter size of the small reflector, and as the diameter becomes larger, the HPBW becomes smaller. There is good agreement between the calculated and measured values.

Sidelobe Level

Figure 9.23 shows the example of the sidelobe level versus the rim height (H_r). The numerical results of the E-plane are shown by a solid line, and that of the H-plane by a broken line. Measured results are also shown. The other antenna parameters are $D_l = 2.0\lambda$, $D_s = 0.5\lambda$, $H_s = 0.5\lambda$, $H_d = 0.25\lambda$, and $D_d = 0.48\lambda$, respectively. Note that the rim plays an effective role in suppressing the sidelobe level, and in the case of H_r more than about 0.2λ, the sidelobe level becomes less than 20 dB.

Antenna Gain

Figure 9.24 shows the antenna gain versus the large reflector's diameter with parameters of the small reflector's diameter. In this figure, the indication of aperture efficiency is

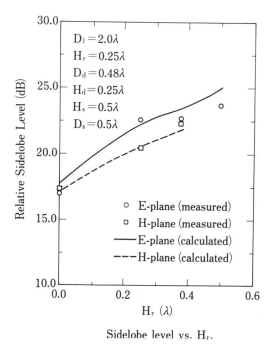

Sidelobe level vs. H_r.

Figure 9.23 Sidelobe level of SBF antenna.

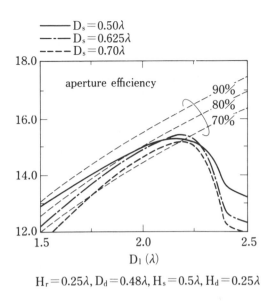

$H_r = 0.25\lambda$, $D_d = 0.48\lambda$, $H_s = 0.5\lambda$, $H_d = 0.25\lambda$

Figure 9.24 Antenna gain (dB) of SBF antenna.

also denoted by broken lines. We can see from this figure that for a size of D_l around 2.0λ, the aperture efficiency is about 80%, while in the case of D_l less than about 1.5λ, more than 90% seems to be expected. The maximum gain of the SBF antenna is about 15 dBi as the size of small reflectors varies from 0.5λ to 0.7λ, and optimum D_l is around 2.24λ almost regardless of the size of the small reflector. According to our studies, the dependence of the size of H_r, H_s, and H_d on the antenna gain characteristics is relatively small.

Quad-Helical Antenna

Since an axial mode helical antenna has good circular polarization characteristics over a wide frequency range, it has been put into practical use as a single antenna or as array elements [33]. With respect to the structure, this antenna can be considered a compromise between the dipole and the loop antennas, and the radiation mode varies with the pitch angle and the circumference of the helix. In particular, the helix with the pitch angle α of 12 to 15 deg and the circumference C_λ of about 1λ has a sharp directivity toward the axial direction of the antenna. This radiation mode is called the axial mode, which is the most important mode in a helical antenna. Several studies were reported on the properties of the axial mode helical antenna with a finite reflector [24,26,31,32].

The current induced on the helix is composed of four major waves, which are two rapidly attenuating waves and two uniform waves along the helical wire. Such waves include the traveling wave and the reflected wave. In a conventional helical antenna, the uniform traveling wave will be dominant when the antenna length is fairly large. On the other hand, in a short-cut helical antenna, the rapidly attenuating traveling wave will be dominant, especially in a two-turn ($N = 2$) helical antenna.

The two-turn helical antenna has a relatively high antenna gain and excellent polarization characteristics for its size. Radiation patterns with respect to E_θ and E_ϕ planes of the two-turn helical antenna are shown in Figure 9.25. In this figure, D is a diameter of the finite circular reflector of the axial mode helical antenna.

As can be seen in Figure 9.26, the antenna gain is affected remarkably by the size of the reflector. In Figure 9.26, the antenna gain is about 9 dBi and the axial ratio is

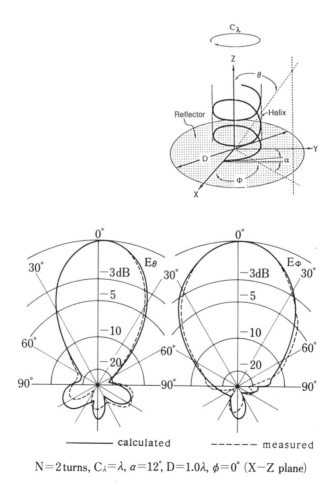

$N = 2$ turns, $C_\lambda = \lambda$, $\alpha = 12°$, $D = 1.0\lambda$, $\phi = 0°$ (X−Z plane)

Figure 9.25 Radiation characteristics of two-turn helical antenna.

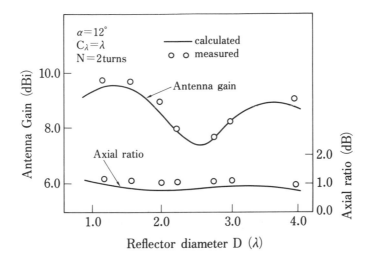

Figure 9.26 Gain and axial ratio of two-turn helical antenna.

about 1 dB, with the reflector diameter D around 1.0λ. Such a two-turn helical antenna has a comparatively high performance in spite of its small size. From the above mentioned consideration, a highly efficient antenna for the INMARSAT-M SES can be realized by applying the two-turn helical antenna to elements of an array antenna.

The quad-helical array antenna is composed of four two-turn helical antennas in a square arrangement whose elements are oriented in the manner shown in Figure 9.27. According to related studies [23], the effect of the mutual coupling between each element of this antenna is not negligible, and this mutual coupling mainly degrades the axial ratio. The axial ratio of a single helical antenna is about 1 dB, as shown in Figure 9.28, but this value is degraded to about 4.5 dB in the case of the array antenna with an array spacing of 0.7λ.

Figure 9.28 shows the experimental values of the antenna gain and axial ratio versus the rim height with an element spacing of 0.7λ and rim diameter of 0.7λ. As can be seen, the best properties of the antenna gain and the axial ratio can be obtained at a rim height of about 0.25λ. The antenna gain is improved by 0.4 dB and the axial ratio is also improved by 3.5 dB compared to that of the quad-helical array antenna without rims.

The performance characteristics of the quad-helical array antenna shown in Figure 9.27 has an antenna gain of about 13 dBi (HPBW:38 deg), an axial ratio of about 1.0 dBi, and an aperture efficiency of about 100%. It appears that the antenna is also suitable for INMARSAT-M.

Microstrip/Slot Array Antennas

Figure 9.29 shows photographs and element arrangements of an MSA and a XSA. The MSA is a nine-element microstrip array antenna whose element spacing is 94 mm (about

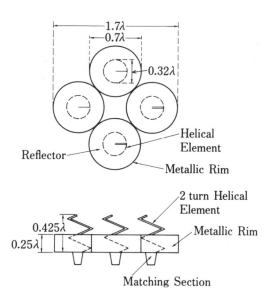

Figure 9.27 Quad-helical array antenna.

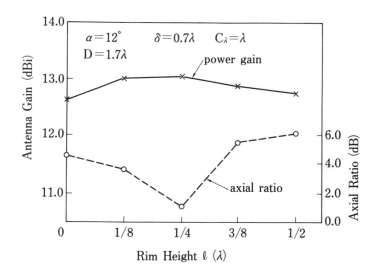

Figure 9.28 Gain and axial ratio of quad-helical array antenna.

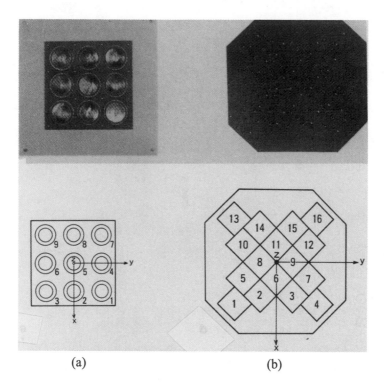

Figure 9.29 (a) Microstrip array; (b) cross-slot array.

a half wavelength at 1.6/1.5 GHz) and whose antenna volume is about 300 × 300 × 10 mm. The XSA is a 16-element cross-slot array antenna with 97-mm spacing and its volume is about 560 × 560 × 20 mm. As shown in these figures, the element arrangement of the MSA is a 3 × 3 square array, while that of the XSA is a modified 4 × 4 square array in order to obtain similar radiation patterns in different cut planes. In both the MSA and XSA, beam scanning is performed by controlling four-bit variable phase shifters attached to each antenna element. These antennas are also applicable to the mobile earth station (MES) antennas for INMARSAT-M as well as INMARSAT-Aero. Figure 9.30 shows the example of INMARSAT-M SES. In this antenna system, a modified SBF antenna, two-axis El/Az mount system, and program tracking are adopted for antenna, mount system, and pointing/tracking method, respectively.

9.4.1.3 Antennas for INMARSAT-C

In the case of the INMARSAT-C antenna system, the simplest and most compact configuration is required; that is, without mount systems and tracking/pointing systems. Therefore,

Figure 9.30 Example of antenna system for INMARSAT-M: Antenna: modified SBF; mount: two-axis (El/Az); fading reduction: polarization shaping method.

the omnidirectional antennas are the most suitable ones for this standard. These are the quadrifilar helix [34], cross-drooping dipole [35], and the microstrip patch antenna [36]. As a shipborne antenna, a quadrifilar antenna is well known because of its good performance of axial ratios in wide angular coverage. Figure 9.31 shows an example of INMARSAT-C terminals for maritime satellite communications. Its general characteristics are a minimum of about 4 dBi in gain, a maximum of about 3 dB in axial ratio, and about 40 cm in height.

9.4.1.4 Portable Terminal's Antennas for INMARSAT-A and -C

Figure 9.32 shows an example of the INMARSAT-C portable terminal (developed by Toshiba Corporation) with four patch antennas [37], which can be connected to a personal computer to exchange messages for about 60 min of receiving and about 30 minutes of transmitting using dry battery operation. The diameters are 34 (w) × 21 (d) × 6 (h) cm in size and 3.8 kg in weight.

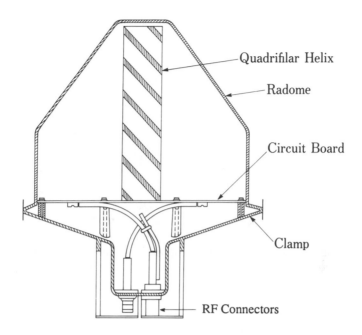

Figure 9.31 Example of antenna system for INMARSAT-C. Antenna: quadrifilar helical.

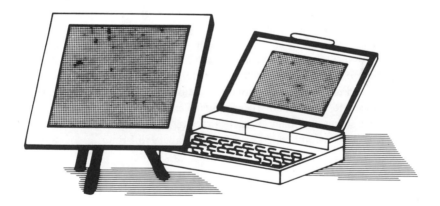

Figure 9.32 Portable terminal for INMARSAT-C. (Courtesy of Toshiba.)

9.4.1.5 Antennas for INMARSAT-Aero

Airborne equipment for aeronautical satellite communications in the INMARSAT system is called the *INMARSAT-Aero terminal.* Two types of high-gain and low-gain antennas are defined in the INMARSAT-Aero terminal. The G/T of high-gain and low-gain antennas is required to be over −13 dBK (gain is about 12 dBi) and −26 dBK (gain is about 0 dBi), respectively. The high-gain antenna is used for public voice/high-speed data (21-kbps) communications, which can be connected to terrestrial public telephone networks. On the other hand, the low-gain antenna is used for low-speed data (600, 1,200, 2,400 bps) communications.

One of the most important issues in the INMARSAT-Aero high-gain antennas is the *key hole,* which is the direction in which the airborne antenna cannot satisfy the required G/T. The phased-array antennas are the best candidates for airborne antennas because of such advantageous characteristics as low profile and mechanical strength; however, they have a fatal disadvantage in the key hole problem. At the present time, two types of phased-array antennas have been used: one is a conformal type, which has two sets of phased arrays on both sides of a fuselage, and the other is a top-mount type, which has a set of phased arrays on the top of a fuselage. The conformal type has the advantage of low air drag because of its very low profile; however, it has a fatal key hole. At present, only a top-mount-type phased-array antenna has satisfied the INMARSAT specifications, which are defined by the INMARSAT Aeronautical System Definition Manual (SDM).

9.4.2 Directional Antennas in the PROSAT Program

In this section, directional shipboard antennas developed for experiments of the PROSAT program are described. Seven European Space Agency (ESA) member states (Belgium, Federal Republic of Germany, France, Italy, Spain, United Kingdom, and Norway) decided to undertake a PROSAT program to promote mobile satellite communications. In the first stage (December 1983 to December 1984) of the PROSAT program, several kinds of antenna systems for maritime satellite communications were developed and evaluated by the European Space Research and Technology Center (ESTEC) using the MARECS satellite. Table 9.3 shows the main characteristics of three antennas, from the PROSAT Phase I Report, which was published by ESA in May 1986 [4].

9.4.3 Directional Antennas in the ETS-V Program

The ETS-V satellite was launched to a geostationary orbit of 150 deg E in August 1987 for the purpose of research and development of mobile satellite communications. In the ETS-V program, the Communications Research Laboratory (CRL) of the Ministry of Posts and Telecommunications, the Electronic and Navigation Research Institute (ENRI) of the Ministry of Transport, (NTT), and Kokusai Denshin Denwa Co. (KDD) carried out the experiments using ships, aircraft, and land mobiles with newly developed directional antennas.

Table 9.3
Antenna Developed in PROSAT Program

Antenna	Peak, Transmit Gain (dB)	Peak G/T (dBk)	Max. EIRP (dBW)	Axial Ratio (dB)	Radome (H, D) (m)	Stabilization
Helix five-turns	12.5	−16	29	3	1.2, 0.84	Gravity elevation on double-gimbaled suspension
Short backfire	15	−12	28	1	1.4, 1.45	Two gyroscope wheels rotating in opposite directions on a platform fixed through a U-point
Array of 16 crossed dipoles	17	−9.5	31.7	0.7	1.25, 1.13	Double-gimbaled mount, gravity stabilized and stiffened by one momentum wheel

9.4.3.1 Shipborne Antenna

Improved Short Backfire Antenna

In the ETS-V program, the electrical characteristics of a conventional SBF antenna have been improved by changing its main reflector from a flat disk to a conical or a step plate, and by adding a second small reflector. Table 9.4 shows the comparison of electrical characteristics between a conventional SBF antenna with a plane main reflector and an improved SBF antenna with a conical main reflector [38].

Table 9.4
Comparison of Electrical Characteristics of Conventional and Improved SBF Antennas

	Plane Reflector SBF Antenna	Conical Reflector SBF Antenna
Effective gain	14.5 dB	15.0 dB
Half-power beamwidth	34 deg	34 deg
Directive gain	14.8 dB	15.5 dB
First sidelobe level	−21.0 dB	−22.5 dB
Axial ratio	−1.3 dB	−1.1 dB
Aperture efficiency:		
Effective gain	65%	76%
Directive gain	75%	85%
Frequency bandwidth for VSWR under 1.5	3%	9% 20%*

*With a small reflector.

9.4.3.2 Airborne Antenna

Phased-Array Antenna

In aeronautical satellite communications, a directional medium gain antenna is considered a key technology, and a major research aim is to realize not only electrical requirements, but also strict mechanical requirements, such as installation, low profile, light weight, and strength [39]. In the ETS-V program, the phased-array antenna [25], developed by CRL, was installed on a commercial jet aircraft for the first time in the world to carry out communication experiments mainly between Tokyo and Anchorage.

Taking account of the electrical and mechanical requirements of aeronautical communications, a phased array with low-profile antenna elements was chosen for a directional antenna. An MSA [40] is chosen as an antenna element because of its very low profile, light weight, and mechanical strength, which satisfy the requirements for airborne antennas [7]. However, one disadvantage is a very narrow frequency bandwidth, usually 2% to 3%. The antenna adopted is a two-frequency resonant element, because it provides a compact array and a simple feed line configuration [37]. However, this type of element has very poor axial ratios. The problem was overcome by using the sequential-array technique [41], where a thin substrate with high dielectric constant is used over a wide frequency bandwidth with excellent axial ratios (Table 9.5). Figure 9.33 shows the antenna installed on the top of a Boeing 747.

9.4.3.3 Land Mobile Antenna

Directional antennas have been expected to provide voice/high-speed data links not only for long-haul tracks but also for small, private cars. From that point of view, cost is a very important factor to be taken into consideration in designing antennas. In the early stage of land mobile satellite communications, a mechanical steering antenna system is considered the best candidate for vehicles; however, it will be replaced by a phased-array antenna in the near future, because a phased-array antenna has many attractive advantages, such as low profile, high-speed tracking, and potential low cost. In recent years, the main research activities have been focused on phased-array antennas and the tracking control method, but first we will note the characteristics of mechanically steered antenna systems.

Antenna

The photograph shown in Figure 9.34 is of a mechanical steering antenna with eight spiral elements, which gives about 15 dBi in system gain. A closed-loop tracking method is adopted. The volume is 30 cm in radius, 35 cm in height, and 1,500g in weight. The array consists of 2×4 elements, and it forms a fan beam having a half-power beamwidth of 21 deg in the azimuth plane and 39 deg in the elevation plane at 1,545 MHz. Its peak

Table 9.5

Characteristics of Airborne Phased-Array Antenna

Characteristic	Measurement
Frequency:	
Transmitter	1,545–1,548 MHz
Receiver	1,647–1,650 MHz
Polarization	Left-handed circular
Gain:	
Transmitter	14.7 dBic (non-scan)
Receiver	13.5 dBic (non-scan)
G/T	−10.8 dBK (non-scan)
Element	Circular patch (one-point feed)
Array	2 × 8 sequential phased array
Substrate	Glass Teflon ($\epsilon_r = 2.6$)
Axial ratio	>2.0 dB (non-scan)
VSWR	>1.4 (non-scan)
Tracking	Step track
Beam step	4 deg
Phase shifter	4-bit digital × 8
Weight	18 kg
Volume	760 (l) × 320 (w) × 180 (h) mm

gain is about 15 dBi, including the feeder loss, and is suitable to track the satellite for mobile satellite communications, because elevation angles to the satellite are not as varied as those of azimuth angles; the antenna beam direction can be shifted in two azimuth directions by switching the pin-diode phase shifters. The difference between the received signals in both directions is used to drive the antenna toward the satellite. The beam-shifting angle is set to about 4 deg.

Tracking System

Figure 9.35 shows a block diagram of the tracking system. The newly developed single-channel tracking system is applied to this mobile antenna. This system is similar to the conical scan tracking system, which shifts the antenna beam direction in order to get the angle difference between the antenna boresight and the satellite direction. In this antenna system, the received signal is modulated by shifting the beam directions, and its modulation period coincides with the beam-shifting period (400 Hz). So, the information on beam-tracking error can be obtained by demodulating the received signals with reference to the beam-shifting signal. We will now note the characteristics of electronically steered antenna systems.

Figure 9.33 View of antenna installed on the top of Boeing 747 (radome removed).

Main Features

Although mechanical tracking antennas have the advantage of wide beam coverage, they have severe disadvantages for small cars. The problems are large size, slow tracking speed, and high cost. Based on the research experiences of the airborne phased-array antenna [25], a new phased-array car antenna has been proposed by CRL [42]. Figure 9.36 is a photograph of the phased-array antenna without a radome, and Figure 9.37 shows a block diagram of the phased-array antenna system.

Antenna Element

Figure 9.38 shows a configuration of the antenna element. An electromagnetic coupled antenna is adopted as an antenna element because it will reduce procedures of assembling the antenna system and it will reduce the cost of the phased-array antenna and thus compete with a conventional mechanical tracking antenna. A feed line is a microstrip line printed on a substrate, which excites the radiating elements by electromagnetic coupling through a coupling aperture. A radiating element, which radiates circularly polarized waves, is also printed on a thin-film substrate.

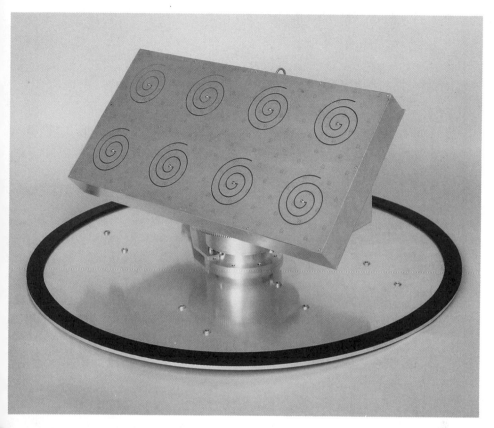

Figure 9.34 Mechanical steering antenna with eight-spiral antennas.

Phased Array

The array consists of 19 elements, which are printed on a very thin film. Figure 9.39 is a photograph of feed lines and phase shifters printed on a substrate installed on the backside of the plate of exciting apertures. The gain of the phased-array antenna in a nonscanned direction (elevation angle is 90 deg) is about 18 dBi, and 10 dBi in a scanned angle of 60 deg (elevation angle is 30 deg). The main characteristics are shown in Table 9.6. Figure 9.40 shows the elevation angular dependence of G/Ts, system noise temperature, and temperature measured at the input port of an LNA. A G/T is about −13 dBK at the elevation angles of 40 to 50 deg.

Briefcase Antenna

The main feature of future mobile satellite communications will be *portability,* which means that a person can directly access a satellite to establish communication channels

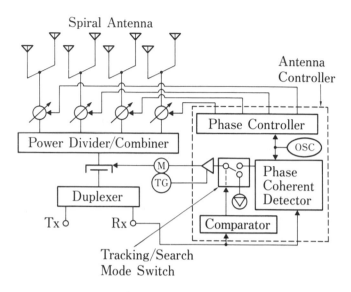

Figure 9.35 Mechanical tracking antenna.

using a very small handheld terminal. Even in the present L-band system, great efforts have been made to develop transportable and portable terminals. Figure 9.41 is a photograph of a briefcase portable terminal [43] developed in the ETS-V program. The main characteristics are:

- Antenna: two microstrip patch antennas;
- Gain: 6 dBic;
- G/T: −21 dBK;
- EIRP: 6 dBW;
- Tracking: hand-pointed; and
- Weight: 13 kg.

The terminal can transmit and receive messages via a satellite with two separate conventional microstrip patch antennas on a lid of a case for transmitting and receiving. The reason for adopting separate antennas is to eliminate a diplexer, which is too large and too heavy to realize a compact and lightweight terminal. The beamwidth of the antenna on the lid is wide enough to point to a satellite by manual pointing.

9.4.4 Directional Antennas in the MSAT-X Program

The National Aeronautics and Space Administration (NASA) and JPL have studied and developed several L-band directional vehicle antennas, mainly for land mobile satellite

Figure 9.36 Vehicle phased-array antenna developed for the ETS-V experiments.

communications [44]. These include a mechanically steered, tilted 1×4 patch array antenna and two electrically steered, planar-phased-array antennas. All three of these antennas feature beams that are narrow in azimuth; hence, they require azimuth steering to keep the beam pointed toward the desired satellite as the vehicle changes its azimuth orientation. These medium-gain antennas provide 9 to 12 dBi gain, reject multipath signals outside their beam pattern, and allow two satellites separated by 30 deg in a geostationary orbital arc to reuse the same frequency to cover CONUS. The following description is mainly derived from [44] and [45]. We commence with the mechanically steered antennas.

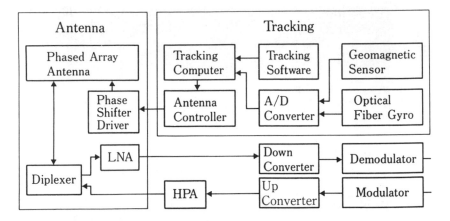

Figure 9.37 Phased array.

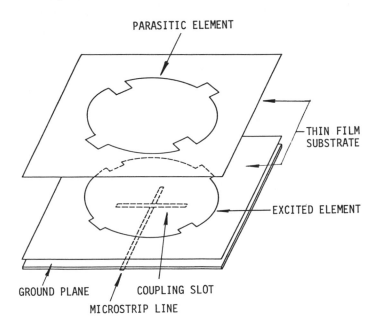

Figure 9.38 Antenna element of the phased array.

Antenna

Two mechanically steered vehicle antennas have been developed at JPL [45,46]. The radiating part is a linear array of four square microstrip patches tilted with respect to the ground plane to provide elevation coverage from 20 to 60 deg with a minimum of

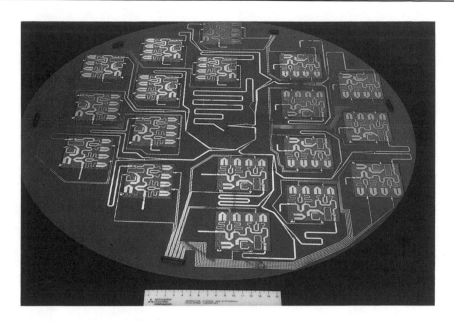

Figure 9.39 Feed lines and phase shifter printed on a substrate.

10 dBi gain. The rotating antenna platform is mounted on a fixed platform that includes the motor drive and pointing system hardware. The success of the breadboard version led to additional antenna development aimed at the reduction of the antenna height from 9 in down to 5 to 6 in and the integration of some of the discrete RF components into the stripline-fed 1 × 4 microstrip patch array. Figure 9.42 shows a side view of the reduced-height mechanically steered antenna.

Table 9.6
Characteristics of Phased-Array Antenna

Characteristic	Measurement
Frequency:	
Receiver	1,530.0–1,559.0 MHz
Transmitter	1,626.5–1,660.5 MHz
Polarization	Left-hand circular
Scanned angle	El 30–90 deg, Az 0–360 deg
Gain	18 dB: (El = 90 deg)
	10 dB: (El = 30 deg)
System temperature	200°K
Axial ratio	4 dB (El = 30 deg)
Volume	60 cm ϕ × 4 cm (H)
Weight	5 kg

562

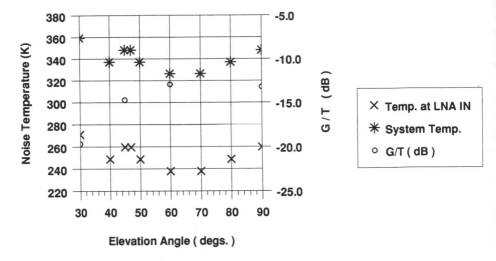

Figure 9.40 Measured G/T and noise temperatures of the array.

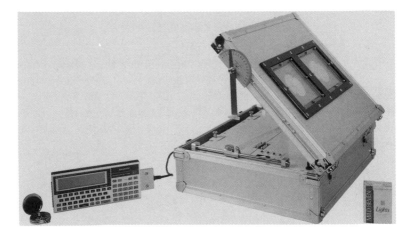

Figure 9.41 Briefcase terminal message communicator with 2 patch antennas.

Tracking

Both antenna systems employ the same monopulse technique, a kind of a close-loop tracking, for tracking the satellite in azimuth. The antenna is divided into two identical subarrays. Since the phase centers of the two halves are physically separate, the phases of their received signals will be offset by an amount proportional to the off-boresight angle. The signals from the two subarrays pass through a sum/difference hybrid whereby their sum and difference are obtained. The difference of *error* signal is then canceled

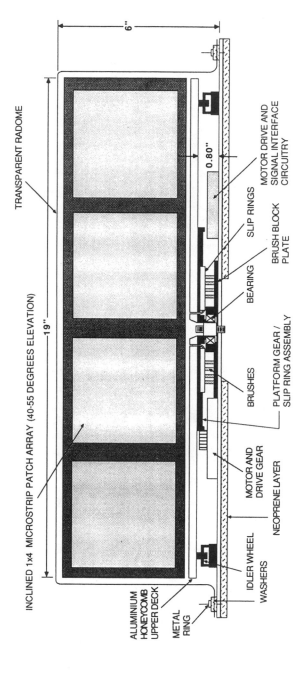

Figure 9.42 Side view of the reduced-height mechanically steered antenna. (Courtesy of JPL.)

through a 1-bit (0 to 180 deg) phase shifter, which will be modulated at a rate of between 200 to 4,800 Hz. It is then passed through a 30-dB ferrite isolator. The difference signal is then combined with the sum signal via a 10-dB coupler in which the difference signal strength is reduced by approximately 10 dB before being combined with the sum signal. The combined output of the coupler is then passed to the single-channel rotary joint. Since the difference signal is modulated before addition to the sum signal, it can be extracted from the composite signal at the receiver by the standard filtering technique. This normalized error signal is then supplied to the pointing control circuitry, which commands a stepper motor to rotate the antenna array in order to maintain pointing toward the source of the incoming signal. The azimuth rate sensor is also used to maintain pointing during the tracking phase when the received signal fades below a prescribed threshold. This constitutes open-loop tracking. We now consider the electrically steered antennas.

Array Antenna

Phased-array antennas were developed principally to provide a thin antenna that can be installed conformal to the top of the vehicle. These antennas are well known for their complexity and high cost. As a result, emphasis was placed on the selection of manufacturing techniques, materials, and component types, in addition to meeting the RF and pointing requirements, to keep the cost down. The breadboard phased-array antennas [47] developed are to meet the following performance goals:

- Frequency: 1,545.0 to 1,559.0 MHz receive and 1,646.5 to 1,660.5 MHz transmit;
- Spatial coverage: 20 to 60 deg elevation and full 360 deg azimuth;
- Gain: 10-dBic minimum above 30 deg in elevation and 8-dBic at 20 deg in elevation;
- Half-power beamwidth: 25 deg in azimuth and 35 deg in elevation;
- Intersatellite signal isolation: 20 dB between two geostationary satellites separated by approximately 35 deg;
- Beam pointing accuracy: +5 deg.

Two phased-array antennas were developed, as shown in Figure 9.43, and they exhibit several common features. Each antenna consists of 19 low-profile radiating elements, with 18 3-bit diode phase shifters. The left model of Figure 9.43 uses a dual resonant stacked circular microstrip element [47] to cover both the transmit and receive bands, while the right model employs a stripline crossed-slot radiator.

Tracking

For the antenna's beam-pointing system, the initial acquisition of the satellite is accomplished by a full azimuth search for the strongest received signal. An angular rate sensor is used to establish an inertial reference point when the acquisition is performed while

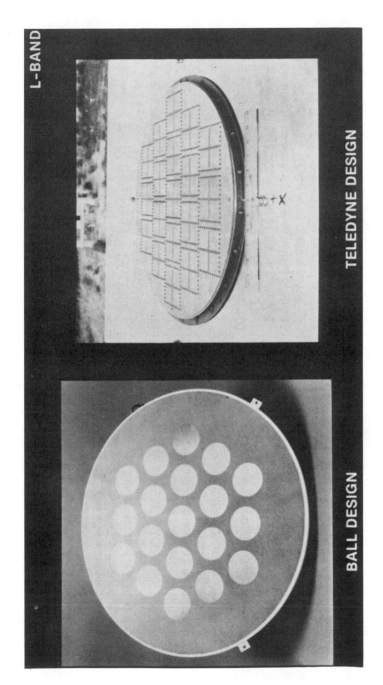

Figure 9.43 Electrically steered phased arrays. (Courtesy of JPL.)

the vehicle is turning. Tests show that the antenna can acquire a reference pilot signal in 2 sec from a random spatial position. After the desired satellite signal has been acquired, the antenna tracks the satellite by a closed-loop sequential lobing technique. In the event of a severe signal fade due to shadowing, the sequential lobing can no longer function properly. In this case, the open-loop angular rate sensor takes over the pointing for approximately a 10-sec period until the sensor drifts away.

9.4.5 Directional Antennas in the MSAT Program

The Communications Research Center (CRC) has developed an antenna for land vehicle operation. Linearly and circularly polarized array antennas have been developed and evaluated in extreme field trials using the MARECS-B satellite. The description in this section is derived from [48].

Adaptive Array

The linearly polarized adaptive antenna array consists essentially of a driven quarter-wave monopole surrounded by concentric rings of parasitic elements all mounted on a ground plane of finite size. The parasitic elements are connected to ground via pin diodes. With the application of suitable biasing voltages, the desired parasitic elements can be activated and made highly reflective. The directivity and pointing of the antenna beam can be controlled both in the elevation and azimuth planes using high-speed digital switching techniques. The use of a circular polarizer in the linearly polarized design can realize an increase in gain at the expense of an increase in antenna height.

The polarizer has an elliptical cross section, a diameter of 40 cm, and a height of 20 cm. It consists of a number of conformal scattering matrices to achieve the 90 deg differential phase shift between two orthogonal polarizations.

A five-ring linearly polarized antenna is shown in Figure 9.44. The antenna incorporates sufficient electronics to control the radiation patterns and pointing on command. It is designed to be mounted on the metallic roof of a vehicle, where the effective ground plane can significantly enhance antenna gain at low elevation angles.

The circularly polarized array, shown in Figure 9.44, is obtained by adding the linearly polarized array. The main characteristics are:

- Frequency: 1,530 to 1,660 MHz;
- Spatial coverage: 15- to 50-deg elevation and 360-deg azimuth;
- Gain: linearly polarized model—9 to 11 dBi and circularly polarized model—10 to 13 dBic;
- Size: linearly polarized model—24 in (d), 2.5 in (h), 11 lb (w) and circularly polarized model—24 in (d), 8 in (h), 16 lb (w).

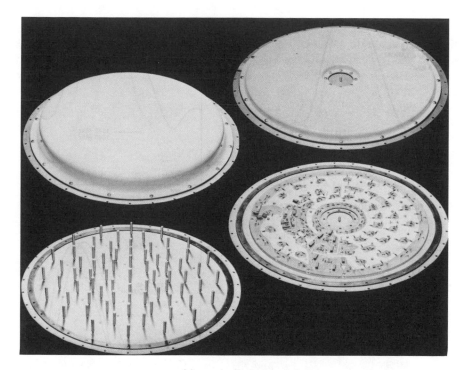

Linearly Polarized Antenna

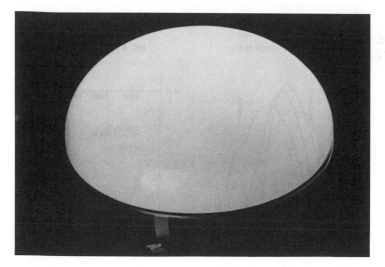

Circularly Polarized Antenna

Figure 9.44 Linearly and circularly polarized antenna developed by CRC. (From IMSC'90 with permission.)

Tracking. The satellite is initially acquired with a closed-loop method of stepping through 16 azimuth beam positions and selecting the beam with the strongest signal. In the event that the signal falls below a given threshold, the acquisition sequence is again initiated until the signal is required. The speed of operation is determined by the terminal C/N_0 ratio, and the signal-to-noise requirements in the control loop bandwidth. Currently, it takes less than 0.1 sec to acquire the satellite after initial phase lock. A dc signal proportional to the RF signal is derived at the terminal receiver. The satellite is tracked by periodically switching on either side of the current beam position and selecting the beam with the strongest signal. A number of algorithms have been devised to minimize any perturbation of the communications signal to less than 1% of the time. The maximum phase transients in azimuth can be kept to less than ± 10 deg over the required angular coverage and operating frequency bands.

9.5 ANTENNA SYSTEMS FOR GPS

9.5.1 General Requirements for GPS Antennas

Several navigation systems using satellites have been used very widely around the world. The oldest system is NNSS, which was originally developed as a U.S. Navy military satellite, but has been open to the civil sector since 1967. The position of a mobile terminal is determined by measuring the Doppler shifts of 400-MHz signals transmitted from the NNSS satellite. The dual frequencies are used in order to compensate for a refraction error due to ionospheric propagation. The Navigation System with Time and Ranging/ Global Positioning System (NAVSTAR/GPS) is the most widely used navigation system at present. A GPS satellite transmits two frequencies, L_1 (1,575.42 MHz) and L_2 (1,227.6 MHz). There are two kinds of positioning applications: an absolute or single positioning and a relative or differential positioning.

The absolute positioning, which provides the absolute three-dimensional position of the GPS receiver with an accuracy of several tens to a hundred meters, is carried out by receiving the signals from four GPSs and decoding the navigation codes, called the *coarse acquisition* (C/A) code. On the other hand, the relative positioning can precisely measure the distance between two GPS receivers by detecting the time delay (usually the phase difference) between the received signals. This system provides accurate land survey-ing over a long distance; for example, an accuracy of less than 1 cm is possible for a base line of several hundred kilometers.

The Global Orbiting Navigation Satellite System (GLONAS) is a system of the former USSR, and now it is in operation using five satellites. The system is very similar to NAVSTAR/GPS, but the frequencies are slightly different: L_1 is 1,597 to 1,617 MHz and L_2 is 1,240 to 1,260 MHz. Requirements for receiver antennas for these navigation systems seem to be similar. On GPS receiver antennas, the following requirements are imposed [49].

Amplitude Radiation Pattern and Gain

An antenna should provide uniform response over approximately the entire upper hemisphere over which the satellites may be visible. This limitation of the coverage is imposed on the antenna because the reception is excluded when GPS satellites are below a specified elevation angle (e.g., 10 deg) in order to avoid severe multipath and tropospheric effects. The hemispherical beam can remove the need for a tracking mechanism for the antenna; therefore, simple antenna systems become available.

The uniform amplitude response over the coverage region is required because the signal level must be sufficient for all angles of the desired view so that the receiver electronics can maintain signal lock with a required signal-to-noise ratio. Beyond this, it is necessary that the antenna be effectively "blind" to all signals arriving from outside the coverage range; that is, the pattern cutoff must be sharp with no backlobes.

For maritime GPS applications, however, antennas are required to have the coverage extending to negative elevation angles to compensate for the vehicle motion due to pitching and rolling. Typically, the gain of -5 dBi or more should be maintained up to -20-deg elevation.

The power flux density of radio waves emitted by a GPS satellite is about 4×10^{-14} (W/m^2) on the surface of the earth. Since most receiving antennas are omnidirectional with gains around 0 dBi or less, LNAs are necessary to amplify the weak received signal. Typical GPS receivers in commercial use seem to have noise figures less than 3 dB and gains between 20 and 40 dB. When the antenna and receiver are located a distance apart and are connected with a long cable, a preamplifier should be implemented at the antenna output in order to compensate for the transmission loss of the cable.

Phase Pattern

In the case of relative positioning systems using direct phase measurements, the phase difference of the antenna outputs corresponding to different directions of the satellites causes considerable position errors, which are unacceptable for accurate land surveying. Within the region of coverage, the antenna should ideally provide uniform response not only in amplitude, but also in phase. The requirement for uniform phase response over the coverage area of the antenna is critical, particularly in phase tracking receivers. The phase measured at the antenna output changes as a signal source moves around the phase center at a constant range. These phase changes produce an apparent displacement of the phase center from its on-axis position, which is dependent on the direction of the signal source as seen from the antenna. In effect, the antenna seems to move, depending on where the signal source is. A true phase center is defined as the center of curvature of an equiphase contour for a defined observation plane, polarization, and frequency [50]. If the phase changes associated with the signal source direction are not considered in the data analysis process, the antenna position determined from the measurements will be incorrect.

A uniform phase response is generally much more difficult to realize than is a uniform amplitude response, particularly in the case of the added constraint of a sharp horizon pattern cutoff. The pattern cutoff (or null) results from interferences of radio waves; consequently, it generates a nonuniform phase response for the antenna. In this respect, small, compact antennas produce a more uniform phase response than longer or more spatially distributed antennas, although this is achieved at the cost of poorer amplitude pattern cutoff characteristics.

Frequencies and Polarization

The GPS satellite transmits two radio waves, L_1 and L_2, with right-hand circular polarizations. In the simple case, a GPS receiver is used either at L_1 or at L_2. For more accurate positioning, however, the dual frequencies are used to compensate the excess delay of radio waves due to ionospheric propagation. The operation in this case requires an antenna that can operate equally well at both frequencies. Omnidirectional antennas introduced in Section 9.3 are potential candidates for GPS reception. Actually, quadrifilar helical antennas (QHA) and MSAs have been most widely used owing to their simplicity, small size, low cost, and satisfactory electrical characteristics. It should be noted, however, that the design parameters of GPS receiver antennas must be determined so that their radiation patterns may be as uniform as possible over the upper hemisphere, which is somewhat different from the omnidirectional antennas for satellite communication described in Section 9.3.

9.5.2 Quadrifilar Helical Antennas

The QHA was invented by Gerst, and some of its characteristics were reported [51]. By changing parameters, these antennas may show widely different radiation patterns, so many manufacturers of the GPS receiver use volutes of different size and construction. The literature [52] is suitable for selecting or designing QHAs for many applications. Data from Kilgus [52] and Adams [24] are particularly useful in obtaining desired amplitude pattern responses; however, owing to the lack of phase data for these antennas, these references are not directly applicable in evaluating the various antenna design choices or in identifying the particular design features that are most critical in determining antenna performance for GPS applications.

Recently, Tranquilla and Best [49] presented a detailed analysis of a QHA using the method of moments technique and various design considerations for GPS applications. In the paper, the phase performance of the antenna as well as the amplitude was taken into account, and the use of the antenna for dual-frequency operation was discussed. Resonant fractional turn volutes are particularly attractive for GPS applications due primarily to their small size and wide beam elevation pattern coverage.

Figure 9.45 shows a model of the QHA with two independent bifilar structures, each having a feed generator at the center of the radial conductor. The parameters of three models examined are summarized in Table 9.7. The design frequency is L_1. Figure 9.46 shows the computed polar amplitude pattern (circular polarization) for these three designs. Note that all three provide adequate upper hemispherical coverage; however, only the shortest volute ($L_{ax} = 0.20\lambda$) has a suppressed backlobe. The figure also shows the phase patterns of both linear polarized components (θ and ϕ) in a fixed observation plane. Figure 9.47 shows the computed phase center location for ϕ linear polarization for the half-wavelength QHA. The longer, slimmer volutes are advantageous in their reduced phase center movement at extremely low elevation angles. This arises due to the small separation distances between helical arms as viewed from near broadside.

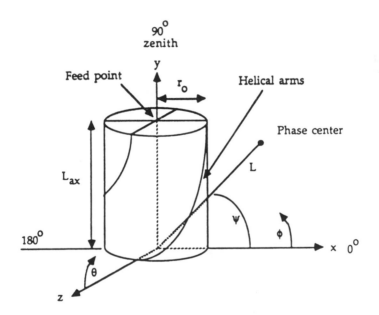

Figure 9.45 Geometry of the quadrifilar helical antenna. (From [49], © 1990 IEEE.)

Table 9.7
Design Parameters of Three Quadrifilar Helical Antennas

	Axial Length (L_{ax}) (λ)	Volute Radius (r_0) (cm)
Antenna 1	0.20	1.60
Antenna 2	0.27	1.39
Antenna 3	0.35	1.04

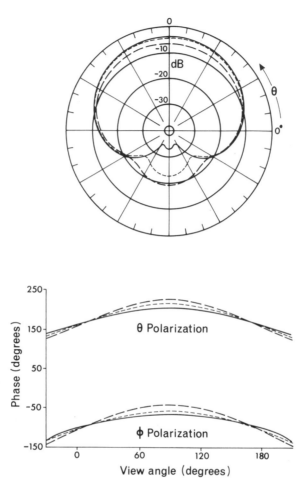

Figure 9.46 Calculated amplitude and phase patterns of the three QHAs of half-wavelength helixes at L_1, with $L_{ax} = 0.2\lambda$ (—), 0.27λ (----), and 0.35λ (—). (From [49], © 1990 IEEE.)

For dual-frequency operation, a single QHA is not available, because it is a resonant-type antenna and is inherently too narrow-band to accommodate both L_1 and L_2. Therefore, the dual-band operation can only be achieved through the incorporation of two antennas into one structure by coaxially mounting them in either an enclosed (L_1 inside L_2) or a *piggyback* (L_1 atop L_2) fashion, as shown in Figure 9.48.

Figures 9.49 and 9.50 show the amplitude and phase patterns of the enclosed and the piggyback types, respectively. For the enclosed volute design, neither the L_1 nor the L_2 performance is as good as with the single volute, the gain reduction being approximately 3.5 and 1.5 dB at L_1 and L_2, respectively, over the upper hemisphere, and a slight increase

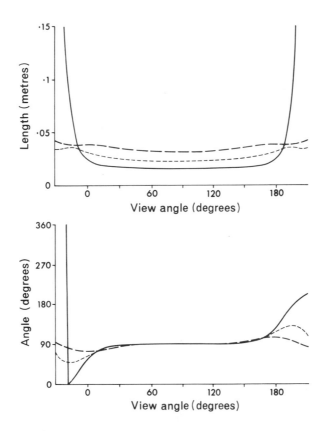

Figure 9.47 The phase center locations of the three QHAs for ϕ polarization component with $L_{ax} = 0.2\lambda$ (—), 0.27λ (----), and 0.35λ (---). (From [49], © 1990 IEEE.)

of the backlobe level at both L_1 and L_2. Phase center performance is also somewhat poorer at off-boresight angles for the compound antenna. From Figure 9.50 it can be seen that the coaxial piggyback arrangement has the greatest difference between L_1 and L_2 performance in the compound arrangement. In particular, the L_1 amplitude and phase response deteriorates significantly (with respect to L_2) at elevation angles below 45 deg. In this case, the radiation from the L_1 antenna is considered to excite the L_2 element and to generate a considerably large backlobe.

9.5.3 Microstrip Antennas

Owing to their low-profile and compact nature, MSAs currently account for almost half of the production of GPS antennas. A number of papers and several monographs [40,53] on MSAs have been published so far. One of the effective design methods for obtaining

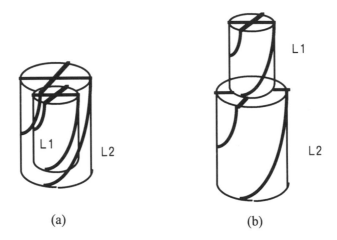

(a)   (b)

Figure 9.48 Dual-frequency quadrifilar helical antennas: (a) enclosed type; (b) piggyback type.

a broadbeam for GPS reception using a typical microstrip antenna is a reduction of the size of its ground plane [54,55].

A circular patch (operating in the TM_{11} mode) with a finite ground plane is shown in Figure 9.51, and the computed radiation patterns using the method of moments are shown in Figure 9.52 [54]. In the upper graph, showing the H-plane pattern, the cross-polarization pattern in the $\phi = 45$-deg plane is also shown. The beamwidth of the H-plane pattern decreases by increasing the ground-plane size. Consequently, the infinite ground plane has the sharpest cutoff pattern. In the E-plane, on the other hand, the beamwidth decreases initially by increasing the ground-plane radius g, but increases for $g > 0.7\lambda$. The infinite ground plane gives the broadest beam, which approaches −6 dB at the horizontal plane. These results show that the assumption of an infinite ground plane in approximate analysis of MSAs will have a serious effect on the correct prediction of the radiation patterns, particularly for angular ranges beyond 45 deg off the main beam.

For a rectangular microstrip patch with a finite ground plane, Huang [55] presented radiation patterns calculated on the basis of slot theory and the uniform GTD. Both the E- and H-plane radiation patterns of a rectangular microstrip patch, whose geometry is illustrated in Figure 9.53, are shown in Figure 9.54 with the measured results. The antenna dimensions in inches are given in the figure, and the operating frequency is 2.295 GHz (wavelength = 5.146 in). The agreement between the measurement and the calculation is quite good. From the figure, it can be seen that the gain reduction for the low elevation angles is relatively large, particularly in the H-plane. These characteristics seem to be similar to those of the circular microstrip patches; therefore, for GPS application, especially for its operation in the low elevation angle, a QHA may be more preferable for an MSA.

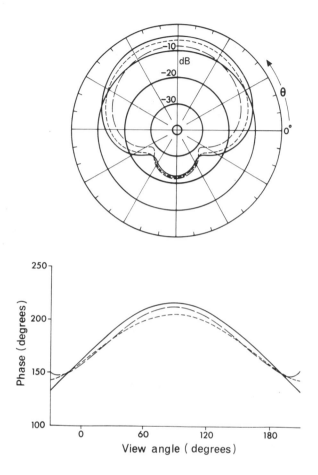

Figure 9.49 Amplitude and phase patterns of the enclosed-type QHA (the L_1 volute is inside the L_2 volute) at L_1 (—–), at L_2 (----), and of a single QHA (—). (From [49], © 1990 IEEE.)

9.5.4 Array Antenna for GPS Reception

The deployment of a directional-array antenna for GPS reception obtains a highly precise pseudorange through the high signal-to-noise ratio of the received signal. The CRL developed a precise relative positioning system using GPS, called PRESTAR (Precise Satellite Ranging) [56]. By measuring the phases of the modulated clock signal and the carriers of L_1 and L_2, PRESTAR provides a three-dimensional relative position (i.e., baseline vectors) of up to several hundred kilometers with an accuracy of 10^{-7}, and also the clock offset between two receivers. Other features are an observation time requirement of less than five minutes and short processing time, and an ability to determine the precise orbit of the GPS satellite from accurate carrier range data. Because the antenna

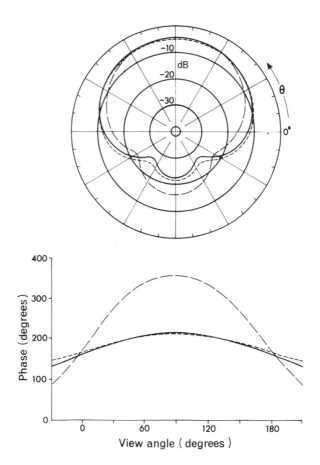

Figure 9.50 Amplitude and phase patterns of the piggyback-type QHA (the L_1 volute is above the top of the L_2 volute) at L_1 (––), at L_2 (----), and of a single QHA (—). (From [49], © 1990 IEEE.)

of PRESTAR is steered mechanically to direct its beam to one of GPS satellites for a moment and then switch to another, PRESTAR is a single-channel sequential system.

The antenna is a dual-frequency microstrip array antenna as shown in Figure 9.55. Eight elements are used for L_1, while six elements are used for L_2. Each element is a single-feed circular microstrip patch with two notches to generate circular polarization. In order to maintain good circular polarization characteristics, the sequential-array technique [41] was adopted, whereby the proper rotation and differential phase shift are given to each element. The numbers written on the elements show the relative rotations and phase shifts to be assigned.

Figures 9.56 and 9.57 show the measured radiation patterns at L_1 and L_2, respectively. From these figures, it can be seen that the excellent circular polarized patterns are

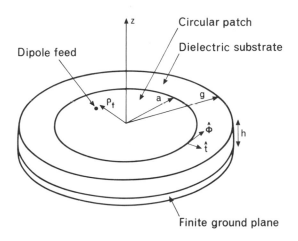

Figure 9.51 Circular microstrip antenna with a finite ground plane. (From [54], © 1986 IEEE.)

obtained by the sequential technique. The measured gains are 18.7 dB at L_1 and 13.1 dB at L_2.

9.6 SATELLITE CONSTELLATION SYSTEMS AND ANTENNA REQUIREMENTS

During the past decade there have been significant developments in the creation of LEO/MEO satellite constellation systems as noted in the introduction, and the practical realizability of such systems is now being commercially demonstrated. Global navigation systems such as GPS and GLONASS are of course well established but are relatively low information rate systems. There are other low information rate systems such as the Advanced Research and Global Observation Satellite (ARGOS) for the time sampled collection of earth environmental data but the main commercial thrust now is to establish satellite constellation systems for voice communication and wide-band data to provide continuous mobile personal services.

The challenge to satellite communication systems designers has been immense as have the projected development and manufacturing costs [57,58]. Antenna designers, in particular, have introduced state-of-the-art electronically controlled spot-beam arrays to satisfy the downlink demands, and the design of compact antenna elements for the mobile handset terminals has emerged as a critical design issue [59]. In this next section the main satellite constellation systems and their system parameters are outlined to highlight the antenna requirements of significance. It also includes some propagation issues relevant to the operation of LEO and MEO satellite systems.

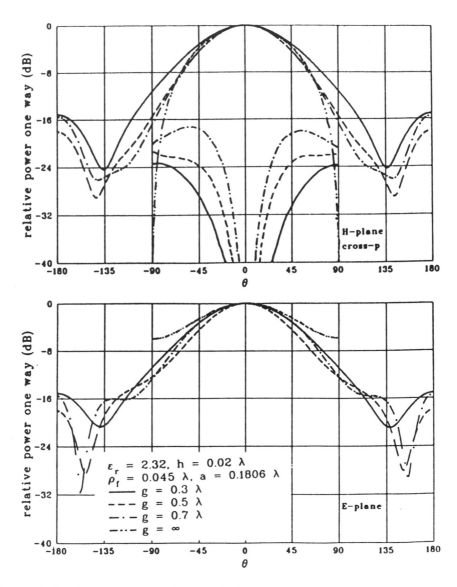

Figure 9.52 The radiation patterns of a circular patch with different ground plane radius g. (From [54], © 1986 IEEE.)

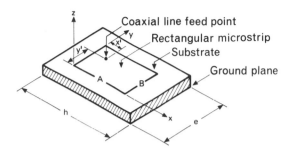

Figure 9.53 Rectangular MSA with a finite ground plane. (From [55], © 1983, IEEE.)

9.6.1 Constellation Systems and Demands on Antenna Design

Iridium, Globalstar, and ICO are the main satellite constellation systems that are designed to provide personal voice communication services. A summary of some of their parameters is given in Table 9.8. These systems have received much attention in both the technical literature [57,60–62] and general interest literature. Other LEO satellite constellation systems include the potentially lower cost PCS Ellipso system, which has elliptical orbits, and Orbcomm, which will offer a service for short messages at low cost.

Several wide-band data-oriented PCS satellite LEO constellation systems are under development, such as SkyBridge and Teledesic, but their higher operational frequencies (ranging over the Ku-, K-, and Ka-bands) permit higher gain antenna arrays to be utilized at the user terminal, thus alleviating many of the problems associated with low-gain handset antennas.

The deployment of low-altitude satellites in conjunction with spot-beam antennas leads to realizable link budgets that are compatible with the use of compact low-gain handset antennas with low-uplink RF transmit power, as shown in Table 9.8, and low path delays. These advantages are accompanied by several propagation and spectral problems. For instance, doppler shifts ranging from 36 kHz at 1.6 GHz and 55 kHz at 2.5 GHz have to be corrected with due regard to the relative direction of flight. Shadowing by buildings is more severe for the LEO and MEO systems than the GEO ones and generates polarization-sensitive reflection and diffraction that manifests itself in multipath effects. For the voice systems of Table 9.8, the channel bandwidths are in general compatible with that obtainable from compact antenna elements but for ICO the separation in uplink and downlink bands makes the provision of a single handset antenna more difficult to realize.

9.6.1.1 Downlink Array Design

Photographs of the Iridium satellite and the ICO satellite in Figures 9.58 and 9.59, respectively, show the outline shape and layout of the spot-beam phased arrays. Full

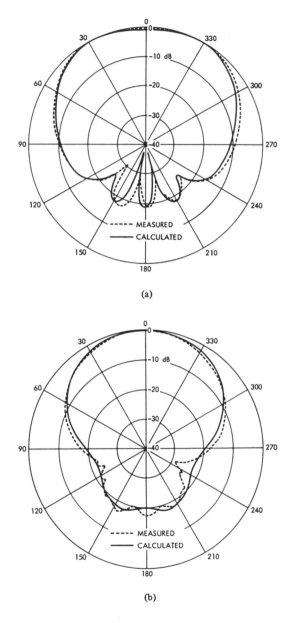

Figure 9.54 E-plane (a) and H-plane (b) radiation patterns of a rectangular MSA. (A = 2.126 in, B = 1.488 in, e = 10.5 in, h = 14.0 in, substrate thickness = 0.125 in, ϵ_r = 2.55, frequency = 2.295 GHz). (From [55], © 1983 IEEE.)

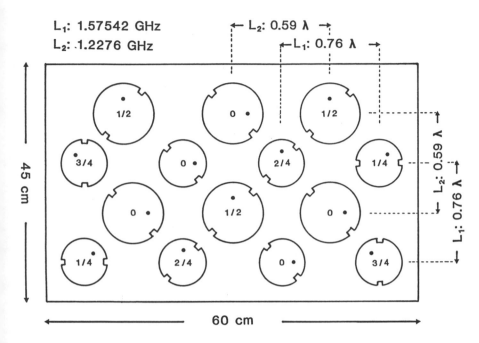

Figure 9.55 Configuration of the microstrip array antenna of GPS receiver PRESTAR.

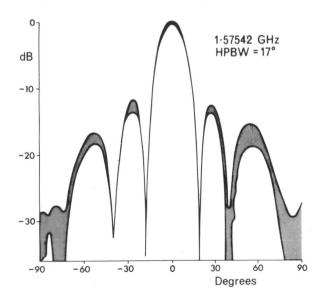

Figure 9.56 The radiation pattern of the PRESTAR at L_1.

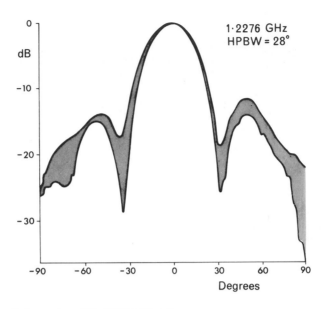

Figure 9.57 The radiation pattern of the PRESTAR at L_2.

Table 9.8
Satellite PCS Systems [57, 60]

	Iridium	*Globalstar*	*ICO*	*Thuraya*
Orbit	LEO	LEO	MEO	GEO
Orbit altitude (km)	780	1414	10355	35786
Number of satellies + spares	66 + 6	48 + 8	10 + 2	1 + 1
Orbit planes and inclination	6 circular 86.5°	6 circular 52°	2 circular 45°	0°
Multiple access	TDMA-FDMA	CDMA-FDMA	TDMA-FDMA	TDMA-FDMA
Target market	Global roamers	Cellular fill-in and fixed	Global roamers	Cellular fill-in and fixed
Spot beams per satellite	48	16	163	256
Uplink freq (GHz)	1.616–1.626	1.610–1.621	1.980–2.010	1.626–1.660
Downlink freq (GHz)	1.616–1.626	2.483–2.495	2.170–2.200	1.525–1.559
User Terminal				
RF power (W)	0.45	0.5	0.625	
G/T (dB/K)	−23.0	−22.0	−23.8	

Figure 9.58 Iridium satellite showing the phased-array antenna panels. (Courtesy Iridium.)

details of the Iridium array, referred to as the main mission antenna (MMA), have been published [63] and confirm the state-of-the-art design that has been realized.

The MMA consists of three fully active phased-array panels, each one producing 16 fixed simultaneous beams, to give a total of 48 beams per satellite. While physical space requirements are not as critical as in the case of the user terminal antenna, the MMA design is constrained to satisfy a stringent specification demanding the simultaneous radiation of multiple carriers into multiple beams with high efficiency and linearity, within a lightweight, space-qualified, temperature-controlled structure. A seamless hand-off process enables users to be transferred from beam to beam and from satellite to satellite to achieve the best link margin at all times. Each array panel consists of more than 100 lightweight patch radiators each of which is connected to a transmit/receive (T/R) module powered by a regulator system embodying redundancy. The beamforming architecture is provided by two orthogonal stacks of Butler matrices and a power divider. Intermodulation products are maintained at least 20 dB below the desired far-field carrier level by employing extensive prediction calculations in the design phase.

9.6.1.2 Generic Handset Antenna Constraints

Personal handset antennas appear to be relatively simple devices yet their design is constrained by many factors. For personal satellite terminal antennas the design is much more critical and even a 1-dB loss or pattern distortion can render the system nonviable.

Figure 9.59 ICO satellite under test in Hughes Laboratories showing the two hexagonal-shaped phased-array antennas. (Courtesy ICO.)

There is, however, a very strong incentive to optimize handset antenna design because any increase in the link budget margin is achieved at much lower cost than that incurred in further modifying the satellite terminal antenna. The design of personal handset antennas for satellite operation is discussed in [59] and the most important issues are:

- Electrically small antenna behavior [64] is important for small-diameter radiators.
- An antenna function independent of the phone body is preferred.
- Requirements of input impedance matching over the prescribed bandwidth exist.

- The radiation pattern polarization and beamshape characteristics need to be maintained over the required bandwidth and in proximity to user.
- Need to minimize noise temperature to achieve the G/T specification.

However, recent research has identified the quadrifilar helical antenna (QHA) and its variants as the best radiating structure for LEO and MEO systems and also in some cases GEO systems, where the margins are inherently small. Compact QHAs have been computed and measured [65] demonstrating that it is also feasible to span both the uplink and downlink bands for the ICO system (Table 9.8) in a single antenna.

A particular advantage of the QHA is the increase in gain due to its hemispherical radiation pattern, which in turn reduces noise pickup. The fundamental relationship between antenna gain and size is well known and for electrically small ground planes the latter is usually part of the radiation structure and contributes to the overall radiator size as previously explained [59]. Given an efficient balun, the QHA is, apart from back-radiation impinging on the case, largely isolated from the handset case ground plane thus decoupling hand effects to some extent. As such the QHA needs to be supported some distance from the handset case to achieve isolation from the latter in which case power loss in the hands is less of a problem than for cellular handset antennas. It thus remains to be seen if it becomes current practice to operate handset QHAs at some distance from both the handset case and head to achieve a clearer view of the satellite and less body perturbation, as opposed to connecting the QHA close to the handset case and sacrificing antenna pattern performance.

Other recent research achievements include the use of ceramics to reduce the inherent bulky QHA size [66,67], and smart QHAs [68], which offer an intelligent diversity system that has the potential to reduce multipath effects without the need for additional antennas. Ultimately it will be the user who will decide [59] the size of the antenna that he or she is prepared to carry around in exchange for a global roamer service. An appreciation of the system concepts for LEO/MEO/GEO satellites and the influence on the antenna design together with some recent practical examples of satcom handset antennas, are presented in the next section.

9.6.2 Handset Antennas for Satellite Systems

9.6.2.1 Basic System Concept

Geostationary systems (GEO) have the satellite in a "fixed" position thus enabling a fixed terminal antenna to be steadily pointing toward the satellite, as is done for TV satellites (DBS). The big path loss caused by the distance (35,786 km) has to be compensated for by a big antenna at the satellite (implying high antenna gain) and perhaps high power (generated by big solar panels).

In the newer satellite systems multilobe antennas are used to combine coverage and capacity and as much as 25–30% of the circumference of the earth can be covered. Wide-lobe terminal antennas are used in recent systems but for somewhat earlier systems bigger terminal antennas are used (such as the cover of a lap top) to meet the power budget requirement. GEO systems are typically supported by telecom companies in the countries below the satellites and the system concept is fairly simple except for the fact that the satellite is very big: 12m antennas are in use and 25m antennas are being studied for future use. One disadvantage when using GEO systems is that the user has to accept a considerable latency due to the distance (0.25-sec time delay one way from one phone on the ground to the other).

Low earth orbit satellite systems (LEO) are the other extreme with satellites at 700–1,400 km giving some 30 dB less path loss due to distance (as compared to GEOs), enabling much smaller antennas to be used in the satellite and also enable a design for better power budget margin.

To keep contact continuous, including a certain redundancy, a considerable number of satellites must be used (such as 66 in the Iridium system, which has been in operation since late 1998) and an advanced handover scheme is necessary. A LEO system includes transmission to many ground stations and possibly also intersatellite connections. Wide-lobe and rather small antennas are used at the terminals. The number of satellites must allow two to three satellites to be visible at any time and outdoor location. This kind of system offers a negligible latency (delay) and enables a bigger capacity than a GEO system using the same bandwidth. A typical LEO satellite (Iridium) is fairly small due both to the small antenna and to the small solar panels, sufficient to fill the moderate power needs.

Medium earth orbit (MEO) systems are a kind of compromise and need fewer satellites than LEO systems in orbits, typically 10,000–12,000 km over the surface of the earth. The latency (time delay) will be a minor problem compared to that of GEO systems. Two examples are the ICO system (which was originally expected to operate in the year 2000), which uses 10 satellites, and the already operating INMARSAT-M using 24 satellites. This larger number of satellites gives redundancy favorable at locations with limited sky visibility due to mountains, vegetation, or houses.

Some recent systems using all three configurations (GEO, MEO, and LEO) in combination with low-gain "omnidirectional" handset antennas. The requirements for power, antenna gain, and so on at the terminals are similar in all three cases, but with a LEO system in particular it is possible to build the system for wider bandwidth, better margins, and so on without the need for excessively large satellite antennas. For GEO systems, only voice transmission is possible for handheld phones with small antennas, while transmission of high rated data, such as 2×64 kb for INMARSAT-M, requires larger antennas, but they are still portable.

9.6.2.2 Frequencies

Most systems are used or intended for telephone and low rate data and operate in a narrow frequency band within the range 1.4–2.5 GHz. This choice of frequency is (besides being an international frequency allocation) favorable for this kind of systems for three reasons:

1. Noise from the earth (man-made noise) is high at low frequencies (<100 MHz) but steadily decreasing at higher frequency. At 1 GHz it is lower than thermal noise.
2. Total noise from the sky is low in the 1- to 3-GHz range but increases with higher frequencies.
3. Effect of rain is negligible but will increase at higher frequencies.

The choice of frequencies can also be illustrated by the propagation attenuation in free space (Friis law), which can be written in three ways to illustrate the influence of the different parameters:

$$P_r/P_t = G_t G_r (\lambda/4\pi r)^2 = A_t G_r/(4\pi r^2) = A_t A_r/(\lambda r)^2 \qquad (9.2)$$

Alternatively antenna gain G and antenna effective area A are used to characterize the antenna connected by the basic relation $G = 4\pi A/\lambda^2$. For a big-dish antenna the area A corresponds to the geometrical surface of the dish. The first formulation implies that at low frequencies where only low-gain antennas are possible the power budget will improve if a low frequency can be used (utilized in early communication radio). The third formulation implies that at limited antenna area (radio links of all kinds) the power budget will be better if a high frequency is used (but also at the expense of the need for accurate direction of the antennas and increased weather sensitivity).

The second formulation is applicable to the actual kinds of communication systems where the receiver antenna gain G_r is low but the transmitter antenna size A_t is critical (cost limited when it comes to 10–12m antennas in the sky). In this case the choice of frequency is not critical for the power budget alone. Another obvious consequence from the second formulation is that the area A_t of the satellite antenna must be bigger at bigger distances r and that can be expressed as a required antenna diameter proportional to the distance to the earth. A GEO satellite may have an antenna diameter of 10–12m, whereas a LEO satellite may have a 25 times smaller antenna diameter ($\approx$35,786/1,400) to fulfill the same power budget. This is unnecessarily small so a LEO system may use smaller power and still have a better power budget margin by using somewhat bigger antennas than the absolute minimum size mentioned above.

There are also LEO systems for low rate data, only they operate at a much lower frequency such as 150 MHz and they also use low-gain terminal antennas (typically a quarter-wave whip antenna). The low frequency enables very small low-cost satellites with little power to be used and still fulfill the power budget. The low frequency greatly improves the power budget assuming low-gain antennas but bandwidth is also very limited as compared to the telephone systems mentioned above. The system cost is much lower than that of a full telephone system and the typical use is to send short messages (below 200 bytes) for fleet management, emergency service, telemetry in remote areas, and so on. One operating system is Orbcom.

In most cases two frequency bands (RX and TX) are used and on the systems to date they are narrow compared to terrestrial cellular bands. In some systems they are widely separated, which sets a number of special requirements on the antennas to keep their size small and still combine double antenna functions in one unit.

9.6.2.3 System Requirements

There are substantial differences between a terrestrial- and a satellite-based telephone system. The power link budget is weak for a satellite system due to the long distance, but on the other hand only line-of-sight propagation is feasible, so the variation of the received power is fairly small and well known. The margin for unexpected degradation can thus be very small—just a few decibels. With the chosen frequencies, the influence of weather is very small. Various ionospheric conditions have a certain influence in terms of a slow rotation of a linear polarized wave. For this and positioning reasons, circular polarization are used, at least on the satellite side. Circular polarization has a special property when reflected in a flat surface; it will change its polarization from RHCP to LHCP. This is a useful tool to improve multipath propagation conditions assuming the terminal antenna has a fairly good purity of its circular polarization (i.e., a low axial ratio). Both flat ground (and also sea surface) and house walls will produce large reflections of this kind. The practical need for a power budget margin for this kind of system will still be experienced when these systems are in widespread use.

As a transmitter the terminal performance is determined by the EIRP, which is transmitter power multiplied with antenna gain G with regard to losses. The gain from the system viewpoint is the minimum gain within the angular coverage area, which is typically defined as 0–80 deg from zenith. For reception the G/T quotient is the key parameter. G/T for the antenna thus depends both on the antenna gain, the losses in the antenna, and the ability of the antenna to avoid picking up noise from the ground below the antenna. For a handheld terminal, a typical EIRP requirement may be 0.5W, whereas the G/T has to be at least −25 dB/K. A terminal for a wider bandwidth requires higher power, higher antenna gain, or higher receiver sensitivity to meet the final signal-to-noise requirements. Practical GEO, MEO, and LEO systems have fairly similar requirements for EIRP and G/T but the satellite antennas are very different and the power budget margins may be different.

The satellite antennas for modern satellites always use a multilobe antenna to improve capacity. The size of the satellite antenna gives the lobe width and thus minimum spot size on the ground. For LEO to GEO systems, the antenna beamwidth can range from 10 deg (LEO) to below 1 deg (GEO). On the ground this corresponds to 100–150 km (LEO) to 600 km (GEO). A LEO system thus can have a "cell size" that is at least comparable to the biggest one of a terrestrial cellular system, whereas a GEO system will have rather big "cells." Within each cell the number of simultaneous calls are limited by the number of available "channels" within the frequency band. A LEO system may thus have a much bigger capacity in terms of simultaneous calls per surface unit at the expense of a more complex system. A GEO system may have several hundreds of beams, while a LEO system may have 30 to 50 antenna beams only per satellite due to the rather broad beams given by the fairly small antennas on the satellites. The minimum satellite antenna size is proportional to the distance to the ground and if a small terminal EIRP is necessary then a big satellite antenna must be used (10–12m for a GEO). To secure the downlink, a slightly higher power must be used in the satellite than in the phone but several channels have to be transmitted simultaneously.

Most satellite phones are dual-mode phones, including a possibility for connection to one or more terrestrial system. Different system solutions currently are on the market as discussed later in Section 9.6.2.10. Besides the obvious customer benefit of being able to automatically choose the alternative having the lowest cost, there are many reasons for the dual-mode function. For instance, a national operator using roaming to a satellite system can offer 100% outdoor coverage from the very beginning before the cellular network is built in sparse populated areas.

9.6.2.4 Terminal Antenna Requirements

Antennas for satellite communication terminals must of course provide a certain minimum gain within the angular coverage area intended for the system and in addition satisfy requirements of needing only a small installation space, integration possibilities, and generally the ability to function in the retracted as well as extended position. Due to space requirements, it is mandatory (or at least highly desirable) to use the same radiating element for transmitting and receiving. In most cases this is fairly easy, but it may require special antenna solutions when the difference between transmitter and receiver band is as big as, for instance, Globalstar having 1.6/2.5 GHz.

For the discussion of the electrical requirement it is assumed that the antenna has a position that can be defined as normally pointing toward zenith ($\Theta = 0°$). The area where the elevation angle Θ is between 0 and Θ_{max} is to be covered with Θ_{max} typically being around 80 deg. Within this "coverage area" certain minimum requirements in terms of gain and axial ratio have to be fulfilled. Note, as a comparison, that for a terrestrial system average gain is usually more important than minimum gain. For the downlink a certain received power is required compared to the receiver noise. The total receiver noise comes from the LNA (low-noise amplifier), internal losses, and noise from sky and surroundings picked up by the antenna. The quotient G/T is the customary measure with a typical minimum requirement for a voice channel being around −25 dB/K. To fulfill this, the gain at low angles ($\Theta > 90$ deg) shall be low to decrease the external noise entering the system from the surrounding ground. The noise from the sky at zenith is typically 10K while the noise from the surroundings (especially the ground) is around 300K. A hemispherical pattern (or slightly less with regard to Θ_{max}) is thus the ideal one, and the use of circular polarization can reduce multipath reflections by 15 dB or more for axial ratios of 2–3 dB. For the uplink a certain minimum EIRP is required that calls for a minimum gain within the desired coverage area.

As a comparison the GPS system solution resembles a MEO system (24 GPS satellites at 22,000-km height) and has rather similar antenna requirements. The GPS system is somewhat more robust as many more satellites can be seen and it can in many cases use simpler antennas.

In some systems simple whip antennas are used at the terminal side and in that case the coverage of Θ is 30–80 deg rather than 0–80 deg. The amount of solid spherical angle close to zenith is rather small, but for a telephone system it is highly undesirable to get too many handovers to another satellite, which would be the case if there was a "sensitivity hole" around zenith. The satellite communication systems using whips are for sending short messages only and then there will not be any interrupted message because of their short duration. The use of a linear antenna for circularly polarized waves obviously will cause 3-dB extra loss, which in special cases may be justified by the simpler solution.

9.6.2.5 Different Uses for Terminal Antennas

Due to the difficult requirements for minimum EIRP and G/T, all working solutions so far for low-gain antenna handset antennas use an extendible antenna that is rather big in order to get sufficient performance. Two different uses are paging and talking. For paging (reception of a call only) the antenna can usually be in the retracted position at least in a free location (outdoors, on a nonconducting table, etc.). Talking in the phone will generally require that the antenna be extended in contrast to cellular systems. Some antennas have a swivel so the antenna can remain in a vertical position, whereas others are just drawn straight out of the phone. Figure 9.60 shows a phone with its extended antenna in talk position.

Because of the bad or nonexistent coverage indoor, inside cars, and so on, auxiliary antennas located outdoors are important accessories. Obvious uses are on cars and small boats. A LNA accompanies the antenna in these cases to allow for the acceptance of the losses in a typical cable length (a few meters). As for cellular antennas, mounting on cars by screws, magnets, and other means is used. Likewise, antennas mounted on the outside of a house are important auxiliary items in areas where satellite phones are the only available systems. Constructions exist where the whole RF part of the phone is integrated with the outdoor antenna, allowing the distance to the user inside the house to be many meters distant. Phone booths with a satellite antenna on the roof will be quite common in countries with a low number of fixed telephones and sparse mobile network. On cars a special kind of antenna in combination with GEO systems has a conical pattern tuned to the height over the horizon where the GEO satellite is to be found in the country in question. Due to the conical shape, the direction of the car will have minor influence as long as the road is reasonably horizontal.

9.6.2.6 Antenna Types

Whip antennas are used in terminals for satellite systems using VHF frequencies and those typically used on lorries. The obvious disadvantages with the linearly polarized

Figure 9.60 Globalstar telephone in use. (Courtesy of Globalstar.)

whip is, of course, the 3-dB loss due to polarization and less suppression of wrongly polarized waves created by ground reflection. For the long wavelength, however, the simple whip implies big practical advantages and a part of the 3-dB polarization loss is regained by the highly simplified antenna structure.

Quadrifilar antennas (see Sections 9.5.2 and 10.3.3) have emerged as the preferred type for handsets based on a number of good virtues:

- Rather constant and well-polarized pattern over angles like Θ = 0–70 deg or Θ = 0–80 deg;
- Low gain for $\Theta > 90°$ giving low antenna noise temperature;
- Reasonable geometry if compared to other antennas giving a similar pattern.

The use of quadrifilar antennas was for a long time, restricted to typically a 0.75λ circumference (corresponding to a 30- to 40-mm diameter in the frequency range used here). On modern phones this is changed to something around 15-mm exterior diameter (or cylinder circumference $\sim\lambda/4$) to make the antenna more handy. The disadvantage with the small diameter is a narrower bandwidth and critical design, which also might give a slightly higher loss. Using a more complicated helical pattern, two separate passbands can be obtained. They are still narrow but can accommodate the transmitting as well as receiving band. Figure 9.61 is an exterior view of one QHA.

Patch antennas are widely used for GPS but not for pocket satellite phones. The coverage for this latter use is rather bad because the radiation at low angles has a poor axial ratio simply because the horizontal polarization implies that the electrical fields are

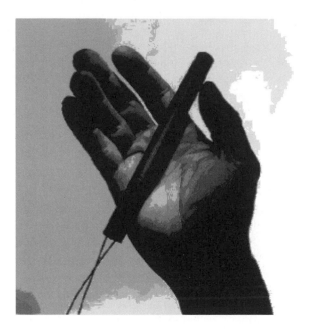

Figure 9.61 Exterior view of a QHA for a satellite phone. (Courtesy of Allgon.)

''shorted'' by the ground plane, which is a functional part of the patch antenna (even if small). Furthermore, a hypothetical patch with sufficient telephone bandwidth generally has to be considerably bigger than a GPS patch, making it quite difficult to integrate in a phone. Roof-mounted patch antennas for cars are available for narrow-band systems. Figure 9.62 is a common GPS patch. Note that the GPS system is very different from a phone system and that the use of patch antennas is restricted to such GPS receivers where a handy shape is worth more than maximum performance. Many GPS receivers use a QHA antenna.

Patch antennas using a higher mode are known for possible use on aircraft as true low-profile antennas having hemispherical coverage. They do not fit in the size of a handheld terminal and are complicated.

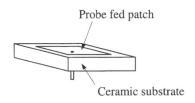

Figure 9.62 Sketch of GPS patch antenna.

There are also other types of antennas creating circular polarization of which some use a combination of two antenna elements to get circular polarization with hemispherical coverage. For handsets, QHAs seem to be the only antenna type reaching the high level of functional standards with some margins. With somewhat less performance, some type of combination antenna or a patch antenna might also be a candidate. Improved satellites (mainly antenna size but also transmitted power) may in the future make other types possible if the constraints on EIRP and G/T are relaxed. This may especially be the case for LEO systems with their intrinsic better power budget margins.

9.6.2.7 General Properties of the QHA

The basic QHA [69] consists of four helical wires or ribbons on a cylinder having the radius a. Note that any number of wires greater than two can be used with the same function. This may be a motif for the term NHA rather than QHA but only $N = 4$ has reached widespread use.

The helical ribbons are fed from a phasing network producing a phase progression of 90 deg between each wire (or $360/N$ in the NHA case). For $N = 4$ the phasing network is readily designed by standard 90- and 180-deg hybrid components of the lumped or distributed type.

The radiation from the QHA geometry can be seen as a combination of vertical and horizontal polarization having a 90-deg phase difference. The function of the radiation can be understood, in a simplified way, as the radiation from a combination of four vertical monopoles together with a circular loop in the middle of the monopoles. The four monopoles radiate vertical polarization as an ordinary monopole but with a 360-deg phase rotation. Like any vertical monopole the radiation toward zenith is zero. A vertical loop on the other hand gives a circular polarization toward zenith and horizontal polarization 90 deg away from zenith. Due to the 90-deg phasing and the small radius, the vertically polarized radiation from the four helices will be much less than the corresponding radiation from a single monopole carrying the same current. The combination of pitch angle and radius determines the relative amplitude of these two horizontal and vertical polarizations and by a suitable choice, a good circularity can be obtained in all directions (at least above the horizon). The condition for circular polarization is approximately $ka \tan(\alpha) \sim 1$ where $k = 2\pi/\lambda$, a = radius, and α = the pitch angle (90 deg corresponds to vertical wires). For a typical phone antenna k is around 32–52 m^{-1} (1.5–2.5 GHz) and a = 6–8 mm, which gives a pitch angle around 75 deg.

Note that each visualized horizontal plane slice of the QHA fulfills the same condition, and the length of the QHA is thus not critical for the polarization part of the radiation properties. On the other hand, the radiation pattern as well as the feeding impedance are more strongly dependent on the length. Some GPS receivers use a quarter-wave QHA while the typical phone antenna is 0.5–0.8 wavelength long. The radiation resistance (as referred to the maximum current somewhere along the helices) will be small due to the small radius of the virtual loop and the counteracting currents of the virtual vertical dipoles. This makes the narrow QHA (i.e., $ka \approx 0.2$–0.35) a narrow-band device. The typical QHA is a resonant antenna so the typical length can be thought of as $\lambda/2$ with an open end or $\lambda/4$ with a shorted end. In contrast to a straight single-wire dipole, the pattern for a QHA having the length λ will not have any zero at $\Theta = 90$ deg due to the combined effect of the twisting and the phase progression. The resonant structure simplifies the matching to a desired impedance (50Ω, etc.), but is not important for the radiation properties.

Radiation properties will be similar for another length but then a condition for a reasonable input impedance may be a tuning coil or capacitor at the end, very analogous to single linear whip antenna. On a typical satellite phone system, the optimization of the gain is very important for making the minimum gain within the desired coverage as big as possible. For fine adjustments of the antenna pattern, the length is important and it turns out that the optimum hemispherical pattern generally occurs when the length is between 0.5 and 0.8 wavelengths. If the QHA is very short ($\lambda/4$ or shorter), the radiation downward will increase, which will deteriorate the noise temperature of the antenna as well as the antenna gain. Quarter-wave QHAs are frequently used for GPS receivers. A QHA that is too long will, on the other hand, have a minimum or even a "hole" at zenith direction. Because QHAs are resonant antennas, the feeding can be made at any point along the helix but a feeding at the bottom is simplest and most common. The bandwidth is mainly limited by the diameter as the radiation resistance (and thus the bandwidth) will decrease proportional to the second power of the circumference expressed in wavelengths. A diameter of 12–16 mm thus seems to be close to the lower limit depending on the type of system. Polarization is determined by the direction of the helices and the phase order of the feeding. Screw direction and phase order must coincide or the antenna will radiate in the wrong direction (i.e., downward).

The feeding network can be done in many ways and for a wide bandwidth (of the feeding network alone) a combination of two 180-deg hybrids (sometimes called baluns) and one 90-deg hybrid or one 180-deg hybrid and two 90-deg hybrids is used. Because of the two bands, the need for bandwidth of the feeding network may be considerable bigger than for each of the antenna bands separately. Distributed networks are used for the same purpose and tend to give lower loss at the expense of a bigger physical space. For very narrow-band QHAs (mainly GPS antennas) a self-phasing QHA has been used. The four helices then can be thought as two pairs of twisted loops each having slightly different resonant frequencies giving a phasing of ±45 deg in each pair within a narrow frequency band. In the self-phasing case, it is important to distinguish between impedance bandwidth (which may be wide) and axial ratio bandwidth, which is very small (typically some 0.1%). Most GPS patches are self-phased by a very slight rectangular shape and a single but slightly off-centered feed.

Figure 9.63 shows a typical phasing network using one 90-deg hybrid and two 180-deg hybrids (balun). In this case the reflection in the QHA as well as possibly wrongly polarized signals received (including ground noise) will be absorbed by the termination

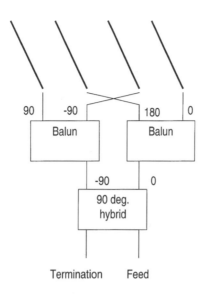

Figure 9.63 Phasing networks showing connections to QHA lines.

connected to the 90-deg hybrid. The possibility of printing the lines on one side of the film substrate only can favor other network solutions. The widely separated uplink and downlink bands are hard to obtain in the original QHA concept using four wires on a small-diameter core, but by manufacturing the QHA as a printed pattern on a flexible film rolled on a cylinder virtually any pattern can printed with good accuracy. Due to the small diameter of this kind of QHA, the total bandwidth is small anyway, but it can be distributed over the two bands by modifying the pattern.

Adding capacitance between the printed conductors is one method that has proven to be useful to achieve a sufficient two band matching, and a meander-shaped pattern is another. Due to the requirements for gain and polarization purity, the design is much more determined by the radiation pattern requirements than by the impedance bandwidth, which over a limited band is more easy to adjust by additional lumped or distributed elements. Two parallel helical and slightly different printed patterns (i.e., 2×4 conductors) have also been used to improve the RX/TX band coverage, and it is especially good in those cases where the RX and TX bands are close to each other but too separated to be treated as one band.

The printed helical pattern is the radiating structure, but there will normally be some nonhelical and more or less undesired radiating parts as well, of which the feeding structure is one. A number of intentionally modified helical printed patterns have been proposed including a short vertical wire connected to the helices. A very important property of the QHA is the good isolation from the ground plane (such as on the roof of a car), and another useful property is the symmetry that can isolate the QHA from a cellular whip.

9.6.2.8 Examples of QHAs

Antenna for Iridium System

Iridium can use a rather narrowband antenna because it uses the same frequency for uplink and downlink (1616–1626.5 MHz). The power budget has a good margin (said to be 16 dB) to simplify use where the coverage is not perfect and to enable fairly low power at good connections (0.1–3.5W). Iridium was the first satellite telephone system with real handsets (operating late 1998). First- and second-generation handsets for Iridium are shown in Figure 9.64. The Iridium system was closed in late spring 2000 for economical reasons.

In the first-generation handset antenna, a combined structure of a microstrip element, a linear element, and a planar element, as shown in Figure 9.65, was employed. The combination of these three elements produces a radiation pattern with wide-angle coverage as Figure 9.66 shows. The microstrip element contributes mostly to the pattern in the vertical direction, whereas the linear and planar elements mainly contribute in the horizontal direction. The linear element produces the vertically polarized wave, while the planar element creates the horizontally polarized wave. A sleeve balun is employed in order to suppress the unnecessary flow of current on the feed cable, so that undesired radiation from the feed cable and the handset unit can be avoided.

In the second generation of Iridium handsets, (Figure 9.64) a QHA element has been used, because it has shown higher antenna performance than the first-stage antenna. Figure 9.67 shows the mechanical outline of such an Iridium antenna using an antenna element that is 100 mm long (active length 77 mm) and having a 17-mm diameter (including radome). The antenna has a swivel joint so it can be used in a vertical position regardless of whether the phone is used with the right or left ear.

A basic QHA with four plain helices is used to cover the rather narrow frequency band. The intended co-polar and the cross-polarized antenna patterns are shown in Figure 9.68 and illustrate the desired near-rectangular pattern. The input VSWR is shown in Figure 9.69, illustrating a fairly wide-band behavior even if the optimized pattern and VSWR performance is obtainable in a narrow band only.

Antenna for Globalstar System

The Globalstar system was put in service in late 1999. It uses two widely separated frequency bands for uplink and downlink (1.6 and 2.5 GHz), which create a design challenge. A meander pattern along the helices together with double capacitive loadings are used to achieve this.

Figure 9.70 shows a mechanical layout. The radome diameter is 15 mm and the total antenna length is 110 mm, of which 77 mm are active. Below the active part of the

Figure 9.64 Iridium handsets: (a) first generation and (b) second generation. (Courtesy Kyocera Corp., Japan.)

antenna, the phasing network is mounted on a small PCB together with the LNA to avoid the losses of any unnecessary piece of cable. The antenna is foldable so it can be used in a vertical position.

The antenna pattern is shown in Figure 9.71 for RX and TX. The difference in shape depends on the fact that the same physical length is used for both bands and the higher RX band has a slight minimum at zenith. The VSWR (Figure 9.72) appears very broadband because the phasing network absorbs power outside of the passbands of the antenna. This is illustrated by Figure 9.73 giving S21 (or the leakage between the antennas), which reveals the basically narrowband behavior.

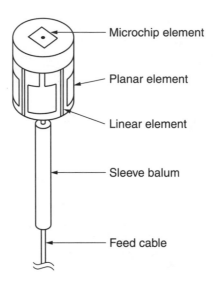

Figure 9.65 Construction details of the first-generation Iridium phone antenna of Figure 9.64(a). (Courtesy of Kyocam.)

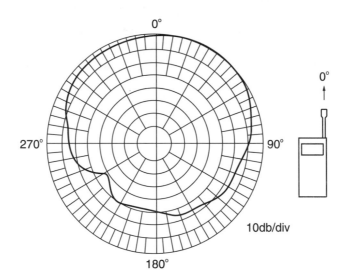

Figure 9.66 Elevation radiation pattern of extended Iridium handset antenna of Figure 9.64(a). (Courtesy of Kyocam.)

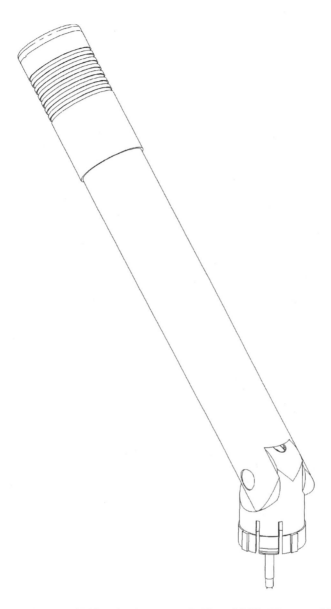

Figure 9.67 Mechanical layout of Iridium handset antenna in Figure 9.64(b). (Courtesy of Allgon.)

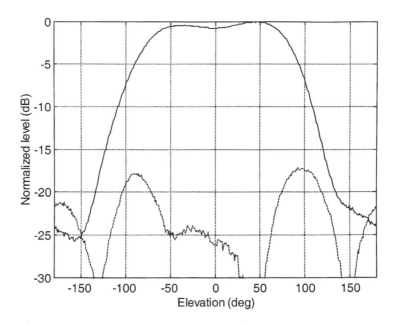

Figure 9.68 Elevation radiation patterns (LHCP and RHCP) of Iridium antenna shown in Figure 9.64(b). (Courtesy of Allgon.)

Antenna for ICO System

An antenna solution for ICO (2 and 2.2 GHz) has, instead of four standard helices, four double helices to get the required double tuning. The radome diameter is 14 mm and the active length is 90 mm ($3\lambda/4$). The mechanical layout is fairly similar to that shown later for Figure 9.76 but without the cellular antenna. Antenna patterns for RX/TX are shown in Figure 9.74 and the VSWR plot is shown in Figure 9.75. The antenna is extended straight out of the handset without a swivel joint. Below the active part of the antenna, the phasing network is mounted on a printed circuit board together with the LNA to avoid the losses of any unnecessary piece of cable.

Antenna for Thuraya System

A number of GEO systems for handheld terminals such as the Thuraya system are in various stages of planning around the world. The active part is 100 mm long with a diameter of 14 mm. It uses the same 4 × 2 helix structure as the ICO antenna and is shown in a Figure 9.76. The VSWR is shown in Figure 9.77, indicating the two-band tuning obtained by the dual helices.

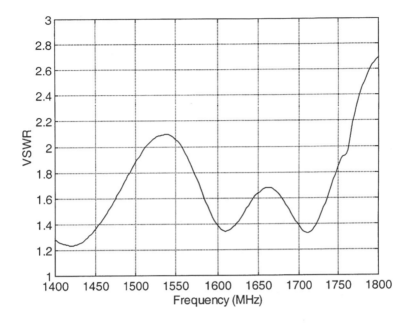

Figure 9.69 VSWR of Iridium antenna shown in Figure 9.67. (Courtesy of Allgon.)

The antenna pattern is shown in Figure 9.78 and it should be remarked that the beamwidth is narrower (110 deg rather than the common 165 deg) than the other described antennas. When the phone is used close to the borders of the coverage, it is assumed that the user will be aware of at least a very coarse direction to the satellite and make a corresponding orientation. The coverage area roughly ranges over a wide area on the earth with among others England, Liberia, and Bangladesh close to the borders.

Figure 9.79 shows a full Thuraya handset where it can be noted that the extended antenna is drawn straight out from the phone body without a swivel. The Thuraya antenna also includes a GSM antenna (see Figure 9.76), which utilizes the difference in symmetry properties between a QHA and a common monopole or axial helix. By the coaxial arrangement, the coupling and the associated losses will be very small. In this case the phone also includes a GPS patch. Because the polarization in this case is different for GPS (RHCP) and Thuraya (LHCP), it is not possible to use the same QHA (which otherwise would have been possible because the frequencies are fairly close).

9.6.2.9 Patch Antennas

It is possible to use a patch antenna if antenna pattern properties can be sacrificed. Any patch antenna (or any small low-profile antenna) on a ground plane (such as the roof of a car) will have poor horizontal polarization near the horizon and thus poor axial ratio at

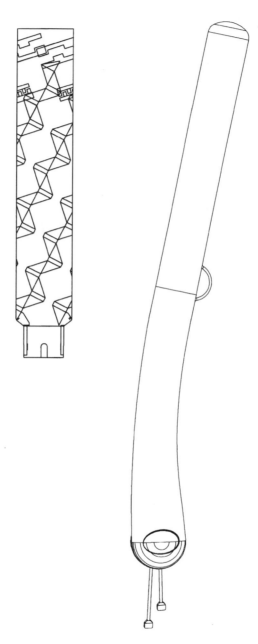

Figure 9.70 Mechanical layout of Globalstar antenna. (Courtesy of Allgon.)

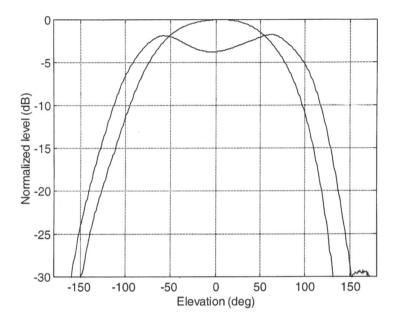

Figure 9.71 Elevation radiation patterns (TX and RX with lower gain at zenith) of Globalstar antenna shown in Figure 9.70. (Courtesy of Allgon.)

low elevations. This will not only decrease the gain for intended circular polarization, but will also allow waves reflected in the ground (and thus having wrong circular polarization) to enter and disturb the performance.

The bandwidth of a patch antenna linearly increases with the volume of the patch (surface multiplied by height) and will also decrease with a higher dielectric constant. Thus a phone application will require a patch that is much larger than, for instance, a typical ceramic GPS patch. The ceramic filling of the GPS patch is generally not useful here because it will decrease the bandwidth without adding anything to the radiation.

9.6.2.10 Auxiliary Antennas for Terrestrial Telephony

Most satellite phones are dual-mode phones that include the possibility of connecting to some terrestrial cellular system. Three system solutions on the market are as follows:

1. Satellite phones with a built-in GSM phone (or other system) in the same unit. Some have two antennas, while other manufacturers utilize the differences in radiation properties to enclose the two antennas together.
2. The satellite phone is made as a docking unit where a special phone (which by itself is usable for terrestrial systems only) is docked for satellite calls.

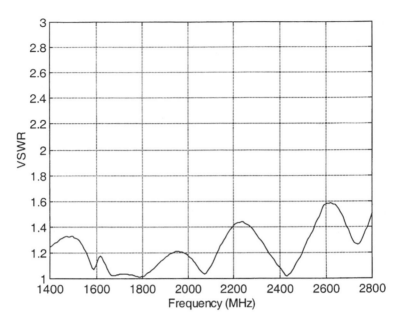

Figure 9.72 VSWR of Globalstar antenna shown in Figure 9.70. (Courtesy of Allgon.)

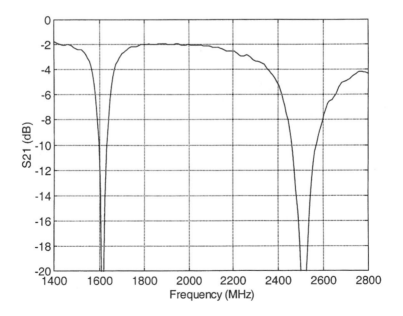

Figure 9.73 S21 response of Globalstar antenna shown in Figure 9.70. (Courtesy of Allgon.)

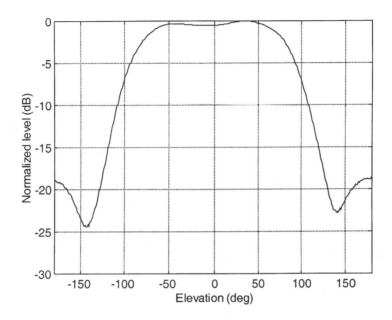

Figure 9.74 Elevation radiation pattern (TX and RX) for ICO handset antenna. (Courtesy of Allgon.)

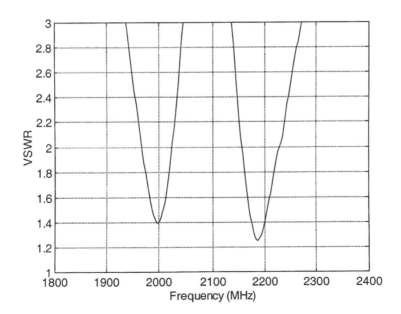

Figure 9.75 VSWR of ICO handset antenna of Figure 9.74. (Courtesy of Allgon.)

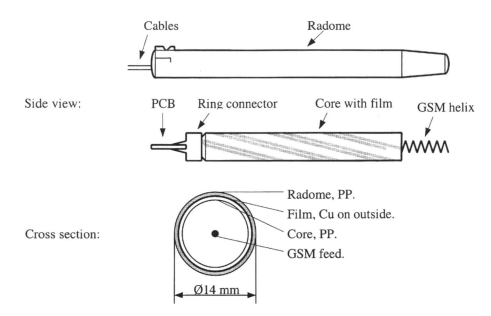

Side view:

Cross section:

Figure 9.76 Mechanical layout of Thuraya handset antenna. (Courtesy of Allgon.)

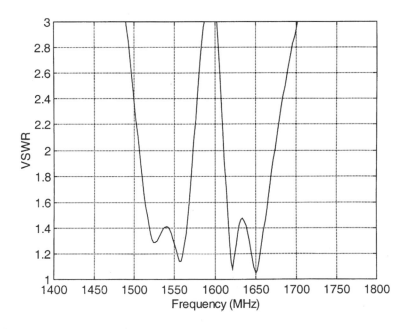

Figure 9.77 VSWR of Thuraya handset antenna of Figure 9.76. (Courtesy of Allgon.)

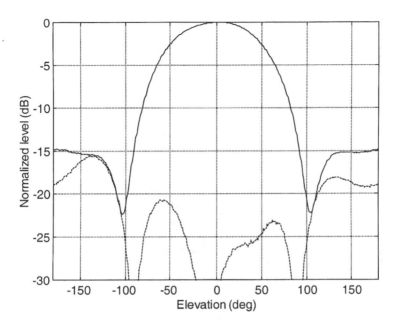

Figure 9.78 Elevation radiation patterns (LHCP and RHCP) of Thuraya handset antenna shown in Figure 9.76. (Courtesy of Allgon.)

3. The terrestrial system is enclosed in a small unit (like an ordinary phone battery), which is hooked on the satellite-only phone from its back. The terrestrial unit is then a kind of plug-in unit that is exchangeable for different countries.

In the first case, it is possible to use a combined antenna where a QHA and a normal cellular antenna (such as a whip with a normal helix at its top) are combined in one unit. Due to the different symmetry properties of the QHA and the whip (with or without a normal mode helix) the leakage between the two antennas will be small.

Figure 9.79 Photograph of Thuraya handset phone. (Courtesy Thuraya.)

REFERENCES

[1] *2nd Mobile Satellite Conf.*, Ottawa, Canada, June 1990.

[2] Hamamoto, N., S. Ohmori, and K. Kondo, ''Results on CRL's Mobile Satellite Communication Experiments Using ETS-V Satellite,'' *Space Communications*, 1990, pp. 483–493.

[3] A series of *MSAT-X Quarterly*, Jet Propulsion Laboratory, No. 1 (1984) to No. 25 (1990).

[4] ''PROSAT Phase I Report,'' European Space Agency, ESA STR-216, May 1986.

[5] Gedney, R. T., and R. J. Scherfler, ''Advanced Communications Technology Satellite (ACTS),'' *Int. Conf. Communications '89* (ICC '89), 1989, pp. 1566–1577.

[6] Takeuchi, M., S. Isobe, N. Hamamoto, S. Ohmori, and M. Yamamoto, ''Experimental Advanced Mobile Satellite Communications System in MM-wave and Ka-Band Using Japan's COMETS,'' *IEEE GLOBECOM '92*, Orlando, Dec. 1992, pp. 443–446.

[7] ''Environmental Conditions and Test Procedures for Airborne Electronics, Electrical Equipment and Instruments,'' Radio Technical Commission for Aeronautics, RTCA Do-160A.

[8] CCIR Recommendation 465–1.

[9] Blake, L. V., Chs. 8 and 9 of *Antennas*, Artech House, 1984.

[10] Bell, D., K. Dessouky, P. Estabrook, M. K. Sue, and M. Bobb, ''MSAT-X Antennas: Noise Temperature and Mobile Receiver G/T,'' *MSAT-X Quarterly*, No. 16, 1988, pp. 12–17.

[11] CCIR, "Propagation Data for Maritime Mobile-Satellite Systems for Frequencies Above 100 MHz," Rep. 884-2, ITU, 1990.

[12] Beckmann, R., and A. Spizzichino, *The Scattering of Electromagnetic Waves from Rough Surfaces*, New York/Oxford: Pergamon Press, 1963.

[13] Karasawa, Y., and T. Shiokawa, "Characteristics of L-Band Multipath Fading Due to Sea Surface Reflection," *IEEE Trans. Ant. Propag.*, Vol. AP-32, No. 6, 1984, pp. 618–623.

[14] Sandrin, W., and D. J. Fang, "Multipath Fading Characteristics of L-Band Maritime Mobile Satellite Links," *COMSAT Tech. Rev.*, Vol. 16, No. 2, 1986, pp. 319–338.

[15] Karasawa, Y., and T. Shiokawa, "A Simple Prediction Method for L-Band Multipath Fading in Rough Sea Conditions," *IEEE Trans. Communications*, Vol. Com-36, No. 10, 1988, pp. 1098–1104.

[16] CCIR, "Fading Reduction Techniques Applicable to Ship Earth-Station Antennas," SG 8, Rep. 1048-1, ITU, 1990.

[17] Kozono, S., and M. Yoshikawa, "Switch and Stay Diversity Effect on Maritime Mobile Satellite Communications," *Trans. Inst. Electron. Comm. Engrs. Japan*, Vol. J64-B, No. 5, 1981.

[18] Karasawa, Y., and T. Shiokawa, "Space and Frequency Correlation Characteristics of L-Band Multipath Fading Due to Sea Surface Reflection," *Trans. Inst. Electron. Comm. Engrs. Japan*, Vol. 67-B, No. 12, 1984, pp. 1347–1354.

[19] Shiokawa, T., and Y. Karasawa, "Ship-borne Antenna Suppressing Multipath Fading in Maritime Satellite Communication," *IEEE Ant. Propag. Society Symp. (AP-S)*, New Mexico, 1982.

[20] Ohmori, S., and S. Miura, "A Fading Reduction Method for Maritime Satellite Communications," *IEEE Trans. Ant. Propag.*, Vol. AP-31, No. 1, 1983, pp. 184–187.

[21] Ohmori, S., et al., "Characteristics of Sea Reflection Fading in Maritime Satellite Communications," *IEEE Trans. Ant. Propag.*, Vol. AP-33, No. 8, Aug. 1985, pp. 838–845.

[22] Giorgio, F., I. Knight, and R. Matthews, "A Maritime Mobile Terminal for Communications Satellite Application," *ICC'74*, 29C1–29C6, 1974.

[23] MSAT-X Technical Brochure, "Low-Cost Omnidirectional Vehicle Antennas for Mobile Satellite Communications," Jet Propulsion Laboratory.

[24] Adams, A. T., et al., "The Quadrifilar Helix Antenna," *IEEE Trans. Ant. Propag.*, Vol. AP-22, March 1974, pp. 173–178.

[25] Taira, S., M. Tanaka, and S. Ohmori, "High Gain Airborne Antennas for Satellite Communications," *IEEE Trans. Aerospace and Electronic Systems*, Vol. 27, No. 2, March 1991, pp. 354–360.

[26] Nakano, H., K. Vichien, T. Sugiura, and J. Yamauchi, "Singly-Fed Patch Antenna Radiating a Circularly Polarized Conical Beam," *Electronic Letters*, Vol. 26, No. 10, May 1990, pp. 638–640.

[27] Yasunaga, M., F. Watanabe, T. Shiokawa, and M. Yamada, "Phased Array Antennas for Aeronautical Satellite Communications," *5th ICAP*, March 1987, p. 1.47.

[28] Howell, J. Q., "Microstrip Antennas," *IEEE Trans. Ant. Propag.*, Vol. 23, January 1975, pp. 90–93.

[29] Kuribayashi, M., and N. Goto, paper at *IECE of Japan*, March 1982, p. 643.

[30] Itoh, K., H. Baba, Y. Ogawa, F. Watanabe, and M. Yasunaga, "L-Band Airborne Antenna Using Crossed Slots," IECE of Japan Technical Report, AP85-101, Jan. 1986.

[31] Ehrenspeck, H. W., "The Short Backfire Antenna, A New Type of Directional Line Source," *Proc. IRE*, Vol. 48, Jan. 1960, pp. 109–110.

[32] Dod, L. R., "Experimental Measurements of the Short Backfire Antenna," Rep. S-525-66-480, Goddard Space Flight Center, Greenbelt, MD., Oct. 1966.

[33] Shiokawa, T., and Y. Karasawa, "Array Antenna Composed of 4 Short Axial-Mode Helical Antennas," *IECE of Japan*, Vol. J65-B, No. 10, 1982, p. 1267.

[34] Kumar, A., *Fixed and Mobile Terminal Antennas*, Norwood, Mass.: Artech House, 1991, Chapter 5.

[35] MSAT-X Technical Brochure, "Low-Cost Omni-Directional Vehicle Antennas for Mobile Satellite Communications," Jet Propulsion Laboratory.

[36] Derneryd, A. G., "Analysis of the Microstrip Disk Antenna Element," *IEEE Trans. Ant. Propag.*, Vol. AP-27, No. 5, Sep. 1979, pp. 660–664.

[37] Long, S. A., "A Dual-frequency Stacked Circular-disk Antenna," *IEEE Trans. Ant. Propag.*, Vol. AP-27, Mar. 1979, pp. 270–273.

[38] Heckert, G. P., "Investigation of L-Band Shipboard Antennas for Maritime Satellite Applications," Automated Marine Int. Rep. NASW-2165, Feb. 1972.

[39] Stricland, P. C., "Low Cost, Elecrtrically Steered Phased Array for General Aviation," *Proc. 2nd Int. Mobile Satellite Conf.*, Ottawa, 1990, pp. 169–171.

[40] James, J. R., P. S. Hall, and C. Wood, *Microstrip Antennas—Theory and Design*, IEEE Electromagnetic Wave Series 12, Peter Peregrinus, 1981.

[41] Teshirogi, T., et al., "Wideband Circularly Polarized Array With Sequential Rotations and Phase Shift of Elements," *Int. Symp. Ant. Propag.*, Tokyo, Aug. 1985, pp. 117–120.

[42] Ohmori, S., K. Tanaka, S. Yamamoto, M. Matsunaga, and M. Tsuchiya, "A Phased Array Tracking Antenna for Vehicles," *Proc. 2nd Int. Mobile Satellite Conf.*, Ottawa, June 1990, pp. 519–522.

[43] Maruyama, S., K. Kadowaki, and Y. Hase, "Experiments on Message Communications With Hand-Held Terminal," *IEICE Trans. B-II*, Vol. J72-B-II, No. 7, July 1989, pp. 269–275.

[44] Woo, K., J. Huang, V. Jamnejad, D. Bell, J. Berner, P. Estabrook, and A. Densmore, "Performance of a Family of Omni and Steered Antennas for Mobile Satellite Applications," *Proc. 2nd Int. Mobile Satellite Conf.*, Ottawa, June 1990, pp. 540–546.

[45] Bell, D., V. Jamnejad, M. Bobb, and J. Vidican, "Reduced Height, Mechanically Steered Antenna Development," *MSAT-X Quarterly*, No. 18, JPL410-13-18, Jan. 1989.

[46] Jamnejad, V., "A Mechanically Steered Monopulse Tracking Antenna for PiFkEx: Overview," *MSAT-X Quarterly*, No. 13, JPL 410-13-13, Jan. 1988.

[47] Huang, J., "L-Band Phased Array Antennas for Mobile Satellite Communications," *IEEE Vehicular Technology Conf.*, Tampa, Florida, May 1987.

[48] Milne, R., "An Adaptive Array Antenna for Mobile Satellite Communications," *Proc. 2nd Int. Mobile Satellite Conf.*, Ottawa, 1990, pp. 529–532.

[49] Tranquilla, J. M., and S. R. Best, "A Study of the Quadrifilar Helix Antenna for Global Positioning System (GPS) Applications," *IEEE Trans. Ant. Propag.*, Vol. AP-38, Oct. 1990, pp. 1545–1550.

[50] Carter, D., "Phase Center of Microwave Antennas," *IRE Trans. Ant. Propag.*, Vol. AP-4, 1956, pp. 597–600.

[51] Gerst, C., and R. A. Worden, "Helix Antenna Take Turn for Better," *Electronics*, Aug. 1966, pp. 100–110.

[52] Kilgus, C. C., "Shaped Conical Radiation Pattern Performance of the Backfire Quadrifilar Helix," *IEEE Trans. Ant. Propag.*, Vol. AP-23, May 1975, pp. 392–397.

[53] James, J. R., and P. S. Hall, eds., *Handbook of Microstrip Antennas*, IEE Electromagnetic Wave Series 28, Peter Peregrinus, 1989.

[54] Kishk, A. A., and L. Shafai, "The Effect of Various Parameters of Circular Microstrip Antennas on Their Radiation Efficiency and the Mode Excitation," *IEEE Trans. Ant. Propag.*, Vol. AP-34, No. 8, Aug. 1986, pp. 969–976.

[55] Huang, J., "Finite Ground Plane Effect on the Microstrip Antenna Radiation Patterns," *IEEE Trans.*, Vol. AP-31, No. 4, July 1983, pp. 649–653.

[56] Sugimoto, Y., N. Kurihara, H. Kiuchi, A. Kaneko, F. Sawada, T. Shirado, and Y. Saburi, "Development of GPS Positioning System PRESTAR," *IEEE Trans. Instrument. Measurement*, Vol. IM-38, No. 2, April 1989, pp. 644–647.

[57] Evans, J. V., "Satellite Systems for Personal Communication," *Proc. IEEE*, Vol. 86, No. 7, July 1998, pp. 1325–1341.

[58] Williamson, M., "Satellite Constellations in the Ascendant," *IEE Review*, Vol. 44, No. 5, Sep. 1998, pp. 209–213.

[59] James, J. R., "Realising Personal Satcom Antennas," *IEE Electronics and Communication Eng. J.*, April 1998, pp. 73–82.

[60] Lisi, M., "Satellite Communications Systems in the 90s and for the Next Millennium," *Microwave Engineering Europe*, May 1999, pp. 15, 16, 18, 20, and 22.

[61] Pattan, B., *Satellite-Based Global Cellular Communications*, New York: McGraw-Hill, 1997.

[62] Siwiak, K., *Radiowave Propagation and Antennas for Personal Communications*, 2nd ed., Norwood, Mass.: Artech House, 1998.

[63] Schuss, J. J., et al., "The IRIDIUM Main Mission Antenna Concept," *IEEE Trans. Antennas and Propagation*, Vol. 47, No. 3, March 1999, pp. 416–424.

[64] Fijimoto, K., et al., *Small Antennas*, Research Studies Press, New York: Wiley, 1987.

[65] Agius, A. A., "Antennas for Handheld Satellite Personal Communications," Ph.D. Thesis, University of Surrey, 1999.

[66] Leisten, O. P., Y. Vardaxoglou, and E. Agboraw, "Simulating the Dielectric-Loaded Quadrifilar Helix Antenna Using a Brute-Force TLM Approach," *Proc. 15th ACES Conf.*, Mar. 15–20, 1999, Vol. 1, pp. 479.

[67] Nicolardis, G., O. Leisten, and Y. Vardaxaglou, "TLM Investigation of Dielectric-Loaded Bifilar Personal Telephone Antennas," *NCAP99*, University of York, Mar. 31–Apr. 1, 1999, pp. 16–19.

[68] United Kingdom Patent Application No. 9803273.3, February 17, 1998.

[69] Kumar, A., *Fixed and Mobile Terminal Antennas*, Norwood, Mass.: Artech House, 1991, Chap. 5.

Chapter 10

Antenna Systems for Aeronautical Mobile Communications

Y. Suzuki

Many different types of antennas are mounted on aircraft today to carry out numerous individual functions. Table 10.1 shows a list of typical antennas and the associated avionic systems, which include navigation, identification, and radar. In this chapter, propagation problems peculiar to aeronautical systems are noted prior to itemizing common requirements for the airborne antennas and the information necessary to design them. After that, examples of current airborne antennas for navigational avionics and identification are introduced, categorized by frequency. Several circularly polarized antenna elements for satellite communications [1] are summarized, along with concise design procedures, in the final section.

10.1 PROPAGATION PROBLEMS

The propagation problems in aeronautical mobile radio links can be analyzed by the same treatment as those in the maritime mobile case, except that the mobile station is flying at high altitude. The problems are, in general, divided into two categories. One is an interferential propagation due to the direct and reflected signals, and the other is diffraction and scattering, but the distinction between these two sources of problems is not clearly defined.

In the interference region, the propagation path is defined within the line of sight, and the field strength on the propagation path is estimated well enough by the calculation

Table 10.1

Antennas for Aircraft

Frequency Band	Equipment or System	Representative Antenna
VLF/LF/MF	Communications, ADF, Loran	Wire, loop, plate
HF	Communications	Wire, tail cap, probe, notch
VHF/UHF	Communications, glide path (glide slope), Locarizer, VOR, marker beacon, homing, telemetry and command, DME/TACAN, ATC transponder	Sleeve monopole, whip, folded dipole, ramshorn, tail cap, annular slot, notch, blade, loop
UHF/SHF	radio altimeter, radar	Horn, reflector, slot (array)

based on a two-ray interference model. When the aircraft flies over land, however, significant fading may not be caused and the propagation can be approximated by that in free space when estimating the local mean value of the field strength.

The region out of sight may be divided into the diffraction and scattering regions. The spherical diffraction waves mainly propagate in the diffraction region and create a field strength that decreases rapidly as the distance increases. After the diffraction waves become very weak, the troposcatter appears and the feeble waves, scattered by the small, random irregularities or fluctuations in the refractive index of the atmosphere, dominate. There was considerable interest in troposcatter propagation during the decade of the 1950s. Many theories have been developed, and the useful performance data have been gathered. With the development of satellite communication systems, however, there is now less need for troposcatter systems.

The field strength on the propagation path for aeronautical mobiles may suffer considerable fluctuation due to the seasonal change of the refractive index of the atmosphere, and the theoretical prediction is difficult. For example, the monthly mean value of refractivity at an altitude of 300m is quoted at about 300 N-units in February, but it changes from 340 to 370 N-units in August. The fluctuation of refractivity causes the change in equivalent earth radius. The refractivity also varies according to altitude and decreases as the altitude increases. So, when the aircraft flies at high altitude, it should be noted that the equivalent earth radius becomes smaller than that assumed on the earth.

10.2 GENERAL REQUIREMENTS AND REMARKS

In this section, the general requirements for airborne antennas are briefly surveyed, along with some useful information necessary to design them. Typical avionic systems that require antennas are illustrated for two aircraft in Figure 10.1.

Airborne antennas must satisfy more severe environmental conditions than the other mobile antennas and the following measurements are typical:

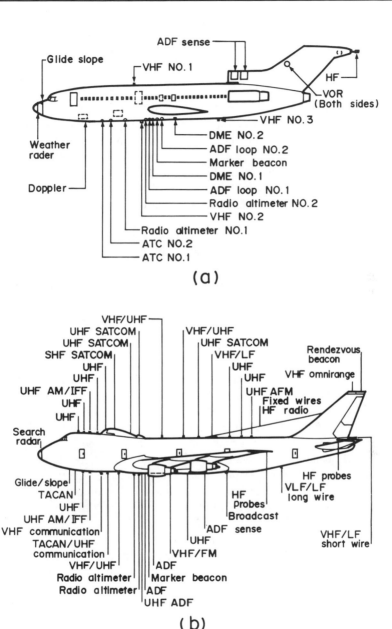

Figure 10.1 Layouts of antennas mounted on aircraft: (a) for B-727 (after [23]); (b) for E-4B.

- Temperature: −85°C to 71°C,
- Vibration: 5 to 2,000 Hz with amplitude from 0.01 to 0.4 in, and
- Acceleration: 1G to 20G.

Also, in order to ensure higher reliability, the antennas must be designed and manufactured ruggedly, without degrading their electrical characteristics, by considering the influence of sand and dust, humidity, salty fog, drops in atmospheric pressure at high altitude, static electricity, and lightning. Static electricity means triboelectric charging of the airframe surface. This is generated by dust or precipitation particles and has been known as P-static. The P-static causes corona discharges, which may produce extreme electrical noise. The corona discharge may also limit the power-handling capacity of the antenna in transmitting operations.

Airborne antennas are also required to have structures that do not increase aerodynamic drag, and this demands light weight, small size, low profile, and conformity. Conformity is defined as constraining an antenna to conform to a surface whose shape is determined by considerations other than electromagnetic (e.g., aerodynamic or hydrodynamic). A conformal antenna should be distinguished from a flush-mounted antenna constructed into the surface on the fuselage, without affecting the shape of that surface.

From the discussions above, it can be said that the main task of airborne antenna designers is realizing rugged antennas that satisfy individual electrical requirements constrained by environmental and aerodynamic conditions. However, the individual electrical requirements are ordinarily different according to each purpose. For example, the antennas for communication or navigation, which are mainly used at frequencies lower than the VHF band, are required to have an azimuthally omnidirectional radiation pattern. In this case, the radiation pattern and other electrical parameters are strongly influenced by the size and shape of an airframe. The current induced on the airframe surface interferes with the original field of the antenna itself, and the overall effect depends on the aircraft's electrical dimensions. A typical variation of a radiation pattern as the frequency is varied, is shown in Figure 10.2 for a tail-cap antenna on a DC-4 aircraft [2]. This example illustrates the frequency dependence of the influence of the airframe on the radiation pattern.

In the low- and medium-frequency bands, below 3 MHz, the wavelengths are more than 100m and are considerably larger than the maximum dimensions of most aircraft; for example, the Boeing 747LR/SR is 70.5m in length. Since the airframe size is generally small compared to the wavelength, the radiation efficiency remains very low. A quality factor Q of the antenna itself will be high, so some tunable matching systems may be necessary for impedance matching over the frequency band.

In the high-frequency band, from 3 to 30 MHz, the wavelength is comparable to the major dimensions for most aircraft. In this case, the airframe becomes a good radiator, so the HF antenna can be designed to maximize the electromagnetic coupling to the airframe. If the antenna succeeds in getting maximum coupling, then the currents on the airframe surface exhibit strong resonant phenomena. The resonant currents can be divided

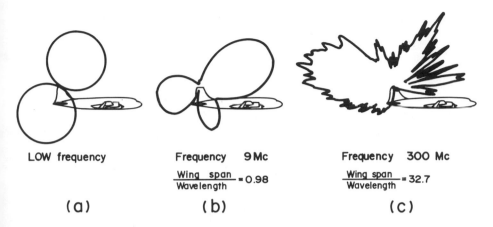

LOW frequency

(a)

Frequency 9 Mc

$$\frac{\text{Wing span}}{\text{Wavelength}} = 0.98$$

(b)

Frequency 300 Mc

$$\frac{\text{Wing span}}{\text{Wavelength}} = 32.7$$

(c)

Figure 10.2 Typical radiation patterns of a tail-cap antenna on a DC-4 aircraft (after [2]): (a) a case of low-frequency range; (b) a case of longitudinal resonance range; (c) a case of transverse resonance range.

into two modes, as shown in Figure 10.3 [2]; one is the symmetric mode and the other is the antisymmetric mode. The antennas should be designed to couple to either of them, so that the system may benefit from this additional degree of design freedom at HF.

In the very high and ultrahigh frequency bands, from 30 to 3,000 MHz, the wavelength is shorter than the airframe size, so the antenna itself or a part of the airframe becomes resonant and the degrees of freedom in both antenna design and installation increase. However, it should be noted that omnidirectional patterns, as required for short-range communication, distance measuring equipment (DME), and air traffic control (ATC), are perturbed by the airframe. The airframes cause shadowing and reflection and can result in major distortions of the primary pattern of the antenna, as shown in Figure 10.2(c). These effects can be analyzed by using the method of moments [3–7], based on a solution of the electromagnetic wave-scattering integral equations, when the airframe is not larger compared with the wavelength. At higher frequency, the diffraction due to the airframe edge can be approximately calculated by using the geometrical optics techniques [8–14].

10.3 ADVANCED CIRCULARLY POLARIZED ANTENNAS

Satellite communication systems require other kinds of antennas, which must radiate circularly polarized waves, and some suitable elements are as follows:

- Crossed dipole;
- Crossed slot;
- Helical;

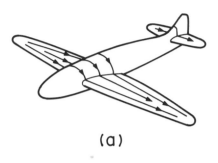

(a)

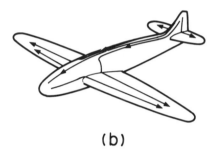

(b)

Figure 10.3 Resonant current modes (after [2]): (a) symmetric mode; (b) antisymmetric mode.

- Conical spiral;
- Cavity-backed planar spiral;
- Quadrifilar helical;
- Microstrip patch.

These elements are the basis of circularly polarized high-gain antennas, phased-array antennas, shaped-beam antennas, and so on. The more recent elements are described below.

10.3.1 Crossed-Dipole Antennas

The resonance of a dipole antenna is obtained when the length is somewhat shorter than a free-space half wavelength. As the thickness is increased, the resonant length is reduced more. The bandwidth is, in general, proportional to the thickness, and the following relationship exists between the unloaded Q (Q_0) and the relative bandwidth (B_r), in which the input VSWR is less than ρ [15]

$$Q_0 B_r = (\rho^2 - 1)/2\rho \tag{10.1}$$

Circular polarization can be produced by a pair of orthogonally positioned dipoles driven in quadrature phase with equal amplitudes. However, this crossed-dipole arrangement cannot provide a good axial ratio off boresight, because the radiation patterns for the straight dipole are different in both principal planes, called the H-plane and the E-plane. This shortcoming can be improved by modifying the straight dipole to a nonstraight version, such as the V-form and U-form. These improved dipoles are called V-type and U-type dipoles. According to some measurements, the U-type provides better electrical performance than the V-type, though the V-type is simpler in mechanical structure and is less complex than the U-type. Figure 10.4(a) shows an inverted V-type crossed-dipole antenna, where the environment metal frame is employed to widen the beamwidth still more and to broaden the pattern in the elevation plane. A quadrature phase condition with equal amplitudes to produce circular polarization is achieved by a hybrid splitter inserted at the input. Figure 10.4(b) shows one example of the radiation pattern, where the pattern was measured at 2,044.25 MHz.

The crossed-dipole can also produce circular polarization without using any external circuits such as the hybrid component. The condition to excite the circularly polarized waves can be established by a balun and the self-phasing of four radiating elements. That is, the balanced outputs, which are in the antiphase with equal amplitudes, are first achieved by the balun. Next, two adjacent elements of the crossed dipole are connected to one-half of the balanced outputs from the balun. The other two adjacent elements are likewise connected to the other half of the balun. As a result, two of the elements are at a 0-deg phase angle and the other two are at a 180-deg phase angle. Finally, the desired 90-deg phase difference is obtained by designing orthogonal elements such that one is larger relative to the desired resonant length to make it inductive, while the other is smaller, to make it capacitive. The resultant crossed-dipole antenna requires no external circuits to produce the circular polarization. This technique is also applicable to the quadrifilar helical antenna to be described later.

10.3.2 Crossed-Slot Antennas

Slot antennas are useful in many applications for high-speed aircraft because they are low-profile in structure and suitable for a flush-mount application. The slot antennas are electromagnetically complementary with the corresponding dipole antennas, so that the radiation pattern is the same as that for the complementary horizontal dipole consisting of a perfectly conducting flat strip. This holds true when both antennas are driven at the same position, but with two differences. One difference is the property that the electric and magnetic fields are interchanged. The other is that the slot electric field component normal to the perfectly conducting sheet is discontinuous from one side of the sheet to the other because the direction of the field reverses. In this case, the tangential component of the magnetic field is, likewise, discontinuous.

The crossed-slot antennas can be said to be also complementary with the corresponding crossed-dipole antenna, although the feeding method for the circular polarization is

Figure 10.4a Inverted-V-type crossed dipole antenna.

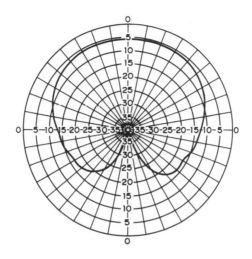

Figure 10.4b One example of radiation pattern at 2,044.25 MHz.

more complicated. Figure 10.5 shows an example of a crossed-slot antenna where the feeding method has been improved [16]. This antenna belongs to a cavity-backed crossed-slot antenna category and is made of copper-clad 3.3-mm-thick Teflon fiberglass with a 2.55 dielectric constant. In this example, the slot length and width are about 112 and 5 mm, respectively, and the cavity dimensions are about $80 \times 80 \times 20$ mm. Each slot antenna, etched on the inside of the cavity, is fed by a balanced feed technique from two feed points on the microstrip feed lines etched on the opposite side of the slot. Although this antenna needs one 90-deg hybrid to produce the circular polarization, this feed

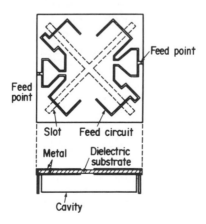

Figure 10.5 Structure of crossed-slot antenna. (After [16].)

technique is effective not only to suppress undesired coupling between the cross-slots, but also to match the input impedance over a wider frequency band.

10.3.3 Quadrifilar Helical Antennas

The quadrifilar helical antenna consists of four helices equally wound circumferentially on a cylinder, and the four terminals are excited with equal amplitude but with relative phases of 0, 90, 180, and 270 deg. It is classified into two categories that are a resonant and nonresonant type. The former is very useful for an aircraft antenna, because of its small size, lack of a ground plane, and insensitivity to nearby metal structures, although the impedance bandwidth is narrow (e.g., the bandwidth for a VSWR less than 2 is typically 3% to 5%). A wider bandwidth can be achieved by using larger diameter wire as a helical arm, as in the case of the dipole antenna. This helical antenna has also been used for various applications requiring a wide beamwidth of 120 to 180 deg over a relatively narrow frequency range.

In order to make the antenna resonate, its four arms must be equal to $m\lambda/4$ in length, where m = 1, 2, 3, 4, . . . , and are generally wound to form a small-diameter helix with an $n/4$ turn, where n = 1, 2, 3, 4, ... The ends of the helices must be open-circuited, when m = odd, and short-circuited, when m = even. These structures may also be regarded as two bifilar helices. Each bifilar helix can be balum-fed at the top through a coaxial line extending to the top along the central axis. In order to produce a circular polarization, some external circuit is required: for example, a 90-deg hybrid is convention-ally employed. The radiation pattern is basically characterized by the radius r_0 and the pitch distance for each element measured along the axis of helix p, defined in Figure 10.6(a) [17]. If these helical parameters are appropriately chosen, the antenna can produce shaped-conical beams having 90- to 240-deg beamwidth, with excellent circular polariza-tion, over the wide angular region. Excellent circular polarization naturally produces high front-to-back ratios. Many measurements indicate that the shaped-conical radiation patterns with full-cone angles from 100 to 180 deg and center minimums down 3 to 20 dB can be realized by an appropriate choice of helical parameters, according to the design curves in Figure 10.7(b) [17].

Figure 10.7(a) shows an example of the resonant quadrifilar helical antennas with $\lambda/4$ coaxial balun, which is designed for use at a frequency in the L-band. A typical radiation pattern, measured with a rotating linearly polarized source at 1,537.5 MHz, is also shown in Figure 10.7(b).

10.3.4 Microstrip Patch Antennas

The typical microstrip patch antenna consists of a metallic ground plane, a thin dielectric or air sheet, and a strip conductor patch on top of the dielectric substrate. It can be made conformal to a metallic surface and can be produced at low cost with photo-etch techniques.

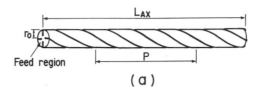

$$(a)$$

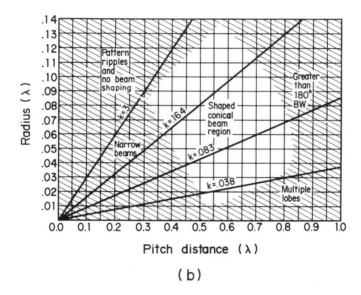

$$(b)$$

Figure 10.6 (a) Quadrifilar helical antenna and design parameters; (b) region of shaped conical beam performances (after [17]).

When low profile, lightweight, small size, and low cost are particularly required, the microstrip patch antennas are very important. In practical applications, typical shapes of the patch radiator are circular and rectangular. Their dimensions are approximately determined by

$$a = a_e \left[1 + \frac{2t}{\pi a_e \epsilon_r} \left\{ \ln\left(\frac{\pi a_e}{2t}\right) + 1.7726 \right\} \right]^{-1/2} \qquad (10.2)$$

$$\therefore a_e = \frac{1.8412\lambda_0}{2\pi\sqrt{\epsilon_r}} \qquad (10.3)$$

for circular patch antennas with the TM_{110} mode excited, and

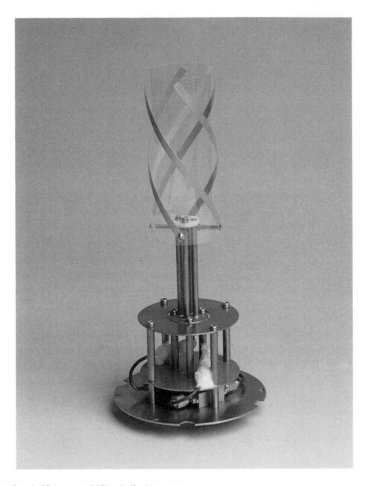

Figure 10.7a One-half-turn quadrifilar helical antenna.

$$L = \frac{\lambda_0}{2\sqrt{\epsilon_r}} - 2\Delta\ell \qquad (10.4)$$

$$\because \epsilon_e = \frac{\epsilon_r + 1}{2} + \frac{\epsilon_r - 1}{2}[1 + 12(t/W)]^{-1/2} \qquad (10.5)$$

$$\because \Delta\ell = 0.412t\frac{(\epsilon_e + 0.3)[(W/t) + 0.264]}{(\epsilon_e - 0.258)[(W/t) + 0.8]} \qquad (10.6)$$

$$\because W = \frac{\lambda_0}{2}\left(\frac{\epsilon_r + 1}{2}\right)^{-1/2} \qquad (10.7)$$

for rectangular patch antennas with the TM_{010} or TM_{100} mode excited; a is the radius of circular patch, L and W are the length and width of the rectangular patch, respectively, t

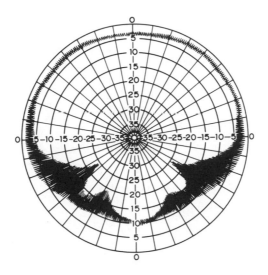

Figure 10.7b Radiation pattern at 1,537.5 MHz.

and ϵ_r are the thickness and dielectric constant of the dielectric sheet (or substrate), respectively, and λ_0 is the resonant wavelength.

However, the microstrip antennas have the serious disadvantage that the frequency bandwidth is narrow and only on the order of 2%. Various impedance matching networks have been investigated [18], but the feed network may become complex and lossy. There are two other techniques for bandwidth extension, as follows. The first technique is to increase the thickness of the substrate and decrease its dielectric constant. Figure 10.8(a) shows an example of this kind of antenna [15], consisting of a paper honeycomb substrate sandwiched between epoxy fiberglass skins. Figure 10.8(b) shows the return loss characteristics, and an approximate 8.4% bandwidth is obtained for a VSWR less than 2 [15]. When circular polarization is required, however, this technique may need some countermeasures [19] against the degradation in the axial ratio, due to the generation of higher order modes.

The second technique is achieved by using the stacked patches, which electromagnetically couple together. One example is shown in Figure 10.9, where the lower patch, which does not appear in this view, is excited with coaxial probes using the four-point feeding technique to improve the axial ratio. The electromagnetic energy from the lower patch couples to the upper parasitic patch that appears in this view. The external circuit to produce the circular polarization consists of a stripline circuit and is attached on the back of the ground plane attached to the lower patch antenna. The radiation patterns, measured by 1,540 and 1,600 MHz, are shown in Figure 10.10(a) and Figure 10.10(b), where the diameters of lower and upper patches are 67.88 and 73.13 mm, respectively, the spacing between them is 20.2 mm, and the substrate thickness and dielectric constant are 3.2 mm and a nominal 2.55, respectively.

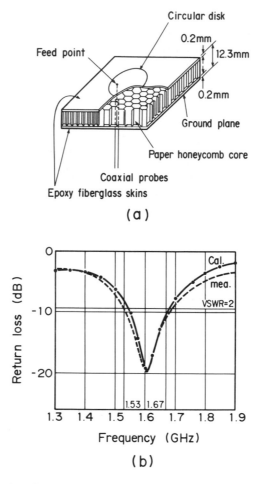

Figure 10.8 (a) Configuration of circular microstrip antenna consisting of epoxy fiberglass skins and paper honeycomb core (after [15]); (b) one example of return loss characteristics (after [15]), where radius = 40.5 mm and $\epsilon_r = 1.21$.

When the receive and transmit frequency bands are completely separate, a dual-frequency technique is also available. This technique assigns the separate frequency bands to each patch antenna so that the antenna system does not need an external duplexer. Figure 10.11 shows an example of a two-layer dual-frequency microstrip patch antenna configuration [16], where the antenna is composed of a 3.2-mm-thick glass-microfiber-reinforced PTFE substrate, with a dielectric constant of 2.3. The upper patch is for transmission, while the lower patch deals with reception. Each patch is individually fed at two points with a phase difference of 90 deg to produce the circular polarization. The

Figure 10.9 Stacked microstrip antenna.

distances from the center to the feed points are about 10 and 20 mm for the upper and lower patch, respectively. In this configuration, the upper layer is a conventional circular microstrip patch antenna with a 66-mm diameter, while the lower patch, with a diameter of 84 mm, functions as a circular microstrip antenna with an electric shielding ring. The electric shielding ring is necessary to avoid any electrical interferences between the various feed lines and consists of 12 short-circuited pins arranged in a circle with a 27-mm diameter.

The above examples all needed some external circuit to produce the degenerate modes for circular polarization. This can, however, be achieved by giving an appropriate perturbation to the patch dimensions and selecting a suitable feed point. Such an antenna can simultaneously radiate a pair of orthogonally polarized waves at some frequency between the resonant frequencies of the two orthogonal modes, because the degenerate modes are, in general, vectorially orthogonal. Figure 10.12 shows typical examples of these circularly polarized microstrip antennas [20], which radiate circularly polarized waves when fed at a point on the chain-dotted lines. In this figure, the conditions, which should be imposed on each patch geometry to produce the circular polarization, are also

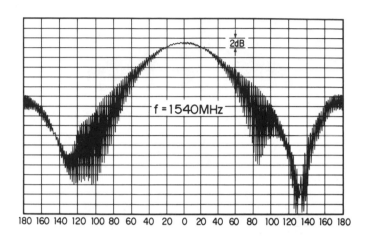

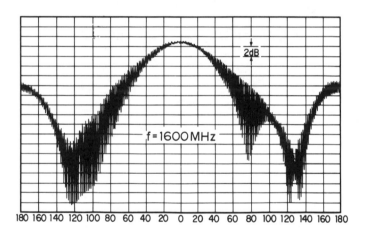

Figure 10.10 Top figure shows radiation pattern at 1,540 MHz; lower figure shows radiation pattern at 1,600 MHz; the diameter of lower patch = 67.88 mm, the diameter of upper patch = 73.13 mm, the spacing between them = 20.2 mm, the thickness of substrate = 3.2 mm, and $\epsilon_r = 2.55$.

presented, where ΔS is the perturbation given to the area of the original patch radiator of area S and Q_0 is the unloaded Q for the original patch antenna. In addition to the above examples, there are various other shapes, including the pentagon proposed by Weinschel [21]. Figure 10.13 shows some of these variants [22], where a and c are defined in Figure 10.13(b) and $b/a = 1.0603$ here. The solid line drawn in each pentagon shows the feed location loci for exciting the left-hand circularly polarized waves, while the broken line shows the feed location loci for exciting the right-hand circularly polarized waves. The

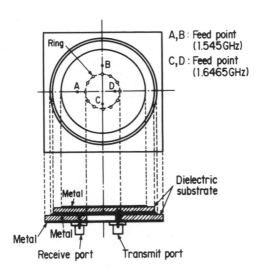

Figure 10.11 Structure of two-layer dual-frequency microstrip antenna. (After [16].)

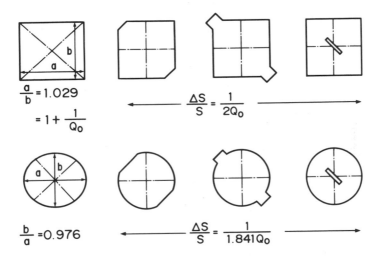

Figure 10.12 Current examples of singly fed circularly polarized microstrip antenna. (After [20].)

symbols f_{c1} and f_{c2} are the two frequencies at which the circularly polarized waves can be excited when $a = 100$ mm, $t = 3.2$ mm, $\epsilon_r = 2.55$, and $\tan \delta = 0.0018$.

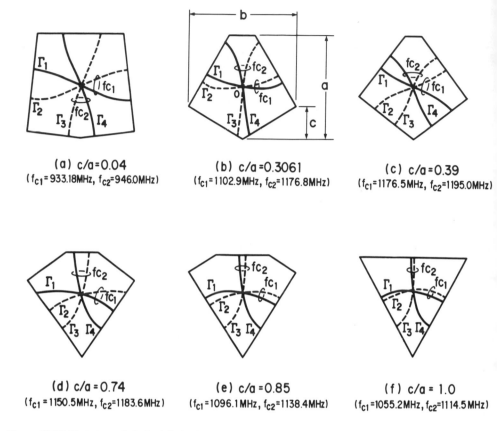

(a) c/a=0.04
(f_{c1}=933.18MHz, f_{c2}=946.0MHz)

(b) c/a=0.3061
(f_{c1}=1102.9MHz, f_{c2}=1176.8MHz)

(c) c/a=0.39
(f_{c1}=1176.5MHz, f_{c2}=1195.0MHz)

(d) c/a=0.74
(f_{c1}=1150.5MHz, f_{c2}=1183.6MHz)

(e) c/a=0.85
(f_{c1}=1096.1MHz, f_{c2}=1138.4MHz)

(f) c/a=1.0
(f_{c1}=1055.2MHz, f_{c2}=1114.5MHz)

Figure 10.13 Variations of singly fed circularly polarized pentagonal microstrip antenna. (After [22].)

REFERENCES

[1] INMARSAT RFP, "Aeronautical High Gain Antenna Subsystem," Request for Proposal (RFP) No. 082, HK/86-2063M/pt, March 1986.

[2] Granger, J.V.N., and J. T. Bolljahn, "Aircraft Antennas," *Proc. IRE*, Vol. 43, May 1955, pp. 533–550.

[3] Harrington, R. F., *Field Computation by Moment Methods*, New York: Macmillan, 1968.

[4] Burke, G. L., and A. J. Poggio, "Numerical Electromagnetic Code (NEC)—Method of Moments, Part 1: Program Description-Theory," Tech. Doc. 116, Naval Electronic Systems Command (ELEX 3041), July 1977.

[5] Richmond, J. H., "A Wire-Grid Model for Scattering by Conducting Bodies," *IEEE Trans. Ant. Propag.*, Vol. AP-14, Nov. 1976, pp. 782–786.

[6] Wang, J. J., and C. E. Ryan, Jr., "Application of Wire-Grid Modelling to the Design of a Low-Profile Aircraft Antenna," *IEEE AP-S Int. Symp. Digest*, June 1977, pp. 222–225.

[7] Knepp, D. L., and J. Goldhirsh, "Numerical Analysis of Electromagnetic Radiation Properties of Smooth Conducting Bodies of Arbitrary Shape," *IEEE Trans. Ant. Propag.*, Vol. AP-20, May 1972, pp. 383–388.

[8] Keller, J. B., "Geometrical Theory of Diffraction," *J. Opt. Soc. Am.*, Vol. 52, Feb. 1962, pp. 116–130.

[9] Kouyoumjian, R. G., and P. H. Pathak, "A Uniform Geometrical Theory of Diffraction for an Edge in a Perfectly Conducting Surface," *Proc. IEEE*, Vol. 62, Nov. 1974, pp. 1448–1460.

[10] Burnside, W. D., M. C. Gilreath, R. J. Marhefka, and C. L. Yu, "A Study of KC-135 Aircraft Antenna Patterns," *IEEE Trans. Ant. Propag.*, Vol. AP-23, May 1975, pp. 309–316.

[11] Yu, C. L., W. D. Burnside, and M. C. Gilreath, "Volumetric Pattern Analysis of Airborne Antennas," *IEEE Trans. Ant. Propag.*, Vol. AP-26, Sept. 1978, pp. 636–641.

[12] Cooke, W. P., and C. E. Ryan, Jr., "A GTD Computer Algorithm for Computing the Radiation Patterns of Aircraft-Mounted Antennas," *IEEE AP-S Int. Symp. Digest*, June 1980, pp. 631–634.

[13] Ryan, C. E., Jr., "Analysis of Antennas on Finite Circular Cylinders With Conical or Disk End Caps," *IEEE Trans. Ant. Propag.*, Vol. AP-20, July 1972, pp. 474–476.

[14] Balanis, C. A., and Y. B. Cheng, "Antenna Radiation Modeling for Microwave Landing System," *IEEE Trans. Ant. Propag.*, Vol. AP-24, July 1976, pp. 490–497.

[15] Suzuki, Y., and T. Chiba, "Designing Method of Microstrip Antenna Considering the Bandwidth," *IECE Japan Trans.*, Vol. E67, Sept. 1984, pp. 488–493.

[16] Yasunaga, Y., F. Watanabe, T. Shiokawa, and H. Yamada, "Phased Array Antennas for Aeronautical Satellite Communications," *IEE ICAP 87*, Part 1, March 1987, pp. 47–50.

[17] Kilgus, C. C., "Shaped-Conical Radiation Pattern Performance of the Backfire Quadrifilar Helix," *IEEE Trans. Ant. Propag.*, Vol. AP-23, May 1975, pp. 392–397.

[18] Pues, H. F., and A. R. Van De Capelle, "An Impedance-Matching Technique for Increasing the Bandwidth of Microstrip Antennas," *IEEE Trans. Ant. Propag.*, Vol. AP-37, Nov. 1989, pp. 1345–1354.

[19] Chiba, T., Y. Suzuki, N. Miyano, S. Miura, and S. Ohmori, "A Phased Array Antenna Using Microstrip Patch Antennas," *Proc. 12th European Micro. Conf.*, Sept. 1982, pp. 472–477.

[20] Hirasawa, K., and M. Haneishi, ed., *Introductory Analysis Design and Measurement of Small or Low-Profile Antennas*, Norwood, Mass: Artech House, 1992.

[21] Weinschel, H. D., "A Cylindrical Array of Circularly Polarized Microstrip Antennas," *IEEE AP-S Int. Symp. Digest*, June 1975, pp. 177–180.

[22] James, J. R., and P. S. Hall, ed., *Handbook of Microstrip Antennas*, IEE Electromagnetic Waves Series 28, London: Peter Peregrinus, 1989.

[23] Hirano, M., ed., *New Radio Wave Technologies*, Part 2, Tokyo: RATESU Co., 1973 (in Japanese).

Appendix
Glossary

K. Fujimoto and J. R. James

The demand for mobile antenna systems has brought with it an ever increasing variety of antennas, as discussed in Chapter 1 and amply illustrated in subsequent chapters. To the uninitiated, the various antennas often appear as distinct types having different fundamental behavior, but this is not necessarily so. In fact, there are surprisingly few generic forms, and even experienced antenna designers may ponder as to the purpose of some seemingly small modification to an otherwise conventional antenna configuration. For instance, does a spiral top-loaded monopole perform very differently from a circular plate top-loaded monopole? The fundamental electromagnetic properties that underlie this perpetual creation of antenna variants has been elaborated on elsewhere [1], where reasons are given explaining why such variety exists. In this glossary, we aim to help the antenna designer collate the likely types of antennas that might fit a given mobile requirement, and, furthermore, to narrow down the choice of antennas. We have kept in mind many special features of mobile antenna systems, which are typically electrically small because of size and shape constraints; furthermore, mobile antennas commonly have to operate on small, ill-defined ground planes and they may also embody additional system functions, such as diversity, polarization selectivity, and beam scanning. As far as possible, the catalog of types in Section A.1 attempts to identify generic forms such as dipoles and loops, but in many cases it is obvious that a simple classification is likely to be ambiguous and serves no useful purpose. The catalog also includes some mention of related issues such as baluns, ground plane effects, and array techniques. Antennas with little or no relevance to mobile systems are excluded, and this includes large parabolic reflectors, large phased arrays, and large broadcast transmitting antennas.

The tables in Sections A.2 to A.5 relate the antenna types to the nature of the mobile terminals, while Section A.6 gives a breakdown in terms of applications and frequency bands. The overlap of information between these sections is international, as are the alternative presentations, but we leave it to the designer to explore how best to sift the data when making an initial choice of antenna for a given mobile system requirement.

A.1 CATALOG OF ANTENNA TYPES

For ease of presentation, diagrammatic sketches are given, together with a literature reference which may be of historical or general interest, or perhaps, may be a more recent research reference. Note that the number in the text corresponds to both the number and the literature reference number up to reference [78].* A few additional references are included and annotated with a, b, c and so on (for example, [7a] is an additional reference to the spiral-loaded antenna [7]).

A.1.1 Dipole Derivatives

The wire dipole [2] and its unbalanced counterpart, the monopole [3], are universally known generic antennas from which have evolved a multitude of variants offering special features such as a more compact size, greater bandwidth, circular polarization, a desired impedance level, or particular physical characteristics. When each arm of the dipole or monopole is a quarter wavelength long, resonant action occurs and the resonant resistance is compatible with conventional transmission line feeders. When the arms are much less than a quarter of a wavelength, matching and efficiency problems are severe, and the feed region radiation can corrupt the overall pattern characteristics. Compactness of size can therefore only be obtained in exchange for some curtailment in performance. When physical size is not a constraint, many desirable characteristics can be achieved by altering the antenna geometry; that is, replacing the dipole or monopole wire with a flat metal plane or metal cone increases the impedance bandwidth, but the cross-polarization is also increased. A selection of dipole derivatives is as follows. The inverted-L antenna [4] readily gives a reduced height h at the expense of accommodating the horizontal sections of length l and decreasing the radiation resistance, bandwidth, and some change in the radiation patterns. The antenna is resonant when $l + h$ is near a quarter wavelength. A more symmetrical form of top loading is [5], while [6] is loaded with a circular plate. The spiral [7] top loading gives good control of resonance, but is a less rigid structure. The spiral can also be mounted on a planar substrate [7a,7b]. The inverted antenna [4] is often referred to as a transmission line antenna when an impedance Z_L loads the line [8]. This loading enables a shorter horizontal line to be deployed, hence saving on space. If Z_L is a short circuit, the structure is rigidly supported, but the horizontal section is then

---

*With the exception of reference [1].

half a wavelength long. The inverted-F antenna [9] is a version of the inverted-L antenna with the additional freedom to tap the input along the horizontal wire to achieve some control of the input impedance [9a]. A parasitic inverted-L section can be coupled to the inverted-F antenna to increase bandwidth [10]. In another version of the transmission line antenna, a boxlike structure has been derived [11]. The inverted-V antenna [12] offers another way of top loading, and when several wires are used, it is called the *umbrella antenna*. The hula-hoop antenna [13] is essentially an inverted-L antenna bent in a circular arc, and as such is more compact, but efficiency is very low. When the thickness of the wire is increased in all the above antennas, we can expect some increase in bandwidth. The Goubau antenna [14] invokes coupled circuits to give a large-impedance bandwidth, and four top-loaded monopoles are used with various coupling arrangements. The input impedance of a monopole can be isolated from the effects of a small ground plane by raising it on a sleeved section [15], but the structure is now a half wavelength in height. Short monopoles commonly contain a series loading coil to achieve resonance [16]. In the extreme, the loading coil itself can act as the radiator, and this is then the helical antenna [17], which gives essentially monopole radiation patterns when dimensioned to function in the normal mode. Antennas composed of compact inductors admit many variations, such as a counterwound version [18], giving wide bandwidth at low efficiency, a means of tapping into the helix to increase the input impedance [19], and a more compact form of transmission line antenna [20]. The quadrifilar helical antenna [21] consists of orthogonal windings excited in quadrature to generate circular polarization. A helical antenna designed to operate in the backfire mode [22] has a conical beam suitable for satellite communication reception on a vehicle. Lumped and continuous impedance loadings [23] have also been designed to modify radiation pattern characteristics. The mutual coupling between two small, wide monopoles can be used to create a wide impedance bandwidth by connecting the radiators by a transmission line Z_0 and magic T circuit [24]. Further freedom in design can be achieved by incorporating active devices in the antenna, assuming that noise and intermodulation constraints can be tolerated [25]. Diodes incorporated in a helical antenna [26] give wide-band electronic tuning at HF and VHF. Orthogonal dipoles [27,27a] have useful polarization diversity characteristics, and considerable progress has been made in designing car window heaters with the capability to function also as antennas [28]. Leaky-cable radiators are used extensively in tunnel communication systems [29] and train telephone systems [29a].

A.1.2 Loop and Slot Derivatives

The small dipole antenna [2] is commonly referred to as an *electric dipole* in recognition of the fact that its fields are similar to those of the Hertzian electrical dipole. In the same way, a small loop [30] has the field characteristics of a Hertzian magnetic dipole. The electric and magnetic Hertzian dipoles exhibit duality in their field components, and this property is retained to a high degree in small dipoles and loops. The latter have, however, differing radiation resistances and Q factors for a given occupancy of space. The loop also behaves differently from a dipole in the presence of a reflecting object, and these properties are extremely useful in the presence of a reflecting object, and these properties are extremely useful in the design of antennas for pagers [30a], which have to function on or near the human body. The loop can also function in a monopolelike mode [31] in the presence of a ground plane. The loop yields a resonant input impedance when its circumference is one wavelength, but it is more typical to use a much smaller loop and tune it with a capacitor [32]. The loop need not be circular; for instance, a rectangular version of [31] and [32] is [33], which is identical to the tuned inverted-L or transmission line antenna [8]. In fact, it could be argued that [7–9,16–20,25] are derivatives of the loop rather than the dipole antenna. Multiturn loops embody mutual inductive coupling [34], but coupled isolated loops [35] offer better control of bandwidth which can be extended somewhat. A resonant slot in a ground plane is a useful radiator [36,37], and the excitation is applied across the slot at a and b. The polarity of the dipolelike fields can be deduced with Babinet's Principle [37a], showing that [36] and [37] are, respectively, dipole and loop radiators with E and H fields interchanged, and satisfying the conducting boundary conditions. A bent slot in a small ground plane [38] gives useful cross-polarization characteristics in multipath conditions, while the combination of a slot and patch antenna gives a compact diversity arrangement [39].

A.1.3 Material Loading

It is well known that most antennas can be detuned to a lower frequency when dielectric material is placed around the radiator. Appreciable size reduction in a monopole [40] and Yagi array [41] can be achieved, but the additional volume occupied by the material and the considerable increase in weight make this technique less attractive in many applications. Partial cladding, as in the skeleton discone antenna [42], is a useful compromise. Ferrite cladding of wire antennas [42a] gives enhanced performance compared to dielectric material, but suitable ferrites are only available at low frequencies. The common ferrite coil antenna [43] has a very low efficiency, but functions well for broadcast reception where the receiver is externally noise-limited. Slot antennas [44] can be size-reduced when fed by a material-loaded cavity, and ceramic material has been used in a compact top-loading monopole [45]. The use of superconducting materials to fabricate electronically small dipoles and loops and their matching components drastically improves efficiency and noise performance at the expense of bandwidth and signal handling capacity [45a]. The use of pyroelectric polymer films to make sensitive probes at microwave frequency has been reported [46].

A.1.4 Printed Elements

The rectangular or circular patch microstrip antenna [47] is the basic printed element. In a sense, it is an extreme case of material loading whereby wave trapping occurs between the conducting patch and the ground plane; cavity action takes place with a leakage of radiation from the patch edges. For the basic rectangular patch, l is about a half wavelength long in the material, while h is typically a few millimeters. The radiation pattern is essentially that of a small loop antenna, but is modified by the contribution of the surface currents on the patch. Higher permittivity substrates give more compact radiators with less efficiency and bandwidth. The conducting patch can take numerous shapes, such as annular, elliptical, and triangular [48]. Higher cavity modes give multilobe patterns. Patch elements can be back-fed with a probe, excited by a coplanar line, excited by capacitive coupling, and so on. Some surface waves are generated in the surrounding substrate. An electric wall across the central region of a rectangular patch can be created by metal rivets [49] and enables the length l to be reduced to a quarter wavelength. This is particularly useful in portable equipment. Given appropriate excitation, patches can be angular segments of disks and annular rings [50]. Circular polarization can be created by generating both orthogonal modes in a symmetric patch, but only one feed is required in patches with appropriate perturbation notches [51,52], at the expense of a narrower bandwidth. Several techniques are available to increase the impedance bandwidth of printed antennas apart from using a low-permittivity substrate. Essentially, the techniques rely on coupled-circuit methods whereby parasitic patches are coupled to the main patch antenna by stacking [53] or planar proximity [54]. Clearly, the additional impedance bandwidth is achieved by an increase in space occupancy. Furthermore, the bandwidth of the radiation characteristics may be a constraint. The positioning of conducting posts in a patch [55] gives further control of the resonant frequency, and slots have similar advantages [55a]. Patches and ground planes can be composed of wire grids, offering many possibilities for multiband operation [56]. Hemispherical coverage with circular polarization has been reported for a circular element [57], and a bent patch on a small ground plane gives adequate cross-polarization for multipath operation [58]. Orthogonally placed quarter-wave elements are used for polarization diversity [59].

A.1.5 Balun Requirements and Imperfect Ground Planes

Coaxial cables are the most common means of interconnecting equipment units, and these unbalanced feeders can be connected directly to any antenna mounted on a ground plane [60]. Balanced antennas without a ground plane, such as a dipole and loop, require a balanced feed. A balun [61] is a device that converts a coaxial cable into a balanced feeder, and it takes a variety of forms. At lower frequencies ferrite-loaded transformers and various component arrangements [62] can be utilized with tolerable losses, while bifilar winding techniques give tight coupling over a wide bandwidth [63]. At microwave frequencies, resonant transmission line lengths in baluns act as wave traps [64] and corporate feed phase inverters [65]. In practice, a ground plane may not be perfectly conducting or flat, and certainly not of infinite extent [66]. Radiation patterns and input impedances are modified by these effects depending on the extent to which the ground plane departs from ideal conditions. This is particularly important for antennas mounted on mobile units that are comparable in size to the antennas. Analytical solutions for a monopole on a sphere or cylinder [66a] indicate the nature of the perturbations in performance, while computational methods [66a] give information about specific geometries. An obstacle in close proximity to an antenna [67] will perturb the radiation patterns and input impedance characteristics. In some cases, conducting materials near an antenna can be used to enhance or improve the antenna performance. Analytical and computational methods [66a] are again a useful indication of practical situations.

A.1.6 Arrays and Diversity Systems

Circularly polarized patch antennas are clustered to form a conical system [68]. A circular patch operating in the TM_{210} mode gives similar characteristics [69], while circular elements with coupled parasitics are used in a 3×3 array [70] with beam control. Flat-plate direct boardcast satellite antennas [71] have been exhaustively addressed worldwide and have reached a high degree of engineering sophistication. Smaller printed arrays are important in airborne satellite receivers [72]. Beam squint techniques have been exploited for tracking systems [73]. Coupled stacked patches can operate in different modes in a novel beam scan arrangement [74], and strip-slot foam-inverted patch elements offer wider bandwidths [74a].

A.1.7 Recent Innovative Concepts

New ways of designing antennas are continually highlighted by international research activity and some of these ideas will influence mobile antenna design. It is too early to state positive advances in this latter respect but mobile antenna designers may benefit from familiarization with some of the recent research findings as follows. Fractal antennas and in particular the Sierpinski antenna [75,75a], have received much publicity. Monopoles [75b], ring arrays [75c], frequency-independent arrays [75d], and random arrays [75e] are some of the antenna structures conceived. Volume fractal monopoles have been demonstrated [75f]. Fractal monopoles enable a good input match to be achieved at certain spot frequencies over a wide bandwidth at the expense of some multilobe radiation pattern behavior. However, one should keep in mind that wider, continuous broadband match and dipole-like patterns can generally be achieved using conventional nonfractal broadband structures [75f]. A nonfractal anisotropic monopole has also been demonstrated [76]. It remains to be demonstrated that the perceived [75g] potential benefits of fractals can be practically realized. Attention continues to be concentrated on compact and/or broadband elements such as a novel compact superdirective helical array [77], a compact circularly polarized microstrip element [78], and numerous variants [78a–e] to mention but a few. Other advances in printed antennas [78f] include dieletric resonator antennas, supercon-ducting microstrip antennas, multifunction printed antennas, and active microstrip antennas. The integration of antennas into circuits and chips [78g] has arisen from phased-array technology and is applicable in some cases to handset antenna design. Photonic band-gap materials [78h] are a new class of periodic dielectric materials that can be used to control electromagnetic wave propagation in substrates. This may prove useful in printed antenna handset design at lower microwave frequencies and enable the loss of radiation into the operator's hand and head to be reduced.

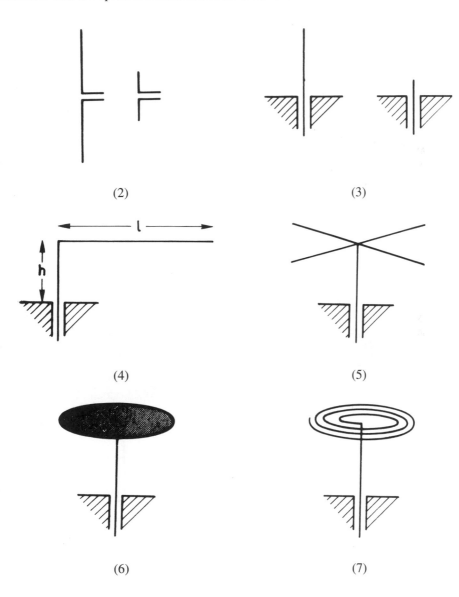

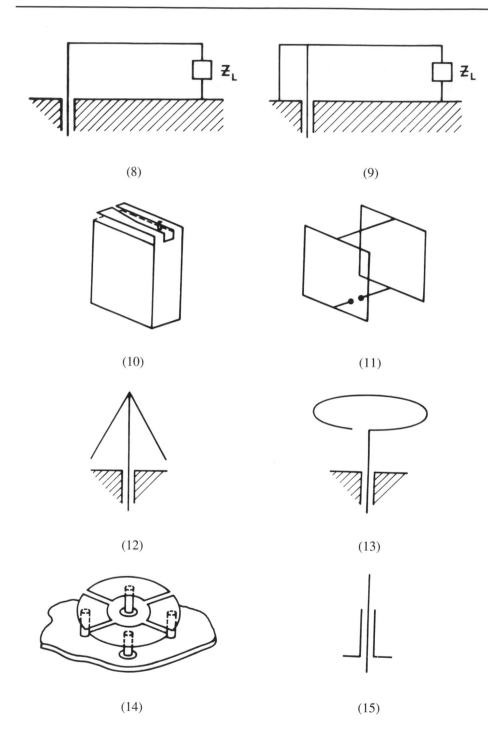

(8)

(9)

(10)

(11)

(12)

(13)

(14)

(15)

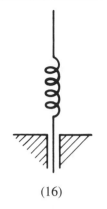

(16)

(17)

(18)

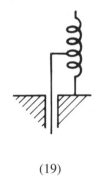

(19)

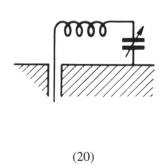

(20)

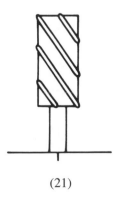

(21)

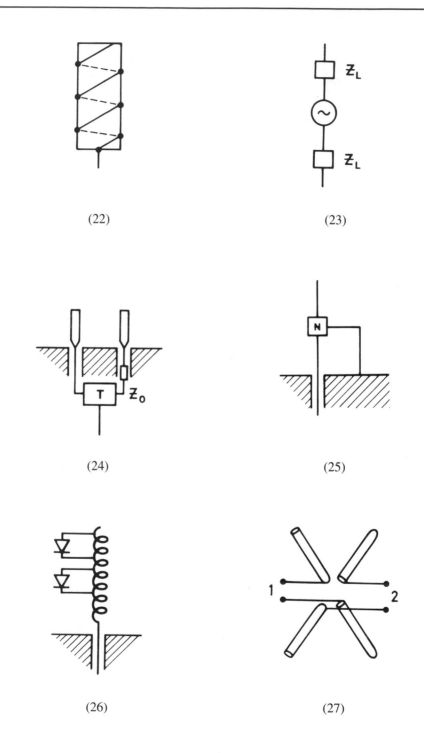

(22)

(23)

(24)

(25)

(26)

(27)

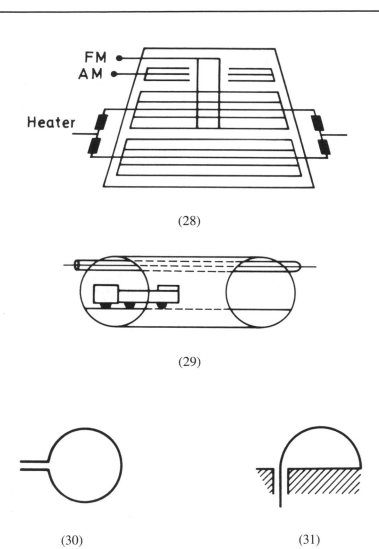

(28)

(29)

(30)

(31)

(32)

(33)

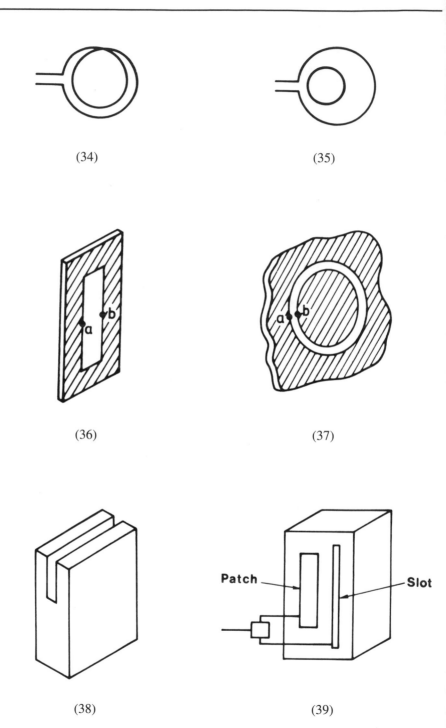

(34)

(35)

(36)

(37)

(38)

(39)

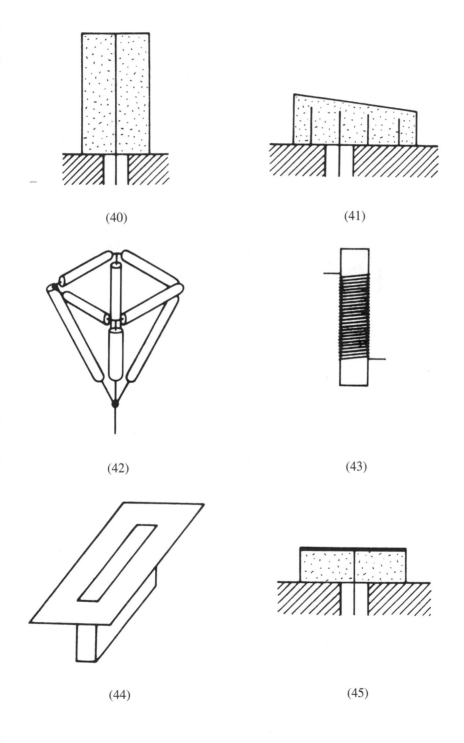

(40)

(41)

(42)

(43)

(44)

(45)

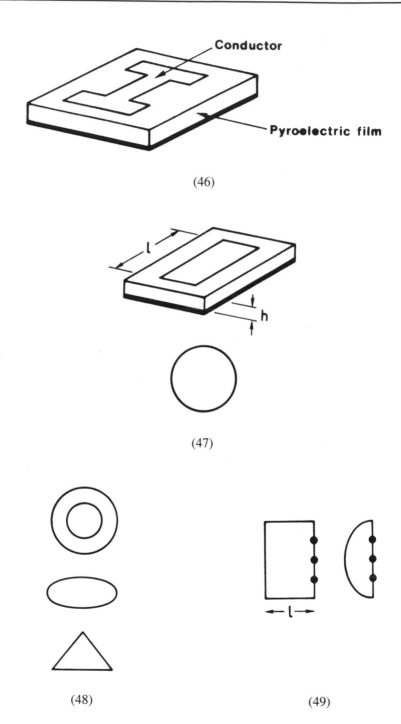

Conductor

Pyroelectric film

(46)

l

h

(47)

(48)

l

(49)

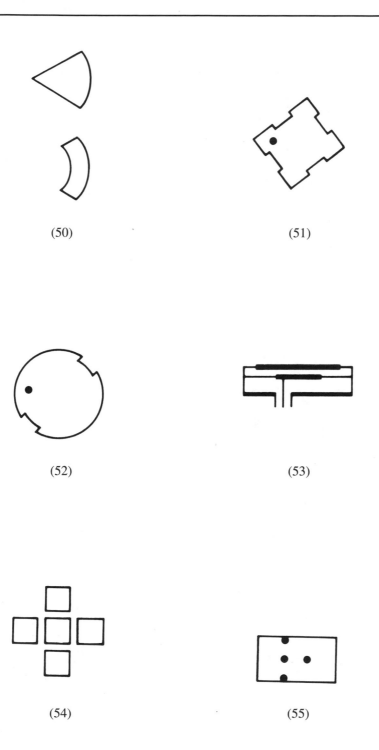

(50)

(51)

(52)

(53)

(54)

(55)

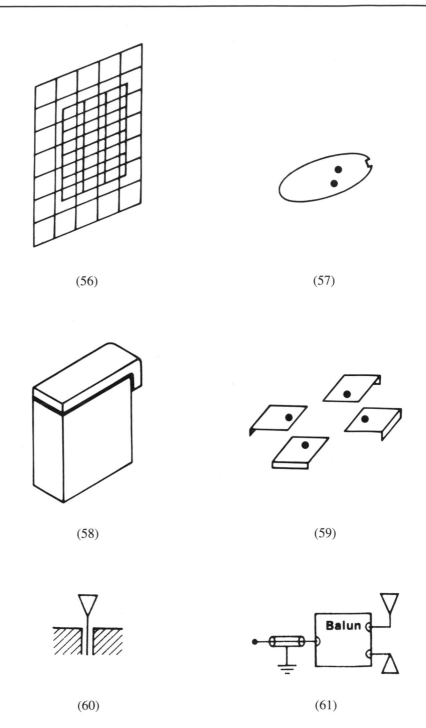

(56)

(57)

(58)

(59)

(60)

(61)

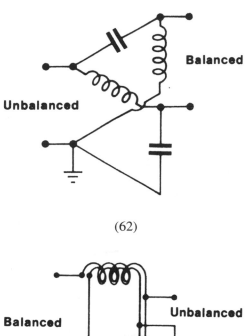

(62)

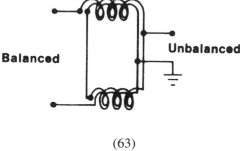

(63)

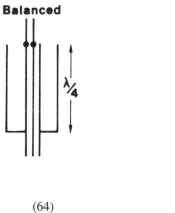

(64)

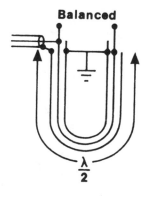

(65)

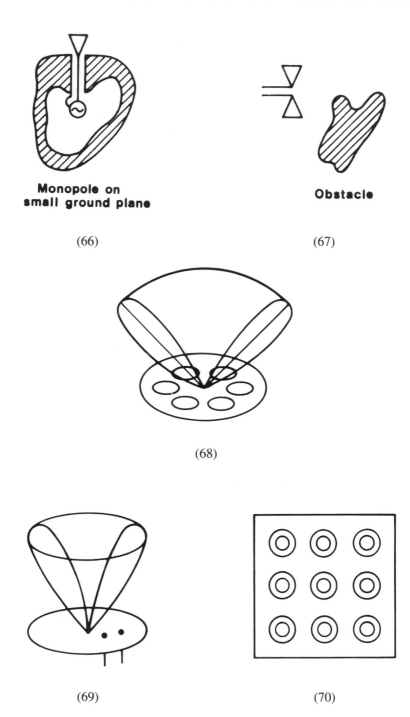

**Monopole on
small ground plane**

(66)

Obstacle

(67)

(68)

(69)

(70)

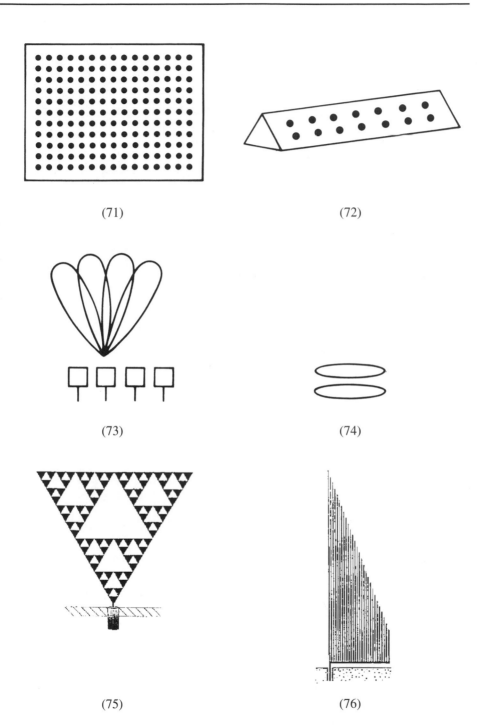

(71)

(72)

(73)

(74)

(75)

(76)

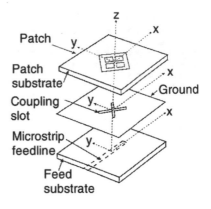

(77) (78)

A.1.8 Key to Symbols and Acronyms Used in Sections A.2 to A.6

For compactness the following cataloging symbols are used to denote the type of mobile system and antenna. The first letter denotes the type of terminal, and the second symbol denotes the antenna type followed by the sequence number.

Examples

C-Mo-3	Automobile monopole antenna number 3
P-S-2	Portable slot antenna number 2
A-A-1	Aeronautical array number 1

Type of Mobile Terminals/Base Station

Automobiles	C	Maritime	M
Portable equipment	P	Aeronautical	A
Train	T	Base station	B
Satellite mobile	S		

Types of Antennas

Monopole	Mo	Printed on glass	G
Dipole	D	Array	A
Inverted-L and F	I	Composite	C
Loop	Lo	Diversity and beam shaping	R
Planar	P	Slot	S
Microstrip	MS	Wire	W
Helix	H	Spiral	Sp
Ferrite	F	Reflector	Re
Meander Line	Me	Ceramic Chip	Cc

Antenna Acronyms

Short backfire	SBF	Inverted-L antenna	ILA
Normal-mode helix antenna	NMHA	Inverted-F antenna	IFA
Microstrip antenna	MSA	Singly fed circularly polarized	SFCP
Planar inverted-F antenna	PIFA		

Next we provide a list of antennas, organized according to mobile terminal.

A.2 LAND MOBILE SYSTEMS

A.2.1 Automobiles

Monopole

Brief View	Type	Technical Data	Cataloging Symbols and Relevant References
	Short monopole	A single element (about 1/300λ for MF and 1/3λ for VHF), used commonly for AM/FM broadcast reception, matching circuits with very high inductive impedance necessary for both AM and FM receivers mounted on the front fender or trunk lid of a car	C-Mo-1 [3,79]
	Short monopole	An element mounted along the pillar, AM/FM broadcast reception, telescopic element, simple element, better performance than other type	C-Mo-2 [79]
	Short monopole	An element mounted on the bumper, inductive loading at the feeding point for the matching, mainly HF and VHF bands	C-Mo-3 [80–82]
	1/4λ monopole	An element mounted on the roof, detachable, mainly used in VHF and UHF bands, almost omnidirectional horizontal pattern	C-Mo-4 [3,79,83]
	On-glass antenna	Simple structure, EM coupling through window glass, directivity about 3 dBd, for car radios and mobile phones in 800 MHz.	C-Mo-5 [84]

Dipoles

	Type	Description	Ref.
	$\lambda/2$ with a sleeve balun	A $\lambda/4$ element with a $\lambda/4$ sleeve balun, used generally in VHF/UHF bands	C-D-1 [2,85,86]
	V-shaped	Horizontal polarization, directional pattern, low-drag structure, VHF/UHF bands, TV reception in cars	C-D-2 [87]

Inverted-L and Inverted-F

	Type	Description	Ref.
	Short inverted-L ($h + l \ll \lambda/4$)	Low-profile, mainly HF bands, narrow bandwidth, matching circuit necessary, vertical and horizontal polarizations	C-1-1 [88,89]
	$\lambda/4$ inverted-F ($h + l \fallingdotseq \lambda/4$) $d \ll \lambda$	Low-profile, VHF/UHF bands, narrow bandwidth, matching impedance obtained by adjusting d and h, both vertical and horizontal polarizations, used on cars, portable units, missiles, and so on.	C-1-2 [88]

Wires printed on glass

	Type	Description	Ref.
	Printed wire	AM/FM radios, printed on the rear window	C-G-1 [90]

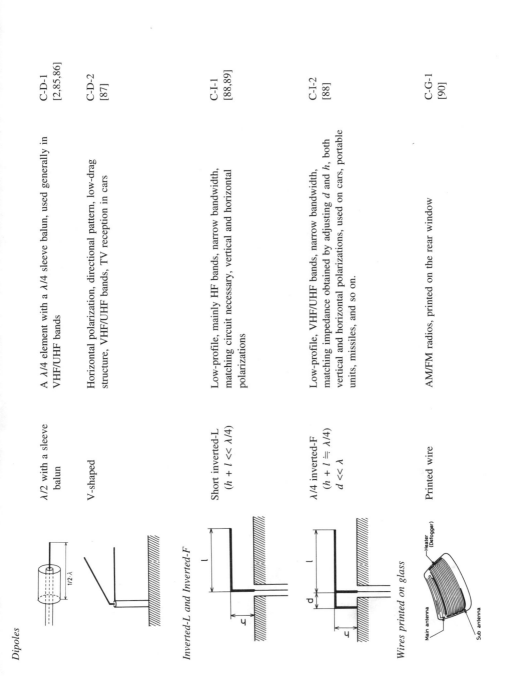

Brief View	Type	Technical Data	Cataloging Symbols and Relevant References
	Printed wire + monopole element	AM/FM radios, TV reception, diversity by a monopole and printed element on the rear window	C-G-2 [90]
	Printed wire	AM/FM radios, diversity by two wire elements printed on the quarter-window and the rear window	C-G-3 [91]
	Printed wires	AM/FM radios, TV reception, diversity by two wire elements printed on both side quarter windows (from [8, Ch. 8]. Reprinted with permission)	C-G-4 [91]
Loops			
	Small loop	A loop built-in side mirror box, MF band, used for door remote control	C-Lo-1 [92]
Planar			
	Printed dipole	Thin, lightweight elements, VHF/UHF mobile telephones, dashboard mount, more than 8% bandwidth, semicircular horizontal radiation pattern (after [94])	C-P-1 [93]

657

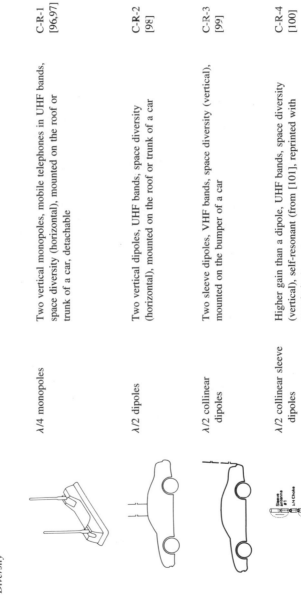

	Dual-frequency-band printed folded dipole antenna	Two frequency bands: 900-MHz and 1.5-GHz bands, thin, lightweight compact structure, semicircular horizontal radiation pattern, directivity (4.5 dBi) (from [95], reprinted with permission)	C-F-2

Diversity

	$\lambda/4$ monopoles	Two vertical monopoles, mobile telephones in UHF bands, space diversity (horizontal), mounted on the roof or trunk of a car, detachable	C-R-1 [96,97]
	$\lambda/2$ dipoles	Two vertical dipoles, UHF bands, space diversity (horizontal), mounted on the roof or trunk of a car	C-R-2 [98]
	$\lambda/2$ collinear dipoles	Two sleeve dipoles, VHF bands, space diversity (vertical), mounted on the bumper of a car	C-R-3 [99]
	$\lambda/2$ collinear sleeve dipoles	Higher gain than a dipole, UHF bands, space diversity (vertical), self-resonant (from [101], reprinted with permission)	C-R-4 [100]

Brief View	Type	Technical Data	Cataloging Symbols and Relevant References
	4-λ/2 elements with reflector	Corner reflector to divide a zone into four, UHF bands, angle diversity	C-R-5
	Dipole loop	Combination of a monopole and a loop element, UHF bands, polarization diversity	C-R-6 [102]
	Cross-dipole	λ/2 dipole elements arrayed perpendicularly, polarization diversity	C-R-7 Chap. 2
	Dual-frequency-band antenna	Sleeve dipoles with parasitic elements, two frequency bands: 900-MHz and 1.5-GHz bands, vertical diversity, trunk lid mount (after [103])	C-R-8
Microstrip patch			
	Circular patch	Circular polarization, low-profile, narrow bandwidth, semisphere coverage pattern, UHF bands, GPS	C-MS-1 [47,103]

Circular patch with a parasitic element	Mobile telephone, 800-MHz bands low-profile, vertical and horizontal polarizations, self-resonant, broadband	C-MS-2 [104,105]
Helix		
Bifilar	Satellite mobile telephone, conical-beam radiation pattern, balanced feed, backfire type, UHF bands	C-H-1 [106]
Composite		
Land mobile/ satellite mobile–compatible antenna	Frequency bands: 870–940 MHz/2.5/2.6 GHz, sleeve antenna and two-wire helical antenna combined, each antenna has different radiation pattern (after [107])	C-C-1
Ferrite		
Cylinder	Very small, narrow bandwidth, LF band, inductive communication, for receiving and transmitting use, bus-roadside communications	C-F-1

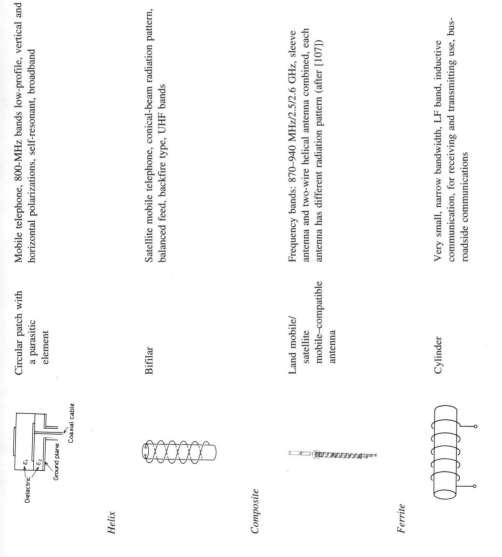

Brief View	Type	Technical Data	Cataloging Symbols and Relevant References

A.2.2 Portable Equipment

Monopole

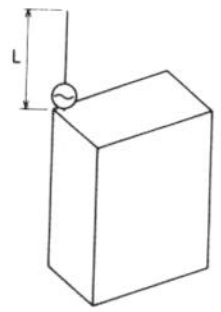

	λ/4 monopole (L = λ/4)	A λ/4 monopole mounted on the body of a unit, simple, popular in VHF and UHF portable units, the body of the portable unit included in the antenna system as a part of radiator	P-Mo-1 [9,108]
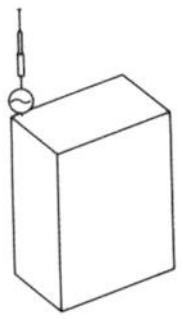	λ/4 (telescopic)	A telescopic λ/4 monopole, convenient for handling, VHF/UHF bands, cordless and portable phones	P-Mo-2
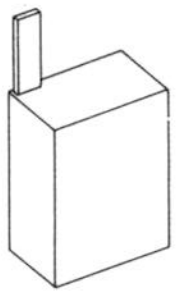	λ/4 (ribbon)	A ribbon-type element mounted on a protable unit, VHF/UHF bands, portable telephones, push-in pull-out type, flexible and thin, convenient for handling	P-Mo-3
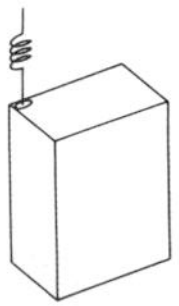	Less than λ/4 (inductance loaded)	A monopole element mounted on a protable unit, VHF bands, inductance-loaded, self-resonant, narrower bandwidth than a monopole	P-Mo-4 [16,81,82]

See P-Mo-I, above	3/8λ, 5/8λ, 7/8λ	A monopole, not a λ/4 length, higher gain than λ/4 monopole, input resistance higher than 50Ω, inductive input reactance, matching circuit necessary, upward radiation pattern, portable telephones, VHF/UHF bands	P-Mo-5 [84]
Dipole	Sleeve (λ/2)	A λ/2 sleeve dipole mounted on a portable unit, Balun not necessary, portable telephones, two-frequency operation in UHF bands possible, less body effect than a λ/4 monopole	P-D-1 [2,108]
IFA	PIFA	A PIFA mounted on or built-into a portable unit, vertical and horizontal polarizations, portable telephones, relay transmitter, UHF bands	P-I-1 [109–111]
Helix	Normal-mode helical antenna	NMHA mounted on a portable unit, shorter than a λ/4 monopole, self-resonant, narrow bandwidth, flexible structure, portable telephone, UHF/VHF bands	P-H-1 [17,112]

Brief View	Type	Technical Data	Cataloging Symbols and Relevant References
Meander Line			
	Meander line	Printed $\lambda/4$ long meander line held on a cylindrical core	P-Me-1 Chap. 7
Slot			
	Bent slot	Slot mounted on the top and sides of a portable unit, the length about $\lambda/2$, bandwidth 3% ~ 8%, vertical and horizontal polarization, self-resonant, portable telephones, semidiversity performance	P-S-1 [38]
	Bent slot	A bent slot mounted on the side of a portable unit, the length about $\lambda/2$, narrow-bandwidth, self-resonant, portable telephones, UHF bands, vertical and horizontal polarization	P-S-2 [38]
Planar			
	Square patch	A square patch with one end shorted, length about $\lambda/4$, UHF bands, narrow bandwidth, vertical polarization, self-resonant, portable telephones, pagers	P-P-1 [103]

Microstrip Patch

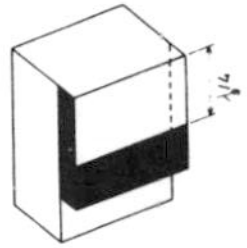 Square patch	A square patch with one end shorted, length about λ/4, UHF bands, narrow bandwidth, vertical polarization, self-resonant, portable telephones, pagers	P-MS-1 [103]

Composite

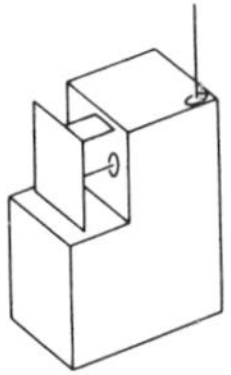 Monopole PIFA	Diversity by a monopole and PIFA, mobile telephones, UHF bands, element spacing of 0.1λ possible with low correlation factor	P-C-1 [113]
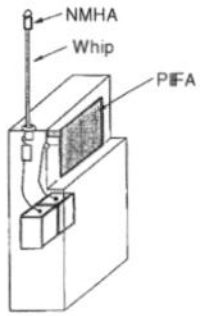 Monopole PIFA NMHA	Monopole and PIFA provide diversity reception; NMHA functions when monopole is retracted	P-C-2 Chap. 7
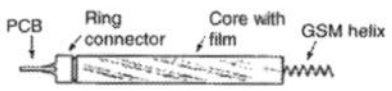 Quadrifilar and helix	Film quadrifilar for personal satcom reception and helix for cellular	P-C-3 Chap. 9

Loop

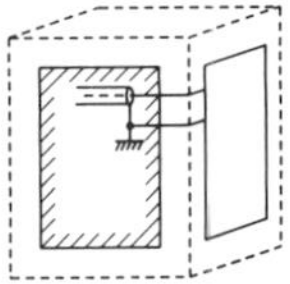 Small rectangular	Electrically small loop, self-resonant, VHF/UHF bands, narrow bandwidth, composite mode: loop plus dipole modes, omnidirectional receiving pattern, sensitivity improved operated near a human body, small volume and lightweight, pager	P-Lo-1 [114]

Brief View	Type	Technical Data	Cataloging Symbols and Relevant References
	Small rectangular (2 turns)	Multiturn small loops, VHF/UHF bands, pagers, almost same performance as the above	P-Lo-2 [115]
	Small rectangular (modified)	A ribbon type element, VHF/UHF, pagers (pencil type), almost same performance as the above	P-Lo-3 [114]
	Parallel plate (magnetic loop)	Aperture between two plates acts as a magnetic loop, 280 MHz, receiving pattern changes according to position of the short point, pager (card type)	P-Lo-4 [116]
Ferrite		Wrist-carried pager, a loop antenna connected in series with a ferrite-cored loop antenna, VHF bands	P-F-1 [117]
		MF bands, very small size, very high permeability, very low efficiency, portable radio	P-F-2 [115,118]

Chip Antenna

| Ceramic chip antenna | Helix embedded in high-permeability ceramic material | P-Cc-1 Chap. 7 |

A.2.3 Trains

| Inverted-L | An inverted element mounted on the roof of the train, 150-MHz bands, communication with nearby stations only | T-I-1 [88,119] |

A.2.4 Base Stations

Dipole	A printed dipole with feeding circuit integrated, 180° beamwidth with corner reflector, UHF bands, telephone, mass production feasible	B-D-1 [103a,120]
Array	Three-dipole element triangle array, pattern shaping, diversity, UHF bands, telephone	B-R-1 [121]
Corner reflector array antenna with printed dipole	16 printed dipole elements, beam tilt, frequency band: 870–940 MHz, beamwidth controllable by adjusting corner angle	B-R-2 Chap. 4

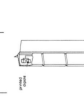

Brief View	Type	Technical Data	Cataloging Symbols and Relevant References
	Microstrip patch array antenna	MSA array, electrical beam tilt downward, frequency band 810–940 MHz, narrow beamwidth	B-R-3
	Dual-frequency band corner reflector	A dipole with a parasitic element and corner reflector, dual frequency bands, narrow beamwidth from 60° ~ 150° by adjusting corner length L (after [19, Ch. 4])	B-R-4
	Sector beam reflector	MSA array and reflector allow choice of beamwidth	B-R-5 Chap. 5
	Pattern diversity antenna with omnidirectional pattern	Four dipole elements combined to perform pattern diversity, each element with omnidirectional pattern, element spacing can be small, mobile telephones	B-R-6 Chap. 4
	Dual beam	MSA array produces two beams for diversity	B-R-7 Chap. 5

	Polarization diversity array antenna	MSA patch array, polarization diversity, thin and compact configuration, beamwidth controllable by adjusting element spacing, mobile telephones (after [28, Ch. 4])	B-R-8
	Precision beam array	Array has precision shaped beam for ETC	B-A-1 Chap. 8

A.3 MARITIME SYSTEMS

	Crossed-drooping dipole	Two crossed-drooping elements, omnidirectional pattern, low gain (4 dBi), L-band, INMARSAT-C	M-D-1
	Quadrifilar helix	Long-structured helix, omnidirectional pattern, low gain (3 dBi), L-band, INMARSAT-C	M-H-1
	SBF	Backfire antenna with disk reflectors, high efficiency, compact, low sidelobe, INMARSAT-M, ETS-V, PROSAT	M-Re-1 Chap. 9
	Helical antenna	Axial mode, wide-band, circular polarization, gain: 8 ~ 15 dBi	M-H-2 Chap. 9

Brief View	Type	Technical Data	Cataloging Symbols and Relevant References
	Yagi-Uda array of cross-dipoles	Endfire array, circular polarization, medium gain: (8 ~ 15 dBi)	M-A-1
	Improved SBF	SBF with two small reflectors, high aperture efficiency, low sidelobe pattern, wide-band	M-Re-2
	Modified SBF	SBF with two small reflectors, high aperture efficiency, low-sidelobe pattern	M-Re-3
	Quad-helix array	Array of four-turn helices, high aperture efficiency, circular polarization	M-A-2 Chap. 9
	Four-element SBF array	High aperture efficiency, circular polarization, gain: 18 ~ 20 dBi	M-A-3
	Cross-dipole array	Array of 16 cross-dipoles, circular polarization, medium gain: ~15 dBi	M-A-4

A.4 AERONAUTICAL SYSTEMS

Dipole

Cross dipole	Inverted V-form with thick elements, Ku-bands, circular polarization with good axial ratio over wide angle, satellite communications	A-D-1
Ramshorn	Horizontal V- or U-form, dragless structure, VHF bands, horizontal polarization, VOR, localizer	A-D-2 [122]

Monopole

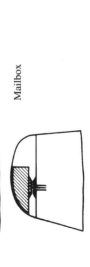

Blade	A flush-mounted antenna of printed stripline elements, vertical polarization, VHF bands, communications (after [123])	A-Mo-1 [123]
Pickax	Tail-cap installation, vertical polarization, VHF bands, communications	A-Mo-2 [124]
Mailbox	Tail-cap installation, vertical polarization, VHF bands, communications	A-Mo-3 [124]

Brief View	Type	Technical Data	Cataloging Symbols and Relevant References
	Loaded wire	A type of transmission line antenna, suitable for lower speed aircraft, horizontal polarization, HF bands, communications	A-W-1
	Trailing wire	VLF/LF bands, suitable for lower speed and large-size aircraft, mainly horizontal polarization, communications	A-W-2 [125,126]
	Probe	A type of isolated-cap antenna, dragless structure, characteristics depend on entire airframe, HF bands, communications	A-Mo-4 [127]
	Tail-cap antenna	Flush-mount, dragless structure, characteristics depend on entire airframe, HF bands, communications	A-S-1 [122]
	Top-loaded monopole	A short monopole with a disk loaded, low-profile structure, narrow bandwidth, LF/MF bands, ADF sense antenna	A-Mo-5 [123]
	Multiband sleeve	A top-loaded sleeve with a sleeve in a fin cap, multifrequency operation in VHF/UHF bands, communications	A-Mo-6 [128]

Bent sleeve		An inverted-L-shaped sleeve, dragless structure, vertical polarization, VHF bands, communications	A-Mo-7 [129]
Partial sleeve		A monopole partially sleeved, tail-cap installation, vertical polarization, VHF bands, communications	A-Mo-8 [124]
Loop	Half-loop loaded with a reactance	Small series-resonant half loop, horizontal polarization, UHF bands, glide path	A-Lo-1 [130]
	Vertical loop	A half loop loaded with a reactance, low-profile structure, polarization parallel to the axis of fuselage, VHF bands, marker beacon	A-Lo-2 [130]
	Loop wound on a ferrite core	Small loop antenna, narrow bandwidth, LF/MF bands, ADF	A-F-1 [123,131]
	Terminated loop	A loop terminated with a resistance, small dimensions, cardioid pattern, VHF/UHF bands, homing	A-Lo-3 [132]

Brief View	Type	Technical Data	Cataloging Symbols and Relevant References
Slot			
	E-slot	Buried structure, cavity backed, horizontal polarization, VHF/UHF bands, VOR, localizer	A-S-2 [122]
	Crossed slot	Flush-mounted, cavity backed, circular polarization, L-bands, satellite communications	A-S-3 [133]
	Annular slot	One-wavelength circumference loop, vertical polarization, flush-mounted, cavity-backed, communications	A-S-4 [124,134]
Notch			
	Standard notch	Buried structure, normally inductive impedance and small radiation resistance, narrow bandwidth, HF/VHF/UHF bands, communications	A-S-5 [129,135]
	Small gap	Body mount (a gap is driven), vertical polarization, LF/MF bands, loran	A-S-6 [130]

Helix	Cylindrical helix	Monofilar or multifilar, a normal, axial, or conical mode, circular polarization without hybrid, L-bands, satellite communications	A-H-1 [136–140]
	Conical helix	HPBW on the order of 100-deg, circular polarization without hybrid gain (4 to 7 dBi), L-band, satellite communications	A-H-2
	Quadrifilar helix	Two bifilar helices, high front-to-back ratio without ground plane, circular polarization with hybrid, narrow band (3% to 5% for VSWR $\leq$ 2), L-band, satellite communications	A-H-3 [141]
Spiral	Conical spiral	Bifilar with self-complementary, circular polarization without hybrid, frequency-independent, L- to Ku-band, satellite communications	A-Sp-1 [142,143]
	Equiangular planar spiral	Bifilar with logarithmic period, cavity-backed, flush-mounted, constant impedance of $60\pi\ \Omega$, circular polarization without hybrid, L- to Ku-bands, satellite communications	A-Sp-2 [144]

Brief View	Type	Technical Data	Cataloging Symbols and Relevant References
	Logarithmic planar spiral	Bifilar with logarithmic period, cavity-backed, flush-mounted, circular polarization without hybrid, L- to Ku-band, satellite communications	A-Sp-3
	Archimedean	Thin-wire bifilar, cavity-backed, flush-mounted, broadband, circular polarization without hybrid, L- to Ku-band, satellite communications	A-Sp-4 [145]

Microstrip Patch

Brief View	Type	Technical Data	Cataloging Symbols and Relevant References
	Standard patch	A square or circular patch, conformal structure, self-resonant, circular polarization with hybrid, L- to Ku-band, satellite communications	A-MS-1 [146,147]
	Thick patch	A patch on a thick substrate with low dielectric constant, conformal structure, higher mode suppression using balanced feeding technique, circular polarization with hybrid, L-bands, satellite communications	A-MS-2 [146–150]
	Stacked patch	A patch stacked with a parasitic element, dual resonance, higher mode suppression using balanced feeding technique, circular polarization with hybrid, L-bands, satellite communications	A-MS-3 [147]
	Stacked-patch dual-frequency operation	Separate stacked configuration with individual feeding, dual frequency, self-duplexing, circular polarization with hybrid, L-bands, satellite communications	A-MS-4 [133]

	SFCP patch	Circular polarization produced by orthogonal modes degenerated, with single feeding, narrow bandwidth, L- and Ku-band, satellite communications	A-MS-5 [146,147,149]
Reflector			
	Parabola	Noncircular periphery mounted on the bottom of fuselage with radome, horizontal polarization, X-bands, weather radar	A-Re-1
	Cassegrain	Dual dishes, mounted on the nose, with radome, either normal pencil beam or modified cosec-square beam, also either circular or horizontal polarization, X-bands, weather radar	A-Re-2
Quadrifilar			
See M-H-1 in A.3	Quadrifilar	Omnidirectional radiation pattern, low gain, low air drag INMARSAT-Aero, GPS	A-H-4
Phased Array			
	Phased array	MSA arrays on two planes, mounted on the top of fuselage, gain: 12 dBi; airplane telephones, INMARSAT-Aero, ETS-V	A-A-1 Chap. 9

A.5 SATELLITE SYSTEMS

Brief View	Type	Technical Data	Cataloging Symbols and Relevant References
Land Mobile Systems			
	Quadrifilar	Two one-and-half turn helices, circular polarization, UHF bands, mobile telephones, satellite mobile systems	S-H-1
	Conical spiral	Spiral elements on a cone, circular polarization, wide bandwidth in 1.5-GHz bands, GPS	S-Sp-1
Maritime Mobile Systems			
See M-H-1 in A.3	Quadrifilar	Omnidirectional pattern, low gain, INMARSAT-C, GPS, AMSC, MSAT, Mobilesat ETS-V, MSAT-X	S-H-2
	Array mechanical tracking	Spiral element array, thin structure, medium gain: 15 dBi, directional pattern, INMARSAT-M, Mobilesat, ETS-V, MSAT-X	S-A-1 Chap. 9
	Phased-array electronic tracking	MSA patch phased array, thin structure, directional pattern, medium gain 15 dBi; ETS-V, MSAT-X	S-A-2 Chap. 9

Type	Description	
Patch array	Four-element patch array, transportable antenna, INMARSAT-C	S-A-3
Patch array	Two-MSA patch array mounted on the lid of briefcase terminal, low-speed data transmission, ETS-V	S-A-4 Chap. 9
Planar-array mechanical tracking	MSA patch array, high gain (40 dBi), Ku-bands (12 GHz), TV and BS reception on a vehicle	S-A-5
Parabolic mechanical tracking	A parabolic antenna, high gain (40 dBi), Ku-band and (12 GHz), TV and BS reception	S-Re-1
Umbrella-type antenna	A deployable parabola antenna, transportable, compact, lightweight	S-Re-2

A.6 TYPICAL ANTENNA TYPES AND THEIR APPLICATIONS

| | | | | *Applications* | | |
Antenna Type	LF	MF 0.3 MHz	HF 3 MHz	VHF 30 MHz	UHF 300 MHz	SHF 3,000 MHz
ILA (long wire) (small, rigid)	Ship, airplane			Train, airplane		
Monopole (1/4λ)		Car (radio, AM)	Car (FM, TV) Portable, car, ship, airplane		Land vehicle, ship	
(3/8λ)				Car	Portable	
(5/8λ)						
Backfire					Ship (INMARSAT)	
IFA			Vehicle		Missile	
Helix (bifilar)					Satellite mobile	
(quadrifilar) (NMHA)				Portable		
(ceramic chip)					Portable	
(quadrifilar)					INMARSAT	
(meander line)					Portable	
Loop (small square)	Car (remote control)		Pager, portable, telephone	Pager		
(multiturn)						
Ferrite	Portable radio (AM)		Pager			
(ceramic chip)					Portable	

Antenna			
PIFA	Portable telephone	Missile	
MSA	Pager	Portable (data) terminal
	Airplane telephone	GPS
		DSRC
Array (MSA) (Wire)	Car (diversity)	Mobile phone (base)
Composite (diversity)		Portable telephone
		Mobile phone (base)
Reflector	Telephone (base)	Airplane (radar)
	Pager (base)	
Spiral		GPS, INMARSAT
Slot	Portable Telephone	Airplane (radar)

REFERENCES

[1] James, J. R., "What's New in Antennas," *Proc. IEE Int. Conf. Ant. Propag.*, ICAP '89, Warwick, England, April 1989, pp. 84–97.

[2] Johnson, R. C., and H. Jasik, Ch. 4 in *Antenna Engineering Handbook*, 2nd edition, New York: McGraw-Hill, 1984.

[3] Ibid.

[4] Prasad, S., and R. W. P. King, "Experimental Study of Inverted L, T and Related Transmission-Line Antennas," *J. Research*, NBS, 65D, Sept./Oct. 1961.

[5] Simpson, T. L., "The Theory of Top-Loaded Antennas: Integral Equations for the Currents," *IEEE Trans.*, Vol. AP-19, 1971.

[6] Weeks, W. L., *Antenna Engineering*, New York: McGraw-Hill, 1968.

[7] Bhojwani, H. R., and L. W. Zelby, "Spiral Top-Loaded Antenna: Characterization and Design," *IEEE Trans.*, Vol. AP-21, 1973, pp. 293–298.

[7a] Fenwick, R. C., "A New Class of Electrically Small Antennas," *IEEE Trans.*, Vol. AP-13, 1965, pp. 379–383.

[7b] Nakano, H., "Research on Spiral and Helical Antennas," *IEEE AP-S Newsletter*, June 1988, p. 20.

[8] King, R. W. P., et al., "Transmission-Line Missile Antennas," *IRE Trans.*, Vol. AP-8, 1960, pp. 88–90.

[9] Hirasawa, K., and K. Fujimoto, "Characteristics of Wire Antennas on a Rectangular Conducting Body," *Trans. IECE Japan*, Vol. J65-B, 1982, pp. 1113–1139.

[9a] Fujimoto, K., A. Henderson, K. Hirasaura, and J. R. James, "Small Antennas," England: Research Studies Press, 1987, distributed by Wiley & Sons, p. 117–122.

[10] Rasinger, J., et al., "A New Enhanced Bandwidth Internal Antenna for Portable Communication Systems," *40th IEEE Vehicular Technology Conf.*, Orlando, May 1990, pp. 7–12.

[11] Tsukiji, R., and Y. Kumon, "On a Modified Transmission Line Type Antenna," *Proc. IEE Int. Conf. Ant. Prop.*, ICAP '91, York, England, April 1991, pp. 38–41.

[12] Gangi, A. F., et al., "The Characteristics of Electrically Short, Umbrella Top-Loaded Antennas," *IEEE Trans.*, Vol. AP-13, 1965, pp. 864–871.

[13] Burton, R. W., and R. W. P. King, "Theoretical Considerations and Experimental Results for the Hula-Hoop Antenna," *Microwave J.*, Nov. 1963, pp. 89–90.

[14] Goubau, G., "Multi-Element Monopole Antennas," *Proc. ECOM-ARO Workshop on Electrically Small Antennas*, G. Goubau and F. Schwering, ed., 6–7 May 1976, Fort Monmouth, pp. 63–67.

[15] Johnson, R. C., and H. Jasik, Ch. 4 in *Antenna Engineering Handbook*, 2nd edition, New York: McGraw-Hill, 1984.

[16] Hansen, R. C., "Optimum Inductive Loading of Short Whip Antennas," *IEEE Trans.*, Vol. VT-24, 1975, pp. 21–29.

[17] Hansen, L. H., "A New Helical Ground-Plane Antenna for 30 to 50 MHz," *IRE Trans.*, Vol. VC-10, 1961, pp. 36–39.

[18] Fujimoto, K., A. Henderson, K. Hirasawa, and J. R. James, "Small Antennas," England: Research Studies Press, 1987, distributed by Wiley & Sons, pp. 236–238.

[19] Ramsdale, P. A., and T. S. M. Maclean, "Active Loop-Dipole Aerials," *Proc. IEE*, Vol. 118, 1971, pp. 1697–1710.

[20] Smith, G. S., "Radiation Efficiency of Electrically Small Multi-Turn Loop Antennas," *IEEE Trans.*, Vol. AP-20, 1972, pp. 656–657.

[21] Kilgus, C. C., "Resonant Quadrifiler Helix Design," *Microwave J.*, Vol. 13, 12 Dec. 1970, pp. 49–54.

[22] Nakano, H., et al., "Generation of a Circularly Polarized Conical Beam From Backfire Helical Antennas," *Proc. Int. Conf. Ant. Propag.*, ICAP '91, York, England, April 1991, pp. 42–45.

[23] Hirasawa, K., and K. Fujimoto, "On Electronically-Beam-Controllable-Dipole Antenna," *IEEE AP-S Int. Symp. Digest*, 1980, pp. 692–695.

[24] Schroeder, K. G., and K. M. Soo Hoo, "Electrically Small Complementary Pair (ESCP) With Interelement Coupling," *IEEE Trans.*, Vol. AP-24, 1976, pp. 411–418.

[25] Hiroi, Y., and K. Fujimoto, "Active Inverted-L Antenna," paper of technical group TGAP 70-29, *IECE Japan*, 1970, pp. 1–12 (in Japanese). See also [19].

[26] Ploussios, G., "An Electronically Tuned HF/VHF Helical Antenna," *Microwave J.*, May 1991, pp. 223, 224, 227, 228, 231, 234, 237, 239, 240.

[27] Taga, T., et al., "Correlation Properties of Antenna Diversity in Indoor Mobile Communication Environments," *IEEE Vehicular Technology Conf.*, 1989, pp. 446–451.

[27a] Gatti, M. S., and D. J. Nybakken, "A Circular Polarized Crossed Drooping Dipole Antenna," *IEEE AP-S Int. Symp. Digest*, Dallas, May 1990.

[28] Lindenmeier, H. K., et al., "Multiple FM Window Antenna System for Scanning Diversity With an Integrated Processor," *IEEE Vehicular Technology Conf.*, Orlando, May 1990, pp. 1–6.

[29] Delogne, P., *Leaky Feeders and Subsurface Radio Communications*, London: Peter Peregrinus, IEE, Vol. 14, 1982.

[29a] Kishimoto, T., and S. Sakai, "LCX Communication Systems," *IECEJ*, Tokyo, 1982.

[30] Johnson, R. C., and H. Jasik, Ch. 5 in *Antenna Engineering Handbook*, 2nd edition, New York: McGraw-Hill, 1984.

[30a] Fujimoto, K., A. Henderson, K. Hirasawa, and J. R. James, "Small Antennas," England: Research Studies Press, 1987, distributed by Wiley & Sons, pp. 100–107.

[31] Johnson, R. C., and H. Jasik, Ch. 5 in *Antenna Engineering Handbook*, 2nd edition, New York: McGraw-Hill, 1984.

[32] Fujimoto, K., A. Henderson, K. Hirasawa, and J. R. James, "Small Antennas," England: Research Studies Press, 1987, distributed by Wiley & Sons, pp. 86–89.

[33] King, R. W. P., et al., "Transmission-Line Missile Antennas," *IRE Trans.*, Vol. AP-8, 1960, pp. 88–90. Same generic form.

[34] Smith, G. S., "Radiation Efficiency of Electrically Small Multiturn Loop Antennas," *IEEE Trans.*, Vol. AP-20, 1972, pp. 656–657.

[35] Technique used by radio amateurs.

[36] Johnson, R. C., and H. Jasik, Ch. 8 in *Antenna Engineering Handbook*, 2nd edition, New York: McGraw-Hill, 1984.

[37] Ibid.

[37a] Silver, S., "Microwave Antenna Theory and Design," McGraw-Hill, 1949, pp. 167–168.

[38] Kuboyama, H., K. Fujimoto, and K. Hirasawa, "UHF Bent-Slot Antenna System for Portable Equipment-I," *IEEE Trans.*, Vol. VT-36, No. 2, May 1987, p. 78.

[39] Ito, K., and S. Sasaki, "A Small Printed Antenna Composed of Slot and Wide Strip for Indoor Communication Systems," *IEEE Int. AP-S Digest*, 1988, pp. 716–719.

[40] James, J. R., and A. Henderson, "Electrically Short Monopole Antennas With Dielectric or Ferrite Coatings," *Proc. IEE*, Vol. 125, No. 9, 1978, pp. 793–803.

[41] James, J. R., "Reduction of Antenna Dimensions by Dielectric Loading," *Electronics Letters*, Vol. 10, 1974, pp. 263–265.

[42] Woodman, K. F., "Dielectric-Clad Discone," *Electronics Letters*, Vol. 13, 1977, pp. 264–265.

[42a] James, J. R., and A. Henderson, "Investigation of Electrically Small VHF and HF Cavity-Type Antennas," *Proc. Int. Conf. Ant. Propag.*, London, 1978, pp. 322–326.

[43] Johnson, R. C., and H. Jasik, Ch. 5 in *Antenna Engineering Handbook*, 2nd edition, New York: McGraw-Hill, 1984.

[44] Johnson, R. C., and H. Jasik, Ch. 8 in *Antenna Engineering Handbook*, 2nd edition, New York: McGraw-Hill, 1984.

[45] Brunner, J. E., and G. Seward, "Low Profile VHF Antenna for Armor," *Proc. ECOM-ARO Workshop on Electrically Small Antennas*, G. Goubau, and F. Schwering, ed., 6–7 May 1976, Fort Monmouth, pp. 153–157.

[45a] Hansen, R. C., "Superconducting Antennas," *IEEE Trans.*, AES, Vol. 26 No. 2, 1990, pp. 345–355.

[46] Lee, T. M., et al., "Microwave Field-Detecting Element Based on Pyroelectric Effect in PVDF," Electronic Letters, Vol. 22, No. 4, 1986, pp. 200–202.

[47] James, J. R., and P. S. Hall, eds., *Handbook of Microstrip Antennas*, Vols. 1 and 2, London: Peter Peregrinus, IEE, 1989.

[48] James, J. R., and P. S. Hall, eds., *Handbook of Microstrip Antennas*, Vols. 1 and 2, London: Peter Peregrinus, IEE, 1989, pp. 25–26.

[49] Penard, E., and J. P. Daniel, "Open and Hybrid Microstrip Antennas," *IEE Proc.*, Vol. 131, pt. H, 1984.

[50] Richards, W. F., et al., "Theoretical and Experimental Investigation of Annular, Annular Sector and Circular Sector Microstrip Antennas," *IEEE Trans.*, Vol. AP-12, 1984, pp. 864–866.

[51] Oswald, L. T., and C. W. Garvin, "Microstrip Command and Telemetry Antennas for Communications and Technology Satellites," *IEE Int. Conf. On Antennas for Aircraft Spacecraft*, London, 1975, pp. 217–222.

[52] Haneishi, M., et al., "Broadband Microstrip Array Composed of Single Feed Type Circularly Polarized Microstrip Element," *IEEE AP-S Int. Symp. Digest*, May 1982, pp. 160–163.

[53] Sabban, A., "New Broadband Stacked Two Layer Microstrip Antenna," *IEEE AP-S Int. Symp. Digest*, Houston, 1983, pp. 63–66.

[54] Kumar, G., and K. C. Gupta, "Non-Radiating Edge and Four Edges Gap Coupled Multiple Resonator Broadband Microstrip Antenna," *IEEE Trans.*, Vol. AP-33, 1985, pp. 173–177.

[55] James, J. R., and P. S. Hall, eds., *Handbook of Microstrip Antennas*, Vols. 1 and 2, London: Peter Peregrinus, IEE, 1989, pp. 1096–1099.

[55a] James, J. R., and P. S. Hall, eds., *Handbook of Microstrip Antennas*, Vols. 1 and 2, London: Peter Peregrinus, IEE, 1989, pp. 1099–1102.

[56] James, J. R., and G. Andrasic, "Superimposed Dichroic Microstrip Antenna Arrays," *Proc. IEE.*, Vol. 135H, No. 5, 1988, pp. 304–312.

[57] James, J. R., and P. S. Hall, eds., *Handbook of Microstrip Antennas*, Vols. 1 and 2, London: Peter Peregrinus, IEE, 1989, pp. 1124–1125.

[58] Microstrip used as bent slot. See [38].

[59] Arai, H., "Flat 4—Direction Antenna for Antenna Pattern Diversity Reception," *Proc. Int. Conf. Ant. Prop.*, ICAP'91, York, England, April 1991, pp. 46–49.

[60] Johnson, R. C., and H. Jasik, Ch. 4 in *Antenna Engineering Handbook*, 2nd edition, New York: McGraw-Hill, 1984.

[61] Johnson, R. C., and H. Jasik, Ch. 43 in *Antenna Engineering Handbook*, 2nd edition, New York: McGraw-Hill, 1984.

[62] Johnson, R. C., and H. Jasik, Ch. 43 in *Antenna Engineering Handbook*, 2nd edition, New York: McGraw-Hill, 1984.

[63] Johnson, R. C., and H. Jasik, Ch. 43 in *Antenna Engineering Handbook*, 2nd edition, New York: McGraw-Hill, 1984.

[64] Johnson, R. C., and H. Jasik, Ch. 43 in *Antenna Engineering Handbook*, 2nd edition, New York: McGraw-Hill, 1984.

[65] Johnson, R. C., and H. Jasik, Ch. 43 in *Antenna Engineering Handbook*, 2nd edition, New York: McGraw-Hill, 1984.

[66] Parhami, P., et al., "Technique for Calculating the Radiation and Scattering Characteristics of Antennas Mounted on a Finite Ground Plane," *Proc. IEE*, Vol. 124, 1977, pp. 1009–1016.

[66a] Fujimoto, K., A. Henderson, K. Hirasawa, and J. R. James, Ch. 4 in "Small Antennas," England: Research Studies Press, 1987, distributed by Wiley & Sons.

[67] Hansen, J. E., ed., *Spherical Near-Field Antenna Measurements*, London: Peter Peregrinus, IEE, 1988, Chapter 2.

[68] Hori, T., et al., "Circularly Polarized Microstrip Array Antenna With Conical Beam," *Natl. Conv. Rec. IECE Japan*, Vol. 655, 1982, (in Japanese).

[69] Hori, T., et al., "Circularly Polarized Broadband Microstrip Antenna Radiating Conical Beam," *Natl. Conv. Rec. IECE Japan*, Vol. 637, 1986, (in Japanese).

[70] Yasunaga, M., et al., "Phased Array Antennas for Aeronautical Satellite Communications," *IEE Int. Conf. Ant. Prop.*, ICAP '87, York, pp. 47–50.

[71] James, J. R., and P. S. Hall, eds., *Handbook of Microstrip Antennas*, Vols. 1 and 2, London: Peter Peregrinus, IEE, 1989, pp. 1112–1121.

[72] James, J. R., and P. S. Hall, eds., *Handbook of Microstrip Antennas*, Vols. 1 and 2, London: Peter Peregrinus, IEE, 1989, pp. 1142–1145.

[73] Hawkins, G. J., et al., "Electronic Beam Squint Tracking for Land Mobile Satellite Terminals," *IEEE Vehicular Technology Conf.*, 1989, pp. 735–741.

[74] Shafai, L., "MSAT Vehicular Antennas With Self Scanning Array Elements," *Proc. Int. Mobile Satellite Conf.*, IMSC '90, Ottawa, June 1990, pp. 523–528.

[74a] Sanford, J., et al., "Shaped Beam Patch Arrays for Mobile Communication Base Stations," *Microwave Engineering Europe*, June/July 1991, pp. 31–33.

[75] Puente, C., et al., "Fractal Multiband Antenna Based on the Sierpinski Gasket," *Electron. Lett.*, Vol. 32, No. 1, Jan. 1996, pp. 1–2.

[75a] Puente, C., et al., " Perturbation of the Sierpinski Antenna to allocate operating bands," *Electron. Lett.*, Vol. 32, No. 24, Nov. 1996, pp. 2186–2188.

[75b] Puente, C., et al., "Small But Long Koch Fractal Monopole," *Electron. Lett.*, Vol. 34, No. 1, Jan. 1998, pp. 9–10.

[75c] Liang, X., et al., "Synthesis of Fractal Patterns from Concentric-Ring Arrays," *Electron. Lett.*, Vol. 32, No. 21, Oct. 1996, pp. 1940–1941.

[75d] Werner, D. H., and P. L. Werner, "Frequency-Independent Features of Self-Similar Fractal Antennas," *Radio Science*, Vol. 31, No. 6, Nov.–Dec. 1996, pp. 1331–1343.

[75e] Kim, F., and D. L. Jaggard, "The Fractal Random Array," *Proc. IEEE*, Vol. 74, No. 9, Sep. 1986, pp. 1278–1279.

[75f] Walker, G. J., and J. R. James, "Fractal Volume Antennas," *Eletron. Lett.*, Vol. 34, No. 16, Aug. 1998, pp. 1536–1537.

[75g] May, M., "Aerial Magic," *New Scientist*, Jan. 1998.

[76] Andrasic, G., et al., "New Anisotropic Monopole Bandwidth Extension Technique," *Electron. Lett.*, Vol. 30, No. 24, Nov. 1994, pp. 1998–2000.

[77] Andrasic, G., and J. R. James, "Height Reduced Superdirective Array with Helical Directors," *Electron. Lett.*, Vol. 29, No. 23, Nov. 1993, pp. 2002–2004.

[78] Huang, C. Y., "Designs for an Aperture-Coupled Compact Circularly Polarised Microstrip Antenna," *IEE Proc. Microw. Antennas Propag.*, Vol. 146, No. 1, Feb. 1999, pp. 13–16.

[78a] Kan, H. K., and R. B. Waterhouse, "Size Reduction Technique for Shorted Patches," *Electron. Lett.*, Vol. 35, No. 12, June 1999, pp. 948–949.

[78b] Huang, C. Y., J. Y. Wu, and K. L. Wong, "Broadband Circularly Polarised Square Microstrip Antenna Using Chip-Resister Loading," *IEE Proc. Microw. Antennas Propag.*, Vol. 146, No. 1, Feb. 1999, pp. 94–96.

[78c] Waterhouse, R. B., "Broadband Stacked Shorted Patch," *Electron. Lett.*, Vol. 35, No. 2, Jan. 1999, pp. 98–100.

[78d] Rowe, W. S. T., and R. B. Waterhouse, "Broadband CPW Fed Stacked Patch Antenna," *Electron. Lett.*, Vol. 35, No. 9, Apr. 1999, pp. 681–682.

[78e] Palit, S. K., and A. Hamadi, "Design and Development of Wideband and Dual-Band Microstrip Antennas," *IEE Proc. Microw. Antennas Propag.*, Vol. 146, No. 1, Feb. 1999, pp. 35–39.

[78f] Lee, K. F., and W. Chen, eds., *Advances in Microstrip and Printed Antennas*, New York: John Wiley and Sons, 1997.

[78g] Gupta, K. C., and P. S. Hall, eds., *Analysis and Design of Integrated Circuit Antenna Modules,* New York: John Wiley and Sons, 2000.

[78h] De Maagt, P., et al., "PBG Crystals: Periodic Dielectric Materials that Control EM Wave Propagation," *Microwave Eng. Europe,* Oct. 1999, pp. 35, 36, 39, 40, 43.

[79] Rudge, A. W., et al., eds., *The Handbook of Antenna Design,* London: Peter Peregrinus, 1986, pp. 1470–1488.

[80] Rudge, A. W., et al., eds., *The Handbook of Antenna Design,* London: Peter Peregrinus, 1986, pp. 1470–1488.

[81] Hansen, R. C., "Optimum Inductive Loading of Short Whip Antennas," *IEEE Trans.,* Vol. VT-24, 1975, p. 21.

[82] Fournier, M., and A. Pomerleau, "Experimental Study of Inductively Loaded Short Monopole," *IEEE Trans. Vehicular Tech.,* Vol. VT-27, 1978, p. 1.

[83] Nishikawa, K., and Y. Asano, "Vertical Radiation Patterns of Mobile Antenna in UHF Band," *IEEE Trans. Vehicular Tech.,* Vol. VT-35, May 1986, pp. 57–62.

[84] Horn, D. W., "Cellular 'On-Glass' Antenna Technology," *Proc. 34th IEEE Vehicular Technology Conf.,* Pittsburgh, 21–23 May, 1984, pp. 65–68.

[85] Johnson, R. C., and H. Jasik, *Antenna Engineering Handbook,* 2nd edition, New York: McGraw-Hill, 1984, pp. 4-18–4-20.

[86] Rudge, A. W., et al., eds., *The Handbook of Antenna Design,* London: Peter Peregrinus, 1986, pp. 1420–1436.

[87] Jasik, H., ed., Ch. 22 in *Antenna Engineering Handbook,* 1st edition, New York: McGraw-Hill, 1961, pp. 24-4–24-5.

[88] Fujimoto, K., A. Henderson, K. Hirasawa, and J. R. James, "Small Antennas," England: Research Studies Press, 1987, distributed by Wiley & Sons, pp. 116–127.

[89] Johnson, R. C., and H. Jasik, *Antenna Engineering Handbook,* 2nd edition, New York: McGraw-Hill, 1984, pp. 27-18 to 27-21.

[90] Toriyama, H., et al., "Development of Printed-On Glass TV Antenna System for Car," *IEEE Vehicular Tech. Conf.,* 1987, pp. 334–342.

[91] Fukumura, H., "Space Diversity FM Radio System for Automobiles," IECE Tech. G, SANE 82-20, 1982, pp. 9–13.

[92] Hirano, M., et al., "Keyless Entry System With Radio Card Transponder," *IEEE Trans. Industrial Electronics,* Vol. IE-35, No. 2 1988, pp. 208–216.

[93] Ebine, Y., and M. Karikomi, "Printed Dipole Antenna With a Parasitic Element," *IEICE Tech. G.,* Vol. AP89-12, 1989. (in Japanese).

[94] Ebine, Y., N. Shimada, and K. Kosaka, "Printed Dipole with Capacity Plate," *IECE Nat. Conv. Rec.,* No. 641, 1986.

[95] Kagoshima, K., T. Takahashi, and Y. Ebine, "Dual Frequency Low Profile Printed Dipole Yagi-Array for a Cabin Mounted Vehicular Antenna," *IEEE AP-S '91,* London, Ontario, June 24-28, 1991, pp. 962–965.

[96] "Vehicular Antenna," Japan Design Patent 793435, Aug. 1990.

[97] "Vehicular Antenna," Japan Design Patent 793436, Aug. 1990.

[98] Lee, W. C. Y., *Mobile Communication Engineering,* New York: Wiley & Sons, 1982, p. 174.

[99] Ibid., p. 175.

[100] Sakitani, A., and S. Egashira, "Analysis of Coaxial Collinear Antenna: Recurrence Formula of Voltages and Admittances at Connections," *IEEE Trans. Ant. Propag.,* Vol. AP-39, No. 1, Jan. 1991, pp. 15–20.

[101] Ebine, Y., and Y. Yamada, "A Vehicular-Mounted Vertical Space Diversity Antenna for a Land Mobile Radio," *IEEE Trans.,* Vol. VT-40, No. 2, 1991, pp. 420–425.

[102] Lee, W. C. Y., *Mobile Communications Engineering,* New York: Wiley & Sons, 1982, p. 165.

[103] Fujimoto, K., et al., "Application in Mobile and Satellite Systems," Chapter 19 in *Handbook of Microstrip Antennas,* J. R. James and P. S. Hall, eds., London: Peter Peregrinus, 1989.

[103a] Nakajma, N., et al., "A Major Angle Corner Reflector Antenna with 180° Beam Width," Natl. Conv. Rec. IECE Japan, Vol. 752, 1985 (in Japanese).

[104] Mishima, H., and T. Taga, "Mobile Antennas and Duplexer for 800 MHz Band Mobile Telephone System," *IEEE AP-S Int. Symp. Digest.*, Quebec, June 1980, pp. 508–511.

[105] Mishima, H., and T. Taga, "Antenna and Dupelexer for New Mobile Radio Unit," *Review of the ECL, NTT*, Vol. 30, No. 2, March 1982, pp. 359–370.

[106] Kumar, A., "Fixed and Mobile Terminal Antennas," Norwood, MA: Artech House, 1991, pp. 116–169.

[107] Terada, N., and K. Kagoshima, "Compatible Mobile Antenna for Mobile Satellite and Cellular Communication System," *IEICE Tech. Rep. on Ant. and Propag.*, A-P91-65, 1991.

[108] Fujimoto, K., A. Henderson, K. Hirasawa, and J. R. James, Ch. 4 in "Small Antennas," England: Research Studies Press, 1987, distributed by Wiley & Sons.

[109] Sato, K., et al., "Characteristics of a Planar Inverted-F Antenna on a Rectangular Conducting Body," *Electronics and Communications in Japan*, Scripta Publishing, Part 1, Vol. 72, October 1989, pp. 43–51.

[110] Taga, T., and K. Tsunekawa, "Performance Analysis of a Built-In Planar Inverted-F Antenna for 800 MHz Band Portable Radio Units," *IEEE Trans.*, Selected Areas in Communication, Vol. SAC-5, June 1987, pp. 921–929.

[111] Taga, T., "Analysis of Planar Inverted-F Antennas and Antenna Design for Portable Radio Equipment," Chapter 5 in *Analysis Design and Measurement of Small and Low-Profile Antennas*, K. Hirasawa and M. Haneishi, eds., Norwood, MA: Artech House, 1991.

[112] Fujimoto, K., A. Henderson, K. Hirasawa, and J. R. James, "Small Antennas," England: Research Studies Press, 1987, distributed by Wiley & Sons, pp. 59–75.

[113] Tsunekawa, K., "Diversity Antennas for Portable Telephones," *Proc. 39th IEEE Vehicular Tech. Conf.*, San Francisco, 1989, pp. 50–56.

[114] Fujimoto, K., A. Henderson, K. Hirasawa, and J. R. James, "Small Antennas," England: Research Studies Press, 1987, distributed by Wiley & Sons, pp. 89–100.

[115] Wheeler, H. A., "Small Antennas," *IEEE Trans. Ant. Propag.*, Vol. AP-23, July 1975, pp. 462–469.

[116] Ishii, N., K. Itoh, "A Consideration on the Numerical Method for a Card-Sized Thin Planar Antenna," Paper of Technical Group, *IEICE Japan*, Vol. AP91-36 (in Japanese), 1991, pp. 9–14.

[117] U. S. Patent 4873527, Oct. 1989.

[118] Sato, S., and Y. Naito, "Dipole Antenna Covered With Ferrite Sleeve," *Trans. IECE Japan*, Vol. J58-B, June 1975, pp. 285–292 (in Japanese).

[119] Watanabe, H., "Electronic Control and Communication System of Shinkansen," *Korona-sha*, Tokyo, 1982, p. 104 (in Japanese).

[120] Nakajima, N., et al., "A Major Angle Corner Reflector Antenna With 180° Beam Width," *Natl. Conv. Rec. IECE Japan*, Vol. 752, 1985, (in Japanese).

[121] Lee, W. C. Y., *Mobile Communication Engineering*, New York: Wiley & Sons, 1982, p. 306.

[122] Johnson, R. C., and H. Jasik, *Antenna Engineering Handbook*, 2nd edition, 1984, pp. 37-18 to 37-33.

[123] IECE Japan, *Antenna Engineering Handbook*, Tokyo, Japan: Ohm-Sha Co., 1980, pp. 319–388 (in Japanese).

[124] Granger, J. V. N., and J. T. Bolljahn, "Aircraft Antennas," *Proc. IRE*, Vol. 43, May 1955, pp. 533–550.

[125] Marin, L., J. P. Castillo, and K. S. H. Lee, "Broad-Band Analysis of VLF/LF Aircraft Wire Antennas," *IEE Trans. Ant. Propagat.*, Vol. AP-26, Jan. 1978, pp. 141–145.

[126] Kossey, P. A., E. A. Lewis, and E. C. Field, "Relative Characteristics of TE/TM Waves Excited by Airborne VLF/LF Transmitters," *AGARD Conf. Prog.*, Vol. 305, 1982, p. 19–1.

[127] Hirano, M., ed., *New Radio Wave Technologies*, Part 2, Tokyo, Japan: RATESU Co., 1973, pp. 276–281 (in Japanese).

[128] Burberry, R. A., "The Rationalization of Aircraft Antennas," *IEE Int. Conf. Antennas for Aircraft and Spacecraft*, June 1975, pp. 204–209.

[129] Burberry, R. A., "Progress in Aircraft Aerials," *Proc. IEE.*, Vol. 109, pt. B, Nov. 1962, pp. 431–444.

[130] Jasik, H., ed., Ch. 22 in *Antenna Engineering Handbook*, 1st edition, New York: McGraw-Hill, 1961, pp. 27-8 to 27-40.

[131] Fujii, K., "Loop Antenna for Aircrafts," Japanese Utility Model 914431, 1970.

[132] Rudge, A. W., et al., eds., *The Handbook of Antenna Design*, London: Peter Peregrinus, 1986, pp. 736–820.

[133] Yasunaga, Y., F. Watanabe, T. Shiokawa, and M. Yamada, "Phased Array Antennas for Aeronautical Satellite Communications," *IEE ICAP '87*, Part 1, March 1987, pp. 47–50.

[134] Cumming, W. A., and M. Cormier, "Design Data for Small Annular Slot Antennas," *IRE Trans. Ant. Propagat.*, Vol. AP-6, April 1958, pp. 210–211.

[135] Blackband, W. T., "A Comparison of the HF Aerials Fitted to XV814," *IEE Int. Conf. Antennas for Aircraft and Spacecraft*, June 1975, pp. 171–177.

[136] Kraus, J., "The Helical Antenna," *Proc. IRE*, Vol. 37, March 1949, pp. 263–272.

[137] King, H. E., and J. L. Wong, "Characteristics of 1 to 8 Wavelength Uniform Helical Antennas," *IEEE Trans. Ant. Propag.*, Vol. AP-28, March 1980, pp. 291–296.

[138] Wong, J. L., and H. E. King, "Empirical Helix Antenna Design," *IEEE AP-S Int. Symp. Digest*, May 1982, pp. 366–369.

[139] Nakano, H., J. Yamauchi, and H. Mikami, "Tapered Balanced Helices Radiating in the Axial Mode," *IEEE AP-S Int. Symp. Digest*, June 1980, pp. 700–703.

[140] Angelakos, D. J., and D. Kajfez, "Modifications on the Axial-Mode Helical Antenna," *IEEE Proc.*, Vol. 55, April 1967, pp. 558–559.

[141] Kilgus, C. C., "Shaped-Conical Radiation Pattern Performance of the Backfire Quadrifilar Helix," *IEEE Trans. Ant. Propag.*, Vol. AP-23, May 1975, pp. 392–397.

[142] Dyson, J. D., "The Characteristics and Design of the Conical Log-Spiral Antenna," *IEEE Trans. Ant. Propag.*, Vol. AP-13, July 1965, pp. 488–499.

[143] Dyson, J. D., "The Unidirectional Equiangular Spiral Antenna," *IRE Trans. Ant. Propag.*, Vol. AP-7, Oct. 1959, pp. 329–334.

[144] Dyson, J. D., "The Equiangular Spiral Antenna," *IRE Trans. Ant. Propag.*, Vol. AP-7, April 1959, pp. 181–188.

[145] Kaiser, J. A., "The Archimedean Two-Wire Spiral Antenna," *IRE Trans. Ant. Propaga.*, Vol. AP-8, May 1960, pp. 312–323.

[146] Hirasawa, K., and M. Haneishi, ed., *Introductory Analysis Design and Measurement of Small or Low-Profile Antennas*. Dedham, MA: Artech House, 1992, pp. 83–159.

[147] James, J. R., and P. S. Hall, eds., *Handbook of Microstrip Antennas*, Vols. 1 and 2, London: Peter Peregrinus, IEE, 1989, pp. 1057–1191.

[148] Chiba, T., Y. Suzuki, N. Miyano, S. Miura, and S. Ohmori, "A Phased Array Antenna Using Microstrip Patch Antennas," *Proc. 12th European Micro.*, Sept. 1982, pp. 472–477.

[149] Weinschel, H. D., "A Cylindrical Array of Circularly Polarized Microstrip Antennas," *IEEE AP-S Int. Symp. Digest*, June 1975, pp. 177–180.

[150] Suzuki, Y., and T. Chiba, "Designing Method of Microstrip Antenna Considering the Bandwidth," *IECE Trans.*, Vol. E67, Sept. 1984, pp. 488–493.

Acronyms and Abbreviations

A/D	analog-to-digital
ADF	automatic direction finder
AF	audio frequencies
AGC	automatic gain control
AM	amplitude modulation
AMPS	Advanced Mobile Phone System
ANSI	American National Standards Institute
AR	axial ratio
ATC	air traffic control
ATCRBS	ATC radar beacon system
ATU	automatic tuning system
BER	bit error rate
BS	broadcast satellite
C/A	course acquisition
C/I	carrier-to-interference ratio
C/M	average power ratio of carrier to multipath wave
CAD	computer-aided design
CAT	computerized axial tomography
CBOCS	Comprehensive Bus Operation Control System
CCIR	International Committee on Radio
CDMA	code division multiple access
CFM	compounded frequency modulation
CIR	carrier-to-interference ratio
CNR	carrier-to-noise ratio
CONUS	continental United States
CPU	central processing unit

CRC	Communications Research Center
CW	continuous wave
D/U	desired-to-undesired signal strength ratio
DBS	direct broadcast satellite
DGPS	Differential GPS
DME	distance measuring equipment
DSRC	Dedicated Short Range Communication System
EGC	equal-gain combining
EIRP	equivalent isotropically radiated power
EM	electromagnetic
EMF	electromagnetic field
ENRI	Electronic and Navigation Research Institute
ERP	effective radiated power
ESTEC	European Space Research and Technology Center
ETC	Electronic Toll Collection System
ETS/V	Engineering Test Satellite-V
FDMA	frequency division multiple access
FDTD	finite difference time domain
FEC	forward error correction
FET	field-effect transistor
FM	frequency modulation
FSK	frequency shift keying
FSPL	free-space path loss
G/T	ratio of system gain to system noise temperature
GEO	geostationary earth orbiting
GLONAS(S)	global orbiting navigation satellite system
GMDSS	global maritime distress and safety system
GPS	global positioning satellite
GSM	Groupe Special Mobile
GTD	geometrical theory of diffraction
HEO	high earth orbit
HP	horizontally polarized
HPBW	half-power beamwidth
IE3D	intregal equation in three dimension
IF	intermediate frequency
IFA	inverted-F antenna
IFF	identification friend or foe
ILA	inverted-L antenna
ILS	instrument landing system
IMT-2000	International Mobile Telephone-2000
INMARSAT	International Maritime Satellite Organization
ISDN	integrated services digital network

ITS	Intelligent Transportation System
ITU	International Telephony Union
JPL	Jet Propulsion Laboratory
LCX	leaky coaxial cable
LEO	low-earth orbit
LNA	low-noise amplifier
loran	long-range navigation
LSI	large-scale integration
MCA	multichannel access
MEG	mean effect gain
MEO	medium earth orbit
MES	mobile earth station
MIC	microwave integrated circuit
MOM	method of moments
MRC	maximal ratio combining
MRI	magnetic resonance imaging
MSA	microstrip antenna
MSS	mobile satellite station
NASA	National Aeronautics and Space Administration
NAVSTAR	Navigation System with Time and Ranging/Global Positioning System
NDB	nondirectional beacon
NMHA	normal-mode helical antennas
NNSS	Navy Navigation Satellite System
PAG	pattern averaging gain
PCN	personal communication network
PCS	personal communication system
PHS	personal handy phone system
PIFA	planar IFA
PIM	passive intermodulation
PSM	polarization shaping method
PSTN	public service telephone network
PTFE	polytetrafluoroethylene
QHA	quadrifilar helix antenna
RAM	random access memory
RF	radio frequency
RFM	random field measurement
RHCP	right hand circular polarization
RT	roadside transceiver
SAR	specific absorption rate
SAW	surface acoustic wave
SBF	short backfire antenna

SDM	System Definition Manual
SDMA	spatial division multiple access
SEC	selective combining
SES	ship earth station
SFCP	singly fed circularly polarized
SMR	special mobile radio
SS	spread spectrum
SSB	single-sideband
SSR	secondary surveillance radar
SWC	switch combining
TDD	time division duplex
TDMA	time division multiple access
TEM	transverse electromagnetic
TLM	transmission line method
TR	transmitting and receiving
UHF	ultrahigh frequency
UMTS	universal mobile telephone system
VCO	voltage-controlled oscillator
VHF	very high frequency
VICS	vehicle information and communication system
VLSI	very-large-scale integration
VOR	VHF omnidirectional range
VP	vertically polarized
VSWR	voltage standing wave ratio
VT	vehicle transceiver
WARC	World Administrative Radio Committee
WCDMA	wideband CDMA
WLL	wireless local loop
XPR	cross-polarization ratio
XSA	cross-slot antenna

About the Authors

Dr. Quirino Balzano
Vice President of Technical Staff
Motorola Inc.
Fort Lauderdale, FL 33322
USA

Dr. C. K. Chou
Motorola Florida Research Laboratories
Motorola Inc.
Fort Lauderdale, FL 33322
USA

Mr. Olle Edvardsson
Research Manager
Allgon Mobile Communications AB
Box 500, Nasvagen 17 SE 184
25 Akersbeigen
Sweden

Professor Kyohei Fujimoto
Professor Emeritus, University of Tsukuba
2-16-9 Kataseyama
Fujisawa 251-0033
Japan

Professor Kazuhiro Hirasawa
Institute of Information Science
University of Tsukuba
Tsukuba 305
Japan

Dr. Jochen Hopf
Institute for High Frequency Systems
University of the Bundeswehr
Munich, Neubiberg D85577
Germany

Professor Jim James
Royal Military College of Science
Cranfield University
Shrivenham, Swindon SN6 8LA
England

Dr. M. A. Jensen
Dept. of Electrical and Computer Engineering
Brigham Young University
Provo, UT 84602
USA

Professor Kenichi Kagoshima
Dept. of Media and Telecommunication Engineering
Ibaraki University
4-12-1 Naka-narusawa
Hitachi, Ibaraki 316-8511
Japan

Professor Yoshio Karasawa
Dept. of Electronic Engineering
The University of Electro-Communications
1-5-1 Choufugaoka
Choufu, Tokyo 182-8585
Japan

Dr. K. W. Kim
P-Com Inc.
31755 Winchester Blvd.
Campbell, CA 95008
USA

Dr. William C. Y. Lee
Vice President Corporate Technology
AirTouch Communications
2999 Oak Road
Walnut Creek, CA 95696
USA

Professor Heinz K. Lindenmeier
Institute for High Frequency Systems
University of the Bundeswehr
Munich, Neubiberg D85577
Germany

Mr. Richard Mumford
Consultant
25 Lordsfield Gardens
Overton, Basingstoke RG25 3EW
England

Dr. Shingo Ohmori
Manager, Planning Section
Communications Research Laboratory
Ministry of Posts and Telecommunications
Koganei-Shi, Tokyo 184
Japan

Professor Yahya Rahmat-Samii
Dept. of Electrical Engineering
University of California Los Angeles
Los Angeles, CA 90095-1594
USA

Dr. Leopold Reiter
Institute for High Frequency Systems
University of Bundeswehr
Munich, Neubiberg D85577
Germany

Dr. Simon Saunders
CCSR
School of Electronic Eng, IT and Maths
University of Surrey
Guildford GU2 5XH
England

Professor Takayasu Shiokawa
Dept. of Electronic Engineering
Tohoku University
Aoba, Aramaki, Sendai 980-0845
Japan

Dr. Yasuo Suzuki
Senior Research Scientist
Research and Development Center
Toshiba Corporation
Kawasaki 210
Japan

Dr. Tokio Taga
Senior Researcher, Supervisor
Radio Communications Systems Laboratories
Nippon Telegraph and Telephone Corporation
Yokosuka 238-03
Japan

Dr. Tasuku Teshirogi
Manager
Systems Division
Anritsu Corporation
5-10-27 Minamiazabu
Minato-ku, Tokyo 106
Japan

Mr. Mark Williamson
Space Technology Consultant
The Glebe House
Kirkby Thore, Cumbria CA10 1UR
England

Dr. Yoshihide Yamada
Dept. of Electronic Engineering
National Defence Academy
1-10-20 Hashirimizu
Yokosuka 239-8686
Japan

Index

Recent Titles in the Artech House
Antennas and Propagation Library

Helmut E. Schrank, Series Editor

Handbook of Antennas for EMC, Thereza MacNamara

Iterative and Self-Adaptive Finite-Elements in Electromagnetic Modeling, Magdalena Salazar-Palma, et al.

Mobile Antenna Systems Handbook, Second Edition, K. Fujimoto and J. R. James, editors

Microstrip Antenna Design Handbook, Ramesh Garg, et al.

Quick Finite Elements for Electromagnetic Waves, Giuseppe Pelosi, Roberto Coccioli, and Stefano Selleri

Radiowave Propagation and Antennas for Personal Communications, Second Edition, Kazimierz Siwiak

Solid Dielectric Horn Antennas, Carlos Salema, Carlos Fernandes, and Rama Kant Jha

Understanding Electromagnetic Scattering Using the Moment Method: A Practical Approach, Randy Bancroft

WIPL-D: Electromagnetic Modeling of Composite Metallic and Dielectric Structures, Software and User's Manual, Branko M. Kolundzija, et al.

For further information on these and other Artech House titles, including previously considered out-of-print books now available through our In-Print-Forever® (IPF®) program, contact:

Artech House
685 Canton Street
Norwood, MA 02062
Phone: 781-769-9750
Fax: 781-769-6334
e-mail: artech@artechhouse.com

Artech House
46 Gillingham Street
London SW1V 1AH UK
Phone: +44 (0)20 7596-8750
Fax: +44 (0)20 7630 0166
e-mail: artech-uk@artechhouse.com

Find us on the World Wide Web at:
www.artechhouse.com